Drug Discovery from Natural Products

RSC Drug Discovery Series

Editor-in-Chief
Professor David Thurston, *London School of Pharmacy, UK*

Series Editors:
Dr David Fox, *Pfizer Global Research and Development, Sandwich, UK*
Professor Salvatore Guccione, *University of Catania, Italy*
Professor Ana Martinez, *Instituto de Quimica Medica-CSIC, Spain*
Professor David Rotella, *Montclair State University, USA*

Advisor to the Board:
Professor Robin Ganellin, *University College London, UK*

Titles in the Series:
1: Metabolism, Pharmacokinetics and Toxicity of Functional Groups
2: Emerging Drugs and Targets for Alzheimer's Disease; Volume 1
3: Emerging Drugs and Targets for Alzheimer's Disease; Volume 2
4: Accounts in Drug Discovery
5: New Frontiers in Chemical Biology
6: Animal Models for Neurodegenerative Disease
7: Neurodegeneration
8: G Protein-Coupled Receptors
9: Pharmaceutical Process Development
10: Extracellular and Intracellular Signaling
11: New Synthetic Technologies in Medicinal Chemistry
12: New Horizons in Predictive Toxicology
13: Drug Design Strategies: Quantitative Approaches
14: Neglected Diseases and Drug Discovery
15: Biomedical Imaging
16: Pharmaceutical Salts and Cocrystals
17: Polyamine Drug Discovery
18: Proteinases as Drug Targets
19: Kinase Drug Discovery
20: Drug Design Strategies: Computational Techniques and Applications
21: Designing Multi-Target Drugs
22: Nanostructured Biomaterials for Overcoming Biological Barriers
23: Physico-Chemical and Computational Approaches to Drug Discovery
24: Biomarkers for Traumatic Brain Injury
25: Drug Discovery from Natural Products

How to obtain future titles on publication:
A standing order plan is available for this series. A standing order will bring
delivery of each new volume immediately on publication.

For further information please contact:
Book Sales Department, Royal Society of Chemistry, Thomas Graham House,
Science Park, Milton Road, Cambridge, CB4 0WF, UK
Telephone: +44 (0)1223 420066, Fax: +44 (0)1223 420247,
Email: booksales@rsc.org
Visit our website at http://www.rsc.org/Shop/Books/

Drug Discovery from Natural Products

Edited by

Olga Genilloud* and Francisca Vicente
Fundación MEDINA, Granada, Spain
*E-mail: olga.genilloud@medinaandalucia.es

RSC Publishing

RSC Drug Discovery Series No. 25

ISBN: 978-1-84973-361-8
ISSN: 2041-3203

A catalogue record for this book is available from the British Library

Published by The Royal Society of Chemistry,
Thomas Graham House, Science Park, Milton Road,
Cambridge CB4 0WF, UK

Registered Charity Number 207890

For further information see our web site at www.rsc.org

Printed in the United Kingdom by Henry Ling Limited, at the Dorset Press, Dorchester, DT1 1HD

Preface

The role of natural products in the treatment of diseases has inspired pharmaceutical scientists in their search for new avenues in drug discovery. The use of medicinal plants is recorded in most ancient archaeological sources, and, today, plant-derived molecules remain a significant fraction of the pharmaceuticals in the clinic. The production of antibiotics by microorganisms has been one of the major breakthroughs in the history of drug discovery in the 20th century, and bacteria and fungi have represented one of the most important sources for novel therapeutic agents exhibiting the most diverse biological actions. In the last decades, in spite of the difficulty of finding novel scaffolds and a fading interest of major companies in maintaining their natural products research, growing number of groups have dedicated numerous efforts to the exploration of alternative sources and the marine environment and the use of marine products has become an extraordinary rich alternative source of new drugs. At present, plants, microorganisms and marine invertebrates represent major sources of natural products for discovering new and novel drugs.

When a vast majority of new leads discovered in the last decades have originated from nature, and success in natural products drug discovery could be expected to be more likely than with synthetic molecules, the complexity of discovering and developing new natural products into new drugs is requiring today the interaction of multiple and new research areas with new and more sophisticated technical requirements. Whereas current research trends in the field suggests an optimistic future for natural products in drug discovery, novel strategies and innovative approaches are still needed for the development of new molecules, all being essential components for generating new promising leads.

RSC Drug Discovery Series No. 25
Drug Discovery from Natural Products
Edited by Olga Genilloud and Francisca Vicente
© The Royal Society of Chemistry 2012
Published by the Royal Society of Chemistry, www.rsc.org

Our intention with this book has been to reflect the current situation in natural products research where the confluence of different disciplines, ranging from the most recent chemical biology, genomics and synthetic biology to the traditional biology and microbiology, are ensuring the concurrence and synergism of expertises with high impact in the discovery of novel molecules. This volume offers an integrated review of the most recent trends in natural products drug discovery with case studies of key lead candidates outstanding for their new chemistry and biology as starting points for development of novel drug candidates. This selection of case studies on key molecules and derivatives from reference authors from academia and industry was established according to the novelty in the chemistry, biology and clinical applications of the compounds.

The book is divided in three major sections comprising 17 chapters covering a wide range of natural product topics.

The first section includes a group of chapters from reference authors in the field that bring recent approaches to exploit natural products chemistry from quite different perspectives, ranging from *de novo* synthesis and semi-synthesis to cope with the compound supply to the engineering of synthetic pathways, genome mining and epigenetic approaches.

Chapter 1, by Francesch and Cuevas, evaluates the impact of ecteinascidins, marine natural products isolated from the colonial ascidian *Ecteinascidia turbinata*, which has become a fascinating target for the synthetic organic chemist. The lead compound, Ecteinascidin-743 (ET-743, Yondelis[®]) is the first marine anticancer agent approved in the European Union for patients with soft tissue sarcoma (STS) and for the treatment of relapsed ovarian cancer. Kirschning *et al.* (Chapter 4) take a different approach, with examples of the elegant application of mutasynthesis, a new emerging strategy combining advanced methods of chemical synthesis with those of molecular biology and microbiology. Different cases of families of compounds are covered within this section from the perspective of their chemistry and their improvement. Three chapters are devoted to review the development of novel classes of metabolites that stand out for their chemistry and bioactivity. Kalesse *et al.* (Chapter 2) focus on reviewing the chondramides and chivosazoles, two metabolites interfering with the actin skeleton, Sussmuth *et al.* (Chapter 3) review the potential of the new Class III lantibiotics as a source of new interesting activities, as emphasized by labyrinthopeptin A2 as reference model, whereas Carter (Chapter 5) provides an updated review of recent biosynthetic modification efforts to enhance the neuroprotective and regenerative potential of the rapamycin family of microbial products.

As an alternative to the chemistry perspective discussed above, the next chapters review the application of engineering of biosynthetic pathways and genome mining as a new ways to reveal the potential to produce novel molecules. Several examples including cases of generation of derivatives by combinatorial biosynthesis are discussed by Salas *et al.* (Chapter 6) and by Donadio *et al.* (Chapter 7). Challis and Oves (Chapter 8) discuss the potential of genome mining in the study of microbial genomes as a source of untapped novel chemistry and new microbial pathways are discussed brilliantly in this chapter in the case of actinomycetes and by Neilan and Kalaitzis (Chapter 9) in the case of cyanobacteria. Complementing the former approaches, Keller and Soukup (Chapter 10) introduce the reader into the application of epigenetic modulation as a new tool to finely tune the expression of fungal metabolites.

The second section of the volume provides a very comprehensive review of some of the new antimicrobial drug discovery screening paradigms that have contributed in the last decade to the discovery of several of the most important microbial natural products.

Chapter 11, by Genilloud and Vicente, provides insights into the advances in high throughput screening technologies, putting special emphasis in the biological tools and whole-cell target-based assay platforms implemented to untap novel natural products scaffolds with novelty in their mode of actions.

This section also include an excellent review by Singh (Chapter 12) of the antisense screening approach, with a comprehensive discussion of the discovery and development of platensimycin and platencin from the different perspectives of their biological characterization, mechanism of action, structural biology, production and biosynthetic studies, structural modifications, structure–activity relationship and total synthesis. This chapter is complemented by the contribution by Roemer *et al.* (Chapter 13), who thoroughly review the identification and characterization of an exemplary set of novel natural antifungals resulting from a chemical genomic approach known as the *Candida albicans* fitness test, with special emphasis in the description of the parnafungins, a new natural product class of antifungal compounds with demonstrated *in vivo* efficacy and novel MOA.

The last section is devoted to investigations of novel microbial natural products leads and their derivatives. It contains four chapters that, in the form of case studies, provide examples of key bioactive molecules with applications in the therapeutic areas of infection diseases, oncology and parasitic diseases.

Novak (Chapter 15) reviews the advances made in the class of pleuromutilins as such, provides some insight into respective registrational trials and possible issues thereof, and finally offers a brief perspective on future opportunities of the class of pleuromutilins.

In the case of tuberculosis, Manjunatha *et al.* (Chapter 14) highlight in their chapter new agents against multi-drug resistant tuberculosis, the most widespread bacterial disease in the world, with nearly 2 billion people infected and one new infection occurring every second. When new chemical entities with novel modes of action are needed for the treatment of this disease, natural products play an important role to respond to this huge medical need for new anti-mycobacterials due to their excellent track record.

Altmann *et al.* (Chapter 16) provide a summary of the basic features of the *in vitro* and *in vivo* biological profile of epothilone B, and the chemistry, biology and pharmacology of ixabepilone and sagopilone, two synthetic analogs of epothilone B in advanced Phase II clinical trials. Moreover, they review some of the most recent developments in the area of analog synthesis and SAR studies, and discuss new findings of the tubulin-bound conformation of epothilones.

The final chapter highlights the contribution of marine chemistry in the field of antimalarial research. Fattorusso *et al.* focus Chapter 17 on four classes of compounds which have emerged as the most promising for future development and for which more detailed data are available: plakortin and related polyketide endoperoxides; isonitriles and analogues; manzamines; lepadins and salinosporamide.

In summary, the book is willing to reflect the current dynamism and the constant evolution of the area that is responding to a renovated interest in natural products with the concurrence of the most recent diverse expertises in chemical and synthetic biology focused on the identification of novel chemical diversity.

While some aspects of drug discovery from natural products have inevitably received more attention than have others, the editors have tried to minimize the duplication of the work discussed and to provide a balanced coverage of key topics.

It is a real pleasure to acknowledge the collective contributions of the individual authors and the much valued assistance received from the staff of the Publications Section of the Royal Society of Chemistry in the preparation of this book.

Olga Genilloud & Francisca Vicente
Fundación MEDINA, Granada, Spain

Contents

Section 1 New Approaches to Exploit Natural Products Chemistry

1.1 Semisynthesis/Synthesis *de novo* of Natural Products to Cope with Supply Issues

RSC Drug Discovery Series No. 25
Drug Discovery from Natural Products
Edited by Olga Genilloud and Francisca Vicente
© The Royal Society of Chemistry 2012
Published by the Royal Society of Chemistry, www.rsc.org

Chapter 3   Class III Lantibiotics – an Emerging Family of Thioether-Containing Peptides   **42**
Bartlomiej Krawczyk, Joanna M. Krawczyk and Roderich D. Süssmuth

Chapter 4   Mutant Manufacturers   **58**
Andreas Kirschning, Simone Eichner, Jekatherina Hermane and Tobias Knobloch

Section 2   New Methodologies and Screening Technologies for the Exploitation of Microbial Resources

Section 3   Novel Microbial Natural Products and Derivatives

Section 1
New Approaches to Exploit Natural Products Chemistry

Section 1.1
Semisynthesis/Synthesis *de novo* of Natural Products to Cope with Supply Issues

CHAPTER 1

Semisynthesis Approach of Ecteinascidin 743 (ET-743, Yondelis®)

CARMEN CUEVAS* AND ANDRÉS FRANCESCH

PharmaMar, S. A., Avenida de los Reyes, 1, Colmenar Viejo, Madrid, 28770, Spain
*E-mail: ccuevas@pharmamar.com

1.1 Introduction

For thousands of years natural products obtained from terrestrial sources have played a very important role in health care and prevention of diseases. However, it was not until the nineteenth century (1804) that scientists (Friedrich Sertürner) isolated active components (morphine) from various medicinal plants (*Papaver somniferum*) and since then terrestrial natural products have been extensively screened for their medicinal purposes. Between the years 1981 and 2006, about a 100 anticancer agents have been developed, of which, 25 are natural product derivatives, 18 are natural product mimics, 11 candidates are derived from a natural product pharmacophore, and 9 are pure natural products.[1] In recent years, the chemistry of natural products derived from marine organisms has become the focus of a much greater research effort. This is due in large part to the increased recognition of marine organisms as a source for bioactive compounds with pharmaceutical applications or other economically useful properties. Because of the physical and chemical conditions in the marine environment, almost every class of marine organism

RSC Drug Discovery Series No. 25
Drug Discovery from Natural Products
Edited by Olga Genilloud and Francisca Vicente
© The Royal Society of Chemistry 2012
Published by the Royal Society of Chemistry, www.rsc.org

possesses the capacity to produce a variety of molecules with unique structural features. These molecules offer an unmatched chemical diversity and structural complexity, together with a biological potency and selectivity. The fact that marine resources are still largely unexplored has inspired scientists from academia and the pharmaceutical industry to intensify their efforts by using novel technologies to overcome the inherent problems in discovering compounds which may have potential for further development as pharmaceuticals or as functional products such as cosmetics, nutritional supplements, and functional foods. These efforts have resulted in the development of around 15 marine natural products in various phases of clinical development, mainly in the oncology area, that includes the PharmaMar compounds: Yondelis®, Aplidin®, Irvalec®, Zalypsis®, PM01183, and PM060184.

1.2   Isolation of Ecteinascidin 743

Ecteinascidia turbinata Herdman (1880) (family Perophoridae) is a colonial ascidian (tunicate) species from the Caribbean and the Mediterranean that belongs to the class Ascidiacea within the subphylum Tunicata (also called Urochordata) possessing a transparent, orange or whitish-colored tunic. Ascidians, or sea squirts, are small, bottom-dwelling soft-bodied marine animals that form colonies comprising many individuals, called zooids. The name 'tunicate' is derived from their characteristic protective covering, or tunic, which functions to a certain extent as an external skeleton and consists of some cells, blood vessels, and a secretion of a variety of proteins and carbohydrates, including tunicin, a cellulose-like polymer – an unusual finding in animals. Within the tunic is the muscular body wall, which controls the opening of the siphons used for feeding. A typical colony consists of a dense cluster of elongated, somewhat club-shaped, zooids connected at their bases by a network of stolons that adheres the colony to the surface of the substrate on which it grows. The tunicate normally lives in coastal shallow waters (0 to 15 m depth) and in lagoons growing on red mangroves roots, rocks, shells, sand, and marine meadows. It is distributed throughout the Caribbean and in the temperate regions of the Atlantic and the Mediterranean. Reproduction is through a sexual cycle in which eggs are fertilized, hatched, and brooded internally and the larvae released to the sea as they reach maturity. Asexual reproduction is by budding of new zooids from the base of an existing zooid or from the stolon mass of the colony. Whereas dispersion of the species is facilitated as a consequence of the larvae being carried to new locales by ocean currents, the role of the stolon constitutes an important adaptive strategy for regeneration and growth, assuring fast colonization and extension of extant colonies over available surfaces of both natural and artificial substrata.

Aqueous ethanol extracts of *Ecteinascidia turbinata* were shown to have antitumor effects in 1969, but isolation and structural characterization of the active compounds was not achieved until 1990 when Rinehart and co-workers reported six new chemical entities called ecteinascidins (ETs), 743 (**1**), 729 (**2**),

745 (**3**), 759A (**4**), 759B (**5**), and 770 (**6**), of which ET-743 was the most abundant representative (0.0001% yield).[2,3] Simultaneously, Wright and co-workers described ET-743 and 729.[4] The novel and unique chemical structures of ecteinascidins, determined by extensive NMR and mass spectral studies, is formed by a monobridged pentacyclic skeleton composed of two fused tetrahydroisoquinoline rings (subunits A and B) linked to a 10-membered lactone bridge through a benzylic sulfide linkage. Most ecteinascidins have an additional tetrahydroisoquinoline or tetrahydro-β-carboline ring (subunit C) attached to the rest of the structure through a spiro ring (Figure 1.1). This is one of the features distinguishing these molecules from saframycins, safracins, and renieramycins, compounds isolated from bacterial sources and sponges.[5]

1.3   Mechanism of Action

In contrast to traditional alkylating agents that bind guanine at the N7 or O6 position in the DNA major groove, ET-743 is the first of a new class of DNA

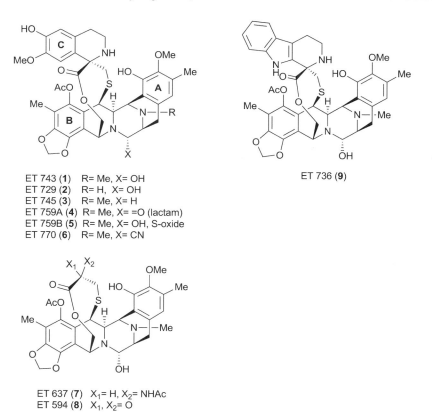

ET 743 (**1**)   R= Me, X= OH
ET 729 (**2**)   R= H, X= OH
ET 745 (**3**)   R= Me, X= H
ET 759A (**4**)   R= Me, X= =O (lactam)
ET 759B (**5**)   R= Me, X= OH, S-oxide
ET 770 (**6**)   R= Me, X= CN

ET 736 (**9**)

ET 637 (**7**)   X₁= H, X₂= NHAc
ET 594 (**8**)   X₁, X₂= O

Figure 1.1   Chemical structure of Ecteinascidin 743 (ET-743, Yondelis®) and natural ecteinascidins.

binding agents with a complex, transcription-targeted mechanism of action. ET-743 binds the exocyclic N2 amino group of guanines in the minor groove of DNA with preference for GC-rich triplets through an iminium intermediate generated *in situ* by dehydration of the carbinolamine moiety present in the monobridge pentacycle skeleton. The binding of ET-743 in the minor groove induces the formation of DNA adducts, which bend DNA towards the major groove.[6,7] The resulting covalent adduct is additionally stabilized through van der Waals interactions and one or more hydrogen bonds between the monobridge pentacycle skeleton with neighboring nucleotides in the same or opposite strand of the DNA double helix, thus creating the equivalent to a functional interstrand crosslink.

The additional subunit C apparently does not participate in DNA binding and it was proposed to protrude out of the DNA, being able to interact with different DNA-binding proteins located in the DNA adduct area. One of these proteins is XPG endonuclease, a member of the nucleotide excision repair (NER) system.[8] Moreover, ET-743 is apparently blocking the trans-activating ability of chimeric proteins such as FUS-CHOP or EWS1-Fli1 modulating the transcription of genes that should be crucial for tumorigenesis in specific cancer subtypes.[9] At the cell cycle level, these events result in a decrease in the rate of progression of the tumor cells through the S phase toward G2 or as a prolonged G2-phase blockade.

1.4   Preclinical Drug Development

Preclinical data generated during the development of ET-743 have provided important insight for the selection and design of the clinical trials. Early *in vitro* studies carried out by PharmaMar and the National Cancer Institute (NCI) in a panel of 60 human tumor cells identified the potent activity (1 pM to 10 nM) of ET-743. The NCI COMPARE analysis with more than 100 standard anticancer agents was negative, indicating a new mechanism of action of ET-743.

ET-743 has been tested in a great variety and number of models against tumors of murine origin (P388 leukemia and B16 melanoma), human sensitive xenografts (melanoma, MEXF 989; non-small-cell lung cancer, LXFL 529; breast, MX-1 early and advanced; and ovarian, HOC 22), and human resistant xenografts (melanoma, MEXF 514; non-small-cell lung cancer, LXFL 629; and ovarian, HOC 18). These first efforts showed that ET-743 has a broad spectrum of antineoplastic activity, with several tumor types showing selectivity; namely, melanoma, non-small-cell lung cancer, and ovarian carcinomas. As a further example of strong activity and long-lasting antitumor effects, the action of ET-743 on human endometrial carcinoma xenografts (HEC-1-B) results in complete regression lasting for more than 125 days. The determination that soft tissue sarcomas (STS) are more sensitive to ET-743 than other solid tumors was not predicted during the preclinical development of the drug. The finding was serendipitous and came from the prevalence of responding or stable STS patients in the clinical trials. As a result, there has been a considerable effort to confirm this finding to supplement the extensive nonclinical profile already characterized for the antineoplastic effect of ET-743 against other solid tumors. In fact, a variety of sarcomas are differentially sensitive to ET-743, showing IC_{50} potencies in the picomolar and sub-picomolar range compared to the nanomolar concentrations established against non-STS solid tumors.[10,11]

1.5 Clinical Studies

Yondelis® (Trabectedin, ET-743) has been designated an orphan drug by the European Commission (EC) and the USA Food and Drug Administration (FDA) for the indications of soft tissue sarcoma (STS) and ovarian cancer.

Soft tissue sarcomas are malignant tumors that originate in the soft tissues of the body. Soft tissues connect, support, and surround other body structures. The soft tissues include muscle, fat, blood vessels, nerves, tendons, and synovial tissues. The annual incidence of STS in Europe is approximately 0.004% (4 in 100 000 people). Five-year overall survival (OS) rates are on the order of 50 to 60%, irrespective of disease stage at diagnosis. Within 2 to 3 years from diagnosis, approximately 30 to 50% of patients develop metastases despite optimal treatment for localized disease.

Ovarian cancer is one of the deadliest gynecological cancers. Unfortunately, detection of ovarian cancer is difficult, and the disease is often diagnosed too late for successful treatment. Although with the current standard of care (debulking surgery followed by platinum-based chemotherapy) most of the patients achieve a complete clinical remission, eventually the majority of them will relapse and die owing to their disease.

In September 2007, Yondelis® received marketing authorization from the European Commission for the treatment of patients with advanced or metastatic soft tissue sarcoma after failure of anthracyclines and ifosfamide, or who are unsuited to receive these agents. In September 2009 Committee for Medicinal Products for Human Use (CHMP) adopted a positive opinion on a variation to extend the indication for Yondelis® for the treatment of patients with relapsed platinum-sensitive ovarian cancer in combination with pegylated liposomal doxorubicin (PLD).[12]

Worldwide, more than 12 000 patients in more than 63 countries have already been treated with this innovative drug and shown a good safety and tolerability profile. The most frequent adverse event appears to be neutropenia, which is reversible. Transaminase elevations were also reported but were transient. No mucositis, alopecia, neurotoxicity, cardiotoxicity, or cumulative toxicities have been observed.

In other indications, such as breast and prostate cancer, Yondelis® is currently being studied in Phase II clinical trials trying to identify the patients that should respond to the drug treatment by measuring levels of the endonuclease XPG. Yondelis® is also being tested in pediatric indications.

1.6   Chemical Synthesis

Ecteinascidia turbinata has been successfully grown and harvested in aquaculture facilities located along the Mediterranean coast. The purification of the active ingredient was then accomplished on an industrial scale, using chromatographic procedures that represent a more practical and environmentally sound practice than harvesting the creature from the wild. Nevertheless, in recent years several synthetic schemes have been developed for industrial production of ET-743 in the quantities and quality required for a drug product (Yondelis®) that will be used in worldwide clinical studies and manufacturing for commercialization.

1.6.1   Synthetic Routes to Ecteinascidin 743 (ET-743, Yondelis®)

To date, three distinct total synthetic routes to Yondelis® (ET-743) have been reported. The pioneer work of E. J. Corey and co-workers[13] provided for the first time a total synthesis of this complex molecule in 1996. This breakthrough scheme resolved one of the main roadblocks to the synthesis – the cyclization to obtain the 10-membered ring – is based on the elegant maneuver of the generation of the short-lived *ortho*-quinone methide and attack by cysteine thiol. Five years later, Fukuyama and co-workers[14] published a second total synthesis of ET-743, based in part on previous efforts targeting members of the saframycins and renieramycins, in which the cyclization reaction takes place in the conditions previously developed by Danishefsky and co-workers[15] to converge with the general approach established by Corey for the later stages of the synthesis. Finally in 2006, a highly convergent total synthesis of ET-743 was reported by Zhu and co-workers[16] as a conclusion of previous investigations on the synthesis of this family of compounds. This third total synthesis has been achieved in 31 steps in the longest linear sequence from 3-methyl catechol. Additionally, two formal syntheses have been described by the Danishefsky[17] and Williams[18] groups when describing the synthesis of an advanced intermediate in the Fukuyama total synthesis and an intermediate of the first formal synthesis, respectively.

The procedures outlined above represent some of the most outstanding work in recent synthetic organic chemistry. However, the long and involved procedures for total synthesis of the molecule represent a tremendous barrier to industrial manufacture of the drug, which is particularly challenging in the face of regulatory requirements for pharmaceuticals. This problem was finally solved with the development of a semisynthetic procedure[19] representing the first industrially feasible route to the manufacture of the drug on a large scale.[20] The procedure uses cyanosafracin B (**11**),[21] an antibiotic available through fermentation of the bacteria *Pseudomonas fluorescens*,[22] as the starting point. This approach is similar to traditional semisynthetic approaches, though in this case the semisynthetic product is a difficult-to-source natural product. Optimization of the fermentation process, followed by its transformation according to Scheme 1.1, provided a robust, easily scaled-up procedure for manufacturing the drug.

The amino and phenol groups of cyanosafracin B (11) were protected as the *tert*-butyloxycarbonyl (BOC) and methoxymethyl ether (MOM) derivatives, respectively, to give 12. The hydrolysis of the methoxy-*p*-quinone and subsequent reduction of the *p*-quinone afforded the unstable hydroquinone, which was treated with bromochloromethane and Cs_2CO_3 to give the methylendioxy ring. Alkylation of the remaining phenol gave fully protected intermediate 13. After simultaneous deprotection of the MOM and BOC groups from 13, cleavage of the amide was accomplished by an Edman degradation[23] by forming first the thiourea with excess phenyl isothiocyanate, followed by treatment with HCl in dioxane to give 14. Protection of the phenol

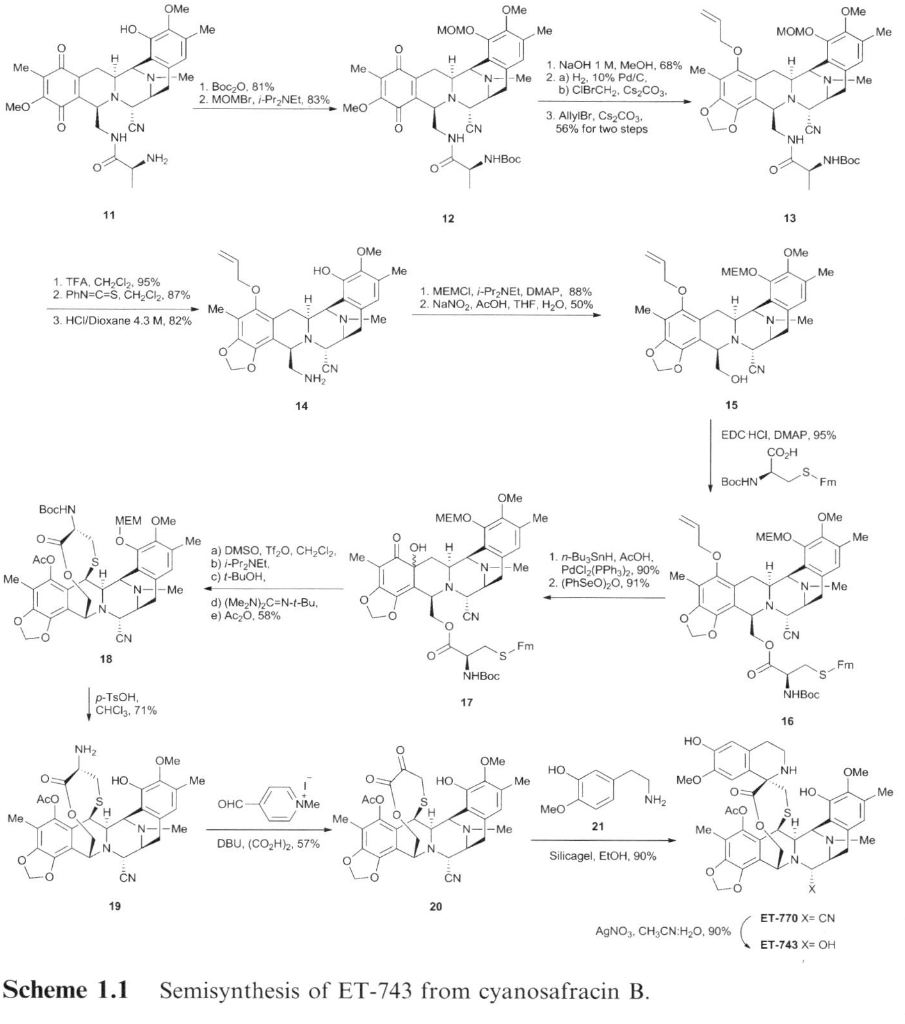

Scheme 1.1   Semisynthesis of ET-743 from cyanosafracin B.

as β-methoxyethoxymethyl ether (MEM) allowed for the diazotization of the primary amine for conversion to alcohol. The critical substitution of the amino by an alcohol function was best performed by treatment with $NaNO_2/AcOH$ yielding **15**, a key intermediate. Next, esterification with a diprotected cysteine ((*S*)-*N*-(*tert*-butoxycarbonyl)-*S*-(9-fluorenylmethyl)cysteine) produces compound **16**. The synthesis of ET-743 was completed using the chemistry from Corey and co-workers on similar substrates. A five-step sequence was used to form α-keto lactone **20**. Deprotection of the allyl group and oxidation of the phenol with benzeneseleninic anhydride[24] effected position-selective angular hydroxylation to give the dihydroxy dienone **17**. Compound **17** was transformed in one flask to the bridged 10-membered lactone **18** by the following operations: (a) reaction of **17** with the *in situ* generated Swern reagent from excess triflic anhydride and dimethyl sulfoxide, (b) addition of *i*-Pr$_2$NEt to form the *ortho*-quinone methide, (c) quenching with *tert*-butanol to destroy excess Swern reagent, (d) addition of excess *N*-*tert*-butyl-*N'*,*N'*,*N''*,*N''*-tetramethylguanidine[25] to convert the 9-fluorenylmethyl thiol ether to the thiolate ion and to promote nucleophilic addition of sulfur to the *ortho*-quinone methide to generate the 10-membered lactone bridge, and (e) addition of excess acetic anhydride to acetylate the resulting phenoxide group. Simultaneous removal of the MEM and BOC protecting groups using *p*-toluenesulfonic acid to achieve α-amino lactone **19** in 71% yield was followed by ketone formation by transamination[26] with the *N*-methylpyridinium-4-carboxaldehyde iodide, 1,8-diazabicyclo[5.4.0]undec-7-ene (DBU), and oxalic acid to yield α-keto lactone **20**. Diastereoselective Pictet-Spengler condensation[27] of **20** with 5-(2-aminoethyl)-2-methoxyphenol (**21**)[28] in the presence of silica gel generated the final tetrahydroisoquinoline system of ET-770 (**6**). Finally, the nitrile group was replaced by the hydroxyl group with excess of $AgNO_3$ in a mixture of $CH_3CN:H_2O$ to produce ET-743 (**1**).

1.6.2 Total Syntheses of other Natural Ecteinascidins

In 2003, this semisynthetic process developed to prepare ET-743 (**1**) by PharmaMar[19] was demonstrated as a versatile methodology that allowed the preparation of other members of the ecteinascidins and analogous compounds (Figure 1.1): ET-729 (**2**), ET-745 (**3**), ET-759B (**5**), ET-637 (**7**), ET-594 (**8**), and ET-736 (**9**).[29] Particularly significant for the synthesis of ET-729 (**2**)[30] are the smooth *N*-demethylation reaction conditions: *meta*-chloroperoxybenzoic acid (*m*CPBA), triethylamine, and trifluoroacetic anhydride (TFAA) used over the fully protected intermediate **22** (obtained in one step from **15**, Scheme 1.1) to give **23** (85% yield) that could allow the preparation of a wide variety of new *N*-derivatives of the ecteinascidins difficult to obtain from the natural source. With compound **23** in hand, deprotection of the allyl group, oxidation of the phenol, and subsequent protection with allyl bromide of the bridgehead amine furnished the *N*-allyl intermediate **24**, which was submitted to desilylation under standard conditions to give the corresponding alcohol. Next, esterification with (*S*)-*N*-[(*tert*-butoxycarbonyl)-*S*-(9-fluorenylmethyl)]cysteine and subsequent cyclization of **25** gave the 10-membered lactone bridge *via* formation of the exo quinone methide followed by nucleophilic addition of the deprotected cysteine and further acetylation of the phenoxide ion. Simultaneous removal of the BOC and MEM protecting groups with *p*-toluenesulfonic acid in CHCl$_3$ afforded α-amino lactone **26**. Finally, transamination and introduction of the 3-hydroxy-4-methoxyphenethylamine (**21**) by Pictet-Spengler reaction gave the tetrahydroisoquinoline ring in excellent yield. Deprotection of the allyl protecting group and replacement of CN by OH with AgNO$_3$ in a mixture of CH$_3$CN:H$_2$O gave ET-729 (**2**), which had identical data upon comparison with that of a natural sample (Scheme 1.2).

The synthesis of other natural ecteinascidins that retain the *N*-Me group can be achieved from common intermediates of the semisynthesis of ET-743 (**1**). The choice of the starting material depends on the complexity of the structures. ET-637 (**7**) was obtained in two steps from α-amino lactone **19** that was treated with Ac$_2$O without base to avoid acetylation of the free phenol in 96% yield,

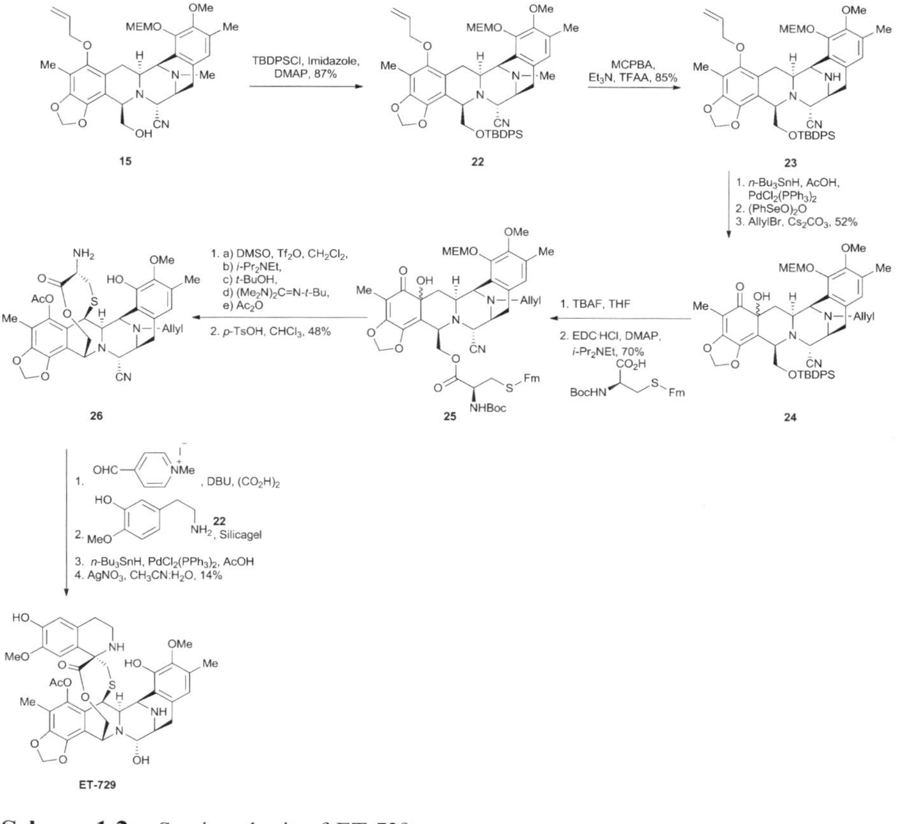

Scheme 1.2 Semisynthesis of ET-729.

but substitution of the CN group by OH was better performed in this case with CuCl in a mixture of THF:H$_2$O in 75% yield since the reaction failed when carried out under the standard conditions with AgNO$_3$. α-keto lactone **20** was used as starting material to prepare ET-594 (**8**) in one step: replacement of CN by OH was carried out in 91% yield under the same conditions used to prepare ET-637 (**7**). Intermediate **20** was also used to synthesize ET-736 (**9**); the tetrahydro-β-carboline ring was introduced under mild conditions by treatment of **20** with tryptamine in AcOH as solvent in excellent yield (99%), following by reaction with AgNO$_3$ (92% yield). Finally, ET-745 (**3**) was directly obtained by reduction of ET-743 (**1**) with sodium cyanoborohydride in 77% yield, whereas ET-759B (**5**) was synthesized in two steps from ET-770 (**6**) by oxidation of the sulfide with *m*CPBA in excellent yield (90%) to give a single stereoisomer, and treatment with AgNO$_3$ to replace the nitrile by the hydroxyl group (Scheme 1.3).

1.7 Conclusions

The biological activity of marine natural compounds is often startling, demonstrating, for example, potent cytotoxic, immunosuppressive, and antibiotic properties – making them of interest to the pharmaceutical industry. However, the low natural abundance of many of these lead structures means

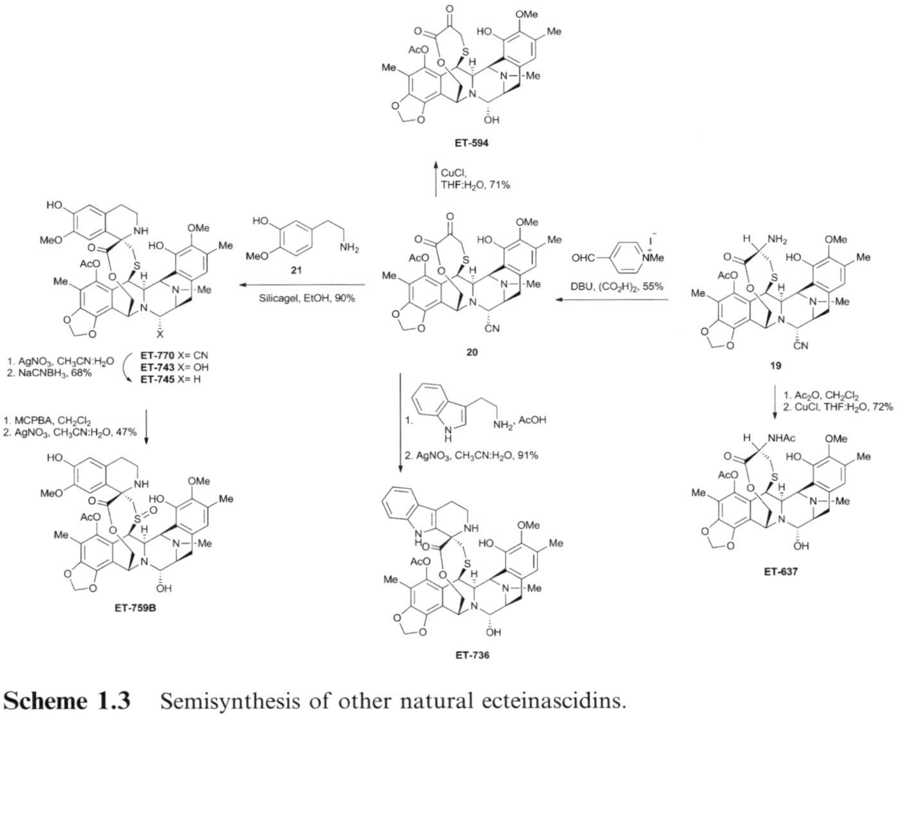

Scheme 1.3   Semisynthesis of other natural ecteinascidins.

that realistic and practical synthetic routes are required to provide material to investigate and exploit further their biological activity. These factors, combined with the exquisite molecular architectures that many of these compounds possess, offer demanding challenges to the modern synthetic chemist.

This dedicated synthetic effort has triumphed in removing the supply problem for ET-743, providing sufficient material for extensive clinical development, and commercialization, as well as facilitating SAR studies for lead optimization.

References

1. D. J. Newman and G. M. Cragg, *J. Nat. Prod.*, 2007, **70**, 461.
2. K. L. Rinehart, T. G. Holt, N. L. Fregeau, J. G. Stroh, P. A. Keifer, F. Sun, L. H. Li and D. G. Martin, *J. Org. Chem.*, 1990, **55**, 4512.
3. K. L. Rinehart, T. G. Holt, N. L. Fregeau, J. G. Stroh, P. A. Keifer, F. Sun, L. H. Li and D. G. Martin, *J. Org. Chem.*, 1991, **56**, 1676.
4. A. E. Wright, D. A. Forleo, P. G. Gunawardana, S. P. Gunasekera, F. E. Koehn and O. J. McConnell, *J. Org. Chem.*, 1990, **55**, 4508.
5. J. D. Scott and R. M. Williams, *Chem. Rev.*, 2002, **102**, 1669.
6. Y. Pommier, G. Kohlhagen, C. Bailly, M. Waring, A. Mazumder and K.W. Kohn, *Biochemistry*, 1996, **35**, 13303.
7. M. Zewail-Foote and L. H. Hurley, *J. Med. Chem.*, 1999, **42**, 2493.
8. A. B. Herrero, C. Martin-Castellanos, E. Marco, F. Gago and S. Moreno, *Cancer Res.*, 2006, **66**, 8155.
9. C. Forni, M. Minuzzo, E. Virdis, E. Tamborini, M. Simone, M. Tavecchio, E. Erba, F. Grosso, A: Gronchi, , P. Aman, P. Casali, M. D'Incalci, S. Pilotti and R. Mantovani, *Mol. Cancer Ther.*, 2009, **8**, 449.
10. W. W. Li, N. Takahashi, S. Jhanwar, C. Cordon-Cardo, Y. Elisseyeff, J. Jimeno, G. Faircloth and J. R. Bertino, *Clin. Cancer Res.*, 2001, **7**, 2908.
11. N. Takahashi, W. W. Li, D. Banerjee, K. W. Scotto and J. R. Bertino, *Clin. Cancer Res.*, 2001, **7**, 3251.
12. European Medicines Agency (EMA). Assessment report for Yondelis. International non-propietary name/Common name: trabectedin. Procedure. No. EMEA/H/C/000773-II/0008. http://www.ema.europa.eu/docs/en_GB/document_library/EPAR_Product_Information/human/000773/WC500045832.pdf. 2010.
13. E. J. Corey, D. Y. Gin and R. S. Kania, *J. Am. Chem. Soc.*, 1996, **118**, 9202.
14. A. Endo, A. Yanagisawa, M. Abe, S. Tohma, T. Kan and T. Fukuyama, *J. Am. Chem. Soc.*, 2002, **124**, 6552.
15. B. Zhou, J. Guo and S. J. Danishefsky, *Org. Lett.*, 2002, **4**, 43.
16. J. Chen, X. Chen, M. Bois-Choussy and J. Zhu, *J. Am. Chem. Soc.*, 2006, **128**, 87.

17. S. Zheng, C. Chan, T. Furuuchi, B. J. D. Wright, B. Zhou, J. Guo and S. J. Danishefsky, *Angew. Chem., Int. Ed.*, 2006, **45**, 1754.
18. D. Fishlock and R. M. Williams, *J. Org. Chem.*, 2008, **73**, 9594.
19. C. Cuevas, M. Pérez, M. J. Martín, J. L. Chicharro, C. Fernández-Rivas, M. Flores, A. Francesch, P. Gallego, M. Zarzuelo, F. de la Calle, J. García, C. Polanco, I. Rodríguez and I. Manzanares, *Org. Lett.*, 2000, **2**, 2545.
20. C. Cuevas and A. Francesch, *Nat. Prod. Rep.*, 2009, **26**, 322.
21. Y. Ikeda, H. Matsuki, T. Ogawa and T. Munakata, *J. Antibiot.*, 1983, **36**, 1284.
22. Y. Ikeda, H. Idemoto, F. Hirayama, K. Yamamoto, K. Iwao, T. Asao and T. Munakata, *J. Antibiot.*, 1983, **36**, 1279.
23. P. Edman, *Acta Chem. Scand.*, 1956, **10**, 761, and references cited herein.
24. H. J. Reich and S. Wollowitz, *Org. React.*, 1993, **44**, 1.
25. D. H. R. Barton, M. Chen, J. C. Jászberényi and D. K. Taylor, *Org. Synth.*, 1997, **74**, 101.
26. T. F. Buckley and H. Rapoport, *J. Am. Chem. Soc.*, 1982, **104**, 4446.
27. E. D. Cox and J.M. Cook, *Chem. Rev.*, 1995, **95**, 1797.
28. E. J. Corey and D.Y. Gin, *Tetrahedron Lett.*, 1996, **37**, 7163.
29. R. Menchaca, V. Martínez, A. Rodríguez, N. Rodríguez, M. Flores, P. Gallego, I. Manzanares and C. Cuevas, *J. Org. Chem.*, 2003, **68**, 8859.
30. J. M. Reid, D. L. Walker and M. M. Ames, *Cancer Chemother. Pharmacol.*, 1996, **38**, 329.

CHAPTER 2

Chondramides and Chivosazoles – Two Metabolites Which Interfere with the Actin Cytoskeleton

LYNETTE A. SMYTH[a], TOBIAS BRODMANN[a] AND
MARKUS KALESSE*[a,b]

[a] Centre for Biomolecular Drug Research (BMWZ), Leibniz Universität
Hannover, Schneiderberg 1b, D-30167 Hannover, Germany; [b] Helmholtz
Centre for Infection Research, Inhoffenstrasse 7, D-38124 Braunschweig,
Germany
*E-mail: markus.kalesse@oci.uni-hannover.de

2.1 Introduction

Natural products have been utilised as an important source of medicines for
more than 2000 years. Although at the beginning, active ingredients were
isolated mainly from plants, there is now a stronger interest in studying the
secondary metabolites of microorganisms.

The pivotal role of natural products in the therapeutic areas of cancer and
infection is apparent. Sixteen out of the twenty most prescribed groups of
antibiotics are derived from natural sources or at least inspired by natural
products. A similar trend can be seen for antitumour compounds. For some
targets all known and approved drugs stem from natural products. The
evolutionary pre-optimisation of natural products to interact with biological
targets often serves as one plausible explanation for this observation.

RSC Drug Discovery Series No. 25
Drug Discovery from Natural Products
Edited by Olga Genilloud and Francisca Vicente
© The Royal Society of Chemistry 2012
Published by the Royal Society of Chemistry, www.rsc.org

Based on this background Gerd Höfle and Hans Reichenbach at the Helmholtz centre for infection research (HZI, formerly known as GBF) have studied natural products from myxobacteria for the last 25 years culminating in the isolation of more than 140 unique structural motifs.

Two remarkable observations form the basis for an explanation of the significance of these natural products. First, myxobacteria produce increased amounts of these secondary metabolites at the cell differentiation stage, when they also produce fruiting bodies. Secondly, they often produce both agonists and antagonists for a particular cellular process. Both observations indicate that secondary metabolites play a key role in the regulation of cellular processes and their importance is not only as a means of chemical protection or warfare.[1,2]

An important target affected by secondary metabolites is the actin cytoskeleton. The challenge with this target is to develop modulators which specifically interfere with one isoform, or in the case of antitumour compounds, which distinguish between cancer cells and normal, non-transformed cells. This often requires the synthesis of the natural product first, followed by the synthesis of modified analogues. Before one can address these challenges the structure of the natural product including its stereochemistry needs to be elucidated. Here, we discuss the structural elucidation and synthesis of two classes of natural products which target the actin cytoskeleton.

The actin cytoskeleton is a fundamental component of cells. The construction and degradation of the actin filaments is crucial for cell motility, phagocytosis and cytokinesis. There are two mechanisms by which compounds can interfere with the actin cytoskeleton; compounds such as chondramide and jasplakinolide lead to the polymerisation of actin G. Rhizopodin and chivosazole, on the other hand, lead to the destabilisation and destruction of actin F polymers.[3] These two possible mechanisms of interference can be achieved using natural products and have fascinated chemists, biologists and physicians alike.

2.2 Chondramides: Stabilisation of actin polymerisation

The chondramides were isolated in 1995 by Reichenbach and Höfle as a group of secondary metabolites from the myxobacterium *Chondromyces crocatus* (Figure 2.1).[4,5] Their structure is similar to the jasplakinolides and is composed of a polyketide and a polypeptide segment embedded in a 18-membered ring macrocycle. In addition to the differing number of chiral centres, two of the chondramides contain a chloride at the tryptophan moiety. Compared to jaspamide, the structures of chondramide show a different ring size (18- versus 19-membered).

2.2.1 Total Syntheses of Chondramide C by Kalesse and Waldmann

The absolute configuration of chondramide C was elucidated simultaneously through total synthesis by Waldmann and Kalesse. The absolute configuration

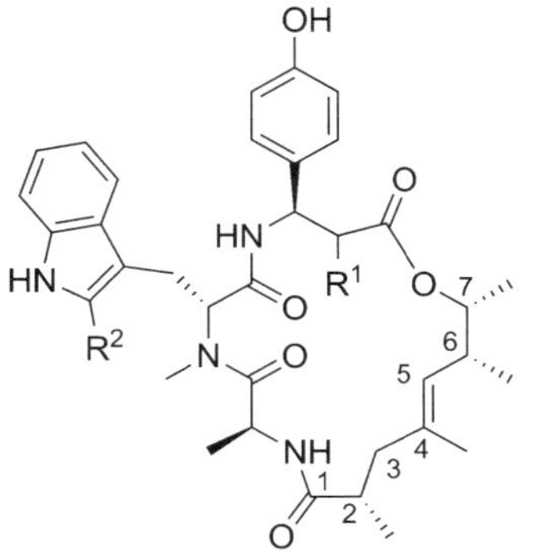

Chondramide

A (**1**): R^1 = OMe, R^2 = H
B (**2**): R^1 = OMe, R^2 = Cl
C (**3**): R^1 = H,   R^2 = H
D (**4**): R^1 = H,   R^2 = Cl

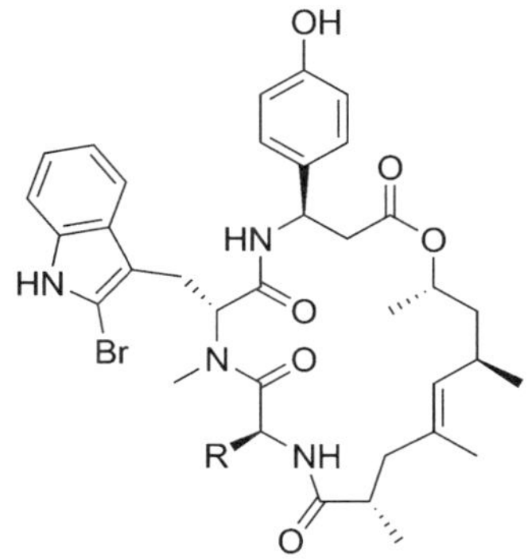

Jaspamide (Jasplakinolide) (**5**): R = CH$_3$
Jaspamide D: (**6**) R = CH$_2$CH$_3$
Jaspamide E: (**7**) R = CH$_2$OH

Figure 2.1   The chondramides and jasplakinolides.

of the stereocentres in the peptide segment and at C2 position could be deduced from the biosynthetic pathway and by comparison with the jasplakinolides.

The synthetic route by Kalesse was planned to allow a flexible construction of four possible diastereomers existing at the C6 and C7 positions. The further retrosynthetic disconnection led to a polyketide and a peptide fragment (Figure 2.2).[6,7]

The synthesis of the polyketide segment began with a four-step sequence to synthesise all possible diastereomers of ester **8**. Standard transformations finally provided the corresponding acids **12a–d** (Scheme 2.1).

Segments **12a–d** were finally coupled to peptide **13** to give compounds **14a–d**. Removal of the PMB protecting group with BCl$_3$ and saponification of the methyl ester provided the precursor for the final cyclisation step.[8] The key macrolactonisation step proved to be more challenging than expected, but after systematic investigations of different conditions and reagents the Shiina conditions allowed for the desired ring closure to accomplish the synthesis of all four chondramides (**3**, **15a–15c**). Through comparison of the spectroscopic data with those from the authentic material the configuration was finally assigned.

In subsequent biological experiments, the ability of chondramide C and the three chondramide derivatives to induce actin polymerisation was tested. In these studies, against five cell lines, chondramide C (**3**) and **15b** were shown to have the highest activity. To rationalise the results, a Monte Carlo conformational analysis was applied. An overlay of the energy-minimised structures of all four substances showed that the peptide segment of **3** and **15b** had a similar conformation. The conformations of the derivatives **15a** and **15c**,

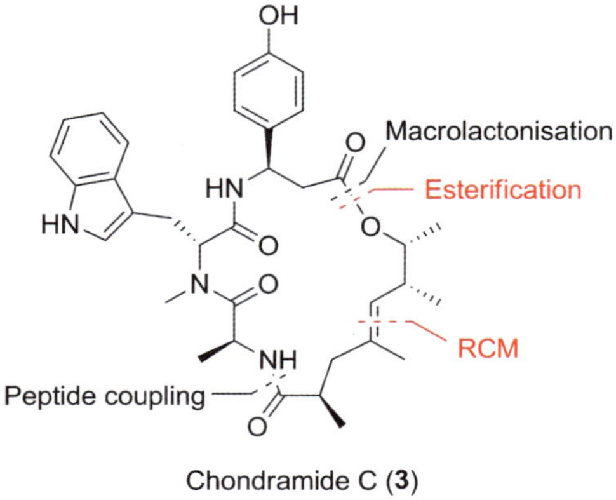

Chondramide C (**3**)

Figure 2.2   Retrosynthetic disconnection of chondramide C by Kalesse and Waldmann.

on the other hand, differed significantly from that of the natural chondramide C. The similar biological activity, despite the inversion of two stereogenic centres at C6 and C7, can be explained by the similar conformation of the peptide segments of the chondramides (Scheme 2.2).

Waldmann's retrosynthesis went back to acid **18**, which should be the substrate for the following esterification and ring-closing metathesis (Scheme 2.3). Compound **18** was constructed using solid-phase synthesis and all four

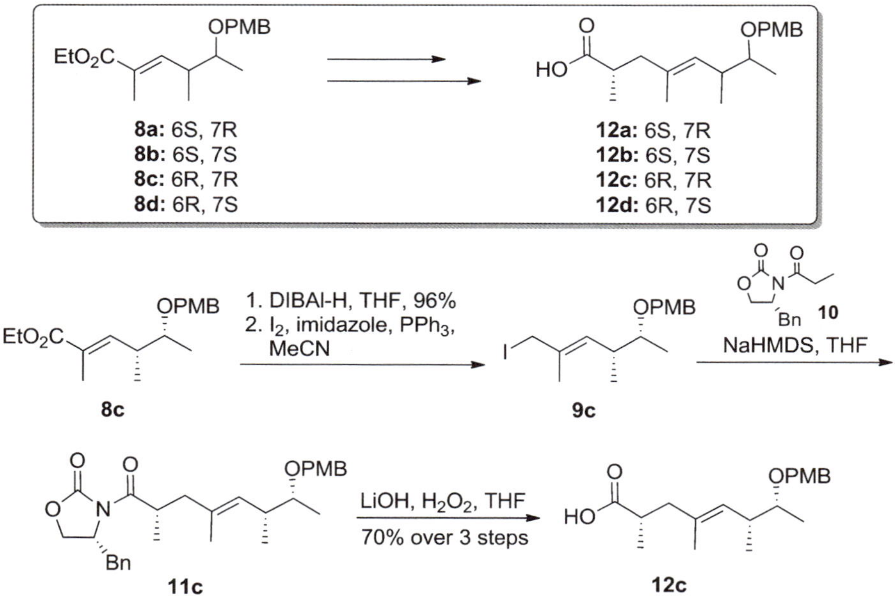

Scheme 2.1   Synthesis of secondary acids **12a–d**.

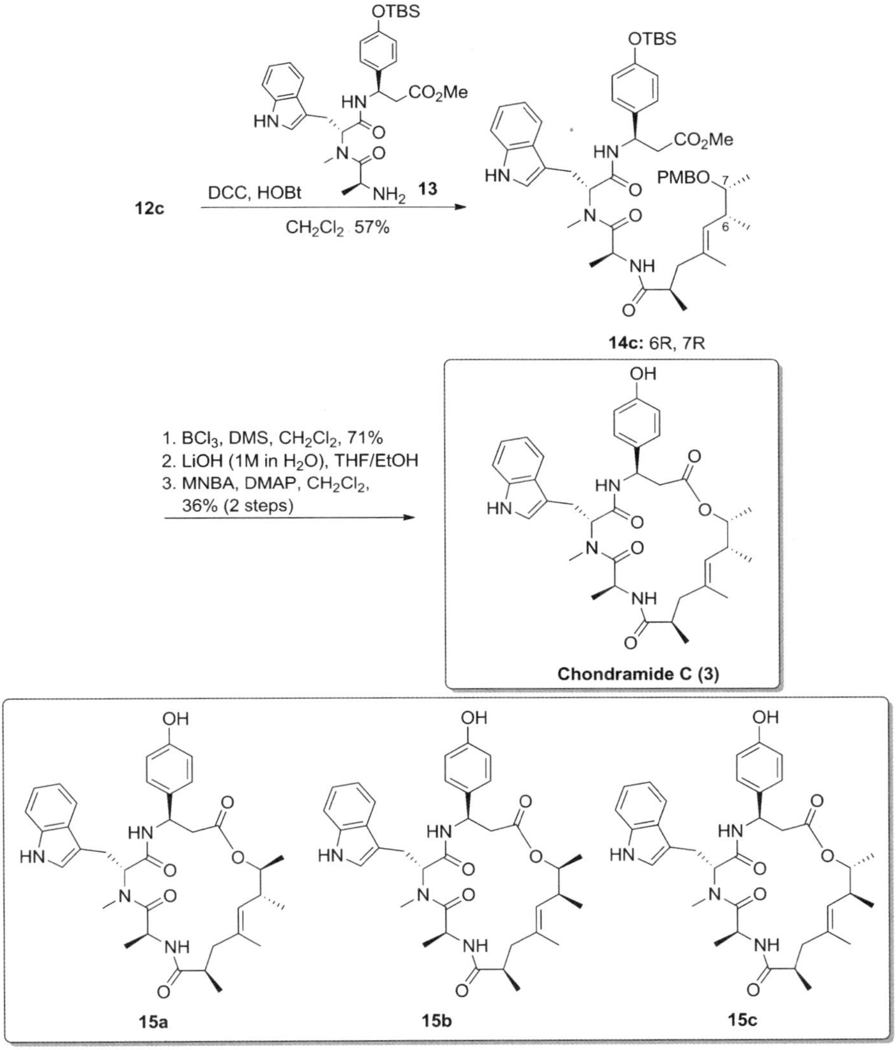

Scheme 2.2 Synthesis of chondramide C (**3**) and derivatives **15a–15c**.

diastereomers were accessed by esterification of **18** with the four diastereomers of homoallyl alcohol **19a–d** to form **20a–d**.[9,10] The key step involved the synthesis of an α-branched trisubstituted double bond. Remarkably, the double bond geometry established during the metathesis reaction depended significantly on the configuration of the C6 and C7 stereocentres. Compound **20b** cyclised to the undesired *Z*-isomer, but compounds **20c** and **20d** provided the desired *E*-isomer. After TIPS deprotection, the configuration of chondramide C could be identified as 2*S*, 6*R* and 7*R* for the polyketide region. The three further derivatives synthesised in these efforts showed

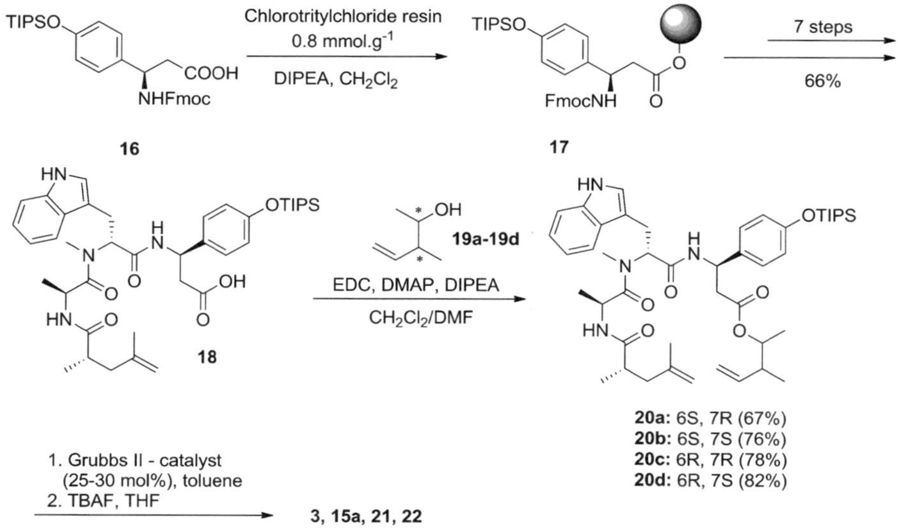

Scheme 2.3   Synthesis of chondramide C (**3**) and derivatives **15a**, **21** and **22**.

reduced activity in the subsequent biological evaluation compared to the natural product (Figure 2.3).

2.2.2   Total Synthesis of Chondramide A by Maier

Maier *et al.* established the synthesis of chondramide A and were able to elucidate the additional configuration at the β-position on the tyrosine residue.[11,12]

Retrosynthetically, chondramide A was constructed from two fragments, ω-hydroxyester **31** and the peptide fragment **35** (Figure 2.4). A Yamaguchi esterification should then lead to the fully functionalised, linear precursor **36**. Following this, intramolecular amide formation and deprotection should then lead to chondramide A (**1**). The key step in the synthesis of the polyketide fragment was a variation of the Mukaiyama aldol reaction established by Kobayashi *et al.*, allowing a trisubstituted double bond and two stereocentres to be built in one step.

The synthesis of the polyketide fragment started from silylenolether **23** which was transformed using a Mukaiyama aldol reaction to provide the *anti*-product **24** in excellent selectivity (Scheme 2.4). A Mitsunobu inversion led to the 6,7-*syn*-diastereomer, which, after removal of the chiral auxiliary and protecting group manipulation, led to the primary alcohol **27**. An asymmetric Evans alkylation with iodide **28** built up the stereocentre at C2 in very good yield and selectivity. Esterification, to give the *tert*-butyl ester and removal of the silyl protecting group led to **31**.

The total synthesis established the configuration of the 3-amino-2-methoxypropanoic acid to be 2*S*, 3*R*. The completion of the synthesis of

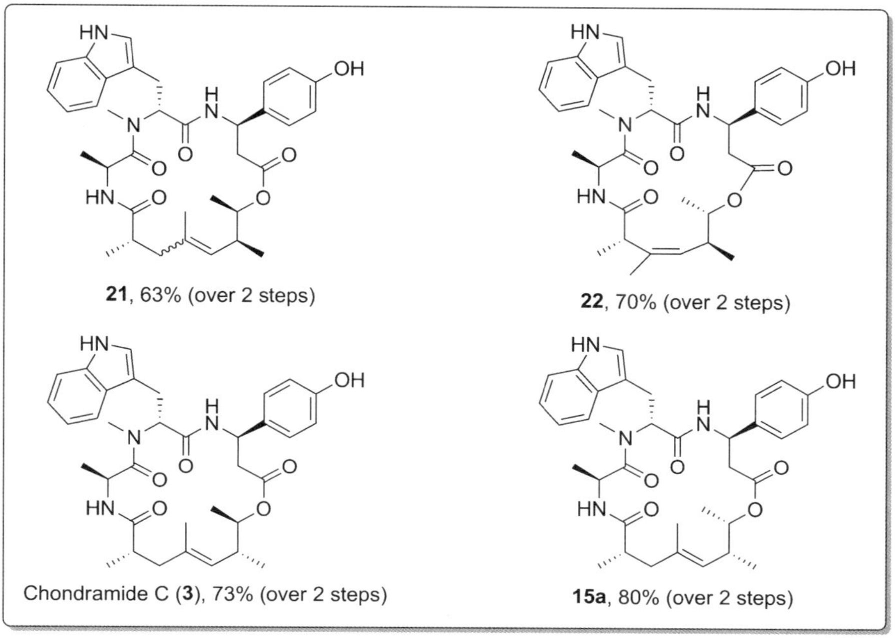

21, 63% (over 2 steps)

22, 70% (over 2 steps)

Chondramide C (**3**), 73% (over 2 steps)

15a, 80% (over 2 steps)

Figure 2.3   Chondramide C (**3**) and derivatives **15a**, **21** and **22**.

chondramide A from Maier *et al.* began with a peptide coupling to connect **32** and **33** (Scheme 2.5). The unusual β-tyrosine moiety with the additional α-methoxy-substituent was synthesised from the 4-TIPS-protected hydroxyphenylacryl acid methyl ester obtained by Sharpless dihydroxylation and Mitsunobu inversion. Saponification of the methyl ester with trimethyl tin hydroxide yielded the free acid **35**, which was then linked to the polyketide

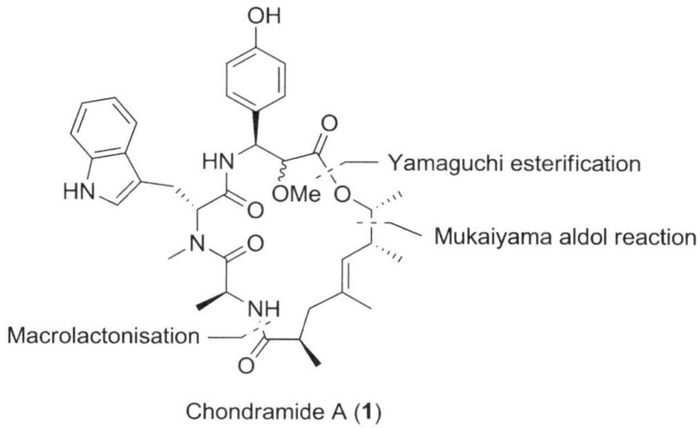

Chondramide A (**1**)

Figure 2.4   Retrosynthesis of chondramide A (**3**).

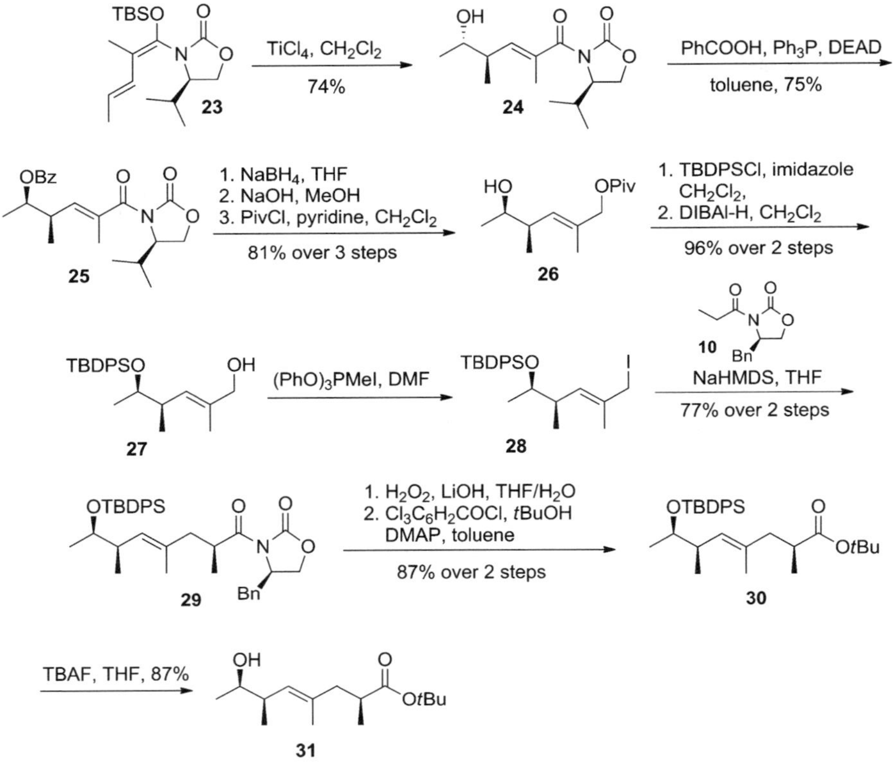

Scheme 2.4   Synthesis of the polyketide fragment **31**.

fragment **31** using a Yamaguchi esterification. Trifluoroacetic acid was used to remove the BOC protecting group and the *tert*-butyl ester in one step and the following macrolactonisation with HOBt and TBTU yielded the protected depsipeptide **36**. In the last step, the TIPS group was removed with TBAF leading to chondramide A (**1**) in 17 linear steps.

2.3   Chivosazoles: Inhibitors of Actin Polymerisation

The chivosazoles are a class of natural compounds made up of seven 31-membered ring macrolides (Figure 2.5, Table 2.1). They were isolated in 1995 by Höfle and co-workers from the myxobacterium *Sorangium cellulosum*. The chivosazoles have a wide range of biological activity, including high antibiotic activity, activity against yeast and filamentous fungi and cytotoxicity against various cancer cell lines (MIC: 9 ng/ml, L 929 and HeLa).

In 2009 Sasse *et al.* published the results of a more detailed study of the cytotoxicity of chivosazole A (**37**) and F (**43**).[13] In this study the anti-proliferative activities of the compounds were tested in further cell lines and

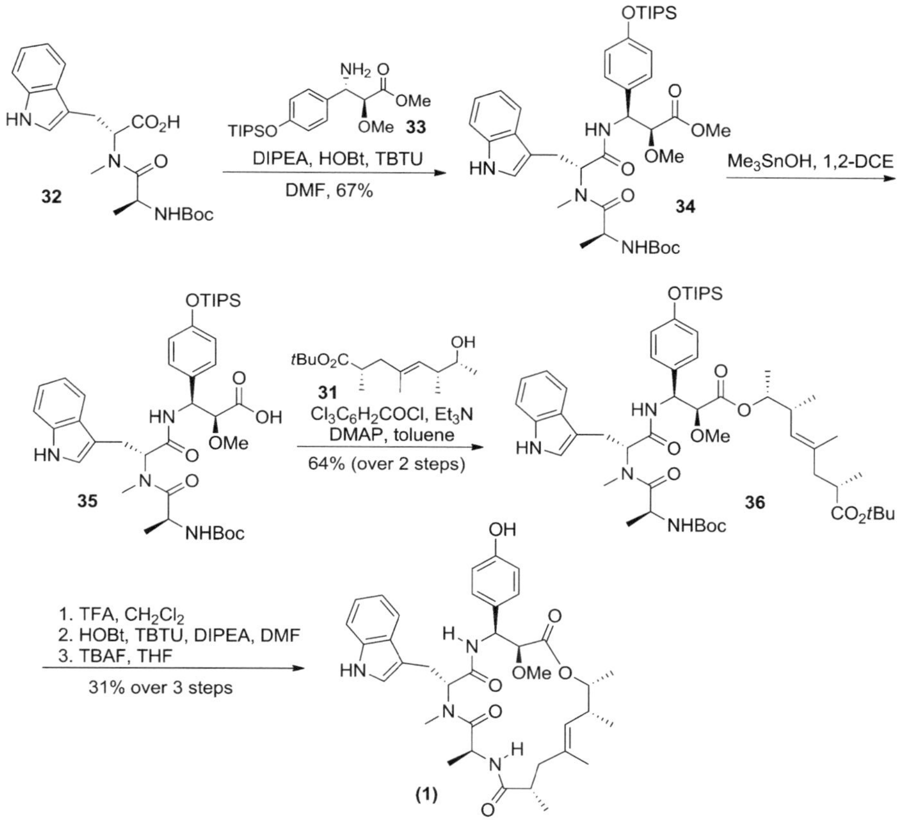

Scheme 2.5 Synthesis of chondramide A (**1**).

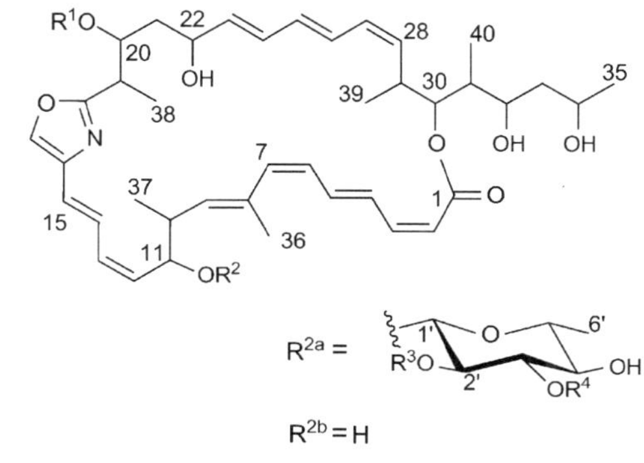

Figure 2.5 Chivosazoles A–F, **37–43**.

Table 2.1   The Chivosazoles A–F, **37–43**.

Chivosazole	*R¹*	*R²*	*R³*	*R⁴*
A (**37**)	CH₃	a	CH₃	CH₃
A₁ (**38**) (6,7-*E*)	CH₃	a	CH₃	CH₃
B (**39**)	H	a	CH₃	CH₃
C (**40**)	CH₃	a	H	H
D (**41**)	H	a	H	CH₃
E (**42**)	H	a	H	H
F (**43**)	CH₃	b	–	–

the results compared with those for rhizopodin and cytochalasin D.[14–21] In all cases the IC_{50} values were found to be in the nanomolar range, therefore in the same order of magnitude as those for rhizopodin and cytochalasin D. The mode of action of chivosazole had already been shown to be due to interaction with the actin cytoskeleton. After the treatment of human epithelial carcinoma cell line A 431 with chivosazole F it was observed that the actin cytoskeleton appeared only to be destroyed in isolated patches. The further observation that many of the cells contained two nuclei indicates inhibition of cytokinesis which also leads to inhibition of proliferation.

The basic structure of the chivosazoles was determined using chivosazole A (**37**) with the aid of mass spectrometry and NMR spectroscopy. The chivosazoles are 31-membered macrolactones comprising of three polyene segments and a side chain bearing a diol at C30. Other structural characteristics include an oxazole ring and a glycosidic side chain at C11. An exception is chivosazole F (**43**), which bears no glycosidic side chain.

Further variation of the chivosazoles is solely through the hydroxyl and methyl ether groups present both on the main structure next to the oxazole ring and on the glycoside. The double bond geometry of the disubstituted double bonds was determined using the coupling constants in the ¹H NMR, which were 10–12 Hz or 14–16 Hz for *Z* and *E* configurations respectively. The configuration of the trisubstituted double bond was determined to be *E* due to the presence of an NOE signal between the C36 methyl group and the proton on C9.

2.3.1   Elucidation of the Absolute and Relative Configuration of Chivosazole A

The configuration of chivosazole A (**37**) was first determined by the Kalesse group using a combination of genetic analysis, conformational analysis and synthesis (Figure 2.6).[22–25]

The relative configurations of the diols at C32 and C34 in chivosazole A (**37**) could be determined using the acetonide method developed by Rychnovsky and Evans.[26,27] The reaction of chivosazole A (**37**) with acetone and *para*-toluenesulfonic acid gave acetonide **44** and the ¹³C NMR shifts showed that the hydroxyl groups in this case have an *anti*-conformation (Scheme 2.6).

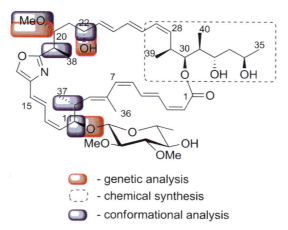

Figure 2.6 Determination of the absolute configuration of chivosazole A (**37**).

To determine the relative configuration of all stereocentres between C28 and C35, the free hydroxyl groups on chivosazole A (**37**) were TBS protected and fragmented using ozonolysis followed by reductive work-up to give **45** (Scheme 2.7). This fragment was studied using spectroscopic methods which showed strong NOE contacts between the proton on the methyl group C39 and the proton at C31, between the protons at C40 and C29 and between protons at C33 and C31.[28]

The proposed configurations were confirmed through chemical synthesis. The synthesis started from commercially available (*S*)-3-hydroxy-2-methyl-propionic acid methyl ester (**46**). Firstly a *syn*-selective Evans aldol reaction with aldehyde **48**, followed by acetal formation, reductive cleavage of the chiral auxiliary and Swern oxidation gave aldehyde **50**.

An *anti*-Felkin-selective Mukaiyama aldol reaction and reduction with Me₄NHB(OAc)₃ gave selectively the *anti*-diol. TBS protection and cleavage of the acetal led to synthetic fragment **54** (Scheme 2.8) which could be compared to the same fragment derived from ozonolysis.

The other five stereocentres (Figure 2.5) were determined through a combination of NMR experiments and molecular modelling. The results were

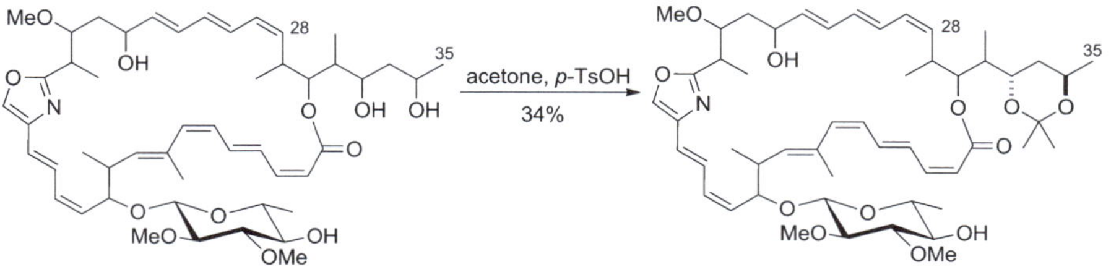

Scheme 2.6 Synthesis of the acetonide **44**.

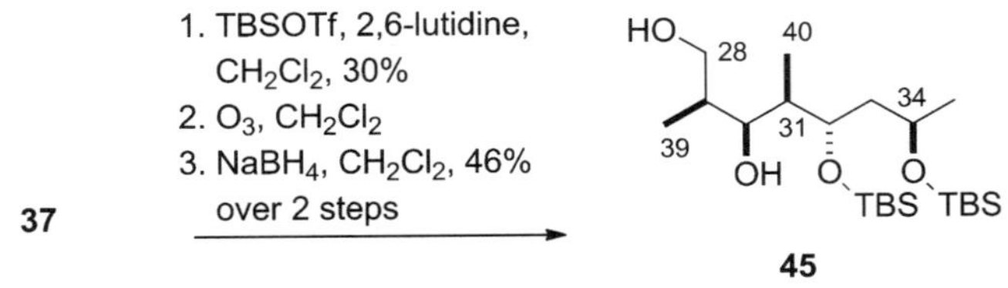

Scheme 2.7 Fragmentation of chivosazole A (**37**) by TBS protection and reductive ozonolysis.

based on a transannular NOE between H7 and one of the diastereotopic protons on C21. Additionally, genetic analysis of the indicative region of the ketoreductase (KR) independently confirmed the configurations at C30, C32 and C34 to be consistent with the analysis performed by standard spectroscopic methods. The advantage of genetic analysis to determine the configuration of polyketides is that it is stand-alone; it depends on no other complementary methods.

2.3.2 Synthesis of Chivotriene, a Chivosazole Shunt Product from *Sorangium cellulosum*

Chivotriene (**55**) is a shunt product of chivosazole biosynthesis and was also isolated at the HZI in Braunschweig from myxobacteria *Sorangium cellulosum*

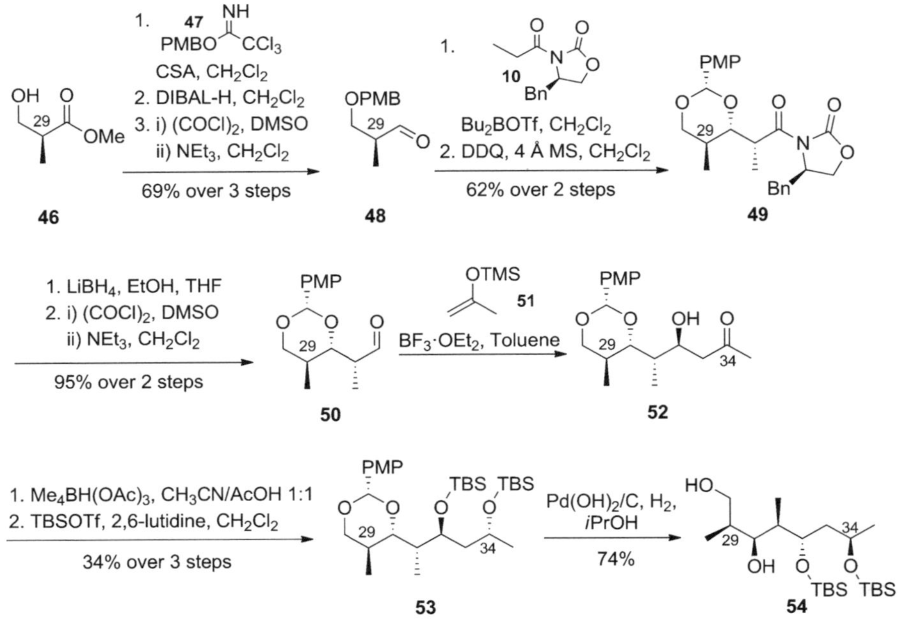

Scheme 2.8 Synthesis of fragment **54**.

So ce12. The structure of chivotriene (**55**) is consistent with that suggested by Perlova *et al.* and can be explained by the hydrolysis and decarboxylation of the biosynthetic product at ACP8 (Scheme 2.9).[29,30]

Comparison of the NMR spectra of chivotriene (**55**) with those of chivosazole A (**37**) allowed the shunt product to be identified as the C20 to C35 fragment of the chivosazole.

The retrosynthesis of the shunt product was developed in analogy to that of the northern fragment of chivosazole A (**37**). Chivotriene (**55**) was taken back to a west and east fragment which could be connected using a Wittig olefination to form the double bond between C8 and C9 (Scheme 2.10). Consequently, chivotriene (**55**) could be assembled from either α,β-unsaturated aldehyde **57** or **59** and phosphonium salt **56** or **58**. The β-hydroxyketone proved to be a further challenge as under the basic conditions of the olefination reaction, elimination of the TBS-ether occurred to give the fully conjugated system.

For the synthesis of both fragments Kalesse *et al.* took advantage of the already established synthetic approach to the northern hemisphere of chivosazole A. The protecting group strategy for the later synthesis of chivosazole F (**43**) was evaluated in the synthesis of the shunt product. Therefore, one of the protecting groups on the triol portion of the molecule, at C13, must be differentiable to the others. The synthesis of the α,β-unsaturated aldehyde **57** began with a Nagao aldol reaction to install the configurations at C5. After TBS protection, the Nagao auxiliary was removed and the Weinreb amide **62** formed (Scheme 2.11).[31,32] It was reacted with ethyl magnesium bromide to give the ethyl ketone and the PMB protecting group removed using DDQ to liberate the alcohol. A final oxidation using MnO$_2$ led to α,β-unsaturated aldehyde **57** required for the Wittig olefination. The corresponding phosphonium bromide **56** was constructed along the same line as for aldehyde **57**. Rather than the final oxidation, alcohol **63** was mesylated and LiBr was used to give the allyl bromide. The final step to form **56** was substitution of the bromide by tributylphosphine.

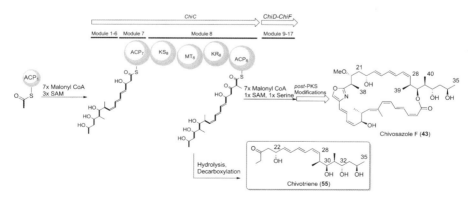

Scheme 2.9   Suggested biosynthesis of chivotriene **55**.

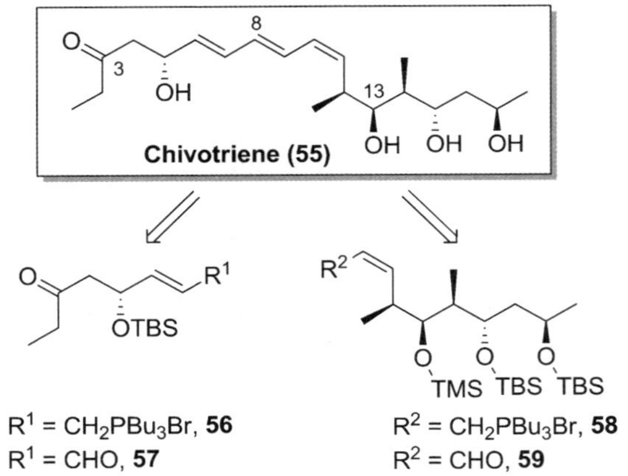

Scheme 2.10   Retrosynthesis of chivotriene (**55**).

The synthesis of the eastern fragments **58** and **59** started from an advanced intermediate **53**, synthesised by Janssen, already described in Section 2.1.3.1. The modifications to this synthesis are shown in Scheme 2.12. After Ando olefination to form ethyl ester **67**, the PMB ether was changed for a TMS protecting group. Reduction with DIBAl-H gave primary allyl alcohol **68** which was oxidised with MnO_2 or converted to the allyl bromide. The phosphonium bromide **58** was achieved from the allyl bromide and used directly in the Wittig reaction.

The Wittig reaction to form triene **68** proved problematic. Using KO^{t}Bu in toluene at 0 °C led to elimination products (Table 2.2) as the main by-products. Furthermore, these reaction conditions only provided non-selective double bond formation between C8 and C9 ($E:Z = 1:1$). Even at lower

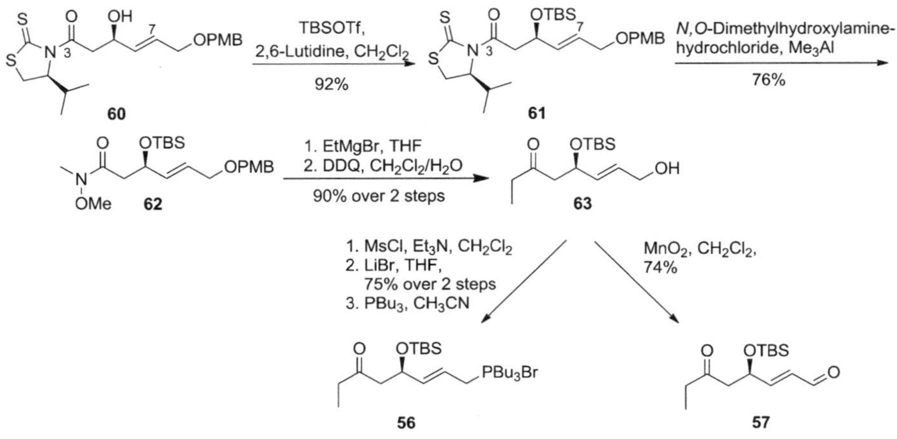

Scheme 2.11   Synthesis of the α,β-unsaturated aldehyde **57**.

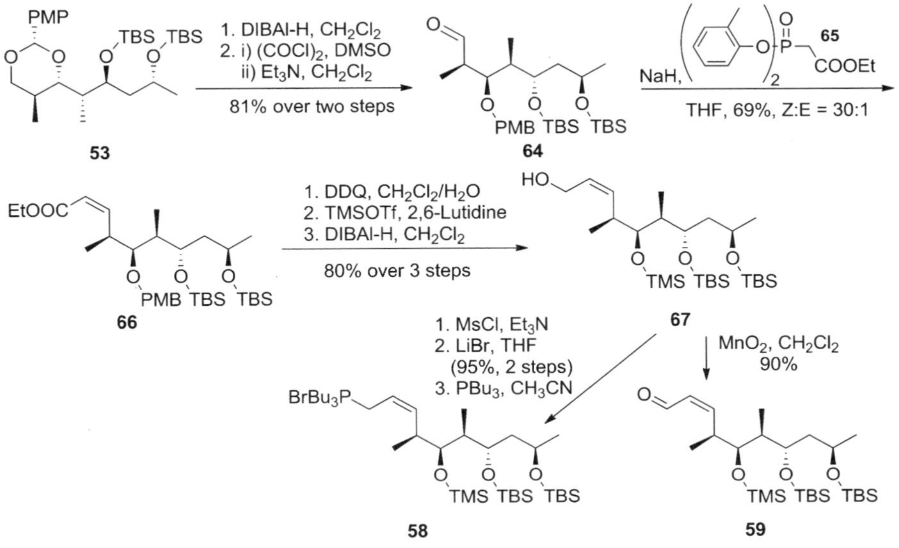

Scheme 2.12   Modified synthesis of the east fragments **58** and **59**.

Table 2.2   Optimisation of the Wittig olefination

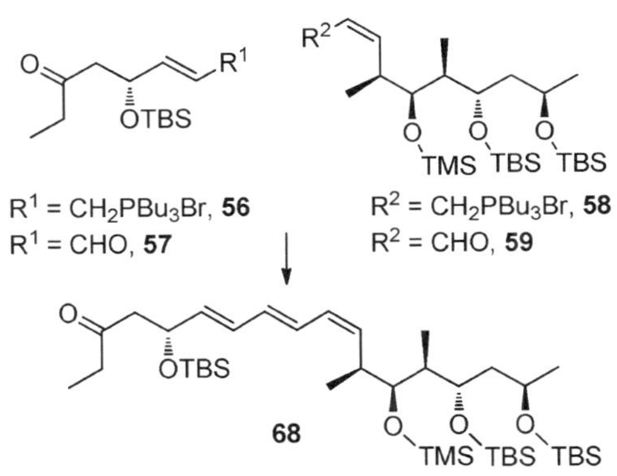

$R^1 = CH_2PBu_3Br$, **56**
$R^1 = CHO$, **57**

$R^2 = CH_2PBu_3Br$, **58**
$R^2 = CHO$, **59**

Entry	Base	R^1	R^2	E/Z	T [°C]	Yield[a] [%]
1	KO^{t}Bu	CHO	CH$_2$PBu$_3$Br	1:1	0	57, 36[b]
2	KO^{t}Bu	CHO	CH$_2$PBu$_3$Br	2:1	−30	78
3	LiO^{t}Bu	CHO	CH$_2$PBu$_3$Br	2:1	−30	71
4	LiHMDS	CHO	CH$_2$PBu$_3$Br	2:1	−30	81
5	KO^{t}Bu	CH$_2$PBu$_3$Br	CHO	7:1	−30	73

[a]Calculated over two steps from the allyl bromide. [b]Elimination products.

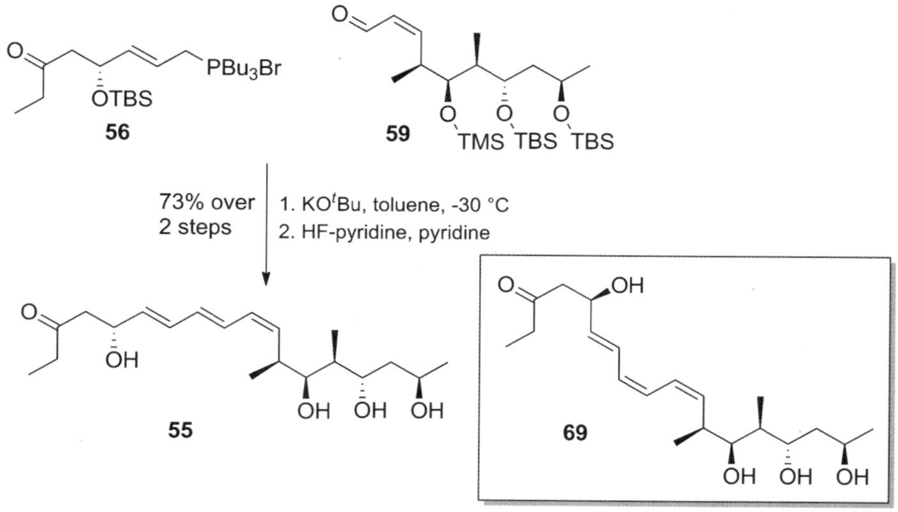

Scheme 2.13   Optimised synthesis of chivotriene **55**.

temperatures, the selectivity ($E{:}Z = 2{:}1$) could not be improved to practicable ratios.

A solution to this problem was to interchange the functionalities of the eastern and western fragments. Gratifyingly, these changes provided the desired product in 73% yield and with an E/Z selectivity of 7:1 (Scheme 2.13). The shunt product could be globally deprotected using buffered HF-pyridine and the two double bond isomers could be separated using HPLC.

Experiments to explore the biological activity of chivotriene (**55**) and the E,Z,Z-isomer **69** were carried out against various cell lines. Both molecules exhibited weak anti-proliferative activity with IC_{50} values in the micromolar range. Surprisingly, the E,Z,Z-isomer **69** showed a higher activity than the isolated shunt product (Table 2.3). The activity in relation to actin polymerisation was investigated, but no effect on the actin filaments could be detected. Therefore, it seems that the shunt product **55** has a different mode of action compared to that of chivosazole.

Table 2.3   Biological activity of chivotriene (**55**) and the E,Z,Z-isomer **69**

	IC_{50} [µg/mL]	
Cell line	*Chivotriene (55)*	*E,Z,Z-Isomer 70*
L-929	50	23
KB-3-1	>100	>100
PC-3	>100	>100
U-937	—	25
HUVEC	100	25

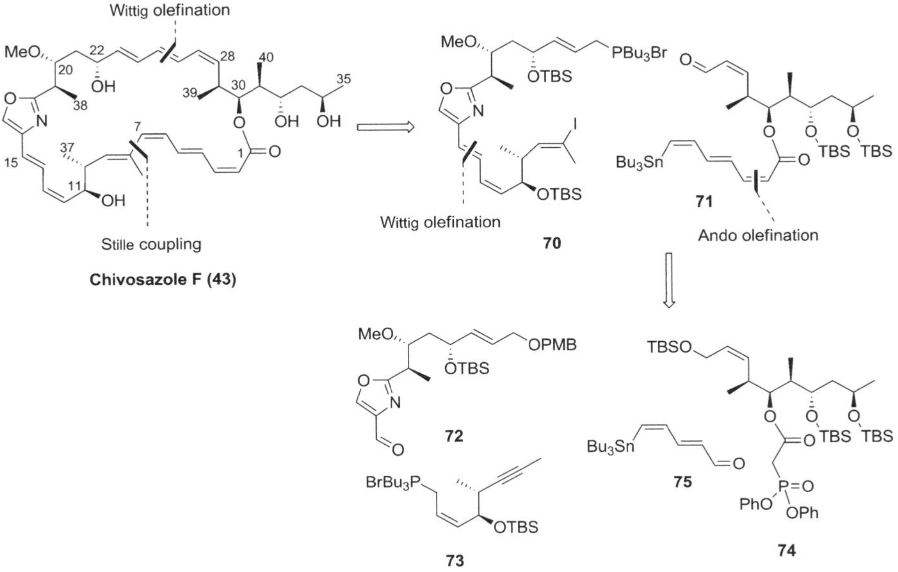

Scheme 2.14 Retrosynthesis of chivosazole F (**43**).

2.3.3 Total Synthesis of Chivosazole F by Kalesse

The retrosynthesis of chivosazole F (**43**) leads to an eastern and a western fragment (Scheme 2.14). These fragments were chosen to construct the sensitive tri- and tetraene segments at a late stage in synthesis. This strategy consequently requires that the fully functionalised eastern and western fragments can be joined in two steps leading to protected chivosazole F.

The *Z,E,Z*-triene of the east fragment **71** could be made from aldehyde **75** by Ando olefination with phosphonate **74**. For the synthesis of phosphonate **74**, the literature-known ester **66** could be used. The PMB deprotection was

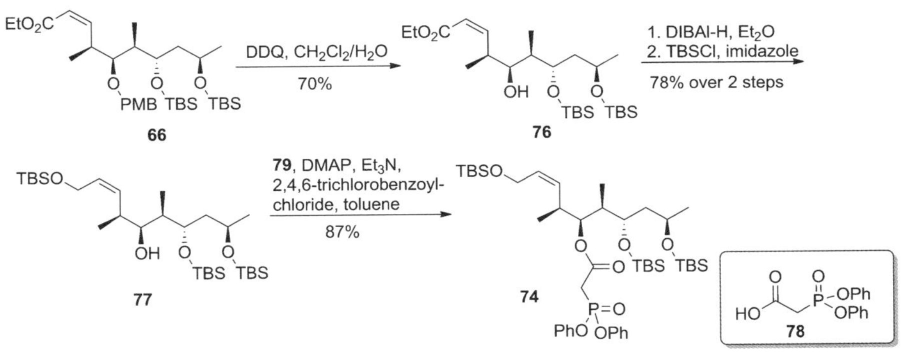

Scheme 2.15 Synthesis of phosphonate **74**.

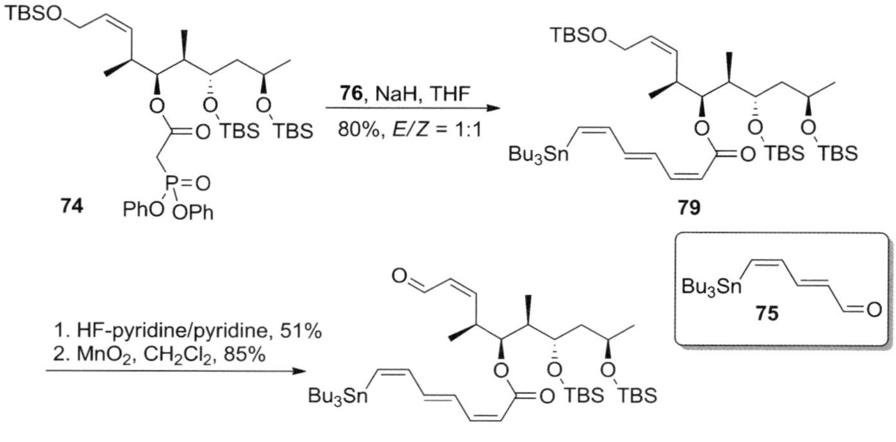

Scheme 2.16   Synthesis of the east fragment **71** of chivosazole F (**43**).

carried out using DDQ in good yield and reduction of the ethyl ester and selective protection of the primary alcohol as the TBS ether led to compound **77** (Scheme 2.15). Esterification with phosphonate **78** under Yamaguchi conditions gave **74** in very good yield.[33]

Phosphonate **74** was reacted with aldehyde **75** in an Ando olefination (Scheme 2.16). This gave a very good yield of the desired alkene as a 1:1 mixture of the *E* and *Z* isomers. The separation of the isomers could be carried out by flash column chromatography after deprotection of the primary alcohol with HF-pyridine. An oxidation with MnO_2 led to the eastern fragment **71** which contained the required aldehyde and stannane for the following Wittig olefination and Stille coupling reactions to form the triene and tetraene respectively.

The western fragment **70** could be synthesised starting from aldehyde **72** and phosphonium bromide **73**. The *anti*-relationship in fragment **72** was established through a Marshall-Tamaru reaction.[34–36] In order to circumvent the modest selectivities of the Marshall reaction with unsaturated aldehydes, Kalesse *et al.* used aldehyde **80** with a triethyl silyl group in the α-position to enhance facial and diastereoselectivity. Using aldehyde **80** and *S*-configured propargylic mesylate **81** in an indium-promoted Marshall reaction, alcohol **82** was obtained in good yield and high selectivity (68%, *anti:syn* = 10:1). Further standard transformations led to phosphonium bromide **73** (Scheme 2.17).

The oxazole fragment **83** was prepared according to the route described by Janssen *et al.*[23] DIBAl-H reduction established the aldehyde required for the Wittig olefination with phosphonium bromide **73** (Scheme 2.18). This olefination installed the C25–C26 double bond in high yield and selectivity (91%, *E:Z* > 10:1).

For the subsequent functional group manipulations it proved beneficial to first remove the PMB group with DDQ and then to install the tri-substituted

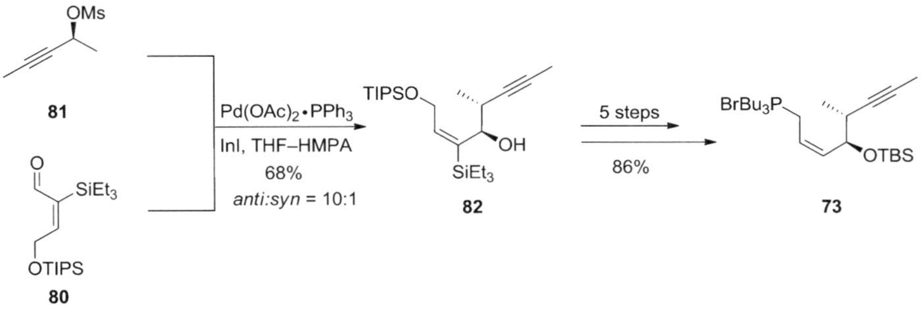

Scheme 2.17 Synthesis of phosphonium bromide **73**.

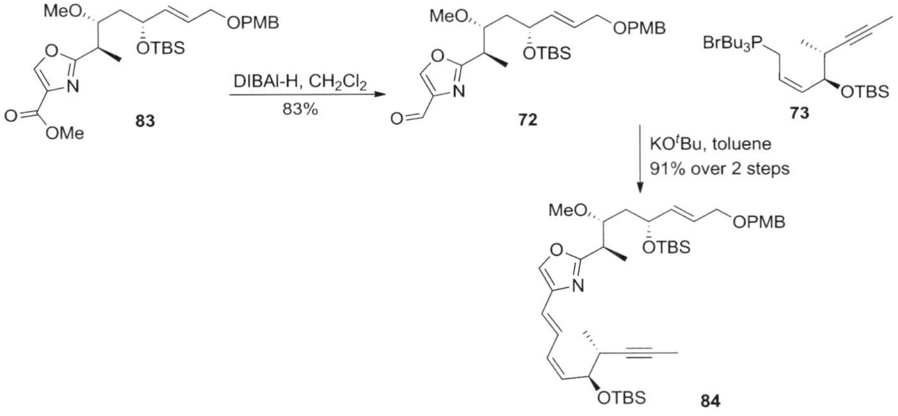

Scheme 2.18 Synthesis of alkyne **84**.

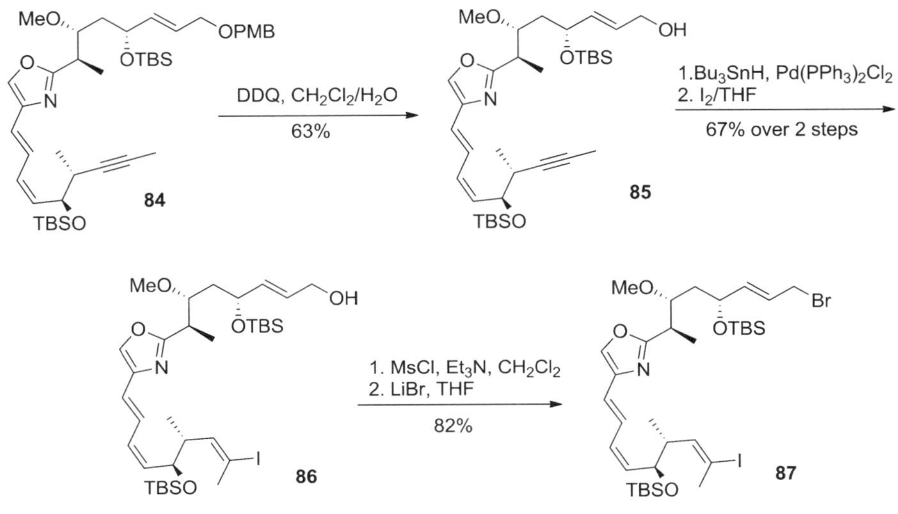

Scheme 2.19 Synthesis of allylbromide **87**.

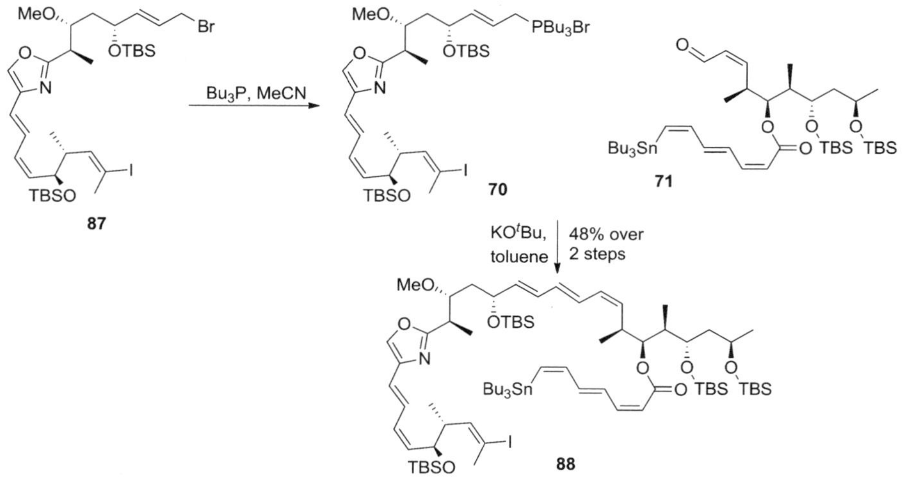

Scheme 2.20   Coupling of west **70** and east **71** fragments using Wittig olefination.

double bond through a palladium-catalysed hydrostannylation reaction with tributyl tin hydride in the presence of the free hydroxyl group (Scheme 2.19). The light- and acid-sensitive vinyl iodide **86** was obtained as a single isomer.

The required bromide was introduced via a displacement reaction of the mesylate with LiBr. Reaction with tributylphosphine led to west fragment **70** which was coupled to the east fragment using Wittig olefination to give the *E* double bond between C25 and C26 (Scheme 2.20)

In the subsequent step, the intramolecular Stille coupling was carried out to give TBS protected chivosazole F **89** as a single product (Scheme 2.21). Problems due to double bond isomerisation in the tetraene segment, as seen in the synthesis of the southern fragment, were not observed using this route. The crude product **89** was globally deprotected using buffered HF-pyridine. The ^{1}H and ^{13}C NMR spectra of synthetic chivosazole F (**43**) were identical to those from the authentic sample.

2.3.4   Synthesis of the Southern Hemisphere by Paterson

Paterson *et al.* independently published a synthesis of the southern fragment of chivosazole F (**43**).[37] Their retrosynthesis led to northern and southern fragments which could be joined using a cross-coupling reaction and an esterification (Scheme 2.22).

It was planned to build the southern fragment **92** from two dienes (**93, 94**) by Stille cross-coupling. In this key step, the bis-halogenated intermediate **93** should be used in two consecutive cross-coupling reactions. The iodide is to be first reacted in a Stille coupling with **94** to build the southern fragment, followed by coupling to the bromide to attach the northern fragment.

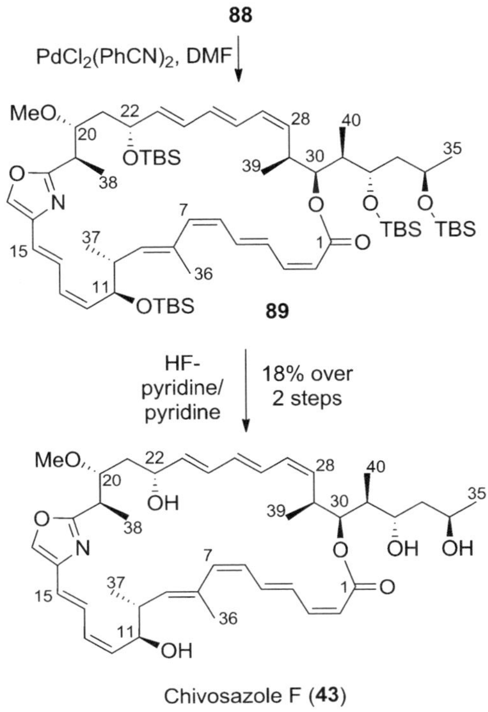

88

PdCl$_2$(PhCN)$_2$, DMF

89

HF-pyridine/pyridine | 18% over 2 steps

Chivosazole F (**43**)

Scheme 2.21 Intramolecular Stille coupling and global deprotection to give chivosazole F (**43**).

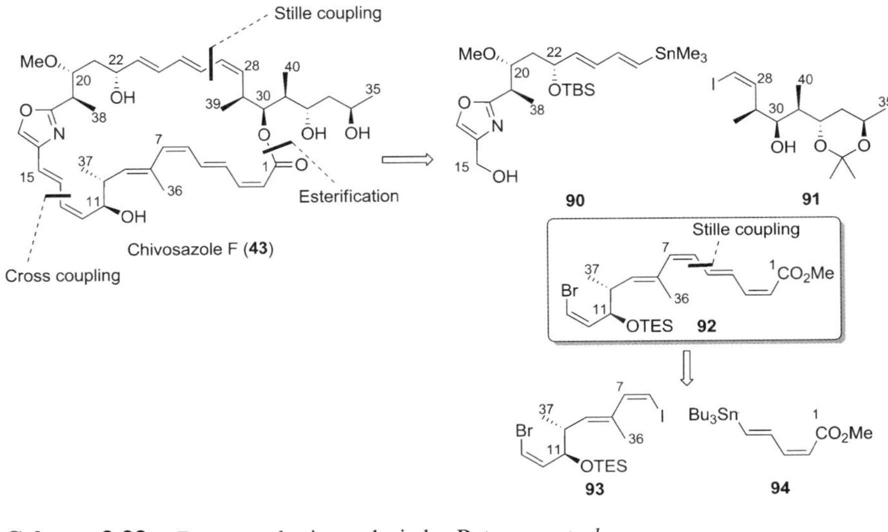

Stille coupling

Chivosazole F (**43**)

Cross coupling

Esterification

90

91

Stille coupling

92

93

94

Scheme 2.22 Retrosynthetic analysis by Paterson *et al.*

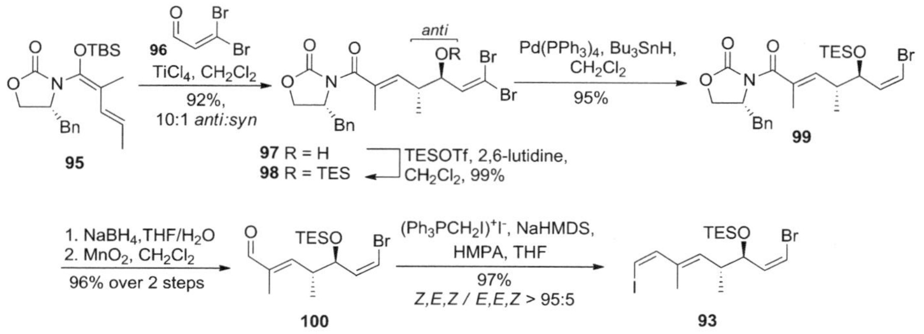

Scheme 2.23   Synthesis of the bis-halogenated triene **93**.

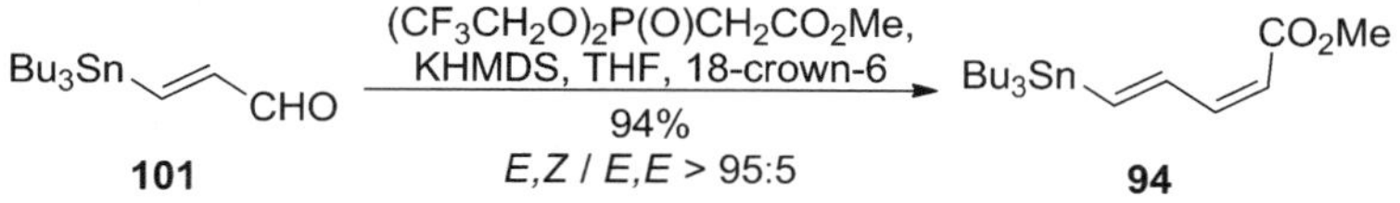

Scheme 2.24   Synthesis of stannane **94**.

For the construction of the bis-halogenated intermediate **93** an asymmetric vinylogous Mukaiyama aldol reaction was envisaged.[38,39] In this reaction, a chiral α-methylated *N,O*-silyl-ketene acetal is reacted with an aldehyde in the presence of a Lewis acid to give the product in high stereoselectivity due to remote 1,6- or 1,6,7-induction from the Evans chiral auxiliary. The product from the reaction of **95** and **96** in the presence of TiCl$_4$ as a Lewis acid gave the product **97** in very good yield (92%) and diastereoselectivity (10:1 *anti:syn*) (Scheme 2.23). The aldol product was then TES-protected and selectively debrominated using tributyl tin and catalytic amounts of palladium(0) to give the (*Z*)-vinylbromide **99**. Reductive removal of the Evans auxiliary and oxidation with MnO$_2$ gave α,β-unsaturated aldehyde **100**, which was used in a Stork-Wittig olefination with excellent *Z/E*-selectivity to give the light-sensitive bis-halogenated triene **93**.[40]

For the Stille coupling, the required stannane **94** could be made using a Still-Gennari olefination from the literature-known α,β-unsaturated aldehyde **101** (Scheme 2.24).[41,42]

The Stille coupling to build the tetraene system used Pd(MeCN)$_2$Cl$_2$ as a catalyst in DMF in the presence of Ph$_2$PO$_2$NBu$_4$ as a tin scavenger (Scheme 2.25). Under these conditions compound **102** was the main product; it was obtained as a mixture of isomers which could not be separated by column chromatography. NMR studies confirmed that the bis-halogen **93** reacted chemoselectively at the vinyl iodide with stannane **94** and optimisation of the Stille coupling to use conditions from Fürstner *et al.*;[43] Pd(PPh$_3$)$_4$, Ph$_2$PO$_2$NBu$_4$ and copper(I)-thiophene-2-carboxylate (CuTC) as an additive in DMF, reduced the *E/Z*-isomerisation and gave the desired southern fragment **92** in a good yield.

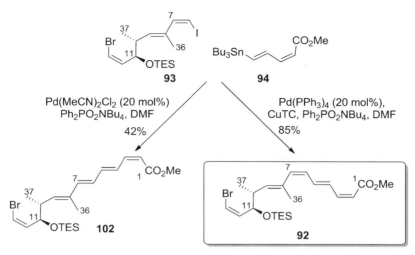

Scheme 2.25   Stille coupling to form the southern fragment **92** of chivosazole F (**43**).

2.4   Conclusion

In summary, the total synthesis of the chondramides confirmed their absolute and relative configurations. The synthesis of additional diastereomers enabled first SAR studies, in which the ability to induce actin polymerisation was compared with the anti-proliferative effects. The results show that the potential antitumour activity is derived through interaction with the actin skeleton. Surprisingly, both chondramide C and isomer **15b** exhibit nearly the same biological activity, which can be explained through a conformational analysis of the isomers. This analysis clearly supports the hypothesis that the polyketide segment acts as a structural element that performs a fine-tuning on the conformation of the amino acids in the molecule.

The first total synthesis of (–)-chivosazole F and independently the synthesis of the southern hemisphere has been achieved through a strategy that establishes the delicate polyene systems using late-stage Wittig reactions and a Stille coupling as the pivotal transformations.

References

1. H. Reichenbach, *J. Ind. Microbiol. Biotechnol.*, 2001, **27**, 149.
2. S. C. Wenzel and R. Müller, *Mol. BioSyst.*, 2009, **5**, 567.
3. G. Hagelueken, S. C. Albrecht, H. Steinmetz, R. Jansen, D. W. Heinz, M. Kalesse and W.-D. Schubert, *Angew. Chem.*, 2009, **121**, 603.
4. B. Kunze, R. Jansen, F. Sasse, H. Höfle and H. Reichenbach, *J. Antibiot.*, 1995, **48**, 1262.
5. R. Jansen, B. Kunze, H. Reichenbach and G. Höfle, *Liebigs Ann.*, 1996, 285.

6. U. Eggert, R. Diestel, F. Sasse, R. Jansen, B. Kunze and M. Kalesse, *Angew. Chem., Int. Ed.*, 2008, **47**, 6478.

7. U. Eggert, R. Diestel, F. Sasse, R. Jansen, B. Kunze and M. Kalesse, *Angew. Chem.*, 2008, **120**, 6578.

8. P. Ashworth, B. Broadbelt, P. Jankowski, P. Kocienski, A. Pimm and R. Bell, *Synthesis*, 1995, 199.

9. H. Waldmann, T.-S. Hu, S. Renner, S. Menninger, R. Tannert, T. Oda and H.-D. Arndt, *Angew. Chem., Int. Ed.*, 2008, **47**, 6473.

10. H. Waldmann, T.-S. Hu, S. Renner, S. Menninger, R. Tannert, T. Oda and H.-D. Arndt, *Angew. Chem.*, 2008, **120**, 6573.

11. A. Schmauder, S. Müller and M. E. Maier, *Tetrahedron*, 2008, **64**, 6263.

12. A. Schmauder, L. D. Dibley and M. E. Maier, *Chem. Eur. J.*, 2010, **16**, 4328.

13. R. Diestel, H. Irschik, R. Jansen, M. W. Khalil, H. Reichenbach and F. Sasse, *ChemBioChem*, 2009, **10**, 2900.

14. F. Sasse, H. Steinmetz, G. Höfle and H. Reichenbach, *J. Antibiot.*, 1993, **46**, 741.

15. T. M. A. Gronewold, F. Sasse, H. Lunsdorf and H. Reichenbach, *Cell Tissue Res.*, 1999, **295**, 121.

16. N. Horstmann and D. Menche, *J. Chem. Soc., Chem. Commun.*, 2008, **41**, 5173.

17. D. C. Lin and S. S. Lin, *Proc. Natl. Acad. Sci. USA*, 1979, **76**, 2345.

18. M. D. Flanagan and S. Lin, *J. Biol. Chem.*, 1980, **255**, 835.

19. S. L. Brenner and E. D. Korn, *J. Biol. Chem.*, 1979, **254**, 9982.

20. S. S. Brown and J. A. Spudich, *J. Cell Biol.*, 1979, **83**, 657.

21. S. MacLean-Fletcher and T. D. Pollard, *Cell (Cambridge, MA, US)*, 1980, **20**, 329.

22. D. Janssen, D. Albert, R. Jansen, R. Müller and M. Kalesse, *Angew. Chem., Int. Ed.*, 2007, **46**, 4898.

23. D. Janssen and M. Kalesse, *Synlett*, 2007, **17**, 2667.

24. T. Brodmann, D. Janssen, F. Sasse, H. Irschik, R. Jansen, R. Müller and M. Kalesse, *Eur. J. Org. Chem.*, 2010, 5155.

25. T. Brodmann, D. Janssen and M. Kalesse, *J. Am. Chem. Soc.*, 2010, **132**, 13610.

26. S. D. Rychnovsky and D. J. Stalitzky, *Tetrahedron Lett.*, 1990, **31**, 954.

27. D. A. Evans, D. L. Rieger and J. R. Gage, *Tetrahedron Lett.*, 1990, **31**, 7099.

28. C. R. Landis, L. L. Luck and J. M. Wright, *J. Magn. Reson., Ser. B.*, 1995, **109**, 44.

29. S. Schneiker, O. Perlova, O. Kaiser, K. Gerth, A. Alici, M. O. Altmeyer, D. Bartels, T. Bekel, S. Beyer, E. Bode, H. B. Bode, C. J. Bolten, J. V. Choudhuri, S. Doss, Y. A. Elnakady, B. Frank, L. Gaigalat, A. Goesmann, C. Groeger, F. Gross, L. Jelsbak, L. Jelsbak, J. Kalinowski, C. Kegler, T. Knauber, S. Konietzny, M. Kopp, L. Krause, D. Krug, B. Linke, T. Mahmud, R. Martinez-Arias, A. C. McHardy, M. Merai, F.

Meyer, S. Mormann, J. Muñoz-Dorado, J. Perez, S. Pradella, S. Rachid, G. Raddatz, F. Rosenau, C. Rückert, F. Sasse, M. Scharfe, S. C. Schuster, G. Suen, A. Treuner-Lange, G. J. Velicer, F.-J. Vorhölter, K. J. Weissman, R. D. Welch, S. C. Wenzel, D. E. Whitworth, S. Wilhelm, C. Wittmann, H. Blöcker, A. Pühler and R. Müller, *Nat. Biotechnol.*, 2007, **25**, 1281.

30. O. Perlova, K. Gerth, O. Kaiser, A. Hans and R. Müller, *J. Biotechnol.*, 2006, 121, , 17 4..

31. D. A. Evans, S. L. Bender and J. Morris, *J. Am. Chem. Soc.*, 1988, **110**, 2506.

32. A. Basha, M. Lipton and S. M. Weinreb, *Tetrahedron Lett.*, 1977, **18**, 4171.

33. J. Inanaga, K. Hirata, H. Saeki, T. Katsuki and M. Yamaguchi, *Bull. Chem. Soc. Jpn.*, 1979, **52**, 1989.

34. Y. Tamaru, S. Goto, A. Tanaka, M. Shimizu and M. Kimura, *Angew. Chem.*, 1996, **108**, 962; Y. Tamaru, S. Goto, A. Tanaka, M. Shimizu and M. Kimura, *Angew. Chem., Int. Ed. Engl.*, 1996, **35**, 878.

35. J. A. Marshall and N. D. Adams, *J. Org. Chem.*, 1998, **63**, 3812.

36. J. A. Marshall and C. M. Grant, *J. Org. Chem.*, 1999, **64**, 8214.

37. I. Paterson, S. B. K. Kan and L. J. Gibson, *Org. Lett.*, 2010, **12**, 3724.

38. S. Shirokawa, M. Kamiyama, T. Nakamura, M. Okada, A. Nakazaki, S. Hosokawa and S. Kobayashi, *J. Am. Chem. Soc.*, 2004, **126**, 13604.

39. M. Shinoyama, S. Shirokawa, A. Nakazaki and S. Kobayashi, *Org. Lett.*, 2009, **11**, 1277.

40. G. Stork and K. Zhao, *Tetrahedron Lett.*, 1989, **30**, 2173.

41. I. Beaudet, J.-L. Parrain and J. P. Quintard, *Tetrahedron Lett.*, 1991, **32**, 6333.

42. B. H. Lipshutz, E. L. Ellsworth, S. H. Dimock and D. C. Reuter, *Tetrahedron Lett.*, 1989, **30**, 2065.

43. A. Fürstner, J. A. Funel, M. Tremblay, L. C. Bouchez, C. Necado, M. Waser, J. Ackerstaff and C. C. Stimson, *Chem. Commun.*, 2008, **25**, 2873.

CHAPTER 3

Class III Lantibiotics – an Emerging Family of Thioether-Containing Peptides

BARTLOMIEJ KRAWCZYK, JOANNA M. KRAWCZYK
AND RODERICH D. SÜSSMUTH*

Technische Universität Berlin, Fakultät II – Institut für Chemie, Strasse des 17. Juni 124, 10623 Berlin, Germany
*E-mail: suessmuth@chem.tu-berlin.de

3.1 Introduction

The term 'lantibiotic' was introduced in 1988 to describe lanthionine-containing peptides displaying antibacterial activities.[1] Lanthionine is a diamino diacid which is reminiscent of the amino acid cystin, with the difference that it contains a thioether moiety in place of the disulfide (Figure 3.1). The most-studied lantibiotic nisin A, produced by *Lactococcus lactis*, was discovered already in 1928 and is one of the oldest known antibacterial agents.[2] It was not until 1971, however, that the structure of nisin was solved and the presence of lanthionine as a constituent was proved. Nisin is still used by the industry as a food preservative due to its high efficiency against Gram-positive bacteria and the lack of human toxicity. Interestingly after 40 years of extensive use no significant resistance was observed.[3,4] Shortly after the structure elucidation, the ribosomal origin of nisin and the mechanism of the biosynthesis were proposed and later experimentally confirmed.[5] Hence, a ribosomally synthesized precursor peptide (designated

RSC Drug Discovery Series No. 25
Drug Discovery from Natural Products
Edited by Olga Genilloud and Francisca Vicente
© The Royal Society of Chemistry 2012
Published by the Royal Society of Chemistry, www.rsc.org

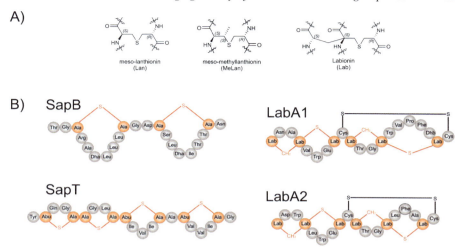

Figure 3.1 (A) Comparison of lanthionine, methyllanthionine and labionin structures. (B) Structures of class III lantibiotics.

sometimes in literature as a prepropeptide) is composed of two segments: the N-terminal region termed leader peptide and the C-terminal region constituting the core peptide, which accommodates amino acids involved in further post-translational modifications (PTMs).[6] As a first step, these PTMs include installation of didehydroamino acids: 2,3-didehydroalanine (Dha) and/or 2,3-didehydrobutyrine (Dhb) obtained by dehydrations of the precursor amino acids: serine and threonine, respectively. In subsequent steps, enzyme-catalysed Michael-type thiol additions of cysteine side chains to unsaturated double bonds form intramolecular bridges: lanthionine (Lan) or methyllanthionine (MeLan) (Figure 3.1). The stereochemistry of Lan and MeLan has been determined as (2*S*,6*R*) and (2*S*,3*S*,6*R*), respectively and is consistent throughout all investigated examples. As the last step of maturation the leader peptide is proteolytically removed to form the mature product. The presence of Lan or MeLan is the unifying feature of all lantibiotics; however, in addition, various other post-translationally introduced modifications were found.[6] Formation of Lan rings is basically mandatory for the biological activity of lantibiotics. In addition, Lan and MeLan increase the proteolytic, thermal and chemical stabilities of the lantibiotic structures. These features together with restraints of conformational flexibility appear beneficial also for other peptidic compounds. It was already demonstrated that introduction of the Lan motif to non-lantibiotic bioactive peptides can be valuable, as described for enkephalin or contryphans.[7–9]

The main reasons of the great interest in lantibiotics by the scientific community are the unique biological activities as demonstrated by the example of nisin. More recently, more than 60 different lantibiotics were described, many of which display very promising and commercially interesting properties, mostly antibacterial.[10] At least three different classification systems were developed to accommodate this growing group of natural products. The first classification was based on the structural and functional aspects and divides lantibiotics into two groups: group A with an elongated, amphiphilic, screw shape and group B, with globular structures.[5,6] Another classification system was based on the amino acid sequence similarities between precursor peptides.[11] The most recent and very universal classification scheme is based on the differences in the arrangement of the biosynthetic gene clusters (Figure 3.2), the architecture of enzymes introducing post-translational modifications and on the presence or absence of antimicrobial activity.[5,12] Within this system three lantibiotic classes were proposed, designated as classes I–III; most recently also the existence of class IV lantibiotics was proposed.[13,14] This classification underlines the common evolutionary origin of members belonging to the same class. Each class has a unique architecture of biosynthetic gene cluster (named *lan* cluster).[6] The essential gene present in all clusters encodes the precursor peptide (designated as *lanA*). The maturation process, i.e. formation of Lan and MeLan, is catalysed by enzymes also encoded within the *lan* gene cluster (referred to as modifying enzymes or lantibiotic synthetases). For class I lantibiotics the dehydration of Ser and Thr to Dha and Dhb is catalysed by the LanB enzyme, and the subsequent conjugate addition of Cys residues to unsaturated didehydroamino acids is catalysed by the enzyme LanC.[6] For class II and III lantibiotics both reactions are performed by one multifunctional enzyme: LanM (class II), and RamC- or LabKC-like enzymes (class III), respectively.[6,14] Another important component of the lantibiotic biosynthesis apparatus is an ATP-binding cassette (ABC) transport system (*lanT*), which is found in almost all lantibiotic gene clusters.[6] The LanT proteins are responsible for secretion of either the final mature product or the post-translationally modified peptide still attached to its leader sequence. These transport systems can either be mandatory or alternative systems can fulfil this role, as observed for example in epidermin biosynthesis. In addition to these components, in the case of class I and II lantibiotics also other genes were identified in boundaries of gene clusters.[15,16] Accordingly, genes encoding a protease (*lanP*) responsible for the proteolytic removal of the leader peptide are often found. Alternatively, the task of proteolytic removal can be fulfilled by other host proteases, or in some cases by ABC-transporting proteins (LanT) harbouring a protease domain. In addition, in most of gene clusters self-immunity genes were also identified coding for either a secondary transport system, lantibiotic-binding proteins or two-component regulatory systems.[6,17,18] Finally, also other genes involved in additional post-translational modifications were identified.[6]

With this respect, gene clusters of class III lantibiotics appear to represent a minimalist design. The only elements present in all characterized clusters of this family are: structural gene(s), one gene encoding the modifying enzyme and two genes coding for the ABC-transporter system. Additionally, a single gene for the response regulatory protein was identified in two cases. Genome-mining analysis suggests also that some of the gene clusters of putative class III lantibiotics might contain a protease gene in the close proximity (unpublished data). This minimalist architecture, e.g. the lack of immunity genes, can be explained in the light of the concomitant lack of antibacterial activity which stands in clear contrast to class I and II. It was suggested that these class III compounds play important morphogenetic functions, as demonstrated for SapB.[19–21] With this respect, the presence of immunity genes as well as a more sophisticated regulatory system or enzymes introducing additional PTMs seem to be unnecessary. Because of the lack of antibacterial activity, class III lantibiotics are sometimes referred to as lantipeptides.[22] One other significant difference between class III and other classes of lantibiotics is manifested in the arrangement of the biosynthetic gene cluster. For class I and II the *lanA* gene is located upstream of the gene coding for the modifying enzyme LanB (class I) or LanM (class II), which is an important factor for maintaining a proper transcription ratio between the structural peptide LanA and the rest of corresponding biosynthetic machinery. For instance the lactocine S structural gene is under the direct control of a promoter and is followed by a leaky terminator which occasionally allows read through of the other biosynthetic components.[23] In the case of class III lantibiotics, the structural gene *lanA* follows the gene encoding the modifying enzyme which suggests a different transcriptional regulation. In fact it was shown in the case of the *ram* gene cluster from *S. lividans* that a strong promoter is located upstream of the *ramC* gene (see Figure 3.2) and a leaky terminator is located downstream of *ramS* terminating most of the transcripts.[24] This points to an equimolar transcription ratio for *ramC* and *ramS*.

Very recently, the existence of a new class of lantibiotics (class IV) was proposed.[14] This class closely resembles class III, with the most prominent difference in the cyclase domain of the synthetase (probably also the architecture of the gene cluster is different (unpublished data)). Up to now only one example of class IV lantibiotics has been described: venezuelin.[13] This peptide was obtained only *in vitro* and no production *in vivo* was detected, probably due to the cryptic nature of the biosynthetic cluster. For this reason a better characterization of class IV lantibiotic biosynthesis systems is required in the future.

For a more comprehensive and general overview on lantibiotics and historical aspects, readers can refer to excellent review articles.[6,5,12,25]

3.2 Class III Lantibiotics

From the current perspective, class III lantibiotics represent the least studied group of lantibiotics, excluding the very recently proposed class IV. Class III lantibiotics were discovered during the course of the identification and characterization of morphogenic genes in *Streptomyces coelicolor*. In contrast to class I and II these peptides do not display any pronounced antibacterial activity. So far only several examples of class III lantibiotics have been described in the literature. These include SapB, SapT, AmfS and the labyrinthopeptins A1 and A2 which are produced by *Streptomyces coelicolor*, *Streptomyces tendae*, *Streptomyces griseus* and *Actinomadura namibiensis*, respectively (Table 3.1).[21,23,27,28] The discovery of these lanthionine-containing peptides without antibacterial activity has significant implications. While the production titres, with the exception of labyrinthopeptins, are extremely low, it is highly likely that many class III lantibiotics were overlooked in common screens and probably many new compounds will be described in the near future as the number of available genome sequences grows every year.

To date, structures of only three class III peptides have been elucidated by means of different analytical techniques. For instance, the structure of SapT has been solved using NMR spectroscopy experiments whereas the structure of labyrinthopeptin A2 has been determined by X-ray crystallography.[26,28] The latter represents one of the rarely available X-ray structures of lantibiotics. The structure of SapB was solved based on Edman degradation and mass spectrometry experiments supported by the genetic data.[21] Interestingly, known structures of SapB, SapT and LabA2 (presented in Figure 3.1) show significant differences in rings topologies underling the structural diversity of this class. Moreover, SapB and LabA1/2 structures show some similarity manifested by the presence of two separate ring systems (Lan in SapB and Lab in labyrinthopeptins).

3.2.1 SapB

SapB was the first discovered and structurally described class III lantibiotic.[21,29] It is also the most studied example, belonging to this group, in terms of physiological function. As described in Section 3.3 of this chapter, SapB has an important function for aerial mycelium formation with a designated role as a

Table 3.1   Overview of known class III lantibiotics

Lantibiotic	Producer	Known: structure/gene cluster	Activity	Reference
SapB	*Streptomyces coelicolor*	yes/yes	surface-active molecule, aerial mycelium formation	21
SapT	*Streptomyces tendae*	yes/no	surface-active molecule, aerial mycelium formation	26
AmfS	*Streptomyces griseus*	no/yes	aerial mycelium formation	27
LabA1	*Actinomadura namibiensis*	yes/yes	antiviral activity	28
LabA2	*Actinomadura namibiensis*	yes/yes	antipain and antiviral activity	28

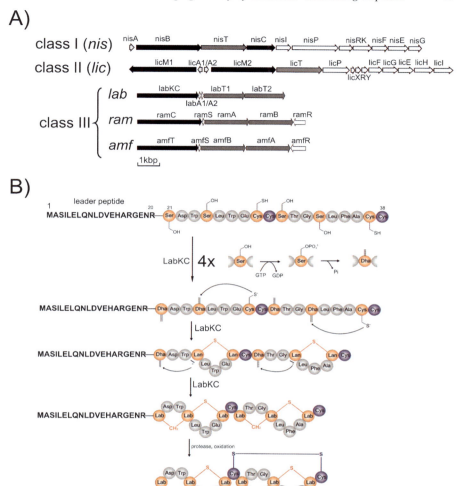

Figure 3.2   (A) Biosynthetic gene clusters of selected lantiobiotics: nis (nisin, class I), lic (lichenicidins, class II), lab (labyrinthopeptins, class III), ram (SapB, class III), amf (AmfS, class III). Genes coding for: precursor peptides in light grey, modifying enzymes in black, transporters in grey. Other genes not relevant for the discussion in white. (B) Proposed mechanism of labyrinthopeptin A2 biosynthesis.

surfactant decreasing the surface tension and allowing the upward growth of the hyphae during the life cycle of *Streptomyces coelicolor*. SapB was isolated in 1988 from *S. coelicolor* cultures but the structure was elucidated significantly later in 2004 by Willey and co-workers.[21] The structure elucidation was based on Edman degradation and mass spectrometric experiments. SapB is composed of 21 amino acid residues and contains two non-overlapping lanthionine rings separated by a dipeptide unit (Figure 3.1). A closer inspection of the amino acid sequence reveals that both rings have a significant hydrophobic character and additionally each of the rings contains one unsaturated Dha residue. Molecular modelling studies suggested an amphiphilic nature of the peptide structure which is in accordance with the surfactant activity of SapB. Before the structure was solved, the biosynthetic gene cluster was identified by a screen for genes triggering aerial mycelium formation.[24] The SapB biosynthetic gene cluster (*ram* cluster) is composed of five genes (designated as *ramABCSR*) (Figure 3.2). The first gene, *ramC*, encodes the modifying enzyme responsible for peptide maturation. Early analysis of this protein revealed that the RamC protein contains a Ser/Thr-kinase like domain.[30] Site-directed mutagenesis studies confirmed the importance of putative active site residues for *in vivo* activity.[30] Furthermore, it was shown that the enzyme is located in the membrane and forms homodimers.[31] This observation gave rise to the premature conclusion that RamC might be a membrane receptor. Later, however, structural data and the identification of a lanthionine cyclase domain revealed that RamC is a lantibiotic synthetase.[32] Recently also the presence of an N-terminal lyase domain was suggested, responsible for phosphate group elimination during installation of didehydroamino acids.[14] The *ramC* gene is followed by *ramS* which codes for the precursor peptide composed of a 21 amino acid long leader peptide and a core part with the same number of amino acids. The genes *ramAB* code for the ATP-binding cassette transporting system. The gene *ramR*, located on the complementary strand of the DNA, encodes a response regulator.[33] The studies devoted to the RamR protein revealed typical domain architecture: an N-terminal receiver domain and a DNA-binding domain at its C-terminus. *In vitro* experiments demonstrated that RamR cannot undergo autophosphorylation (in contrast to other known regulatory response proteins).[34] Nevertheless, non-phosphorylated RamR forms a homodimer which binds the *ramC* promoter region.[34] These observations suggest that the typical phosphorylation-dependent mechanism is not responsible for the activation.

It is also worth noting that a closely related homologous gene cluster was also found in *S. lividans* sharing 99.7% identity with the *ram* cluster of *S. coelicolor*.[24] The discovery and structure elucidation of SapB had serious implication. This peptide was the first described lantibiotic without antibacterial activity, which indicates that the same class of compounds which may be assigned to a common antibacterial defence system of the expression host apparently can also serve other physiological purposes.[32] It also gave rise to a new class of lantibiotics with distinctive modifying enzymes.

3.2.2 SapT

Shortly after the structure elucidation of SapB, another lantibiotic, SapT, from *S. tendae* was isolated and the structure was solved.[26] The compound was identified in the course of purification of streptofactin and restored aerial mycelium formation in non-sporulating *S. coelicolor* mutants.[26,35] The structure was solved using NMR spectroscopy in conjunction with ESI tandem mass spectrometry and Edman degradation. SapT is a 21-residue peptide but despite the same length and similar *in vivo* activity the sequence and structure differs significantly from SapB (Figure 3.1). The structure of SapT contains three MeLan rings and one Lan ring, as presented in Figure 3.1. Nevertheless, as for SapB, molecular modelling studies suggested an amphiphilic nature of the molecule, supporting the proposed surfactant activity. Additionally, as of the time of writing the biosynthetic gene cluster has not yet been identified. With this respect it cannot be ruled out that SapT belongs to a different class of lantibiotics and further investigations are necessary.

3.2.3 AmfS

In the course of attempts to identify genes restoring the aerial mycelium formation in an A-factor-deficient *S. griseus* strain, displaying an aerial mycelium-negative phenotype, a gene cluster designated as *amf* was identified.[27,36,37] This gene cluster shows high similarity to the *ram* cluster identified later. Subsequent studies showed that the peptidic product of the *amfS* gene is responsible for the aerial mycelium formation as demonstrated by knockout mutants.[27] The same studies also suggested that the AmfS peptide acts intracellularly as a signalling molecule based on the wild type morphology of a double *amfAB* knockout mutant lacking the extracellular export of AmfS. Up to now, however, neither a molecular weight has been determined nor has the putative lantibiotic peptide AmfS been subjected to rigorous structure elucidation.

3.2.4 Labyrinthopeptins (LabA1 and LabA2)

Labyrinthopeptins A1 and A2 (LabA1 and LabA2) were identified for the first time at Sanofi-Aventis in 1988 in cultures of the desert bacterium *Actinomadura namibiensis*.[38] More than two decades later the structure of LabA2 was determined by X-ray crystallography (Figure 3.3 and 3.1) by our groups since NMR spectroscopic studies failed in the structure elucidation.[28]

Labyrinthopeptin A2 has a globular structure that primarily consists of hydrophobic, aliphatic and aromatic amino acids and due to the presence of Glu it might posses a slightly acidic character. The most characteristic structural feature of LabA2 is a unique carbacyclic side chain linkage composed of the new post-translationally modified triamino triacid named labionin (Lab) (Figure 3.3).[28] Lab partially integrates structural features of Lan, extended by a methylene bridge to another amino acid. Five different rings, ring A, B, A′, B′ and C are formed to yield the highly side chain bridged structure of LabA2. Each pair of rings A/B and A′/B′ is built within a framework of the triamino triacid labionin and they all share a quaternary

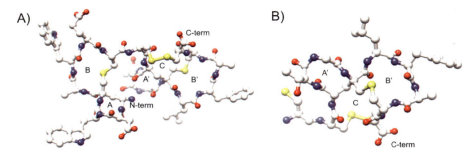

Figure 3.3 (A) X-ray structure of LabA2. (B) Close up view on labionin rings A'B' and C.

α-carbon. Another rare but not uncommon modification for lantibiotics, constituting ring C is a disulfide bond formed between a C-terminal Cys18 residue and Cys9.

The stereochemistry of Lab has been determined as (2*S*,4*S*,8*R*). Interestingly, the stereochemistry of the C-terminal lanthionine-like ring is the same as for the commonly occurring lanthionine (3*S*,6*R*). Only recently, our group succeeded in developing a GC-MS based method for the unambiguous detection of Lab and its differentiation from Lan.[39] The importance of such method lies in the fact that standard techniques used for the characterization of lantibiotics, e.g. Edman degradation and tandem MS, may simply overlook possible labionins. With this method it is now possible to evaluate if labionin is a common structural feature which occurs more frequently in certain bacterial producers. Recent efforts of our group are also directed towards the total synthesis of labyrinthopeptins in order to obtain access to new derivatives not accessible by biosynthetic manipulations. First success has been achieved by the synthesis of a Lab precursor.[40]

Degenerated primer probes used to screen a genomic DNA cosmid library of *A. namibiensis* allowed identification of the labyrinthopeptin (*lab*) biosynthesis gene cluster (6.4 kb) which only encodes five genes.[28] Upstream, the gene *labKC* codes for a multifunctional protein with an N-terminal lyase domain,[14] Ser/Thr protein kinase domain and a C-terminal putative lanthionine cyclase domain. Hence, as this enzyme appeared of central importance for the formation of labionin, the biosynthesis of LabA2 was reconstituted *in vitro* using heterologously expressed modifying enzyme and synthetic substrate peptide.[41] The proposed reaction mechanism is presented in Figure 3.2. Interestingly as a phosphate donor the LabKC enzyme utilizes GTP in contrast to previous reports on class II modifying enzymes which are active in the presence of ATP. The *labKC* gene is followed by *labA1* and *labA2*, coding for precursor peptides of labyrinthopeptin A1 (40 amino acids, the structure had been deduced from the DNA sequence) and labyrinthopeptin A2 (38 amino acids), respectively. The genes *labT1* and *labT2* code for ABC transporters with a putative export function. Although a disulfide bridge (C ring) is present in labyrinthopeptins, a gene coding for a thiol-disulfide oxidoreductase, which could be involved in its formation, like in the case of sublancin 168 biosynthesis,[42] was not identified – suggesting spontaneous disulfide formation after secretion. Likewise, a gene coding for a protease (LabP) could not be identified within the borders of the sequenced cosmid. Finally, the *lab* gene cluster does not display any additional regulatory sequence or immunity genes.

3.2.5 Other Class III Lantibiotics

The occurrence of other class III lantibiotics was suggested by gene cluster analysis of two other Streptomyces strains, i.e. *S. scabies* and *S. avermitilis*.[43] In addition, Table 3.2 presents other hypothetical class III lantibiotics producers identified during genome mining in the NCBI database (www.ncbi.nlm.nih.gov).

3.3 Biosynthesis of Class III Lantibiotics

The general scheme for the biosynthesis of lantibiotics is the same for all lantibiotic classes, with the most prominent difference manifested in the sequences and structures of the modifying enzyme(s). The biosynthesis commences with the formation of Dha and Dhb from Ser and Thr residues

Table 3.2 Putative class III lantibiotics producers (identified by NCBI protein database BLAST search using LabKC sequence as a query) with the number of precursor peptides encoded in the gene cluster and the NCBI accession number of the modifying enzymes

Organism	ID	Number of putative precursor peptides
Streptomyces lividans TK24*	ZP_05522301.1	1
Verrucosispora maris AB-18-032	YP_004406242.1	1
Streptosporangium roseum DSM 43021	YP_003341890.1	3
Kribbella flavida DSM 17836	YP_003384295.1	1
Kribbella flavida DSM 17836	YP_003380339.1	1
Saccharopolyspora erythraea NRRL 2338	YP_001106424.1	1
Actinosynnema mirum DSM 43827	P_003099120.1	1
Streptomyces scabiei 87.22	YP_003486142	1
Stackebrandtia nassauensis DSM 44728	YP_003509339.1	1
Streptomyces avermitilis MA-4680	NP_828679.1	1
Thermomonospora curvata DSM 43183	YP_003301182.1	1
Streptomyces scabiei 87.228	YP_003486614.1	1
Streptomyces griseus subsp. *griseus* NBRC 13350	YP_001823909.1	1
Actinosynnema mirum DSM 43827	YP_003102521.1	2
Lactobacillus delbrueckii subsp. *bulgaricus* ATCC 11842	YP_618260.1	1
Anaerococcus prevotii DSM 20548	YP_003142327.1	3
Catenulispora acidiphila DSM 44928	YP_003114944.1	1

in the ribosomally synthesized precursor peptide. With this respect residues are first phosphorylated providing a leaving group for the subsequent elimination of phosphate (PO_4^{3-}) accompanied by abstraction of the α-proton.[44] In the subsequent step lanthionine rings are formed catalytically by a Michael-type addition of the Cys thiol to the unsaturated double bond of Dha or Dhb.[45] The crystal structure of nisin cyclase revealed the presence of a zinc atom in the active site and further studies proved the importance of this metal for the reaction.[45,46] As mentioned above in case of class I lantibiotics, these reactions are catalysed by two separate enzymes (LanB and LanC) and for class II and III lantibiotics only one multifunctional enzyme is involved. The mechanism of labionin formation relevant for the biosynthesis of labyrinthopeptins is still unknown. Precursor amino acids for Lab comprise a Ser-(Xxx)$_2$-Ser-(Xxx)$_{3/5}$-Cys motif, which upon dehydration of serines mechanistically would afford a double Michael-type addition (Figure 3.2). It has been postulated that during the thiol addition of the Cys side chain to the unsaturated double bond of Dha the α-carbon is not protonated; instead a carbanion is formed undergoing the subsequent addition to the second double bond located two residues apart.[41] Further investigation of this reaction, however, is necessary to unveil details of the mechanism.

The uniqueness of the biosynthesis of class III lantibiotics is manifested in the distinctive domain architecture of the modifying enzymes. As mentioned above the modifying enzyme LabKC is composed (according to homology comparisons) of three domains: lyase, Ser/Thr protein kinase and cyclase. With this respect each enzymatic activity required for the formation of the mature product can be assigned to a well-defined protein domain. In contrast, in the case of class I and II lantibiotic synthetases the lyase and kinase activities cannot be unambiguously assigned to specific regions of the protein based only upon the sequence analysis. The requirement of the Ser/Thr kinase activity *in vivo* was also proven by site-directed mutagenesis of the RamC enzyme in the producer *S. coelicolor*. Hence, substitution of highly conserved active site residues abolished the aerial mycelium formation of *S. coelicolor*.[30] Most recently the presence of the lyase domain for related class IV enzyme VenL was proven, and the reaction mechanism was investigated *in vitro*.[13,14] The conserved catalytically important residues were also shown to be present in class III enzymes, suggesting similar enzymatic mechanisms. In contrast, the catalytic activity of the cyclase domain was neither investigated for LabKC nor for RamC. It was argued that the absence of the active site residues, coordinating zinc atom (in contrast to other classes), might indicate that this domain is inactive.[14] It has, however, to be emphasized that catalysis of the cyclization step is most likely necessary to maintain the proper stereo- and regioselectivity of the reaction since thioether formation can occur spontaneously. For instance LabA2 contains four Cys residues, two of which are involved in labionin formation (Cys8 and Cys17). Cys9 and Cys18 form the disulfide bond (ring C). A spontaneous cyclization reaction would most likely not result in one defined structure as observed. The proposed scheme of LabA2 biosynthesis is presented in Figure 3.2. In this context, it is of particular interest to investigate the cyclization reaction not only with respect to the labionin formation but also for other class III lantibiotics (e.g. SapB), as it can reveal a new mechanism of lanthionine formation. It would be also valuable to understand the features of modifying enzymes which discriminate between the lanthionine and labionin formation.

Although the role of the leader sequence is not entirely clear, it has been already shown on multiple examples that it is required for the proper maturation.[47] The leader peptide can also prevent the activation of the peptide inside the cell. In this case, activation occurs by removal of the leader peptide during or after secretion.[6] Very recently the importance of the leader peptide in the biosynthesis of labyrinthopeptin LabA2 was investigated.[48] It was shown that the leader needs to be attached to the core peptide for proper processing of the precursor peptide by the modifying enzyme LabKC. Moreover, *in vitro* investigation of the processing of different leader peptide mutants allowed identification of a structural motif responsible for the recognition by the enzyme. This motif seems to be conserved in leader peptides from other class III lantibiotics, as shown in Figure 3.4.[43] It was also proposed that the C-terminal part of the leader acts as a spacer necessary for the proper positioning of the residues undergoing modifications. The secondary structure prediction on the leader peptide revealed the possibility of an α-helix formation, which is in agreement with previous observation for class I and II lantibiotics.[48]

As a final comment, it is noteworthy that enzymes displaying homology to class III lantibiotic synthetases are widely spread as judged by NCBI protein database analysis. Interestingly, many BLAST search hits correspond to class III lantibiotic synthetase-like enzymes which are not located in typical lantibiotic gene clusters – thus suggesting the involvement in the production of novel compounds.

3.4 Bioactivities and Physiological Relevance

Most of the lantibiotics from class I and II demonstrate strong antibacterial activities, which together with the ribosomal origin, attracts great interest from the scientific community. In contrast, no significant antimicrobial activity was described for class III lantibiotics. Most likely these compounds play other important physiological functions for the producing organisms. For example

LabA2	M A S I L E L Q N L D V - - E H A R G E N R - -
LabA1	M A S I L E L Q D L E V - - E R A S S A A D - -
SapB	- M N L F D L Q S M E T P K E E A M G D V E T G
AmfS	- M A L L D L Q A M D T P A E D S F G E L R T G
S.avermitilis	- M A L L D L Q T M E S - - D E H T G G G G A -

Figure 3.4 Alignment of leader peptides of class III lantibiotics. The conserved motif is marked with a black line.

bacteria belonging to the Streptomyces genus display a complex lifecycle. One of the most significant features is the ability of cellular differentiation within the colony. At certain time points in the lifecycle in addition to substrate mycelium, a colony can start to form aerial hyphae growing upwardly and being able to differentiate into spores. The first described and characterized class III lantibiotics produced by *Streptomyces* were shown to play morphogenetic functions and be involved in this process as in case of SapB, SapT and AmfS.[19,20,21,26,27] With respect to the role in the formation of aerial hyphae, the most extensively studied example is SapB, which was described already in 1988 as a spore-associated protein.[29] It was demonstrated that purified SapB can restore aerial mycelium formation in bald mutants displaying an aerial mycelium negative phenotype.[19] The biophysical examination confirmed the hypothesis that SapB can act as a surfactant, decreasing the surface tension of aqueous support and allowing bacteria to grow upwards.[43] The surfactant function was in agreement with the proposed amphiphilic nature of the peptide. A very similar function was described for SapT from *S. tendae*, which is also able to restore aerial hyphae formation of *S. coelicolor bld* mutants.[26] Additionally, SapT displays antibacterial activity at higher concentrations against *B. cereus*.[26] However, the physiological relevance of this observation is still unclear. Experiments described above lead to the conclusion that although SapB and SapT differ significantly on the structural level, their function is similar.

The function of AmfS was investigated by the preparation of a *S. avermitilis* mutant with a nonsense mutation in the *amfS* gene coding for the hypothetical precursor peptide. This mutant displays significantly reduced aerial mycelium formation.[27] Further examination of the double knockout mutant of transporters (*amfAB*) showed that this mutant can form aerial mycelium but does not produce AmfS extracellularly (as shown by complementary assay with an *amfS* knockout mutant). This observation led the authors to suggest the intracellular activity of AmfS which is in contrary to results obtained for SapB and SapT. It has to be noticed that AmfS has not been isolated and the structure of AmfS has not yet been elucidated. Therefore an extracellular surfactant activity cannot be ruled out. Interestingly, initiation of an aerial mycelium formation is not the only proposed function for AmfS. For instance the disruption of *amfS* gene had a significant influence on streptomycin production, suggesting its role as a regulatory molecule for antibiotic production.[27] However, the function of SapB and SapT as regulatory molecules was suggested to be unlikely and other effects, beside the ones caused solely by surfactant activity, can be explained as a response to the environmental changes occurring in aerial mycelium.[43] The reader can refer to a more comprehensive excellent review article describing the morphogenic surfactants.[43]

Unfortunately, similar studies on the physiological importance of labyrinthopeptins for the producing host *A. namibiensis* have not yet been conducted. The main reasons are the poor sporulation behaviour of the strain and the absence of the genetic transformation protocols for the producer strain. Nevertheless, labyrinthopeptins isolated from *A. namibiensis* display the most prominent and probably most exciting bioactivities among class III lantibiotics known to date. Biological characterization showed that LabA2 displays an anti-viral activity against *Herpes simplex* and an activity in a spared nerve injury mouse model of neuropathic pain (NP).[28] The antiallodynic effect of LabA2 *in vivo* is of particular importance mainly due to the limited efficacy of currently available drugs and the significant fraction of the population suffering from chronic pain. Activity against NP has been already reported for naturally occurring peptides, e.g. toxins from spiders or cone snails.[49–52] The most prominent example of a peptide in pain treatment is ziconotide (Prialt®), which has been approved for the therapy of severe chronic pain.[53] These peptides are reminiscent of labyrinthopeptins with respect to complex ring topologies. The discovery of remarkable and unexpected bioactivities of labyrinthopeptins implies that other class III lantibiotics might also possess interesting properties and could serve as a new source of compounds with a pharmacological potential. In this context, it could be also of great importance to utilize the labionin scaffold as a building block for new biologically active compounds in a similar fashion as was already done with lantionine.[7–9] However, in order to achieve this goal the synthetic route or the universal biosynthetic approach has to be developed.

Acknowledgements

This work was supported by grants from the Deutsche Forschungsgemeinschaft (DFG SU239/8-1) and the Cluster of Excellence 'Unifying Concepts in Catalysis' coordinated by the Technische Universität Berlin. B.K. thanks the Berlin International Graduate School of Natural Sciences and Engineering (BIG-NSE).

References

1. N. Schnell, K.-D. Entian, U. Schneider, F. Gotz, H. Zahner, R. Kellner and G. Jung, *Nature*, 1988, **333**, 276.
2. C. Piper, P. D. Cotter, R. P. Ross and C. Hill, *Curr. Drug Discov. Technol.*, 2009, **6**, 1.
3. J. Delves-Broughton, P. Blackburn, R. J. Evans and J. Hugenholtz, *Antonie Van Leeuwenhoek*, 1996, **69**, 193.
4. K. L. Blake, C. P. Randall and A. J. O'Neill, *Antimicrob. Agents Chemother.*, 2011, **55**, 2362.
5. G. Bierbaum and H.-G. Sahl, *Curr. Pharm. Biotechnol.*, 2009, **10**, 2.
6. C. Chatterjee, M. Paul, L. Xie and W. A. van der Donk, *Chem. Rev.*, 2005, **105**, 633.
7. Y. Rew, S. Malkmus, C. Svensson, T. L. Yaksh, N. N. Chung, P. W. Schiller, J. A. Cassel, R. N. DeHaven and M. Goodman, *J. Med. Chem.*, 2002, **45**, 3746.

8. H. Kessler, *Angew. Chem., Int. Ed. Engl.*, 1982, **21**, 512.
9. M. R. Levengood and W. A. van der Donk, *Bioorg. Med. Chem. Lett.*, 2008, **18**, 3025.
10. D. Field, C. Hill, P. D. Cotter and R. P. Ross, *Mol. Microbiol.*, 2010, **78**, 1077.
11. C. Piper, P. D. Cotter, R. P. Ross and C. Hill, *Curr. Drug Discov. Technol.*, 2009, **6**, 1.
12. J. M. Willey and W. A. van der Donk, *Annu. Rev. Microbiol.*, 2007, **61**, 477.
13. Y. Goto, B. Li, J. Claesen, Y. Shi, M. J. Bibb and W. A. van der Donk, *PLoS Biol.*, 2010, **8**, e1000339.
14. Y. Goto, A. Okesli and W. A. van der Donk, *Biochemistry*, 2011, **50**, 891.
15. A. Peschel, N. Schnell, M. Hille, K. D. Entian and F. Götz, *Mol. Gen. Genet.*, 1997, **254**, 312.
16. A. Kuipers, J. Wierenga, R. Rink, L. D. Kluskens, A. J. M. Driessen, O. P. Kuipers and G. N. Moll, *Appl. Environ. Microbiol.*, 2006, **72**, 7626.
17. T. Stein, S. Heinzmann, I. Solovieva and K.-D. Entian, *J. Biol. Chem.*, 2003, **278**, 89.
18. K.-ichi Okuda, , S. Yanagihara, K. Shioya, Y. Harada, J.-ichi Nagao, Y. Aso, T. Zendo, J. Nakayama and K. Sonomoto, *Appl. Environ. Microbiol.*, 2008, **74**, 7613.
19. J. Willey, R. Santamaria, J. Guijarro, M. Geistlich and R. Losick, *Cell*, 1991, **65**, 641.
20. R. D. Tillotson, H. A. Wösten, M. Richter and J. M. Willey, *Mol. Microbiol.*, 1998, **30**, 595.
21. S. Kodani, M. E. Hudson, M. C. Durrant, M. J. Buttner, J. R. Nodwell and J. M. Willey, *Proc. Natl. Acad. Sci. USA*, 2004, **101**, 11448.
22. Y. Shi, X. Yang, N. Garg and W. A. van der Donk, *J. Am. Chem. Soc.*, 2011, **133**, 2338.
23. M. Skaugen, C. I. Abildgaard and I. F. Nes, *Mol. Gen. Genet.*, 1997, **253**, 674.
24. B. J. F. Keijser, G. P. van Wezel, G. W. Canters and E. Vijgenboom, *J. Bacteriol.*, 2002, **184**, 4420.
25. C. van Kraaij, W. M. de Vos, R. J. Siezen and O. P. Kuipers, *Nat. Prod. Rep.*, 1999, **16**, 575.
26. S. Kodani, M. A. Lodato, M. C. Durrant, F. Picart and J. M. Willey, *Mol. Microbiol.*, 2005, **58**, 1368.
27. K. Ueda, K.-I. Oinuma, G. Ikeda, K. Hosono, Y. Ohnishi, S. Horinouchi and T. Beppu, *J. Bacteriol.*, 2002, **184**, 1488.
28. K. Meindl, T. Schmiederer, K. Schneider, A. Reicke, D. Butz, S. Keller, H. Gühring, L. Vértesy, J. Wink, H. Hoffmann, M. Brönstrup, G. M. Sheldrick and R. D. Süssmuth, *Angew. Chem., Int. Ed.*, 2010, **49**, 1151.
29. J. Guijarro, R. Santamaria, A. Schauer and R. Losick, *J. Bacteriol.*, 1988, **170**, 1895.
30. M. E. Hudson, D. Zhang and J. R. Nodwell, *J. Bacteriol.*, 2002, **184**, 4920.

31. M. E. Hudson and J. R. Nodwell, *J. Bacteriol.*, 2004, **186**, 1330.
32. S. Kodani, M. E. Hudson, M. C. Durrant, M. J. Buttner, J. R. Nodwell and J. M. Willey, *Proc. Natl. Acad. Sci. USA*, 2004, **101**, 11448.
33. T. J. O'Connor, P. Kanellis and J. R. Nodwell, *Mol. Microbiol.*, 2002, **45**, 45.
34. T. J. O'Connor and J. R. Nodwell, *J. Mol. Biol.*, 2005, **351**, 1030.
35. M. Richter, J. M. Willey, R. Süssmuth, G. Jung and H.-P. Fiedler, *FEMS Microbiol. Lett.*, 1998, **163**, 165.
36. K. Ueda, K. Miyake, S. Horinouchi and T. Beppu, *J. Bacteriol.*, 1993, **175**, 2006.
37. K. Ueda, C.-W. Hsheh, T. Tosaki, H. Shinkawa, T. Beppu and S. Horinouchi, *J. Bacteriol.*, 1998, **180**, 5085.
38. J. Wink, R. M. Kroppenstedt, G. Seibert and E. Stackebrandt, *Int. J. Syst. Evol. Microbiol.*, 2003, **53**, 721.
39. A. Pesic, M. Henkel and R. D. Süssmuth, *Chem. Commun.*, 2011, **47**, 1359.
40. G. M. Sambeth and R. D. Süssmuth, *J. Pept. Sci.*, 2011, **17**, 581.
41. W. M. Müller, T. Schmiederer, P. Ensle and R. D. Süssmuth, *Angew. Chem., Int. Ed.*, 2010, **49**, 2436.
42. R. Dorenbos, T. Stein, J. Kabel, C. Bruand, A. Bolhuis, S. Bron, W. J. Quax and J. M. Van Dijl, *J. Biol. Chem.*, 2002, **277**, 16682.
43. J. M. Willey, A. Willems, S. Kodani and J. R. Nodwell, *Mol. Microbiol.*, 2006, **59**, 731.
44. C. Chatterjee, L. M. Miller, Y. L. Leung, L. Xie, M. Yi, N. L. Kelleher and W. A. van der Donk, *J. Am. Chem. Soc.*, 2005, **127**, 15332.
45. B. Li, J. P. J. Yu, J. S. Brunzelle, G. N. Moll, W. A. van der Donk and S. K. Nair, *Science*, 2006, **311**, 1464.
46. B. Li and W. A. van der Donk, *J. Biol. Chem.*, 2007, **282**, 21169.
47. T. J. Oman and W. A. van der Donk, *Nat. Chem. Biol.*, 2010, **6**, 9.
48. W. M. Müller, P. Ensle, B. Krawczyk and R. D. Süssmuth, *Biochemistry*, 2011, **50**, 8362.
49. S. P. Park, B. M. Kim, J. Y. Koo, H. Cho, C. H. Lee, M. Kim, H. S. Na and U. Oh, *Pain*, 2008, **137**, 208.
50. N. J. Saez, S. Senff, J. E. Jensen, S. Y. Er, V. Herzig, L. D. Rash and G. F. King, *Toxins*, 2010, **2**, 2851.
51. R. J. Lewis, *Prog. Mol. Subcell. Biol.*, 2009, **46**, 45.
52. B. G. Livett, D. W. Sandall, D. Keays, J. Down, K. R. Gayler, N. Satkunanathan and Z. Khalil, *Toxicon*, 2006, **48**, 810.
53. A. Schmidtko, J. Lötsch, R. Freynhagen and G. Geisslinger, *Lancet*, 2010, **375**, 1569.

CHAPTER 4

Mutant Manufacturers

ANDREAS KIRSCHNING*, SIMONE EICHNER,
JEKATHERINA HERMANE AND TOBIAS KNOBLOCH

Institut für Organische Chemie und Biomolekulares Wirkstoffzentrum
(BMWZ) der Gottfried-Wilhelm-Leibniz Universität Hannover, Schneiderberg
1B, 30167 Hannover, Germany
*E-mail: andreas.kirschning@oci.uni-hannover.de

4.1 Introduction

4.1.1 The Quest for Natural Product Libraries

Quo vadis, pharmaceutical research? The current situation in the development
of new drugs is problematic because the big pharmaceutical companies are
lacking new drug candidates in their research pipelines. It can be debated
whether a focus on combinatorial concepts and a reduced reliance on the
experience and wisdom of medicinal chemists, or to a greater degree, the
closure of natural product research units is responsible for this critical
development in drug research.

In fact, regarding natural products as being too complex, too difficult to
handle and too expensive in pre-commercial development has become a
widespread attitude. Thus, simple automation technologies, i.e. robots,
designed to copy and replace the classical synthetic chemist, have infiltrated
the laboratories. In medicinal chemistry, large libraries of fully synthetic
compounds were generated and screened, disregarding the large potential of
natural products – compounds already receptor-optimized during evolution.
The consequence was disappointing: Although large in number but without
significant structural variety, these libraries were incapable of generating

RSC Drug Discovery Series No. 25
Drug Discovery from Natural Products
Edited by Olga Genilloud and Francisca Vicente
© The Royal Society of Chemistry 2012
Published by the Royal Society of Chemistry, www.rsc.org

enough new lead compounds suitable to be developed to medicaments. The result of this approach, known as 'Combinatorial Chemistry', is well known.

It became evident that it was not just producing thousands of analogues but rapidly finding the counterpart to a biological target, i.e. the key to the lock. Natural products, being evolutionary validated ligands, provide excellent chances to fulfill the structural requirements of the receptor. During biological evolution, Nature has only tested and developed a fraction of possible protein receptors. Due to statistical considerations, a full evaluation of all possible structures is not even feasible for Nature.[1] Small ligands, i.e. natural products, are the adaptors to these macromolecular structures, being specifically developed for this world of receptors. Of equal importance is their ability to pass through biological phase layers such as membranes, a highly desirable key characteristic for drug development.[2]

Since evolution is an unfocussed process, not all structural elements in a natural product are essential for its biological activity. This circumstance is similarly reflected by cases in which the biological target for a natural product aimed at in a drug development programme is seemingly very different from all potential targets occurring in the natural environment of its producing organism. A case in point are natural products used as immunosuppressants, e.g. rapamycin. Consequently, there is a quest to develop new or better concepts for the creation of natural product libraries, facilitating structural alterations at any part of a given complex secondary metabolite in order to identify its essential pharmacophores and hence optimise and simplify the structure for its intended area of application.

4.1.2 The Role of Synthetic Chemistry in the Development of Natural Products as Drugs

The chemical sciences are based on the creation of new molecules and materials with desired properties. In principle, organic synthesis allows accessing every possible organic molecule. The last three decades have seen an enormous growth in the development of new reagents and catalysts. As a result, new chemical transformations have been rendered possible, allowing even total synthesis (*de novo* synthesis) to be considered for the industrial production of complex natural products and their derivatives.

For example, the epothilone derivative ZK-EPO 1 is the result of a drug development programme carried out by Bayer-Schering that included the total synthesis of an epothilone library consisting of several hundred derivatives (Figure 4.1). Although ZK-EPO 1 is an inspiring example of success for modern organic chemistry, the extraordinary high chemical and structural complexity of many natural products limits the options to access compound libraries via semisynthesis and total synthesis in order to provide material for structure–activity relationship (SAR) studies.

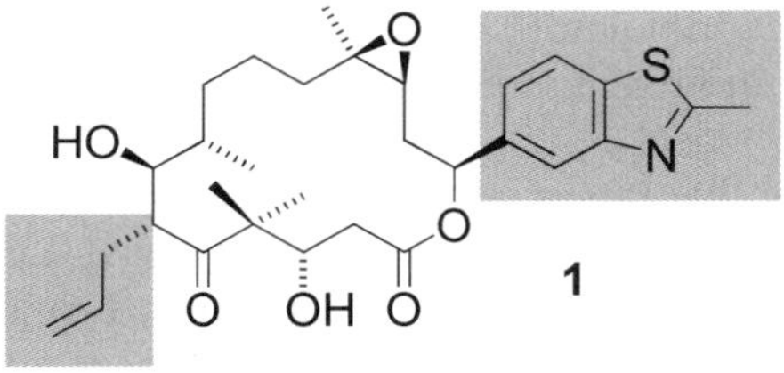

1

Figure 4.1 Bayer-Schering's epothilone derivative **1** (ZK-EPO, synthetic changes with respect to epothilone are marked in grey) is prepared in an industrial scale.

4.1.3 Biosynthetic Tools to Create Natural Product Libraries

Besides the rather traditional synthetic concepts of total synthesis and semisynthesis, new strategies have appeared on the scene that rely on manipulations of microbial secondary metabolites not after but in the course of their biosynthesis. The classical method for introducing structural diversity into microbial secondary metabolites is the utilisation of precursor-directed biosynthesis (PDB).[3] Here, typically a derivative of a biosynthetic precursor to the natural product of interest is given to a culture of the microbial producer's wild-type strain (Scheme 4.1, case B), thus competing with the endogenous natural precursor(s) for further processing by the biosynthetic machinery (Scheme 4.1, case A). Commonly, this approach yields mixtures of natural and unnatural secondary metabolites which often require tedious work-up protocols to yield pure material. Since the natural, endogenous precursor is commonly better processed by the biosynthetic machinery, yields for the new derivatives are often rather low.

A state of the art derivative of this strategy circumvents these disadvantages by employing mutant organisms blocked in the biosynthesis of endogenous key precursors on a genetic level (Scheme 4.1, case C). According to Rinehart and Gottlieb,[4] mutational biosynthesis or in short 'mutasynthesis'[5] comprises the generation of mutants of a producer organism blocked in the formation of a key biosynthetic building block required for assembly of the end product.[6]

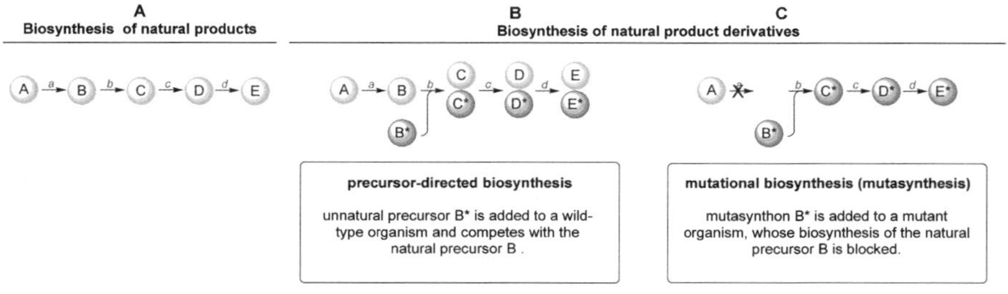

Scheme 4.1 Schematic descriptions of natural product biosynthesis (A), precursor-directed biosynthesis (B) and mutasynthesis (C).

Administration of potentially altered building blocks, so-called mutasynthons, to the blocked mutant may result in new metabolites which are isolated and biologically evaluated.[7] Considering that the tools required for genetic engineering and sequencing and methods for genetic manipulations have dramatically improved over the past two decades, mutasynthesis can nowadays be seriously considered as an elegant approach to access libraries of structural derivatives of pharmacologically important natural products.

While simple gene knockouts are needed for the creation of the required blocked deletion mutants, the crucial point of this concept is the efficiency and selectivity of enzymatic catalysis. In order to allow the generation of unnatural natural products and for the concept of mutasynthesis as a whole to work, the enzymes involved are required to exhibit broad substrate tolerance when presented with non-natural precursors. While the prerequisites of concomitant selectivity and flexibility seem contradictory at first glance, the examples discussed in this article show that a practical compromise between these characteristics of enzymatic machineries exists and can be exploited accordingly.

When viewed from the perspective of a synthetic chemist, the application of mutasynthesis for the creation of natural product libraries is particularly attractive when the following criteria are fulfilled:

1. The secondary metabolite should have excellent biological activity in a pharmaceutically relevant field, or alternatively it could serve as a tool for studies of cell biology. Otherwise the laborious efforts of sequencing and genetic manipulation for its own sake cannot be justified.
2. The secondary metabolite should be structurally complex so that mutational biosynthesis becomes the superior approach over total synthesis.
3. The genetic manipulation should be able to have effect on a structural element in the secondary metabolite not accessible by semisynthesis.
4. The fermentation with modified precursors should yield sufficient amounts of new metabolites for full structural elucidation of the new derivative, e.g. by NMR spectroscopy, and for biological evaluation.

4.2 Mutasynthesis: Illustrative Examples from the Literature

The first important mutasynthetic experiment was conducted in 1991 with *Streptomyces avermitilis*, the producer of the potent anthelmintic drug avermectin **4a** (Scheme 4.2). The structural complexity precludes a facile access to new derivatives by total synthesis. By employing a blocked mutant unable to generate the natural polyketide synthase (PKS)-starter unit *S*-2-methylbutyric acid **3** from the biosynthetic precursor isoleucine **2**, new avermectins modified at C25 were obtained.[8] The loading domain of the avermectins' core biosynthetic machinery turned out to show a very broad

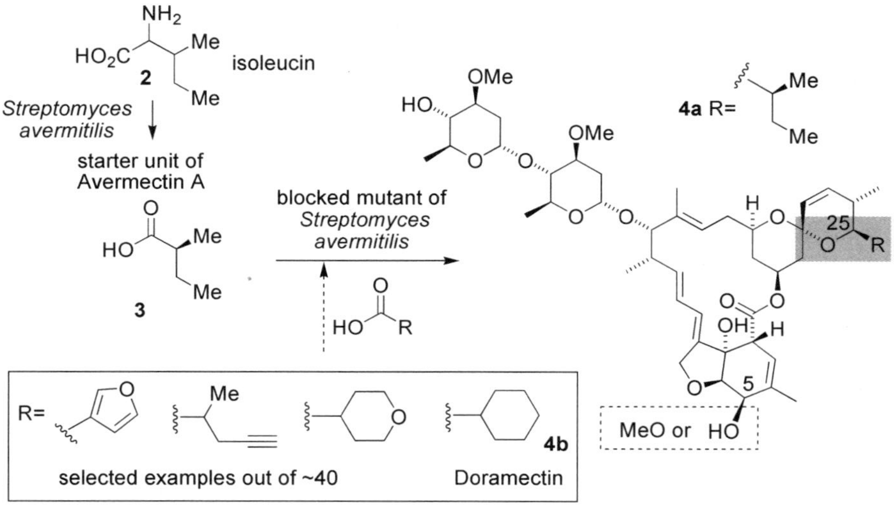

Scheme 4.2   Mutasynthetic preparation of avermectin derivatives and doramectin **4b**.

substrate spectrum. Almost 40 different carboxylic acids were successfully incorporated by the mutasynthesis approach and in addition these studies yielded the first commercial drug obtained by mutasynthesis – doramectin **4b**.

Based on these findings, Reynolds and co-workers developed an elegant access to doramectin **4b**[9] via combinatorial biosynthesis. They transferred the gene cluster that codes for the biosynthesis of cyclohexylcarbonyl-coenzyme A (CoA) from the ansatrienin producer *Streptomyces collinus* into a blocked mutant of *Streptomyces avermitilis*, eliminating the need for supplementation of the fermentation broth with cyclohexylcarboxylic acid.

The glycopeptide vancomycin is a broad spectrum antibiotic in clinical use. The increasing occurrence of resistance, particularly among nosocomial bacterial strains, has evoked intense research activities towards new vancomycin-based derivatives. Semisynthesis did mainly address the glycon part of the molecule, while the peptide aglycon was kept unaltered.[10] The classical synthetic methods were complemented by Süssmuth, Wohlleben and co-workers using a mutasynthetic strategy.[11] They utilised different blocked mutants of the actinomycete *Amycolatopsis balhimycina*, the producer of the structurally related antibiotic balhimycin **5**. Inactivation of the biosynthesis of the precursor β-hydroxytyrosine allowed structural modifications of the resulting 3-chloro-β-hydroxytyrosine subunits, so that structural variants of the tricyclic aglycon, such as the fluorinated product **6**, could be accessed (Scheme 4.3).

Further diversity was achieved by utilisation of a second mutant blocked in the biosynthesis of 3,5-dihydroxyphenylglycine, thereby allowing different modifications of the AB-macrocycle (see **7a–d**) (Scheme 4.4).[12]

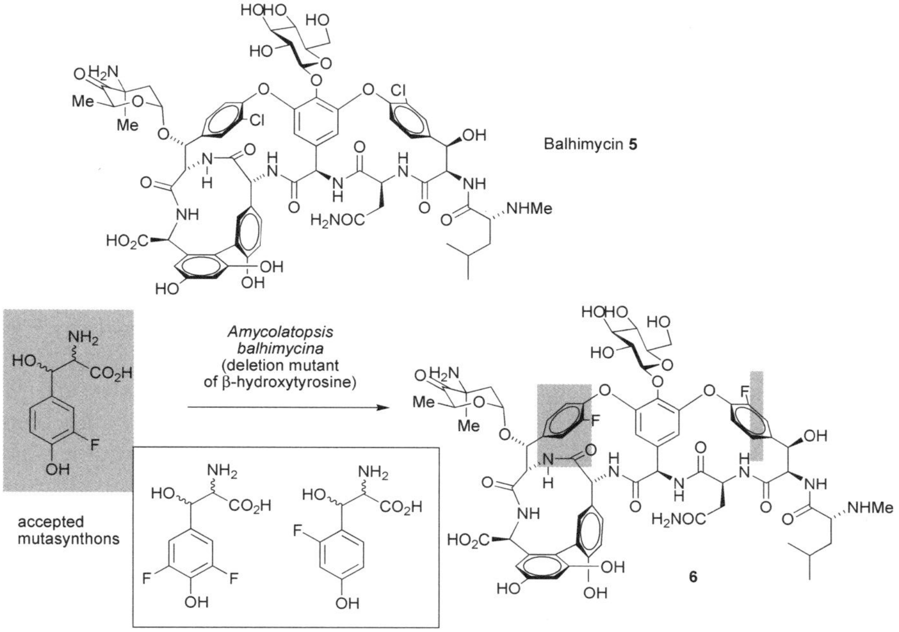

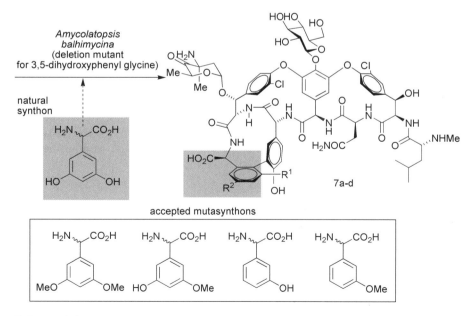

Scheme 4.3 Mutasynthetic preparation of balhimycin derivatives, e.g. compound **6**.

Scheme 4.4 Mutasynthetic preparation of balhimycin derivatives, e.g. compounds **7a–d**.

4.3 Ansamycin Antibiotics: a Showcase for Combined Mutasynthetic/Semisynthetic Strategies

4.3.1 Mutasynthetic/Semisynthetic Hybrid Concepts

As discussed above, mutasynthesis has become a concept for the generation of derivatives of complex natural products which are otherwise very difficult to access by semi- or total synthesis. In principle, structural diversity can further be broadened when mutasynthesis is combined with chemical synthesis and/or a second blocked mutant of the same or another organism is used. As will be exemplified for the ansamitocins, three combined strategies have already been pursued in our group. The first concept is based on the preparation of a new natural product derivative by mutasynthesis, which is subsequently followed by semisynthesis (Scheme 4.5, case D). Alternatively, an advanced biosynthetic intermediate can be produced using a 'late stage' blocked mutant. This advanced intermediate serves as a precursor for a semisynthetic diversification step (Scheme 4.5, case E). Thirdly, the advanced intermediate generated by the 'late stage' blocked mutant is modified by a second, 'early stage' blocked mutant, and finally optionally followed by semisynthesis. These two successive steps may also be performed in a reversed mode (Scheme 4.5, case F).

4.3.2 Mutasynthesis Towards New Ansamycin Antibiotics

The ansamycins comprise a group of macrolactam antibiotics that contain an aryl, naphthyl or quinone chromophore incorporated into a polyketide-type ansa chain that is attached in meta orientation on the chromophore. Famous examples are rifamycin, geldanamycin (**12**) and maytansin.[13]

The maytansinoids exhibit cytotoxic activity, evident in the growth inhibition of different tumour cell lines and human solid tumours at very low concentrations (10^{-3} to 10^{-7} μg/mL).[14] Their antimitotic mode of action is based on interaction with β-tubulin, preventing polymerisation of tubulin and thereby promoting depolymerisation of microtubules. The maytansinoids currently attract high clinical interest as 'warheads' in tumour-targeted immunoconjugates.[15]

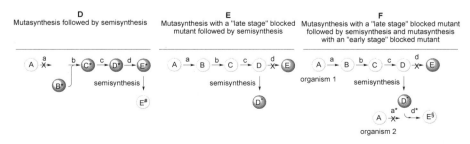

Scheme 4.5 Concepts of combined muta- and semisynthetic strategies (D–F).

In contrast to the parent compound maytansin, isolated from plant material, the closely related ansamitocins, e.g. ansamitocin P3 (**11**), are produced by the bacterium *Actinosynnema pretiosum*. The biosynthesis of ansamitocins involves several unusual steps, among them the utilisation of the aromatic biosynthetic starter 3-amino-5-hydroxybenzoic acid (AHBA) (**8**) supplied by a dedicated biosynthetic pathway[16] (Scheme 4.6) as well as the involvement of a separate amide synthase acting after polyketide synthase processing to generate proansamitocin **9**. This core structure is subsequently further modified by a set of post-PKS tailoring enzymes.

Geldanamycin (**12**) is produced by *Streptomyces hygroscopicus* var. *geldanus*. It is a potential antitumour drug[17] which binds to the N-terminal ATP-binding domain of heat shock protein 90 (Hsp90) and inhibits its ATP-dependent chaperone activities.[18] Most geldanamycin derivatives reported to date are 17-aminated compounds and were obtained by semisynthesis.[19] Besides us, two other groups have recently utilised blocked mutants of the microbial source of geldanamycin to prepare several new derivatives.[20–22] The biosynthesis of geldanamycin (**12**) is closely related to ansamitocin P3 (**11**) as it is PKS-based and utilises AHBA **8** as starter unit and a standalone amide synthase for cyclisation. Next, the cyclised product progeldanamycin **10** is further

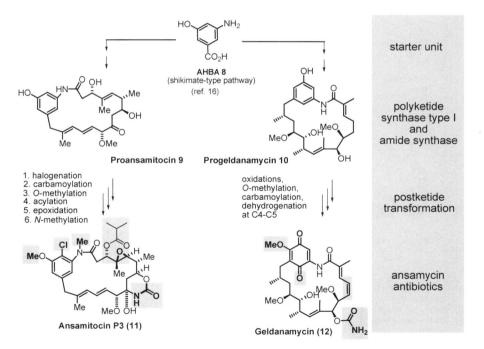

Scheme 4.6 Principle biosyntheses of ansamycin antibiotics ansamitocin P3 (**11**) and geldanamycin (**12**) (post-PKS tailoring modifications are marked in grey; the timing of tailoring events in geldanamycin biosynthesis is not fully elucidated to date; see also Scheme 4.9).

'decorated' by a set of individual post-PKS tailoring enzymes that differ from those occurring in *Actinosynnema pretiosum* (Scheme 4.6).

The biosynthetic uniqueness of AHBA **8** was the ideal starting point to generate a strain of *Actinosynnema pretiosum* blocked in the biosynthesis of AHBA. By doing so, one does not interfere with the essential primary metabolic pathways in the microorganism. Gratifyingly, the AHBA loading domain of the ansamitocin PKS turned out to accept and process a structurally diverse number of 3-aminobenzoic acids **13–28**, as shown in Scheme 4.7.[23,24] In a similar manner, the geldanamycin producer *Streptomyces hygroscopicus* var. *geldanus* could also be genetically manipulated and exploited in mutasynthetic fermentations by supplementing 3-aminobenzoic acid derivatives **13–15**, **20**, **23** and **29–36**.[25]

In many cases not only the expected final product of biosynthesis but instead several new ansamitocin derivatives were formed and could be isolated. These metabolites mainly differed in the extent of post-PKS tailoring modifications. For instance, feeding of 5-hydroxymethyl-3-aminobenzoic acid **21** is a particular illustrative example (Scheme 4.8). Out of eight metabolites detected

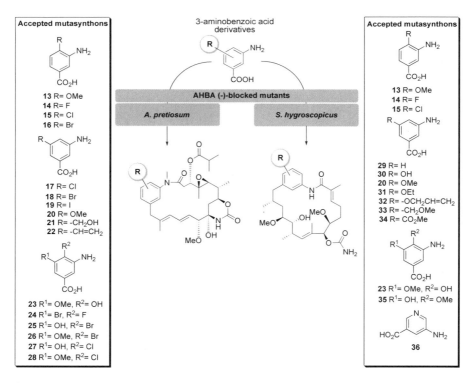

Scheme 4.7 Mutasynthetic generation of an ansamitocin and a geldanamycin library, respectively, by feeding aminobenzoic acid derivatives to AHBA(–) blocked mutants.

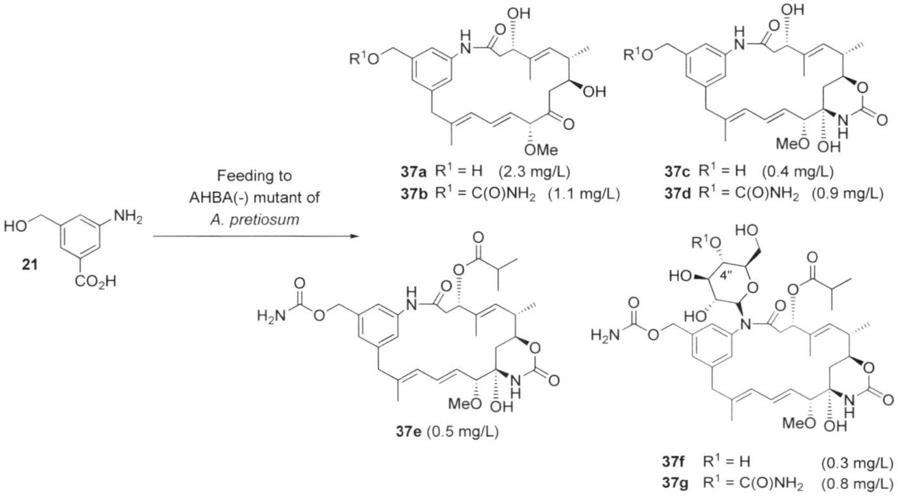

Scheme 4.8 New ansamitocin derivatives **37a–g** after mutasynthetic supplementation of AHBA(−) *Actinosynnema pretiosum* with hydroxymethylaminobenzoic acid (**21**).

by high-resolution mass spectrometry (HRMS), seven new ansamitocin derivatives **37a–g** were isolated, characterised and their anti-proliferative activity was determined. Besides modified proansamitocin **37a**, also metabolites **37b–e** containing common ansamitocin post-PKS tailoring modifications were isolated from the fermentation broth of AHBA(−) *A. pretiosum*. Surprisingly, also the benzylic hydroxy group underwent carbamoylation and unexpectedly two *N*-glycosylated ansamitocins **37f** and **37g** were formed (Scheme 4.8).[26]

Besides the principle relevance for synthetic natural product chemistry, these syntheses also paved the way for the biological evaluation of these new metabolites, thereby elucidating at which level of post-PKS tailoring modification ansamitocins **37a–e** start to exert their anti-proliferative activity. Clearly, when the ester side chain at C3 is being incorporated the very strong biological activity of the ansamitocins is introduced (Table 4.1), while *N*-glycosylation leads to a severe loss of biological potency.

The effect of these new ansamitocin derivatives, e.g. for chloro-demethoxy AP-3 generated from mutasynthon **15** on the microtubule can be visualized in green by means of a double antibody staining procedure while cell nuclei and chromosomes were stained in blue with DAPI (Figure 4.2). Clearly, the mode of action of new ansamitocin metabolites generated by mutasynthesis on the microtubules resembles that of the parent natural product **11**.

Similar experiments were pursued to generate novel geldanamycin derivatives. In cases when sufficient amounts (above 0.2 mg/L of fermentation broth) were formed the new derivatives obtained by mutasynthesis were isolated, characterised and finally their anti-proliferative activity was tested. In

Table 4.1   Anti-proliferative activity IC_{50} [nmol/L] of **37a-e** and **37f–g** in comparison to AP-3 **11**. Human cell lines: U-937 (lymphoma), A-431 (epidermal carcinoma), SK-OV-3 (cervical carcinoma), PC-3 (prostate carcinoma), MCF-7 (breast carcinoma), HUVEC (endothelial – human umbilical vein endothelial cells)

| | *Cell line* | | | | | |
Product	*U-937*	*A-431*	*SK-OV-3*	*PC-3*	*MCF-7*	*HUVEC*
AP-3 **11**	0.01	0.08	0.05	0.06	—	0.02
37a–d	>700	>700	>700	>700	>700	>700
37e	1.5	8.5	6.8	14.0	12.4	6.7
37f	348	>500	>500	>500	>500	>500
37g	659	>500	452	>500	>500	>500

Table 4.2, five representative examples are listed. In these tests the new aza-geldanamycin derivative **43** was completely inactive, while phenol **42**, a non-quinone geldanamycin derivative, exerted similar anti-proliferative activity compared to geldanamycin **12**. It has been demonstrated that benzoquinone Hsp90 inhibitors depend on reductive activation to the hydroquinone by the enzyme NAD(P)H/quinone oxidoreductase 1 (NQO1).[27–30] The activity of this enzyme differs in different patients, so that derivatives such as the phenol derivative **42** should show binding to the ATP-binding pocket of Hsp90 regardless of individual patients since the need for activation by NQO1 is absent for derivative **42**. Additionally, the quinone moiety of geldanamycin is held responsible for undesired hepatotoxic side-effects for which the Michael addition of the thiol moiety of glutathione to the quinone is made responsible.[31] This problem is suppressed in phenolic geldanamycin derivatives such as **42,** making the mutasynthetic strategy for its generation highly attractive.

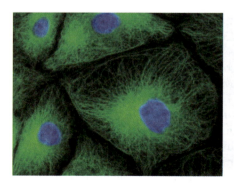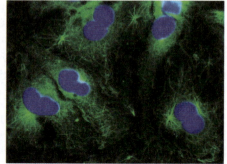

Figure 4.2   Pokoroo cells (PtK2): left: control; right: cells treated with chloro-desmethoxy AP-3 generated from mutasynthon **15** show complete destruction of microtubules.

Table 4.2 Anti-proliferative activities IC$_{50}$ [nmol/L] of **38–43** and comparison with geldanamycin (**12**)

Cell line; origin	38	39	40	41	42	43	12
KB-3-1; cervical carcinoma	109	470	73	123	34	>8000	53
U-937; lymphoma	188	93	62	142	21	2600	9
PC-3; prostate carcinoma	—	300	42	118	36	2800	18
SK-OV-3; ovarian carcinoma	113	1060	54	264	46	4000	125
MCF-7; breast carcinoma	38	320	18	123	120	870	---
A-431; epidermoid carcinoma	60	840	62	228	17	1900	18

It is noteworthy that when an AHBA(–) mutant of *Streptomyces hygroscopicus* var. *geldanus* was supplemented with hydroxymethylaminobenzoic acid (**21**) again a large spectrum of new metabolites was formed, isolated and characterized (Scheme 4.9).[32] The first five metabolites **44a–44e** provide

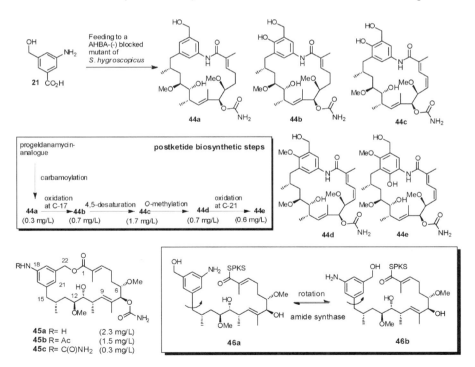

Scheme 4.9 New geldanamycin derivatives **44a–e** after mutasynthesis using 3-hydroxymethylaminobenzoic acid (**21**); sequence of events of post-PKS tailoring transformations and formation of unprecedented 20-membered macrolactones **45a–c**.

information on the sequence of events in the post-PKS tailoring biosynthesis of geldanamycin (**12**). After ring-closure of the progeldanamycin-analogue **44a** by the amide synthase, carbamoylation, oxidation at C17, 4,5-desaturation and *O*-methylation take place before the biosynthesis is terminated by oxidation at C21. Formation of three additional mutaproducts **45a–c** is the most striking result from these mutasynthetic studies. These are 20-membered macro-lactones, compared to the 19-membered macrolactam ring in geldanamycin (**12**). Obviously, the amide synthase responsible for the macrolactamisation is able to macrolactonise PKS-bound alcohols **46** as long as they are more nucleophilic than the arylamino group in *seco*-progeldanamycin.

4.3.3 Mutasynthesis Followed by Semisynthesis

The concept of mutasynthesis can beneficially be extended by combining it with semisynthesis (see Scheme 4.5, case D). This combination can ideally be realised if a new, chemically useful functionality is incorporated into the final product via mutasynthesis, which can subsequently be further modified. The main obstacle of this approach is associated with the fact that mutasynthesis often does not provide sufficient amounts of new product, which is a crucial prerequisite for subsequent semisynthetic efforts.

One of the first simple examples is presented in Scheme 4.10. Formation of AP3 derivative **48** devoid of peripheral functionalization at the aromatic ring could not be achieved by feeding of aminobenzoic acid **29** to the AHBA(–)

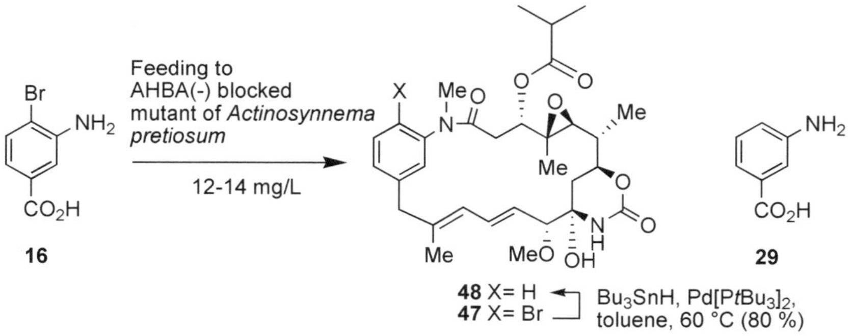

Cell line; origin					
KB-3-1; cervical carcinoma	*U-937; lymphoma*	*PC-3; prostate carcinoma*	*SK-OV-3; ovarian carcinoma*	*A-431; epidermoid carcinoma*	*A-549; lung carcinoma*
48 0.022	0.015	0.020	0.040	0.080	0.095
11 0.11	0.0035	0.035	0.030	0.050	0.095

Scheme 4.10   Preparation of the AP3 derivative **48** and anti-proliferative data (IC$_{50}$ [ng/mL]) of **48** compared to **11**.

blocked mutant of *Actinosynnema pretiosum*.[33] However, 3-amino-4-bromo-benzoic acid **16** yielded the bromo-AP3 derivative **47** in very good yield. The following palladium-catalysed debromination gave AP3 derivative **48** which surprisingly exerts similar anti-proliferative activity against several cancer cell lines, as does ansamitocin P3 **11** itself.

We regarded bromo-AP3 **47** as an excellent precursor for other Pd-catalysed reactions such as the Stille and the Sonogashira cross-coupling reactions. Thus, alkenylation of the aromatic moiety was achieved with TMS acetylene followed by silyl removal to yield ansamitocin P3 derivative **50** (Scheme 4.11). Likewise, the Stille coupling served to equip the ansamitocins with a vinyl group, including functionalised examples such as **49**, to yield **51** and **52**, respectively.[23,34] All three new ansamitocin derivatives **50–52** show strong cytotoxic activity, though weaker than the parent natural product against the classical set of cell lines.

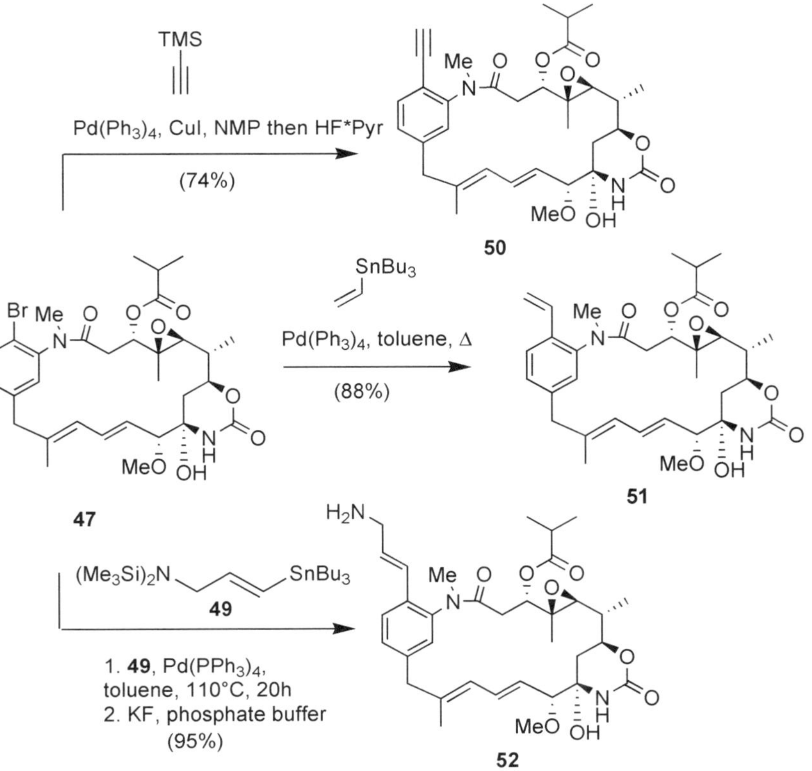

Scheme 4.11 Semisynthesis based on Pd-catalysed cross-coupling reaction with bromo-AP3 **47**.

4.3.4 Mutasynthesis with 'Late Stage' Blocked Mutants Followed by Semisynthesis

The second concept of combining biosynthesis with semisynthesis utilises blocked mutants of a natural product producer whose biosynthesis is blocked at a late stage, e.g. in the post-PKS tailoring part (Scheme 4.5, case E). Thus, two genes of *Actinosynnema pretiosum*, namely those coding for the carbamoyl transferase and the halogenase, were blocked resulting in a mutant that mainly produced the amide synthase ring closing product proansamitocin **9** apart from some minor by-products **53–55** (Scheme 4.12).

Specifically, proansamitocin **9** (106 mg/L), its 10-epimer **53** (3.5 mg/L), the *O*-methylated proansamitocin **54** (2.3 mg/L) and alcohols **55a,b** (1:1; 7.6 mg/L), likely being formed via oxidation product **56**, were isolated. None of these metabolites exert anti-proliferative activities.

Proansamitocin **9** serves as starting point for several semisyntheses, allowing modifications of functional groups situated in the ansa chain of ansamitocin (Scheme 4.13).[34] Firstly, the keto group at C9 was reduced to furnish the two diastereomers **57a,b** (3:1), or alternatively the *O*-persilylated proansamitocin derivative **60** was transformed into the methyl-branched proansamitocins **61a,b** (5:1 for MeLi or 10:1 for MeMgBr, respectively) via conformer **62**. Due to the fact that the isobutyrate at C3 is essential for the strong anti-proliferative activity of the ansamitocins, a synthetic sequence for its selective introduction was developed. This generally applicable sequence is exemplified for the preparation of ester **59** starting from diastereomeric alcohol **57a**.

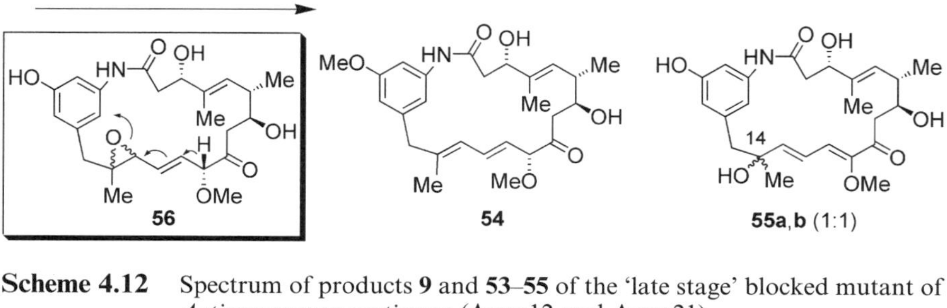

Scheme 4.12 Spectrum of products **9** and **53–55** of the 'late stage' blocked mutant of *Actinosynnema pretiosum* (Δasm12 und Δasm21).

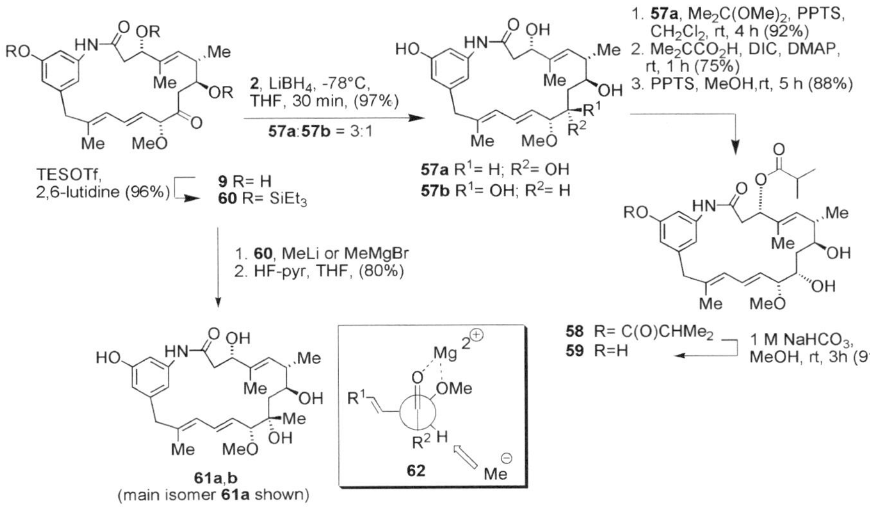

Scheme 4.13 Selected semisyntheses with proansamtocin **9**.

4.3.5 Combining Mutasynthesis Using 'Late and Early Stage' Blocked Mutants With Semisynthesis

An extension of the concept of combining metabolites obtained from mutants blocked at a late stage of the biosynthesis with semisynthesis includes a second mutasynthetic step as a means for further structural diversification

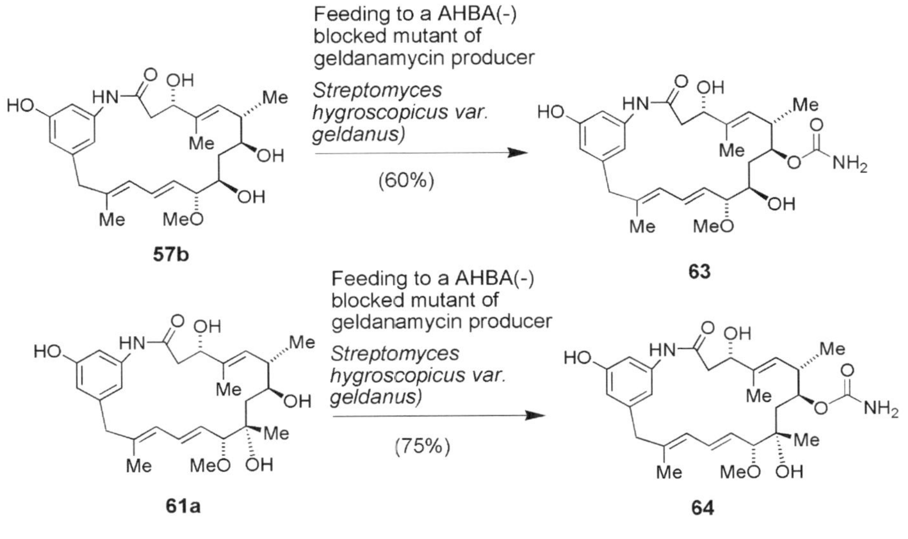

Scheme 4.14 Mutasynthesis with semisynthetically modified proansamitocins **57b** and **61a**.

(Scheme 4.5, case F). In principle, the second blocked mutant can be based on the same organism originally utilised to create the 'late stage' blocked mutant. However, it should be blocked at an earlier stage in the biosynthesis, while alternatively the second blocked mutant may also be based on a different organism coding for a different natural product. This concept can be put on a broader foundation by including a semisynthetic step into the synthesis, either between both fermentations or by terminating both subsequent fermentations with a final synthetic step.

Two straightforward examples containing a second mutasynthetic step after a first fermentation and semisynthesis sequence are listed in Scheme 4.14.[34] When semisynthetically modified proansamitocins **57b** and **61a** were supplemented to a culture of the AHBA(–) blocked mutant of the geldanamycin producer *Streptomyces hygroscopicus* var. *geldanus*, the carbamoylated metabolites **63** and **64**, were formed in good yields. Among the five post-PKS tailoring enzymes in geldanamycin biosynthesis, only the carbamoyl-transferase in *S. hygroscopicus* var. *geldanus* accepted the semisynthetically modified proansamitocin derivatives. The selectivity was very pronounced in that the hydroxy group at C7 was carbamoylated, a site of activity identical to that of the ansamitocin carbamoyl transferase.

Similarly, the reversed sequence was probed. Proansamitocin **9** was first carbamoylated using the AHBA(–) blocked mutant of the geldanamycin producer *Streptomyces hygroscopicus* var. *geldanus* to yield the cyclic carbamate **65** (Scheme 4.15). Afterwards, the selective introduction of the isobutyroyl group at C3 was performed as described in Scheme 4.13, furnishing ansamitocin derivative **66**. Noteworthy, this ansamitocin derivative is not accessible in a single biosynthesis step due to the substrate limitations of the enzymes involved in the established sequence of post-PKS tailoring transformations.

In addition, the carba analogues **68a,b** could be accessed by semisynthesis starting from protected proansamitocin **60**. Thus, Claisen reaction gave diastereomeric lactons **67a,b** after fluoride-induced desilylation. Finally, acylation at C3 was achieved as described for the preparation of carbamate **66** to furnish its deaza analogue. From an SAR point of view it is noteworthy that carba analogue **86a** with an α-orientation of the hydroxyl group at C9 shows anti-proliferative activity in the lower nanomolar range.

4.4   Conclusions

Besides the well-established approaches of semisynthesis and total synthesis, mutational biosynthesis or mutasynthesis has emerged as a third strategy for the generation of natural product libraries. This strategy combines advanced methods of chemical synthesis with those of molecular biology and microbiology required for the modification of biosynthetic cascades and handling of their hosts. With the combination of semisynthesis and mutasynthesis, the synthetic chemist has gained a powerful synthetic tool for

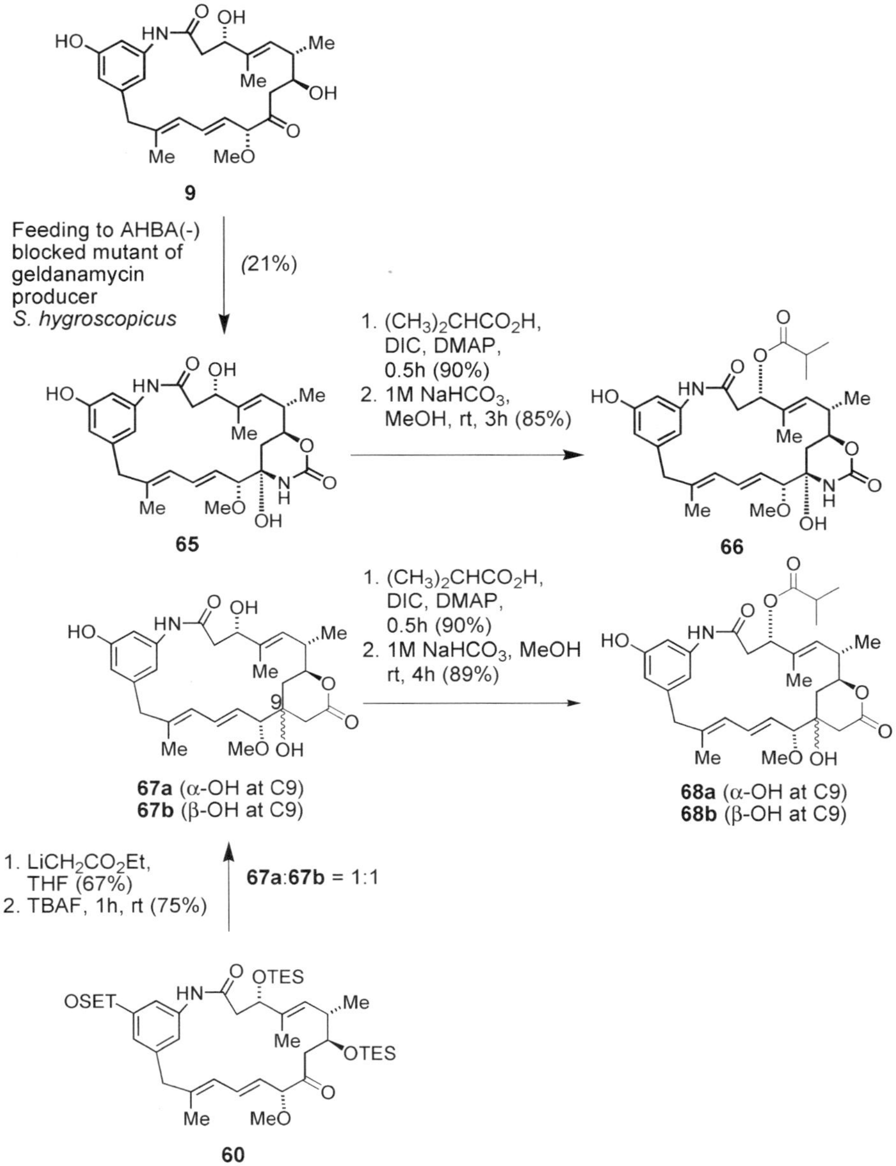

Scheme 4.15 Combined muta- and semisynthetic syntheses of ansamitocin derivate **66** and its deaza analogues **68a,b**.

modifying complex natural products at almost every position. The examples listed in this chapter clearly show that enzymes responsible for the biosynthesis of secondary metabolites are far from being highly substrate specific. The optimisation of substrate specificities which should include evolutionary techniques and the predictable construction of artificial biosynthesis cascades by sophisticated metabolic engineering will certainly expand the scope of mutasynthesis in the near future. For this, a close collaboration of the fields of metabolic engineering and synthetic chemistry will be necessary.

Acknowledgements

Our contributions to this field of research were only possible by a group of excellent co-workers including M. Brünjes, G. Dräger, T. Frenzel, K. Harmrolfs, S. Mayer, A. Meyer, F. Taft and B. Thomaszewski and excellence in the NMR department consecutively led by E. Hofer, D. Albert and today by J. Fohrer. Biological testing was performed by F. Sasse (Helmholtz Center of Infectious Disease (HZI), Braunschweig, Germany). We are particularly grateful to H. G. Floss (University of Washington, Seattle, USA) for initial and continuous generous support including provision of blocked mutants. Finally, we thank the Fonds der Chemischen Industrie and the Deutsche Forschungsgemeinschaft (grant Ki 397/7-1 and Ki 397/13-1) for financial contributions.

References

1. R. S. Bon and H. Waldmann, *Acc. Chem. Res.*, 2010, **43**, 1103.
2. (a) D. J. Newman, G. M. Gragg and K. M. Snader, *J. Nat. Prod.*, 2003, **66**, 1022; (b) Newman and G. Cragg, *J. Nat. Prod.*, 2007, **70**, 461.
3. I. Sattler, R. Thiericke and S. Grabley, in *Drug Discovery from Nature* eds. S. Grabley and R. Thiericke, Springer Verlag, Berlin, Heidelberg, 1999, chapter 11, p. 191.
4. This concept was mentioned for the first time in (a) A. J. Birch, *Pure Appl. Chem.*, 1963, **7**, 527; in a later paper this concept was specified: (b) W. T. Shier, K. L. Rinehart Jr. and D. Gottlieb, *Proc. Nat. Acad. Sci. USA*, 1969, **63**, 198.
5. (a) K. L. Rinehart, Jr., *Pure Appl. Chem.*, 1977, **49**, 1361; (b) K. L. Rinehart Jr, *Jpn. J. Antibiot.*, 1979, **32**, Suppl., S32-46.
6. (a) Reviews: S. Weist and R. D. Süssmuth, *Appl. Microbiol. Biotechnol.*, 2005, **68**, 141; (b) A. Kirschning, F. Taft and T. Knobloch, *Org. Biomol. Chem.*, 2007, 3245.
7. M. A. Gregory, H. Petkovic, R. E. Lill, S. J. Moss, B. Wilkinson, S. Gaisser, P. Leadlay and R. M. Sheridan, *Angew. Chem.*, 2005, **117**, 4835; *Angew. Chem., Int. Ed.*, 2005, **44**, 4757. M. Ziehl, J. He, H.-M. Dahse and C. Hertweck, *Angew. Chem.*, 2005, **117**, 1226; *Angew. Chem., Int. Ed.*, 2005, **44**, 1202.
8. C. J. Dutton, S. P. Gibson, A. C. Goudie, K. S. Holdom, M. S. Pacey, J. C. Ruddock, J. D. Bu'Lock and M. K. Richards, *J. Antibiot.*, 1991, **44**, 357.

9. T. A. Cropp, D. J. Wilson and K. A. Reynolds, *Nature Biotechnol.*, 2000, **18**, 980.

10. (a) T. Doi, A. Kinbara, H. Inoue and T. Takahashi, *Chem. Asian J.*, 2007, **2**, 188; (b) S. S. Printsevskaya, S. E. Solovieva, E. N. Olsufyeva, E. P. Mirchink, E. B. Isakova, E. D. Clercq, J. Balzarini and M. N. Preobrazhenskaya, *J. Med. Chem.*, 2005, **48**, 3885; (c) J. Yang, D. Hoffmeister, L. Liu, S. Fu and J. S. Thorson, *Bioorg. Med. Chem.*, 2004, **12**, 1577; (d) D. A. Thayer and C.-H. Wong, *Chem. Asian J.*, 2006, **1**, 445.

11. (a) R. D. Süssmut and W. Wohlleben, *Appl. Microbiol. Biotechnol.*, 2004, **63**, 344; (b) S. Weist, B. Bister, O. Puk, D. Bischoff, S. Pelzer, G. J. Nicholson, W. Wohlleben, G. Jung and R. D. Süssmuth, *Angew. Chem.*, 2002, **114**, 3531; () *Angew. Chem., Int. Ed.*, 2002, **41**, 3383.

12. S. Weist, C. Kittel, D. Bischoff, B. Bister, V. Pfeifer, G. J. Nicholson, W. Wohlleben and R. D. Süssmuth, *J. Am. Chem. Soc.*, 2004, **126**, 5942.

13. Reviews: (a) J. M. Cassady, K. K. Chan, H. G. Floss and E. Leistner, *Chem. Pharm. Bull.*, 2004, **52**, 1; (b) A. Kirschning, K. Harmrolfs and T. Knobloch, *C. R. Chim.*, 2008, **11**, 1523; (c) H. G. Floss, T.-W. Yu and K. Arakawa, *J. Antibiot.*, 2011, DOI: 10.1038/ja.2010.139.

14. S. Funayama and G. A. Cordell, in *Studies in Natural Product Chemistry* ed. A.-U. Rahman, Elsevier Science B. V., Amsterdam, 2000, vol. 23, pp. 51.

15. (a) H. K. Erickson, W. C. Widdison, M. F. Mayo, K. Whiteman, C. Audette, S. D. Wilhelm and R. Singh, *Bioconjugate Chem.*, 2010, **21**, 84; (b) P. D. Senter, *Curr. Opin. Chem. Biol.*, 2009, **13**, 235.

16. T.-W. Yu, L. Bai, D. Clade, D. Hoffmann, S. Toelzer, K. Q. Trinh, J. Xu, S. J. Moss, E. Leistner and H. G. Floss, *Proc. Natl. Acad. Sci. USA*, 2002, **99**, 7968.

17. C. Prodromou, S. M. Roe, R. O'Brien, J. E. Ladbury, P. W. Piper and L. H. Pearl, *Cell*, 1997, **90**, 65.

18. (a) P. Workman, *Curr. Cancer Drug Targets*, 2003, **3**, 297; (b) L. Neckers and K. Neckers, *Expert Opin. Emerg. Drugs*, 2005, **10**, 137; (c) L. Whitesell and S. L. Lindquist, *Nat. Rev. Cancer*, 2005, **5**, 761.

19. Y. L. Janin, *J. Med. Chem.*, 2005, **48**, 7503.

20. W. Kim, J. S. Lee, D. Lee, X. F. Cai, J. C. Shin, K. Lee, C.-H. Lee, S. Ryu, S.-G. Paik, J. J. Lee and Y.-S. Hong, *ChemBioChem*, 2007, **8**, 1491.

21. PCT Int. Appl. 2008, p 53, Pub. No.: US 2008/0188450 A1.

22. H. G. Menzella, T.-T. Tran, J. R. Carney, J. Lau-Wee, J. Galazzo, C. D. Reeves, C. Carreras, S. Mukadam, S. Eng, Z. Zhong, P. B. M. W. M. Timmermans, S. Murli and G. W. Ashley, *J. Med. Chem.*, 2009, **52**, 1518.

23. F. Taft, M. Brünjes, H. G. Floss, N. Czempinski, S. Grond, F. Sasse and A. Kirschning, *ChemBioChem*, 2008, **9**, 1057.

24. (a) F. Taft, M. Brünjes, T. Knobloch, H. G. Floss and A. Kirschning, *J. Am. Chem. Soc.*, 2009, **131**, 3812; (b) A. Meyer, M. Brünjes, F. Taft, T. Frenzel, F. Sasse and A. Kirschning, *Org. Lett.*, 2007, **9**, 1489; (c) T. Kubota, M. Brünjes, T. Frenzel, J. Xu, A. Kirschning and H. G. Floss, *ChemBioChem*, 2006, **7**, 1221.

25. S. Eichner, H. G. Floss, F. Sasse and A. Kirschning, *ChemBioChem*, 2009, **10**, 1801.
26. T. Knobloch, K. Harmrolfs, F. Taft, B. Thomaszewski, F. Sasse and A. Kirschning, *ChemBioChem*, 2011, **12**, accepted.
27. W. Guo, P. Reigan, D. Siegel, J. Zirrolli, D. Gustafson and D. Ross, *Cancer Res.*, 2005, **65**, 10006.
28. W. Guo, P. Reigan, D. Siegel, J. Zirrolli, D. Gustafson and D. Ross, *Mol. Pharmacol.*, 2006, **70**, 1194.
29. L. R. Kelland, S. Y. Sharp, P. M. Rogers, T. G. Myers and P. Workman, J, . *Natl. Cancer Inst.*, 1999, **91**, 1940.
30. A. C. Maroney, J. J. Marugan, T. M. Mezzasalma, A. N. Barnakov, T. A. Garrabrant, L. E. Weaner, W. J. Jones, L. A. Barnakova, H. K. Koblish, M. J. Todd, J. A. Masucci, I. C. Deckman, R. A. Galemmo Jr. and D. L. Johnson, *Biochemistry*, 2006, **45**, 5678.
31. M. Minami, M. Nakamura, Y. Emori and Y. Minami, *Eur. J. Biochem.*, 2001, **268**, 2520.
32. S. Eichner, T. Eichner, H. G. Floss, J. Fohrer, E. Hofer, F. Sasse, C. Zeilinger and A. Kirschning, *J. Am. Chem. Soc.*, 2012, **134**, 1673–1679.
33. K. Harmrolfs, M. Brünjes, G. Dräger, H. G. Floss, F. Sasse, F. Taft and A. Kirschning, *ChemBioChem*, 2010, **11**, 2517.
34. S. Eichner, T. Knobloch, H. G. Floss, J. Fohrer, K. Harmrolfs, J. Hermane, A. Pieper, F. Sasse, P. Spiteller, F. Taft and A. Kirschning, *Angew. Chem. Int. Ed.*, 2012, **51**, 752–757.

CHAPTER 5

Progress in Enhancing the Neurotrophic Effects of Natural FKBP Ligands

GUY T. CARTER*

Carter-Bernan Consulting, LLC., 350 Phillips Hill Road, New City, NY 10956, USA
*E-mail: gtc9531@gmail.com

5.1 Introduction

Natural products that bind to the FK506-binding protein (FKBP) subclass of immunophilin proteins, including FK506 and rapamycin, have been recognized as potent modulators of signaling pathways in cells.[1] The anti-proliferative effects of these compounds are well known and have resulted in pharmaceutical products for the treatment of transplant rejection and certain types of cancer. The most remarkable aspect of the anti-proliferative effects exhibited by these compounds is their molecular mechanism. First demonstrated for FK506 and subsequently for rapamycin, the compounds first bind FKBP-12 and then recruit the respective targets of their inhibitory action, calcineurin and mTOR, the mammalian target of rapamycin.[2,3] Both FK506 (1) and rapamycin (3) have richly decorated macrocyclic structures that present multiple potential binding domains. As highlighted in Figure 5.1, the pipecolate residue linked to the 'tricarbonyl' moiety is at the core of the FKBP-binding region. This rare structural motif is fully conserved in only two other families of secondary metabolites: meridamycin (4) and antascomicin

RSC Drug Discovery Series No. 25
Drug Discovery from Natural Products
Edited by Olga Genilloud and Francisca Vicente
© The Royal Society of Chemistry 2012
Published by the Royal Society of Chemistry, www.rsc.org

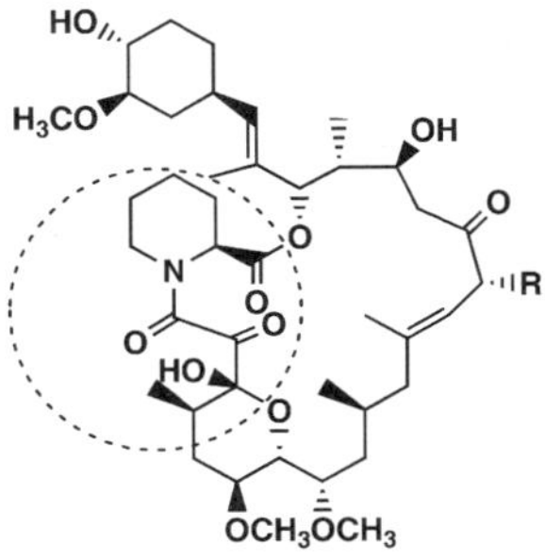

1 FK506 R = CH₂CH=CH₂
2 FK520 R = CH₂CH₃

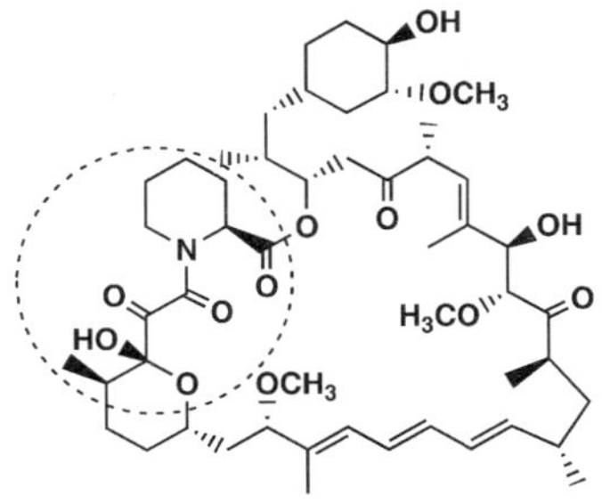

rapamycin
3

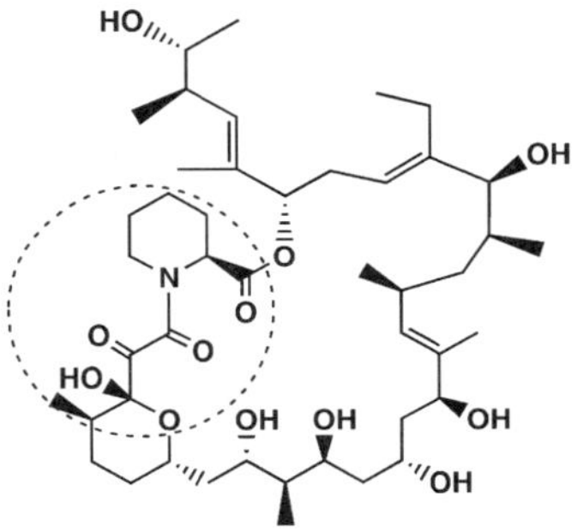

meridamycin
4

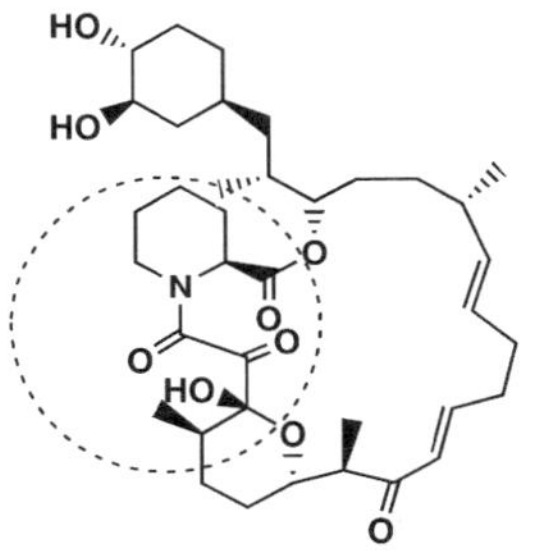

antascomicin A
5

Figure 5.1   Macrocyclic hybrid PKS/NRPS-derived natural products known to bind FKBPs. The dashed circle outlines the core structure that binds FKBPs.

(5).[4,5] Comparatively little information has been published on the biological properties of meridamycin and antascomicin; however, they are not believed to be immunosuppressive.

The therapeutic focus of this chapter centers on the potential neurological benefits of this unique family of secondary metabolites.[6] Unmodified FK-506, but not rapamycin, was shown to have potent neuroprotective effects in an *in vivo* stroke model in 1994.[7] Following this publication, a plethora of reports appeared that sought to address the mechanism of the neuroprotective effect. A seminal paper by Steiner and co-workers in 1997 highlighted the effectiveness of certain experimental compounds to separate the neuroprotective action from immunosuppression, which is a critical requirement for any therapeutically useful agent.[8] In the Steiner report it was demonstrated that non-immunosuppressive analogs of FK506 (L-685,818) **6** and rapamycin (WAY-124,466) **7** as well as small molecule mimics of the FKBP-binding region, e.g. **8** (Figure 5.2) retained the neurotrophic actions of the parent compounds.[9–11] The small molecule-medicinal chemistry approach has been quite extensively pursued wherein simplified compounds were designed that contain an FKBP-binding region alone.[12] Synthetic analogs of this type have shown nanomolar potencies *in vitro*, efficacy in some animal models, and have included a number of clinical candidates. The earlier work in this area is thoroughly reviewed in an excellent report by Hamilton and Thomas.[13] Such *de novo* synthetic approaches remain attractive, although the lack of clinical success has dampened enthusiasm. Much less attention has been paid to the creation of analogs of the natural product binders of FKBPs with respect to enhancing their neurotrophic effects while diminishing anti-proliferative activity.

Clearly the FKBP subclass of immunophilins plays a pivotal and perhaps definitive role in determining the biological effects of these natural ligands. At least 20 FKBPs have been identified in the human genome and are classified according to their respective molecular weights, (i.e. FKBP-12 is approximately 12 kD, etc.). These proteins all share peptidyl prolyl *cis/trans* isomerase activity that is inhibited by the natural ligands. The biological implications of this inhibitory action remain to be fully defined. There have been several reports on the biological roles of these proteins in cells and their potential value as targets of drug action.[1] In addition to facilitating protein folding,

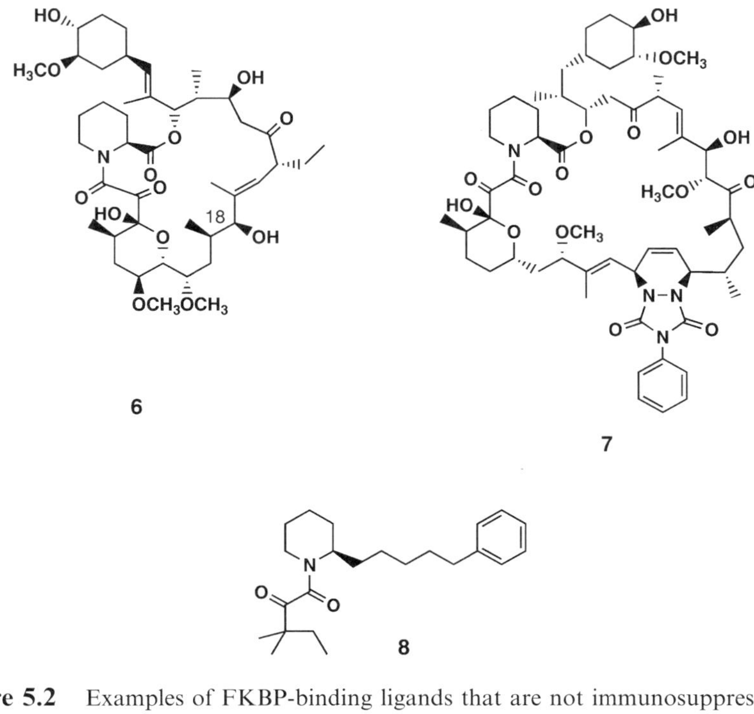

Figure 5.2 Examples of FKBP-binding ligands that are not immunosuppressive.

trafficking and assembly, certain FKBPs play a role in regulation of neuronal growth. Since the late 1990's the evidence has been marshaled by Gold and others implicating FKBP-52 as the key immunophilin that mediates the neurotrophic effects of these natural ligands. Perhaps the most compelling results are derived from experiments done with cells in which individual FKBPs are specifically eliminated either through knockout of production or monoclonal antibody precipitation. Gold has shown that FK506 retains its neurotrophic effects in FKBP-12 knockout mice, whereas in FKBP-52 knockouts the neurotrophic effects of FK506 are abolished.[14] Furthermore there have been reports that implicated FKBP-38 in neuroprotection. In one study it was demonstrated that neurotrophic small molecule inhibitors specifically bound FKBP-38 and that rapamycin had much reduced affinity for FKBP-38 compared to FKBP-12.[15] Therefore it is clear that there are (at least) two mechanisms whereby the interactions of immunophilins with natural ligands produce effects in biological systems. As has been amply demonstrated, FK506 and rapamycin form binary complexes with FKBP-12, which serve to orient the remaining solvent-exposed portion of the natural ligands for optimal inhibition of another enzyme, calcineurin or mTOR, respectively. The resulting tripartite complex is no longer functional in signal transduction and this results in immunosuppression and other anti-proliferative effects. In these two cases, if the natural ligand is modified such that it no longer binds to the second protein effectively, then the biological effects of FKBP binding alone become predominant. It is these latter interactions that appear to result in neurotrophic effects, although the mechanistic details are poorly understood. It is becoming increasingly clear, however that inhibition (or knockdown) of certain FKBPs can ameliorate that neurotoxic effects associated with such conditions as Parkinsons Disease, perhaps through blocking the aggregation of α-synuclein.[16] For the lesser-studied natural products of this family, little is known about the specificity of FKBP binding or whether a ternary complex including another protein is involved in their actions.

This brief chapter is devoted to recent efforts to enhance the neuro-protective/regenerative potential of the macrocyclic, hybrid PKS/NRPS-derived family of microbial products exemplified by four sets of structurally related compounds: FK506/520, rapamycin, meridamycin and antascomicin.

5.2 FK506/520

5.2.1 Synthetic and Biosynthetic Analogs

FK506 (tacrolimus) was the first natural immunophilin ligand to be implicated in neuroprotection. Seminal research by the Schreiber lab led to the synthesis of the hybrid macrocyclic compound 506BD **9** (Figure 5.3) in which the calcineurin-binding domain of FK506 was replaced with a synthetic linker.[17] Compound 506BD retained the immunophilin-binding activity but was no longer immunosuppressive in vitro. Uncoupling of neurotrophic activity from

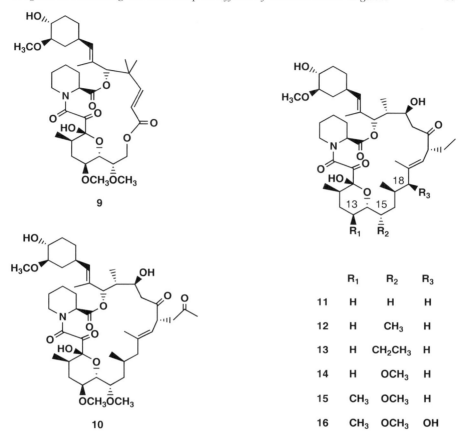

	R₁	R₂	R₃
	R_1	R_2	R_3
11	H	H	H
12	H	CH_3	H
13	H	CH_2CH_3	H
14	H	OCH_3	H
15	CH_3	OCH_3	H
16	CH_3	OCH_3	OH

Figure 5.3 Analogs of FK506/520 that show reduced immunosuppressive potency.

immunosuppression could also be accomplished by much simpler chemical modifications. An early example was the previously cited L-685,818 **6** that was obtained by selective oxidation of FK520 **2** (aka ascomycin, an analog of FK506 with an ethyl group replacing the allyl function at C-21). L-685,818 retained strong affinity for FKBP-12 but showed greatly diminished calcineurin binding and the concomitant reduction in immunosuppressive activity.[18] The structure of the complex was determined by X-ray analysis.[9] The three-dimensional structure of the complex showed essentially identical binding of **6** with FKBP-12 as was found for FK506. This led to the conclusion that any differences in pharmacology must be due to the solvent-exposed regions of the ligands. This is entirely consistent with the structural modification made in **6**, which is now known to be in the region that binds calcineurin.

More recently, a simple keto derivative, FK1706 **10**, synthesized by oxidation of the allyl side chain, was reported to retain potent binding to

FKBP-12 and 52, while showing greatly reduced potency and efficacy in anti-proliferation assays.[19] FK1706 was highly effective at enhancing NGF-induced neurite outgrowth through interaction with FKBP-52 and was proposed to have therapeutic potential for neurological disorders. The compound was effective in inducing neuroregeneration in a rat bilateral nerve crush model of erectile dysfunction.[20,21] It is not clear that this model is sufficiently refined to differentiate the effects of this non-immunosuppressive compound from those of the parent FK506.

In the only published example of genetic engineering approaches to medicinal chemistry in the FK506/520 series, the calcineurin-binding region of FK520 was modified with the aim to reduce immunosuppressive activity.[22] The methoxy groups at positions 13 and 15 (Figure 5.3) were known from structural studies to engage in favorable binding interactions with specific amino acid residues in calcineurin.[23] These particular methoxy functions in FK506/520 are introduced at the stage of polyketide chain elongation through specific acyltransferase (AT) domains, rather than being added post-PKS by tailoring enzymes.[24] Therefore, in order to remove the methoxy functions at C-13 and/or C-15, the specificity of the respective acyltransferases had to be changed. Starting with a strain engineered to produce 13-desmethoxy FK520, alterations in the AT specificity for the 15 position were introduced. Three recombinant pathways were produced in which the methoxymalonyl-coding AT (for C-15) was replaced by a malonyl-specific AT, a methylmalonyl-specific AT or an ethylmalonyl-specific AT domain. These changes resulted in the production of compounds **11**, **12**, and **13**, in which the 15-methoxy group was replaced by a proton, a methyl group, or an ethyl group, respectively. In the course of this work the 13-desmethoxy (**14**), 13-methyl-13-desmethoxy (**15**), and 13-methyl-13-desmethoxy-18-hydroxy (**16**) analogs of FK520 were also prepared. The relative binding affinity of these compounds for FKBP-12 and -52, calcineurin inhibition, and T-cell suppressive potencies were measured. All of the analogs retained strong binding affinity for the FKBPs, while only **14** retained potency against calcineurin and in T-cell suppression. Compounds **15** and **16** were further evaluated in a neurite outgrowth assay that simulates neuroregeneration. Interestingly **16**, which incorporates the 18-hydroxyl group of L-685,818 (**6**), was the most potent analog in these assays and also showed indications of efficacy in a rodent model for peripheral nerve regeneration.

5.2.2 Future Prospects

Each of the non-immunosuppressive analogs derived from the FK506/520 parents has shown positive results in assays indicative of neuroprotection and/or neuroregeneration. Both FK1706 **10** and compound **16** were efficacious in rodent models for neuroregeneration. These highly promising results argue for a more thorough investigation of the efficacy and safety of these analogs. To date no additional reports have appeared to detail these activities. Mechanistic questions also remain unanswered for these analogs: is FKBP-52 binding solely responsible for the induction of neurological effects, or is there an alternate binding partner replacing calcineurin?

5.3 Rapamycin

5.3.1 Synthetic Analogs

As noted, the first non-immunosuppressive analog of rapamycin reported to have neuroprotective activity was WAY-124,466 (**7**), which is a Diels-Alder adduct formed by a [4 + 2] cycloaddition.[10] As shown in Figure 5.4, this process has been revisited in the synthesis of new rapalogs in which highly

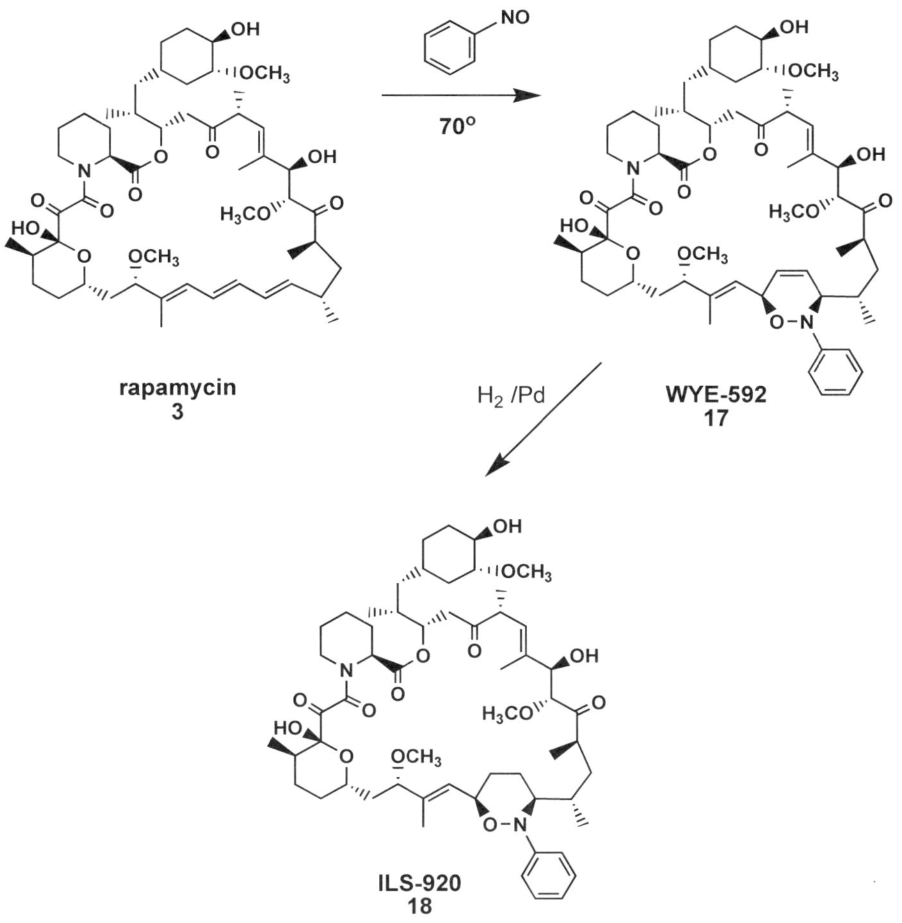

Figure 5.4 Formation of WYE-592 and ILS-920.

reactive nitrosobenzene dieneophiles are condensed with the rapamycin chromophore. These reactions are remarkably regio- and stereoselective and result in moderate yields of one predominant product. The two most studied analogs in this series are WYE-592 **17** and ILS-920 **18**; the latter derived from selective reduction of the new olefinic bond formed in the cyclization. This reduction step became important when the cycloaddition reaction was found to be reversible to the extent of about 3% in solution, resulting in the liberation of free rapamycin. Both WYE-592 and ILS-920 show modestly enhanced potency in neurite outgrowth and neuronal survival assays over the parent rapamycin, while exhibiting greatly reduced inhibition of T-cell proliferation. The conclusion from these experiments was that disruption of the triene system as well as the steric bulk introduced in the addition of the nitrosophenyl moiety effectively abolished the affinity of the compounds for mTOR.[25]

WYE-592 and ILS-920 were used to explore the mechanism of the neurotrophic effects of this group of compounds.[25] Comparative affinities for FKBP-12 and -52 were determined that clearly showed stronger binding to FKBP-52 (10- to 200-fold) while both rapamycin and FK506 preferentially bound FKBP-12. The structural basis for the change in binding specificity remains to be delineated. This result was independently reinforced by the results of a series of affinity capture experiments. In these experiments affinity matrices were prepared with the compounds and used to precipitate putative protein targets from F-11 dorsal root ganglia/neuroblastoma cells. Both WYE-592 and ILS-920 preferentially pulled down FKBP-52 over the more abundant FKBP-12 in these hybridoma cell lines. In the same experiments a previously unrecognized binding partner for the compounds was also identified. This protein was identified as the β-1 subunit of L-type voltage-dependent calcium channels. The full significance of the β-1 subunit binding remains to be established, although it was demonstrated that ILS-920 selectively inhibits the L-type calcium channel.

ILS-920 was subsequently evaluated in rodent models of stroke in which the middle cerebral artery is either transiently or permanently occluded. In these models ILS-920 was shown to be efficacious even when administered 6 to 24 hours after the induced ischemia.[26] The most promising result was the dramatic improvement in neurological deficits when the compound was administered 24 hours post occlusion. One of the major difficulties encountered in the treatment of stroke patients is the lack of therapeutic options once a critical time has elapsed following the ischemic event (3–6 hours). It is in that initial brief period that thrombolytic (clot-busting) therapies are effective. These preliminary results in the rodent model suggest that it will be possible to develop therapies for stroke patients that have a remedial effect many hours after the initial stroke. ILS-920 entered clinical trials in 2009. A full account of the discovery of ILS-920 has recently been published.[27]

5.3.2   Future Prospects

If the safety data from the clinical study of ILS-920 look promising, then it seems compelling that additional members of this class of rapamycin analogs should be prepared and clinically evaluated. In preliminary reports additional substituted nitrosophenyl analogs were mentioned that showed enhanced potency in the neuroprotection models that remain to be further explored. Structural studies with FKBP-52 would also be instructive in the determination of the optimal binding conformations that favor selective binding over FKBP-12. The results of such studies would provide guidance for the next series of FKBP-52-selective analogs. Genetic engineering of the polyketide-derived macrocycle to reduce immunosuppression and fine-tuning of FKBP binding are obvious next steps.

5.4   Meridamycin

5.4.1   Discovery of 3-Normeridamycin

Meridamycin represents an attractive starting point for the investigation of neuroprotective FKBP-binding natural products, since it is not inherently immunosuppressive. The Merck group originally reported the isolation and structure determination of meridamycin in 1995.[4] The compound was recognized as a structural relative of rapamycin and FK506 and was characterized as an antagonist of their actions. Wyeth began investigations of this group of FKBP-binding macrocycles with the disclosure of a previously unrecognized prolyl analog, 3-normeridamycin **19** in 2006 (Figure 5.5).[28] 3-Normeridamycin was shown to be effective in an assay for neuroprotection that measured the ability of a compound to restore dopamine uptake in damaged neurons. In this account the FKBP-binding activity of the compound was reported as was its lack of activity in blocking cellular proliferation. Further biological characterization revealed that 3-normeridamycin promoted neurite outgrowth in cultured cortical cells, dorsal root ganglia, and F-11 cells.[29] Some positive effects were also observed in the mouse MPTP (1-methyl-4-phenyl-1,2,3,6-tetrahydropyridine) model of Parkinson's disease.[30]

The production of 3-normeridamycin by fermentation of *Streptomyces* sp. LL-BB0005 (NRRL 30748) was studied in a series of medium and strain improvement experiments.[31] The wild-type strain produced 20-fold greater titers of meridamycin **4** over 3-normeridamycin **19**. In addition, the original strain produced an overwhelming amount of the polyether antibiotic nigericin. These two factors made the production of sufficient quantities of 3-normeridamycin for further evaluation very challenging. Through classical mutagenesis, fermentation improvement, and careful chemical analysis, a strain was selected that produced **19** as the predominant congener, fully five-fold higher in titer than **4**. This improved strain was used for the production of 3-normeridamycin for *in vivo* studies in rodents.

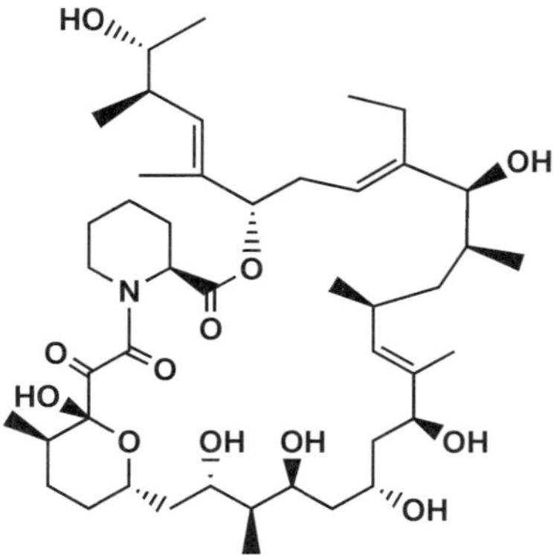

meridamycin
4

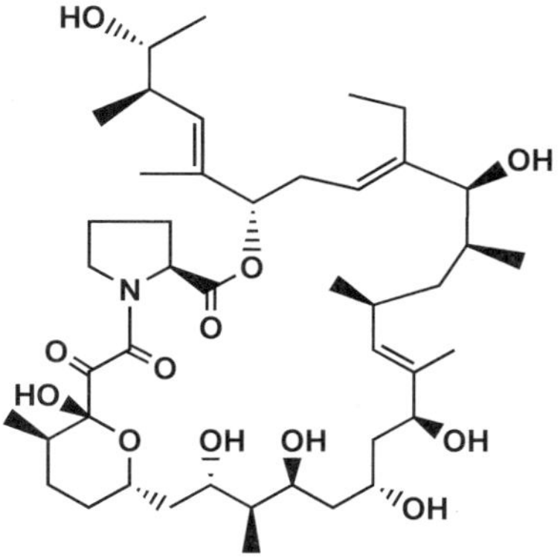

3-normeridamycin
19

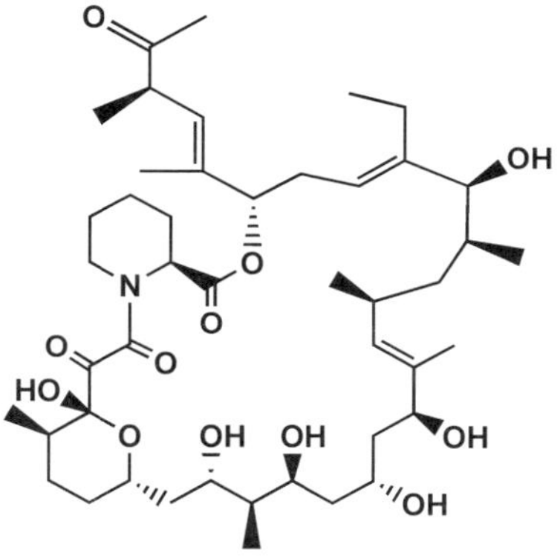

20

Figure 5.5 Meridamycin and analogs.

In preparation for a biosynthetic medicinal chemistry effort, two groups reported the isolation and characterization of the meridamycin biosynthetic gene cluster in 2006.[32,33] Although the two results initially appeared to be divergent, subsequent work confirmed the suggestions of the Cambridge group which led to a consensus organization of the gene cluster.[34] In this follow up work the Wyeth group developed an *E. coli/Streptomyces* shuttle bacterial artificial chromosome (BAC) conjugation vector, on which the entire *mer* cluster was captured. This BAC technology holds the promise of facilitating the biosynthetic manipulation and expression of meridamycin and its analogs. The only biosynthetic analog reported at this time is a 36-keto analog of meridamycin **20** derived from the knockout of the responsible ketoreductase gene in the pathway.[32] The **20** metabolite was formed in quite low yield and was not completely characterized or tested for neuroprotective activity.

5.4.2 Future Prospects

Meridamycin has shown positive indications in preliminary neuroprotection assays, as noted above. Very little research has been done with this subset of natural product immunophilin binders. Given the initial positive assay results and the availability of a biosynthetic platform for modification, this family of compounds is ripe for further development of structure–activity relationships and *in vivo* evaluation. No papers have appeared that describe the FKBP-binding specificity for meridamycins and the question remains open whether meridamycin mediates protein–protein binding in a manner similar to FK-506 and rapamycin.

5.5 Antascomicin

The antascomicin family of natural FKBP-binding ligands is the least abundant of the subgroups discussed thus far. According to the original (and to date only) report on the isolation of compounds in this series, only one of 12 000 strains screened produced the antascomicins, while meridamycin was found in 60 of them.[5] Five congeners were found, as shown in Figure 5.6, with antascomicin B **21** as the major component. It is noteworthy that the producing strain was classified as a *Micromonospora* sp., as opposed to all of the previously discussed producing organisms that are *Streptomyces* strains.

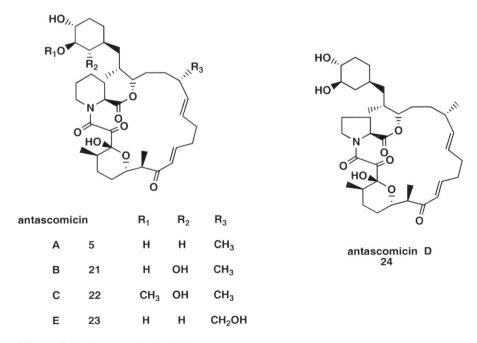

antascomicin		R₁	R₂	R₃
A	5	H	H	CH₃
B	21	H	OH	CH₃
C	22	CH₃	OH	CH₃
E	23	H	H	CH₂OH

antascomicin D
24

Figure 5.6 Antascomicin A–E.

Structurally the antascomicins more closely resemble FK-506/520 than rapamycin or meridamycin, most notably having in common the smaller (23-membered) macrocycle. A prolyl analog, antascomicin D 24, was isolated in approximately 10% of the yield of B. This is the only prolyl congener in the series defined by FK506/520 and antascomicin. The congeners (with the exception of D which is an order of magnitude less potent) bind FKBP-12 with essentially the same avidity as rapamycin or FK506; however, none are active in inhibiting T-cell proliferation. Presumably due to competitive FKBP-binding, the antascomicins are antagonists of rapamycin and FK506. Curiously the antagonistic potency toward rapamycin is 100-fold stronger than against the more closely related FK506.

Regrettably, no additional work on this series has been published by the former Sandoz group and no new reports of the isolation of related compounds have appeared. In the absence of additional research with the original producing strain or the natural products, it is gratifying to note the progress being made in the total synthesis of the compounds. It appears that the Ley group at the University of Cambridge is in the forefront of efforts to produce the fully elaborated compounds.[35] These and other synthetic efforts devoted to this class of natural product were recently reviewed by Ley and co-workers.[36]

5.6   Nocardiopsins

Nocardiopsins A 25 and B 26 (Figure 5.7), natural products that contain a partially elaborated FKBP-binding motif, were isolated from a marine

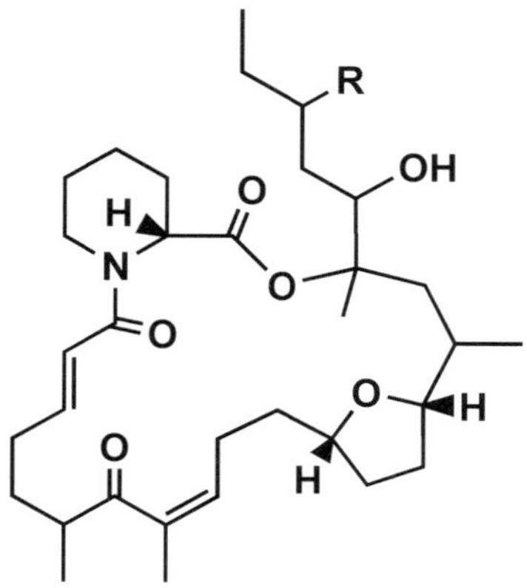

25 nocardiopsin A R = =O
26 nocardiopsin B R = OH

Figure 5.7   Nocardiopsins A and B.

actinomycete of the genus *Nocardiopsis*.[37] These compounds, whose structures have yet to be fully characterized, appear to be biosynthetically related to the other natural FKBP ligands. Their discovery was evidently achieved through a chemistry-driven approach and the compounds display no antifungal or antibiotic activities. Some affinity for FKBP-12 was established in isothermal titration calorimetry experiments, although the binding is considerably weaker than rapamycin in this regard. Non-saline fermentation conditions were developed that greatly enhanced the production of the nocardiopsins over other metabolites, thus paving the way for larger-scale production of the compounds.

The discovery of the nocardiopsins is notable for a number of reasons. Most significantly their isolation demonstrates the continued unexplored reservoir of natural products that exists. Nocardiopsins represent the first new class of this series of compounds to be reported in more than 15 years. Their production by a marine-derived actinomycete also provides further support for the ongoing investigation of microbes in the oceans.

It remains to be seen what biological properties are ascribed to the nocardiopsins. Their structural relationships to other FKBP-binding natural products and relative simplicity suggest that these may well be excellent tools to study binding specificities within the series.

5.7   Virtual Rapalogs

Computational methods were used to design rapamycin analogs that would have optimal affinity for FKBP-12 and not bind mTOR.[38,39] These studies were completed in 1999, but now appear to be particularly relevant. The basic premise of the first study was to pare away as much of the mTOR-binding region as possible without drastically reducing the optimal binding conformation for FKBP-12. The potential role of FKBP-52 in mediating neuroprotection was acknowledged in this work; however, it was believed that the FKBP-binding sites of the two proteins were highly similar. Most notable was that the deletions were proposed in a biosynthetically rational sense. Thus two-carbon (ketide) units were systematically removed from that area of the macrolide encoded by rapB (Figure 5.8) and the resultant hypothetical structures evaluated by virtual screening for FKBP-12 binding. The best candidates from this approach are **27** and **28**, in which diene and triene units respectively have been deleted, as shown in Figure 5.8. In the second paper a smaller template was proposed. Staring with the core of the FKBP-binding domain, a series of biosynthetically plausible ketide linkers was designed and evaluated for FKBP-binding affinity *in silico*. Structures **29** and **30** represent the optimal structures derived from this approach.

Given the advances made in genetic engineering of polyketides in general as well as the specific research on the rapamycin biosynthetic pathway, these virtual rapalogs are feasible targets.[40] In particular, compounds **27** and **28** that would result from modular deletions are excellent candidates for biosynthetic

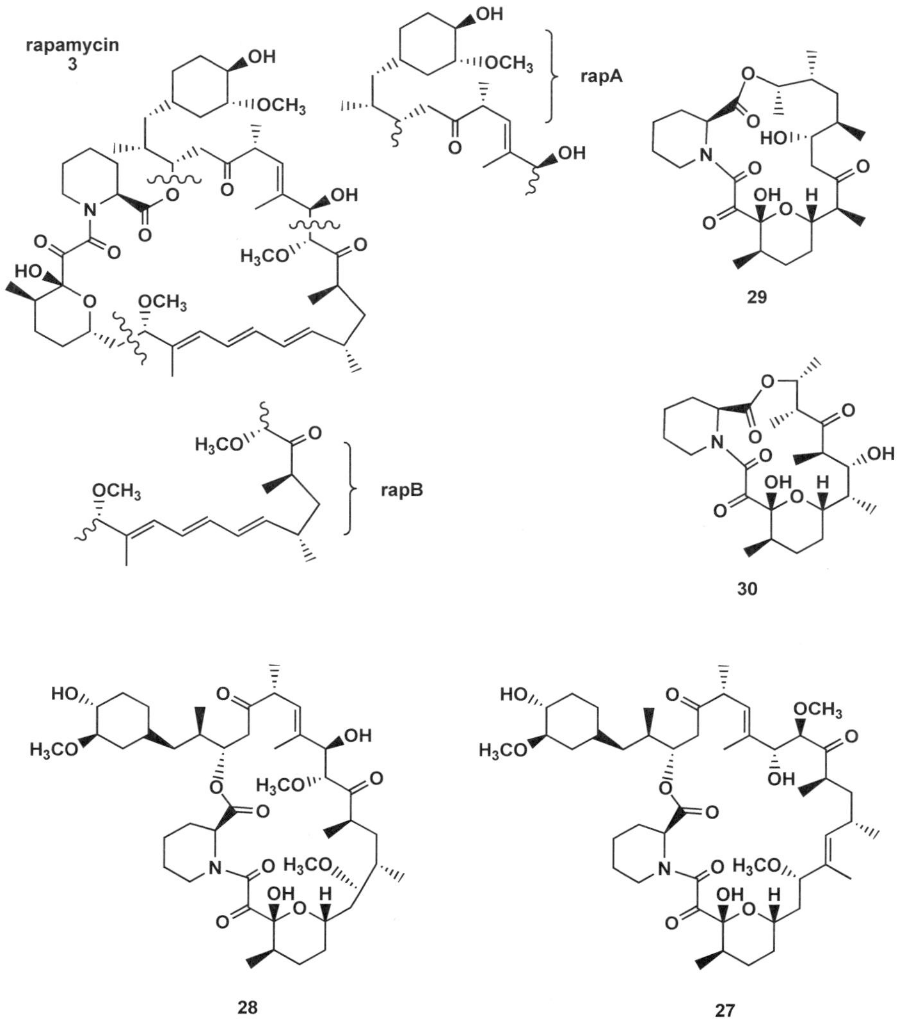

Figure 5.8 Virtual rapalogs designed by biosynthetically rational deletions. The segments of the macrolide encoded by rapA and rapB are illustrated.

engineering. While there are many hurdles to overcome in making productive pathways to elaborate these compounds, the rationale is highly appealing.

5.8 Summary and Perspective

Neurodegeneration and nerve damage in general remain among the most challenging medical conditions for which no effective therapies are available. The early promise of these natural product ligands for FKBP as a route to mediating neurodegeneration appears to have lost momentum. It is remarkable that simple one-step reaction products of rapamycin and FK520 (e.g. **17** and **10**) induce significant enhancement of neurological function and lack antiproliferative effects. Meridamycins share this favorable profile, and appear to be non-toxic. Less is known regarding the biology of antascomicins and even less about the nocardiopsins, other than they have shown some affinity for FKBPs.

Further research in this area is essential to the understanding of the processes by which FKBPs mediate nerve growth. Additional structural studies aimed at probing the differential binding of the various macrocyclic analogs with FKBPs are needed to aid in the design of the next generation of analogs. Mechanistic questions remain for all members of the family. Are ternary complexes of FKBP/natural ligand/protein universal? What are the unknown binding partners? Can this binding specificity be engineered to recruit other proteins?

It also seems highly likely that additional clinical candidates would be generated in the course of such research; given the enormous unmet medical need, the development of such agents is critical. The timing to employ biosynthetic approaches for analog creation, especially for rapamycin, seems ideal. The convergence of enabling technologies with our growing knowledge of FKBP-mediated neurotrophic activities make this area particularly attractive for continued work.

Acknowledgement

I would like to thank Professors Roger Linington and Phil Crews of the University of California, Santa Cruz for their assistance while I was a Visiting Scientist in the Chemistry Department in 2011.

References

1. C. B. Kang, H. Ye, S. Dhe-Paganon and H. S. Yoon, *Neurosignals*, 2008, **16**, 318, and references therein.
2. J. Liu, J. D. Farmer Jr., W. S. Lane, J. Friedman, I. Weissman and S. L. Schreiber, *Cell*, 1991, **66**, 807.
3. J. Choi, J. Chen, S. L. Schreiber and J. Clardy, *Science*, 1996, **273**, 239.
4. G. M. Salituro, D. L. Zink, A. Dahl, J. Nielsen, E. Wu, L. Huang, C. Kastner and F. J. Dumont, *Tetrahedron Lett.*, 1995, **36**, 997.
5. T. Fehr, J. J. Sanglier, W. Schuler, L. Gschwind, M. Ponelle, W. Schilling and C. Wioland, *J. Antibiot.*, 1996, **49**, 230.
6. C. Christner, T. Herdegen and G. Fischer, *Mini-Rev. Med. Chem.*, 2001, **1**, 377.
7. J. Sharkey and S. P. Butcher, *Nature*, 1994, **371**, 336.
8. J. P. Steiner, M. A. Connolly, H. L. Valentine, G. S. Hamilton, T. M. Dawson, L. Hester and S. H. Snyder, *Nature Medicine*, 1997, **3**, 421.

9. J. W. Becker, J. Rotonda, B. M. McKeever, H. K. Chan, A. I. Marcy, G. Wiederrecht, J. D. Hermes and J. P. Springer, *J. Biol. Chem.*, 1993, **268**, 11335.

10. T. D. Ocain, D. Longhi, R. J. Steffan, R. G. Caccese and S. N. Sehgal, *Biochem. Biophys. Res. Commun.*, 1993, **192**, 1340.

11. D. A. Holt, J. I. Luengo, D. S. Yamashita, H. J. Oh, A. L. Konialian, H. K. Yen, L. W. Rozamus, M. Brandt and M. J. Bossard, *J. Am. Chem. Soc.*, 1993, **115**, 9925.

12. T. Mori and T. Uchida, *Curr. Enzyme Inhib.*, 2010, **6**, 46.

13. G. S. Hamilton and C. Thomas, *Adv. Med. Chem.*, 2000, **5**, 1.

14. B. G. Gold, V. Densmore, W. Shou, M. M. Matzuk and H. S. Gordon, *J. Pharmacol. Exp. Ther.*, 1999, **289**, 1202.

15. F. Edlich, M. Weiwad, D. Wildemann, F. Jarczowski, S. Kilka, M.-C. Moutty, G. Jahreis, C. Lucke, W. Schmidt, F. Striggow and G. Fischer, *J. Biol. Chem.*, 2006, **281**, 14961.

16. G. Gerard, A. Deleersnijder, V. Daniels, S. Schreurs, S. Munck, V. Reumers, H. Pottel, Y. Engelborghs, C. Van den Haute, J.-M. Taymans, Z. Debyser and V. Baekelandt, *J. Neurosci.*, 2010, **30**, 2454.

17. P. K. Somers, T. P. Wandless and S. L. Schreiber, *J. Am. Chem. Soc.*, 1991, **113**, 8045.

18. E. J. Dumont, M. J. Staruch, S. L. Koprak, J. J. Siekerka, C. S. Lin, R. Harrison, T. Sewell, V. Kindt, T. R. Beattie, M. Wyvratt and N. H. Sigal, *J. Exp. Med.*, 1992, **176**, 751.

19. R. D. Price, T. Yamaji, H. Yamamoto, Y. Higashi, K. Hanaoka, S. Yamazaki, M. Ishiye, I. Aramori, N. Matsuoka, S. Mutoh, T. Yanagihara and B. G. Gold, *Eur. J. Pharmacol.*, 2005, **509**, 11.

20. N. Hyashi, T. X. Minor, R. Carrion, R. Price, L. Nunes and T. F. Lue, *J. Urol.*, 2006, **176**, 824.

21. A. J. Bella, N. Hayashi, R. E. Carrion, R. Price and T. F. Lue, *J. Sex. Med.*, 2007, **4**, 341.

22. W. P. Revill, J. Voda, C. R. Reeves, L. Chung, A. Schirmer, G. Ashley, J. R. Carney, M. Fardis, C. W. Carreras, Y. Zhou, L. Feng, E. Tucker, D. Robinson and B. G. Gold, *J. Pharmacol. Exp. Ther.*, 2003, **302**, 1278.

23. M. E. Cardenas, R. S. Muir, T. Breuder and J. Heitman, *Eur. Mol. Biol. Organ. J.*, 1995, **14**, 2772.

24. C. R. Reeves, L. M. Chung, Y. Liu, Q. Xue, J. R. Carney, W. P. Revill and L. Katz, *J. Biol. Chem.*, 2002, **277**, 9155.

25. B. Ruan, K. Pong, F. Jow, M. Bowlby, R. A. Crozier, D. Liu, S. Liang, Y. Chen, M. L. Mercado, X. Feng, F. Bennett, D. von Schack, L. McDonald, M. M. Zaleska, A. Wood, P. H. Reinhart, R. L. Magolda, J. Skotnicki, M. N. Pangalos, F. E. Koehn, G. T. Carter, M. Abou-Gharbia, E. I. Graziani, *Proc. Natl. Acad. Sci. USA*, 2008, **105**, 33.

26. K. Pong, E. I. Graziani, S. Liang, D. Liu, Y. Chen, C. Gonzales, H. B. McIlvain, P. H. Reinhart, M. N. Pangalos, B. Ruan, J. Skotnicki, F. Koehn, G. Carter, R. Magolda, M. Abou-Gharbia, M. M. Zaleska and A.

Wood, *2007 Society for Neuroscience Annual Meeting*, Nov. 6, 2007, San Diego, CA. Poster #598.7/Z13.

27. E. I. Graziani, in *Accounts in Drug Discovery*, ed. J. Barrish, P. Carter, P. Cheng and R. Zahler, Royal Society of Chemistry, Cambridge, 2011, p. 316.
28. M. Y. Summers, M. Leighton, D. Liu, K. Pong and E. I. Graziani, *J. Antibiot.*, 2006, **59**, 184.
29. D. Liu, H. B. McIlvain, M. Fennell, J. Dunlop, A. Wood, M. M. Zaleska, E. I. Graziani and K. Pong, *J. Neurosci. Meth.*, 2007, **163**, 310.
30. D. Liu, M. M. Monaghan, A. Sung, H. B. McIlvain, K. Valpreda, M. M. Zaleska, A. Wood, P. H. Reinhart, M. N. Pangalos, M. Summers, F. Koehn, G. Carter, E. I. Graziani and K. Pong, *2007 Society for Neuroscience Annual Meeting*, Nov. 6, 2007, San Diego, CA. Poster #598.6/Z12.
31. K. Kulowski, *C. R. Hutchinson Symposium*, June 20, 2008, Madison, Wisconsin.
32. M. He, B. Haltli, M. Summers, X. Feng and J. Hucul, *Gene*, 2006, **377**, 109.
33. Y. Sun, H. Hong, M. Samborskyy, T. Mironenko, P. F. Leadlay and S. F. Haydock, *Microbiology*, 2006, **152**, 3507.
34. H. Liu, H. Jiang, B. Haltli, K. Kulowski, E. Muszynska, X. Feng, M. Summers, M. Young, E. Graziani, F. Koehn, G. T. Carter and M. He, *J. Nat. Prod.*, 2009, **72**, 389.
35. D. E. A. Brittain, C. M. Griffiths-Jones, M. R. Linder, M. D. Smith, C. McCusker, J. S. Barlow, R. Akiyama, K. Yasuda and S. V. Ley, *Angew. Chem., Int. Ed.*, 2005, **44**, 2732.
36. M. L. Maddess, M. N. Tackett and S. V. Ley, in *Progress in Drug Research*, Vol. 66 (Natural Compounds as Drugs, Volume II)ed. F. Petersen and R. Amstutz, Birkhauser Verlag, AG Basel, 2008, p 13.
37. R. Raju, A. M. Piggott, M. Conte, Z. Tnimov, K. Atexandrov and R. J. Capon, *Chem. Eur. J.*, 2010, **16**, 3194.
38. H. Adalsteinsson and T. C. Bruice, *Biorg. Med. Chem.*, 2000, **8**, 617.
39. H. Adalsteinsson and T. C. Bruice, *Biorg. Med. Chem.*, 2000, **8**, 625.
40. K. J. Weissman and P. F. Leadlay, *Nat. Rev. Microbiol.*, 2005, **3**, 925.

Section 1.2
Engineering Natural Product Synthetic Pathways and Genome Mining

Biosynthesis of Indolocarbazole Alkaloids and Generation of Novel Derivatives by Combinatorial Biosynthesis

CARMEN MÉNDEZ[a], FRANCISCO MORIS[b]
AND JOSÉ A. SALAS*[a]

[a] Departamento de Biología Funcional and Instituto Universitario de Oncología del Principado de Asturias (IUOPA), Universidad de Oviedo, 33006 Oviedo, Spain; [b] EntreChem S.L., Edificio Científico Tecnológico, Campus El Cristo, 33006 Oviedo, Spain
*E-mail: jasalas@uniovi.es

6.1 Introduction

Indolocarbazole alkaloids constitute an important class of natural products, which display a wide range of biological activities, including antibacterial, antifungal, antiviral, antitumor, hypotensive or neuroprotective properties. They have been isolated from actinomycetes, cyanobacteria, slime molds and marine invertebrates.[1,2] A large number of indolocarbazole derivatives have been produced by chemical synthesis or semi-synthesis,[2–5] and several of them have entered clinical trials for the treatment of diverse types of cancer, Parkinson disease or diabetic retinopathy.[6]

Structurally, members of this family are characterized by a core consisting of either an 'open' bisindolylmaleimide (e.g. arcyriarubin B), or a 'closed'

RSC Drug Discovery Series No. 25
Drug Discovery from Natural Products
Edited by Olga Genilloud and Francisca Vicente

indolo[2,3-*a*]carbazole (e.g. tjipanazole F2, rebeccamycin, staurosporine) (Figure 6.1). Most of the latter compounds are in fact derivatives of the indolo[2,3-*a*]pyrrolo[3,4-*c*]carbazole ring system, to which a sugar residue is often attached. In the glycosides, the carbohydrate can be attached through a single N-glycosidic bond (as in rebeccamycin and AT2433), or it can be joined by two bonds consisting of an N-glycosidic bond and an *N,O*-ketal (as in staurosporine and K-252a). Another key structural difference between the different indolocarbazoles resides at the pyrrole moiety, consisting of an imide function (as in rebeccamycin and AT2433) or an amide function (as in staurosporine and K-252a) (Figure 6.1). These differences seem to be essential

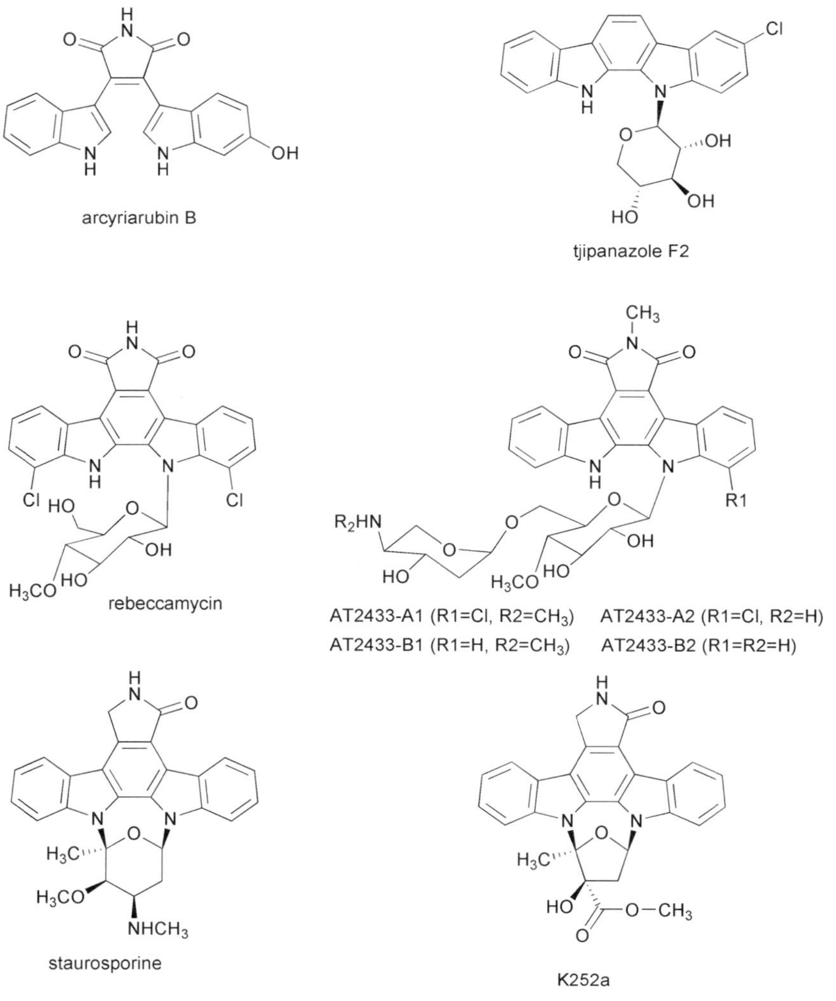

Figure 6.1   Chemical structure of indolocarbazoles.

for target selectivity, so that indolocarbazoles can be grouped into two different classes on the basis of their activity: one represented by rebeccamycin, which are potent stabilizers of topoisomerase I-DNA covalent complexes, and another represented by staurosporine, which includes potent inhibitors of protein kinases A, C and K.

The biosynthetic origin of indolocarbazoles was studied by feeding isotope-labeled precursors to the staurosporine and rebeccamycin producers (*Lentzea albida*, formerly *Streptomyces staurosporeus*, and *Lechevalieria aerocoloni-genes*, formerly *Saccharothrix aerocolonigenes*, respectively).[7–11] It was concluded that the indolocarbazole core derives from two tryptophan units with the carbon skeleton incorporated intact, and the sugar moiety from glucose and methionine. Feeding experiments using tryptophan derivatives showed that indolepyruvate was an intermediate in rebeccamycin biosynthesis, while tryptamine was not, and on the contrary this last compound seemed to be a precursor in staurosporine biosynthesis.[8,9] Differences in the experimental conditions used and in the metabolism between the two strains could be responsible for the apparently conflicting results obtained with the two strains.

6.2 Isolation and Characterization of Gene Clusters for Rebeccamycin and AT2433

Rebeccamycin and AT2433 are two closely related indolocarbazoles characterized by containing a core with an imide function, which is glycosylated by a monosaccharide or a disaccharide chain respectively. This sugar moiety is attached through a single N-glycosidic bond, one of these sugars being 4-*O*-methyl-D-glucose. AT2433 differs from rebeccamycin in having an *N*-methylated and asymmetrically halogenated indolocarbazole core, and by containing an aminodideoxypentose.

The first indolocarbazole gene cluster characterized was that of rebeccamycin (*reb* cluster). It was cloned using as a probe the *ngt* gene, which was the first rebeccamycin gene isolated.[12] This gene was identified as a consequence of bioconversion experiments. It was observed that indolocarbazole J-104303 (identical to the rebeccamycin aglycone but possessing hydroxyl groups instead of the chlorine atoms) was converted into its *N*-glucosylated derivative by the rebeccamycin producer *L. aerocolonigenes* ATCC39243. To identify the gene(s) causing this modification, a gene library of genomic DNA from this producer strain was constructed in *Streptomyces lividans*, and a clone was isolated capable of carrying out this bioconversion. Upon sequencing of the DNA insert, it was found to contain a gene (designated as *ngt* gene) coding for a glycosyltransferase that probably was involved in rebeccamycin biosynthesis, although no definite proof was reported at that time. Later on this gene was used as a probe to identify a cosmid clone from a library of *L. aerocolonigenes* genome, which upon heterologous expression into *Streptomyces albus* conferred the ability to produce rebeccamycin.[13] This result was significant since it unambiguously proved the involvement of the cloned genes in rebeccamycin biosynthesis, and made possible the first characterization of a complete gene cluster governing the biosynthesis of an indolocarbazole alkaloid.[13] Afterwards, the same cluster was reported by two different laboratories.[14,15] Sequencing of the rebeccamycin gene cluster revealed that it spanned 17.6 kb with 11 genes, apparently organized in four transcriptional units (Figure 6.2a).

The AT2433 gene cluster (*atm* cluster) has also been cloned and sequenced from *Actinomadura melliaura* by using degenerate primers designed to amplify homologous genes to *rebD* and *rebP*, which encode the chromopyrrolic acid synthase and a putative P450 oxidase respectively.[16] The sequence spans 45 kb and contains 35 genes. The genetic organization of AT2433 and rebeccamycin gene clusters are quite similar and both contain genes for the biosynthesis of the aglycone, biosynthesis and attachment of sugars, regulation and secretion. Interestingly, the AT2433 cluster contains an extra gene *atmA*, absent in the rebeccamycin cluster, that has been proposed to participate in the biosynthesis of the aglycone (see below).

Based on similarities of the deduced gene products with proteins in databases, the chemical structures of compounds produced by recombinant strains co-expressing different sets of genes, products accumulated by different mutants, and *in vitro* assays of enzyme activities, a pathway for the biosynthesis of these indolocarbazoles has been proposed (Figure 6.2b).[17–27]

6.2.1 Tryptophan Halogenation

The first step in rebeccamycin and AT2433 biosynthesis is the conversion of L-tryptophan to 7-chloro-L-tryptophan, which serves as unique monomer precursor. This reaction is catalyzed by a two-component system formed by reductase/halogenase enzymes that has been characterized *in vitro* using the halogenase RebH from the rebeccamycin pathway.[21–23] *In vitro* halogenation by RebH requires the addition of flavin reductase RebF, which catalyzes the NADH-dependent reduction of FAD to provide $FADH_2$ for the halogenase. The rebeccamycin cluster contains genes for both components (*rebH* and *rebF*) while in the AT2433 cluster the flavin reductase gene is absent. This component can be substituted by other reductases from the host, since *S. albus* expressing the rebeccamycin gene cluster in the absence of *rebF* is able to produce rebeccamycin.[20] Chloride salts such as NaCl are the source of chlorine for RebH-catalyzed halogenation.[21]

6.2.2 Formation of the First Bisindole Intermediate ('Dimerization')

The next step in the biosynthesis is the formation of a chromopyrrolic acid intermediate through the fusion of two tryptophans or halogenated tryptophan moieties. This process requires the tandem action of two enzymes:

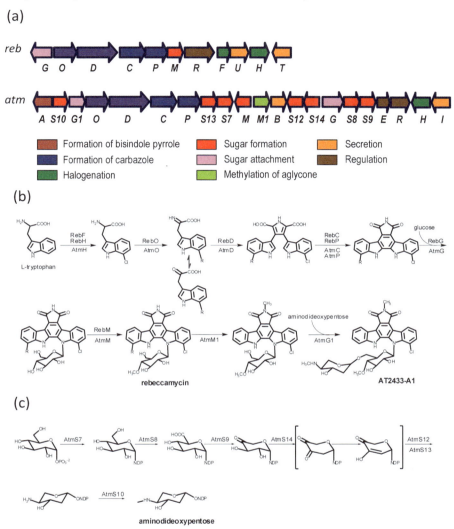

Figure 6.2 Biosynthesis of indolocarbazoles with an imide function. (a) Genetic organization of the gene clusters for rebeccamycin (*reb*) and AT2433 (*atm*). (b) Proposed biosynthesis pathway for rebeccamycin and AT2433. (c) Proposed biosynthesis pathway for the biosynthesis of aminodideoxypentose.

an L-tryptophan oxidase (RebO and AtmO in rebeccamycin and AT2433 biosyntheses, respectively) and a heme protein with chromopyrrolic acid synthase activity (RebD and AtmD). Indications on the role of RebD in the early steps of the pathway came from experiments with mutants in the rebeccamycin pathway and from strains expressing sets of rebeccamycin genes.[14,20] A *rebD*-disrupted mutant of *L. aerocolonigenes* was not able to produce any indolocarbazole-related compounds.[14] Also, it was found that the *rebO* and *rebD* genes were both essential for the production of chromopyrrolic acid (CPA), a known natural product, which was very similar to other metabolite produced by a *rebP*-disrupted mutant of *L. aerocolonigenes*,[14] and by an *S. albus* strain co-expressing *rebH*, *rebO* and *rebD*.[20] More recently, the RebO and RebD enzymes have been purified and characterized.[17–19] RebO is a flavin-dependent L-amino acid oxidase that catalyzes the oxidative deamination of 7-chloro-L-tryptophan to 7-chloroindole-3-pyruvic acid, with the production of hydrogen peroxide.[18] RebD is the first member of a novel subfamily of heme-containing oxidases with both catalase and CPA synthase activities, and performs the condensation and the oxidative coupling of indole-3-pyruvate, in the presence of an exogenously added nitrogen source, to CPA.[17,19] Interestingly, the AT2433 gene cluster contains a gene (*atmA*) that would code for a protein similar to glutamine-dependent amidotransferases. It has been proposed that AtmA catalyzes nitrogen installation to provide an imine derivative of indolepyruvate.[16] AT2433 is the only naturally occurring indolocarbazole to have an asymmetrically halogenated indolocarbazole core. It has been postulated that formation of this asymmetrically halogenated aglycone may reside on the unique substrate specificities of AtmO and AtmA.[16]

6.2.3 Formation of the Indolopyrrolocarbazole Core

Following formation of mono- (as in AT2433) or di-halogenated-CPA (as in rebeccamycin), the 'open' chromopyrrolic acid intermediate is converted into a 'closed' indolocarbazole aglycone containing an imide function at the pyrrole moiety as in the rebeccamycin core or in arcyriaflavin A (the non-chlorinated rebeccamycin aglycone). A pair of enzymes is required for carrying out this process: a cytochrome P450 oxidase (RebP and AtmP in rebeccamycin and AT2433 biosynthesis, respectively) and a flavin-dependent hydroxylase (RebC and AtmC). It has been shown that efficient production of the indolopyrro-locarbazole core in *S. albus* requires the co-expression of genes for CPA formation (*rebO*, *rebD*) and *rebC* and *rebP*.[20] In the presence of *rebH* gene, the rebeccamycin aglycone is obtained. In the absence of *rebC* it is also possible to obtain indolopyrrolocarbazole cores but as a mixture of three derivatives differing only at the C7 position of the pyrrole: arcyriaflavin (7-oxo), staurosporine aglycone (7-deoxo) and 7-hydroxy-staurosporine aglycone.[20] In agreement with these results, *rebP*- and *rebC*-disrupted mutants of *L. aerocolonigenes* have been found to accumulate (respectively) 11,11′-dichloro-CPA or a mixture of rebeccamycin derivatives differing at C7.[14] These authors also reported that the 7-deoxo or 7-hydroxy were not rebeccamycin intermediates, since they were not converted into rebeccamycin in bioconversion experiments. Consequently, it appears that RebP is responsible for the decarboxylative oxidation needed to convert an 'open' (dichloro-) CPA into a 'closed' indolopyrrolocarbazole, and that the monooxygenase RebC is needed for efficient completion of this reaction and for the determination of the C-7 oxidation state in the final product. More recently the crystal structure of RebC has been elucidated.[24] Based on these studies it has been suggested that the role of RebC is to sequester a reactive intermediate produced by RebP and to react with it enzymatically, preventing its conversion to a set of degradation products that includes, at low levels, the desired product.

The AT2433 aglycone is *N*-methylated. Sequence analysis of the AT2433 gene cluster reveals a gene *atmM1* coding for protein containing a thiopurine *S*-methyltransferase domain, which supports this protein as the putative candidate for the AT2433 maleimide *N*-methyltransferase.[16]

6.2.4 Biosynthesis, Attachment and Modification of Sugars

Formation of the aglycone is followed by its glycosylation. Rebeccamycin contains a D-glucose, while AT2433 is glycosylated by a disaccharide of D-glucose and an aminodideoxypentose. These saccharide residues are linked to the aglycone through a single C–N bond. D-Glucose is provided from primary metabolism, while aminodideoxypentose biosynthesis is coded by the AT2433 gene cluster. Comparison of this cluster with that encoding calicheamicin (an antitumor compound also containing an aminodideoxypentose moiety) has allowed the identification of seven candidate genes for the biosynthesis of aminodideoxypentose in the AT2433 cluster and to propose a pathway for the biosynthesis of this unusual sugar (Figure 6.2c).[16] *N*-Glycosylation of the aglycone is carried out by the *N*-glycosyltransferases RebG and AtmG in rebeccamycin and AT2433, respectively.[13,16] By expressing in *S. albus* a plasmid containing rebeccamycin genes, including *rebG*, it was confirmed that *rebG* coded the *N*-glycosyltransferase.[13] Similarly, bioconversion experiments using an *E. coli* strain overexpressing *atmG* confirmed the same role for AtmG in AT2433 biosynthesis.[16] Glycosylation appears to take place only after the formation of a 'closed' planar indolocarbazole ring system, but not on 'open' CPA intermediates: co-expression in *S. albus* of *rebODG* genes or feeding CPA to a strain expressing *rebG* does not result in the production of any glycosylated derivatives.[20] However, co-expression of *rebODCPG* or feeding arcyriaflavin to a strain expressing *rebG* successfully produced a glycosylated compound.[20] In AT2433 biosynthesis, glycosylation of the aglycone is followed by the incorporation of the aminopentose. The *O*-glycosyltransferase AtmG1 has been postulated for being responsible of this enzymatic step.[16]

As final steps in the biosynthesis, several methylations occur decorating the final indolocarbazole structure, acting on the sugars or on the imide function, as was mentioned before for AT2433. Rebeccamycin is *O*-methylated at the sugar residue. Co-expression of *rebO*, *rebD*, *rebC*, *rebP*, *rebG* and *rebM* leads to the production of dideschloro-rebeccamycin, indicating that the methyltransferase RebM methylates the sugar at the 4-hydroxy position.[20] This activity has been confirmed by *in vitro* assays, which showed that RebM displayed substrate promiscuity by being able to accept a wide range of alternate substrates both as donor and as acceptor substrates.[25,26] More recently, a crystal structure of RebM has been reported.[27] The RebM binding pocket implicates a predominance of nonspecific hydrophobic interactions consistent with the above-mentioned ability of RebM to methylate a wide range of indolocarbazole surrogates. Two residues, His140 and Asp166, have been shown to be important for catalysis by site-directed mutagenesis.[27] Based upon homology to RebM first, and then confirmed by *in vitro* assays, it was found that AtmM was the required glucose-4'-*O*-methyltransferase in the AT2433 pathway.[16] Interestingly, AtmM displayed a slightly broader substrate scope than RebM.

6.3 Isolation and Characterization of Gene Clusters for Staurosporine and K-252a

Staurosporine and K252A are two indolocarbazoles that contain an amide function at the pyrrole moiety, and are glycosylated by a monosaccharide attached through two *N*-glycosidic bonds to the aglycone (Figure 6.1). They differ at the glycosidic moiety, which is a L-ristosamine in staurosporine and a dihydrostreptose in K-252a.

The staurosporine gene cluster (*sta* cluster) (Figure 6.3a) was isolated by two different laboratories from *Streptomyces* sp. TP-A0274 and from *S. longisporoflavus* DSM 10189, and they were independently expressed into *S. lividans* and *S. albus*, conferring on them the ability to produce staurosporine, and thus confirming the involvement of the cloned DNA region in staurosporine biosynthesis.[28,29] Both *sta* clusters were essentially identical, with an identical gene organization, with the similarity of the different gene products being higher than 95% for all genes compared.[29] In *Streptomyces* sp. TP-A0274 it was reported that the gene cluster consisted of 14 *orfs* including those involved in deoxysugar biosynthesis (*staABEJIKMAMB*), transcriptional regulation (*staR*), aglycone formation (*staODP*) and attachment of the sugar moiety (*staGN*).[28] However, no *rebC* homologs were reported at that time. Later on, a careful examination of the DNA sequences from *S. longisporoflavus* and from *Streptomyces* sp. TP-A0274 revealed the existence at the right end of the clusters of a *rebC* homolog that was named *staC*.[20] Interestingly, the staurosporine cluster lacks genes coding for putative transmembrane proteins, which are present in other indolocarbazole gene clusters and that have been proposed to be involved in secretion mechanisms of these drugs.[30]

The K-252a cluster (*ink* cluster) has also been cloned and sequenced from *Nonomuraea longicatena* by using as a probe an internal fragment of *vioB*, a *Chromobacterium violaceum* gene coding for an enzyme that catalyzes tryptophan decarboxylation and condensation reactions in violacein biosynthesis (Figure 6.3a).[31] Involvement of the cloned DNA region in K-252a

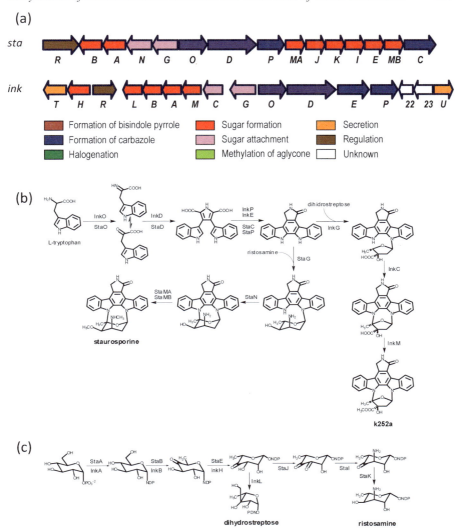

Figure 6.3 Biosynthesis of indolocarbazoles with an amide function. (a) Genetic organization of the gene clusters for staurosporine (*sta*) and K-252a (*ink*). (b) Proposed biosynthesis pathway for staurosporine and K-252a. (c) Proposed biosynthesis pathway for the biosynthesis of L-ristosamine. (d) Proposed biosynthesis pathway for the biosynthesis of dihydrostreptose.

biosynthesis was confirmed by expressing this DNA region into *S. albus* and the subsequent production of this indolocarbazole.[31] Sequence analyses revealed the presence of 14 *orfs* putatively required for K-252a biosynthesis, and which show a rather similar gene organization to those of other indolocarbazole genes clusters. The cluster contains genes for indolocarbazole core formation, glycosylation and sugar methylation, as well as a regulatory gene and two resistance/secretion genes.[31]

Based on similarities of the deduced gene products with proteins in databases, on co-expression of sets of genes in heterologous hosts and on analysis of the compounds produced by the recombinant strains, a pathway for the biosynthesis of these indolocarbazoles has been proposed (Figure 6.3b).[20,28,29,31,32]

6.3.1 Formation of the Aglycone

Similarly to what occurs in rebeccamycin and AT2433 biosynthesis, the first events in staurosporine and K-252a biosynthesis consist in the formation of the indolocarbazole aglycone. However, these two groups of compounds show differences at the C7 position of the aglycone with a carbonyl function in rebeccamycin and AT2433, which are absent in staurosporine and K-252a. The *sta* and *ink* clusters contain a set of genes (*staODCP* and *inkODEP*, respectively) that are responsible for the formation of the indolocarbazole skeleton.[20,28,31] StaO/InkO initiates the biosynthesis pathway by catalyzing the reaction of L-tryptophan to the imine form of indole-3-pyruvic acid, and then StaD/InkD catalyze the coupling of two molecules of indole-3-pyruvic acid imine to generate CPA. Formation of this intermediate was confirmed by co-expressing *inkO* and *inkD* in *S. albus*, being the resultant recombinant strain able to produce CPA.[32] Then, the action of StaP/InkP together with StaC/InkE leads to the formation of the indolocarbazole core. By expressing in *S. albus* different sets of genes, including *rebOD* and different combinations of *rebC*, *rebP*, *StaC* and *staP*, it was demonstrated that *rebP* and *staP* were functionally equivalent, and that *rebC* and *staC* determine different oxidation states at position C7 of the aglycone.[20] Thus, substitution of *rebC* by *staC* leads to the formation of the staurosporine aglycone, which is identical in its structure to the natural product K252c, or staurosporine aglycone. The crystal structure of the StaP-CPA complex has been reported.[33] Based on that structure, a catalytic mechanism for the StaP enzyme has been proposed and later explained.[34]

6.3.2 Biosynthesis, Attachment and Modification of Sugars

Staurosporine contains an L-ristosamine that is *O*- and *N*-methylated, while K-252a contains an *O*-methylated dihydrostreptose, being both sugars linked to the aglycone through two C–N bonds. Genes for the biosynthesis and modification of these two sugars are localized in the corresponding gene clusters (*staABEJIKMAMB* and *inkABHLM*, in the staurosporine and K-252a clusters, respectively), as well as a glycosyltransferase gene (*staG/inkG*) and a P450 oxygenase (*staN/inkC*) required for the formation of two glycosidic bonds.[28,29,31] Based on similarities with proteins in data bases, biosynthetic pathways for these two sugars have been proposed (Figure 6.3c).[28,31] The staurosporine glycosylation events were reconstituted in *S. albus* using a two-plasmid system.[29] The first plasmid ('aglycone plasmid') contained a set of genes from the *reb* and *sta* gene clusters that, when put together, conducted the biosynthesis of the staurosporine aglycone (*rebO*, *rebD*, *staC*, *rebP*). This plasmid also harbored two other genes that were the candidates for the double attachment of the sugar moiety: *staG* (coding for a glycosyltransferase) and *staN* (coding for a cytochrome P450 enzyme). The second plasmid ('sugar plasmid') included a DNA fragment from the chromosome of *S. longisporoflavus* with a set of genes (*staEJIKMAMB*) coding for activities specifically required for the biosynthesis of L-ristosamine. Co-expression of both plasmids in *S. albus* resulted in staurosporine production, which confirmed the role of the cloned genes in glycosylation steps of staurosporine biosynthesis.[29] Deletion of *staN* from the 'aglycone plasmid' resulted in the production of the glycosylated product holyrine A, a known compound in which L-ristosamine is singly attached through an *N*-glycosidic bond to the aglycone, which confirmed the role of StaN in the formation of the second C–N bond. Holyrine A also differs from staurosporine in the absence of the two sugar methylations (despite the presence of both methyltransferase genes *staMA* and *staMB* in the 'sugar plasmid'), indicating that both sugar methylations were the last steps in staurosporine biosynthesis, taking place only after the second linkage is established. Similar conclusions were also drawn from experiments made with a *staN*-disrupted mutant of *Streptomyces* sp. TP-A0274.[35]

6.4 Generation of Novel Glycosylated Indolocarbazoles

Increasing evidence suggests the existence of flexibility in secondary metabolite glycosyltransferases regarding the sugar donor. Taking advantage of this flexibility a number of novel glycosylated derivatives from staurosporine have been generated. Using the two plasmid system for reconstitution of staurosporine biosynthesis described above (see Section 6.3.2), several glycosylated derivatives of staurosporine were generated. For this purpose a family of 'sugar cassette plasmids', each one directing the biosynthesis of different L-deoxysugars and D-deoxysugars[36–38] were used. These experiments led to the formation of novel glycosylated staurosporines containing L-rhamnose (L-Rha), L-olivose (L-Olv), L-digitoxose (L-Dig) or D-olivose (D-Olv) attached either through single C–N bonds (in all the cases; indicated by 1N in Figures 6.4 and 6.5) or double C–N bonds (in the case of the former three sugars but not for D-olivose; indicated by 2N in Figures 6.4 and 6.5)[29] (Figure 6.4). In a similar way novel glycosylated derivatives of rebeccamycin were also generated with this system. In this case it was necessary to replace the *staC* in the 'aglycone plasmid' by the *rebC* gene since one of the structural differences between rebeccamycin and staurosporine resides in the oxygenation state of the molecule and these two genes are responsible for these differences. Rebeccamycin derivatives were formed with the same structural features as those of the staurosporine derivatives, i.e. compounds containing L-rhamnose

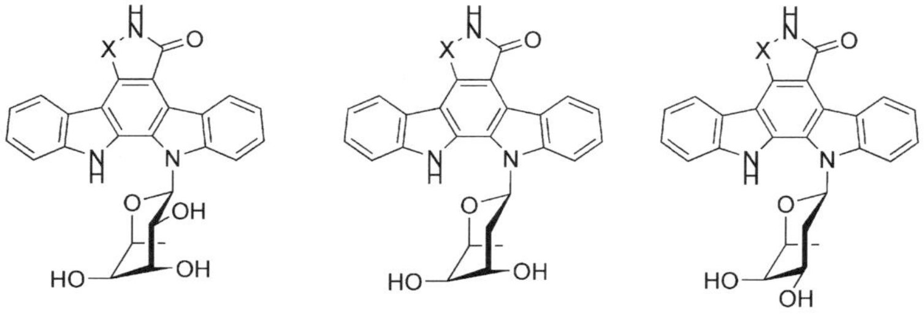

Reb-L-Rha^{-1}N (X = C=O)
Sta-L-Rha^{-1}N (X = CH_2)

Reb-L-Olv^{-1}N (X = C=O)
Sta-L-Olv^{-1}N (X = CH_2)

Reb-L-Dig^{-1}N (X = C=O)
Sta-L-Dig^{-1}N (X = CH_2)

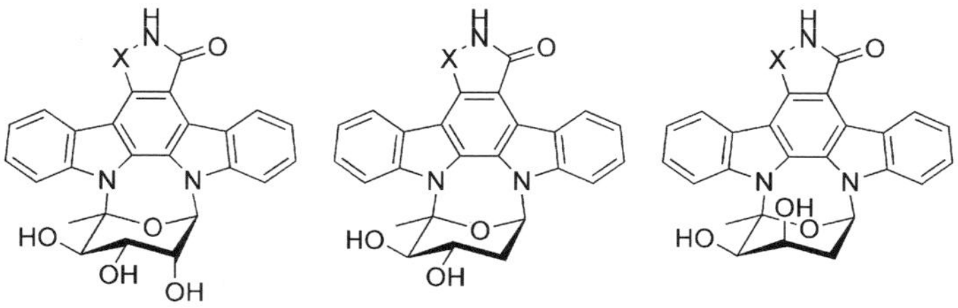

Reb-L-Rha^{-2}N (X = C=O)
Sta-L-Rha^{-2}N (X = CH_2)

Reb-L-Olv^{-2}N (X = C=O)
Sta-L-Olv^{-2}N (X = CH_2)

Reb-L-Dig^{-2}N (X = C=O)
Sta-L-Dig^{-2}N (X = CH_2)

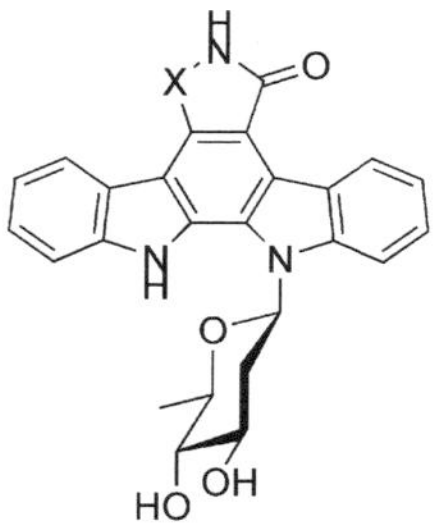

Reb-D-Olv^{-1}N (X = C=O)
Sta-D-Olv^{-1}N (X = CH_2)

Figure 6.4 Novel glycosylated derivatives of rebeccamycin and staurosporine generated by combinatorial biosynthesis. Abbreviations: Sta, staurosporine; Reb, rebeccamycin; L-Rha, L-rhamnose; L-Olv, L-olivose; L-Dig, L-digitoxose; D-Olv, D-olivose. Compounds in which the sugars are attached to the aglycon through a single C–N bond are indicated by 1N, while those attached through a double C–N bond are indicated by 2N.

(L-Rha), L-olivose (L-Olv), L-digitoxose (L-Dig) or D-olivose (D-Olv) attached either through single C–N bonds (in all the cases; indicated by 1N in Figures 6.4 and 6.5) or double C–N bonds (in the case of the former three sugars but not for D-olivose; indicated by 2N in Figures 6.4 and 6.5).[39] Through bioconversion experiments it has been also shown that the RebG rebeccamycin glycosyltransferase can glycosylate a set of indolocarbazole surrogates, and evidence for a remarkable lack of regioselectivity of this transferase in the presence of asymmetric substrates has been provided.[40]

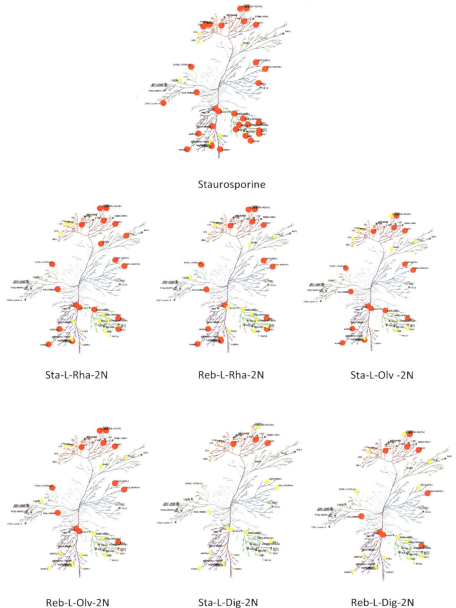

Figure 6.5 Dendrogram showing the human kinome and its inhibition by staurosporine and six novel glycosylated indolocarbazoles. Big red dots represent >50% inhibition at <10 nM, medium yellow dots represent >50% inhibition at <100 nM and small dots represent 50% inhibition at >100 nM. Abbreviations: Sta, staurosporine; Reb, rebeccamycin; L-Rha, L-rhamnose; L-Olv, L-olivose; L-Dig, L-digitoxose. 2N, sugar is attached through a double C–N bond.

Interestingly, the RebM *O*-methyltransferase can modify *in vitro* many of the glycosylated compounds produced by RebG.[40] The RebM *O*-methyltransferase also shows an interesting peculiarity regarding the methyl group donor. RebM is capable of using a non-natural cofactor analogue of *S*-adenosyl-L-methionine (a novel synthetic cofactor bearing a pendant 50-amino acid *N*-mustard) as the methyl group donor to generate a number of new rebeccamycin analogs.[41] This could be another way to increase the molecular biodiversity in indolocarbazoles for drug discovery.

6.5 Assaying for Protein Kinase Inhibition

Kinases, the enzymes that catalyze phosphorylation events, have been implicated in hundreds of different diseases, including cancer, inflammatory diseases and neurological disorders. Compounds that control their activity, therefore, hold rich promise for drug development. The human genome encodes 518 kinases (the human kinome), many of whose functions remain unknown. The total roster of phosphorylation events is even less understood. Many of the human genome's 30 000-odd proteins are phosphorylated at one time or another, often at multiple sites, making the phosphoproteome a vast entity, with perhaps hundreds of thousands of unique members. Since the early times of staurosporine, the quest for selective inhibitors has been a mainstay in kinase drug development. In 2000 it was demonstrated that many commercially available kinase inhibitors were not nearly as specific as they were previously thought to be. Cited more than 1000 times to date, the study underscores the difficulty of drawing conclusions based on inhibitor studies: these reagents can be powerful probes of signal transduction pathways, yet unforeseen effects on non-target kinases can yield misleading results.[42] Such limitations have practical impact, both in the laboratory and in the clinic. The wider the range of pathways a kinase inhibitor disrupts, the more likely the compound is to affect normal cell processes adversely. Comprehensive data on what pathways a potential drug affects will help not only to screen out potential drug candidates, but also to yield a better understanding of a compound's potential uses, side-effects and weaknesses. Determining selectivity during kinase drug discovery is very important and requires a combination of biochemical and cellular approaches. One issue is the extent to which comparison of *in vitro* measurements of compound activity at various isolated targets accurately reflects selectivity in cells, even with proteins from the same family, such as kinases. Inherent differences in kinase expression levels, enzyme kinetics and responsiveness of downstream effectors along individual pathways may lead to selective phenotypic outcomes from apparently unselective inhibitors. Nevertheless, *in vitro* selectivity profiling using panels of isolated kinases remains a valuable tool for comparing overall specificity.[43,44] Taking these issues in mind, 14 indolocarbazole compounds with a strong resemblance to staurosporine were submitted to a microfluidic shift mobility kinase assay, offered by Caliper Life Sciences, a CRO specialized in developing bioassays to

Table 6.1 Kinase inhibition IC_{50} values (nM) for selected pairs of kinase and indolocarbazole

Kinase	*Sta-L-Rha-2N*	*Reb-L-Rha-2N*	*Sta-L-Olv-2N*	*Reb-L-Olv-2N*	*Sta-L-Dig-2N*	*Reb-L-Dig-2N*
AurA	—	>10	5.0	>10	>10	6.7
AurB	—	—	6.8	—	—	4.5
Chk1	2.4	1.0	—	4.9	>10	—
Dyrk1a	—	4.0	—	—	>100	>10
Ftl3	0.6	0.6	0.5	0.6	—	0.4
FGFR1	—	8.9	>10	—	>100	>10
HGK	—	0.8	—	—	>10	>10
Ikkb	—	0.2	—	<0.03	>100	>10
Jak2	0.4	0.7	0.6	0.7	0.5	1.2
KDR	—	3.7	—	0.5	>10	>10
SYK	—	1.0	—	1.1	>10	2.3

aid drug discovery efforts. The 14 indolocarbazole compounds included compounds in which the sugar is attached to the aglycon through a single C–N bond, as in rebeccamycin, and also those in which the sugar is attached through a double C–N bond, as in staurosporine. Figure 6.5 graphically represents the results on the human kinome dendrogram of six novel glycosylated indolocarbazoles and its comparison with staurosporine. In the first pass screening, 54 kinases were tested at concentrations 1 μM, 100 nM and 10 nM. At the highest concentration (1 μM) most kinases were heavily inhibited, and only at 100 nM and especially at 10 nM were differences notable. Staurosporine, according to its promiscuity, shows high activity for most kinases, but the novel indolocarbazoles ranged from moderately selective to highly selective (at 10 nM Sta-L-Dig-2N inhibits only one kinase at >70%, Reb-L-Dig-2N inhibits five kinases and Reb-L-Olv-2N inhibits six kinases).[39] At 10 nM or less, while staurosporine was very active against 29 of the kinases tested, compound Sta-L-Rha-2N was active against 19 kinases, compounds Reb-L-Rha-2N and Sta-L-Olv-2N were active only in 15 kinases; Reb-L-Olv-2N was active against 12 kinases and Sta-L-Dig-2N and Reb-L-Dig-2N were very active against 3 and 8 kinases, respectively. IC_{50} values for a panel of selected pairs of kinase and indolocarbazoles are shown in Table 6.1. These data shows that one round of combinatorial biosynthesis is able to provide a first generation of analogs much more selective than staurosporine, while keeping the potency of inhibition in those kinases that are inhibited.

Acknowledgements

This work has been supported by grants from the Spanish Ministry of Science and Innovation (PET2008-0271) and the Red Temática de Investigación Cooperativa de Centros de Cáncer (Ministry of Health, Spain; ISCIII-RETIC RD06/0020/0026).

References

1. G. W. Gribble and S. J. Berthel, in *Studies in Natural Product Chemistry*, ed. Atta-ur-Rahman, Elsevier, Amsterdam, Vol. 12 (Stereoselective Synthesis, Part H), 1993.
2. H. J. Knolker and K. R. Reddy, *Chem. Rev.*, 2002, **102**, 4303.
3. M. Prudhomme, *Eur. J. Med. Chem.*, 2003, **38**, 123.
4. C. Sánchez, C. Méndez and J. A. Salas, *Nat. Prod. Rep.*, 2006, **23**, 1007.
5. C. Marminon, F. Anizon, P. Moreau, B. Pfeiffer, A. Pierré, R. M. Golsteyn, P. Peixoto, M. P. Hildebrand, M. H. David-Cordonnier, O. Lozach, L. Meijer and M. Prudhomme, *Mol. Pharmacol.*, 2008, **74**, 1620.
6. M. S. Butler, *Nat. Prod. Rep.*, 2005, **22**, 162.
7. C. J. Pearce, T. W. Doyle, S. Forenza, K. S. Lam and D. R. Schroeder, *J. Nat. Prod.*, 1988, **51**, 937.
8. K. S. Lam, S. Forenza, T. W. Doyle and C. J. Pearce, *J. Ind. Microbiol.*, 1990, **6**, 291.
9. S. W. Yang and G. A. Cordell, *J. Nat. Prod.*, 1997, **60**, 788.
10. S. W. Yang, L. J. Lin, G. A. Cordell, P. Wang and D. G. Corley, *J. Nat. Prod.*, 1999, **62**, 1551.
11. S.-W. Yang and G. A. Cordell, *J. Nat. Prod.*, 1997, **60**, 236.
12. T. Ohuchi, A. Ikeda-Araki, A. Watanabe-Sakamoto, K. Kojiri, M. Nagashima, M. Okanishi and H. Suda, *J. Antibiot. (Tokyo)*, 2000, **53**, 393.
13. C. Sánchez, I. A. Butovich, A. F. Braña, J. Rohr, C. Méndez and J. A. Salas, *Chem. Biol.*, 2002, **9**, 519.
14. H. Onaka, S. Taniguchi, Y. Igarashi and T. Furumai, *Biosci. Biotechnol. Biochem.*, 2003, **67**, 127.
15. C. G. Hyun, T. Bililign, J. Liao and J. S. Thorson, *ChemBioChem*, 2003, **4**, 114.
16. Q. Gao, C. Zhang, S. Blanchard and J. S. Thorson, *Chem. Biol.*, 2006, **13**, 733.
17. A. R. Howard-Jones and C. T. Walsh, *Biochemistry*, 2005, **44**, 15652.
18. T. Nishizawa, C. C. Aldrich and D. H. Sherman, *J. Bacteriol.*, 2005, **187**, 2084.
19. T. Nishizawa, S. Grüschow, D. H. Jayamaha, C. Nishizawa-Harada and D. H. Sherman, *J. Am. Chem. Soc.*, 2006, **128**, 724.
20. C. Sánchez, L. Zhu, A. F. Braña, A. P. Salas, J. Rohr, C. Méndez and J. A. Salas, *Proc. Natl. Acad. Sci. USA*, 2005, **102**, 461.
21. E. Yeh, S. Garneau and C. T. Walsh, *Proc. Natl. Acad. Sci. USA*, 2005, **102**, 3960.
22. E. Yeh, L. C. Blasiak, A. Koglin, C. L. Drennan and C. T. Walsh, *Biochemistry*, 2007, **46**, 1284.
23. E. Yeh, L. I. Cole, E. W. Barr, J. M. Bollinger Jr., D. P. Ballou and C. T. Walsh, *Biochemistry*, 2006, **45**, 7904.
24. K. S. Ryan, A. R. Howard-Jones, M. J. Hamill, S. J. Elliott, C. T. Walsh and C. L. Drennan, *Proc. Natl. Acad. Sci. USA*, 2007, **104**, 15311.

25. C. Zhang, C. Albermann, X. Fu, N. R. Peters, J. D. Chisholm, G. Zhang, E. J. Gilbert, P. G. Wang, D. L. Van Vranken and J. S. Thorson, *ChemBioChem*, 2006, **7**, 795.
26. C. Zhang, R. L. Weller, J. S. Thorson and S. R. Rajski, *J. Am. Chem. Soc.*, 2006, **128**, 2760.
27. S. Singh, J. G. McCoy, C. Zhang, C. A. Bingman, G. N. Phillips Jr. and J. S. Thorson, *J. Biol. Chem.*, 2008, **283**, 22628.
28. H. Onaka, S. Taniguchi, Y. Igarashi and T. Furumai, *J. Antibiot.*, 2002, **55**, 1063.
29. A. P. Salas, L. Zhu, C. Sánchez, A. F. Braña, J. Rohr, C. Méndez and J. A. Salas, *Mol. Microbiol.*, 2005, **58**, 17.
30. J. A. Salas and C. Méndez, *Curr. Opin. Chem. Biol.*, 2009, **13**, 152.
31. S. Y. Kim, J. S. Park, C. S. Chae, C. G. Hyun, B. W. Choi, J. Shin and K. B. Oh, *Appl. Microbiol. Biotechnol.*, 2007, **75**, 1119.
32. C. S. Chae, J. S. Park, S. C. Chung, T. I. Kim, S. H. Lee, K. M. Yoon, J. Shin and K. B. Oh, *Bioorg. Med. Chem. Lett.*, 2009, **19**, 1581.
33. M. Makino, H. Sugimoto, Y. Shiro, S. Asamizu, H. Onaka and S. Nagano, *Proc. Natl. Acad. Sci. USA*, 2007, **104**, 11591.
34. Y. Wang, H. Chen, M. Makino, Y. Shiro, S. Nagano, S. Asamizu, H. Onak and S. Shaik, *J. Am. Chem. Soc.*, 2009, **131**, 6748.
35. H. Onaka, S. Asamizu, Y. Igarashi, R. Yoshida and T. Furumai, *Biosci. Biotechnol. Biochem.*, 2005, **69**, 1753.
36. L. Rodríguez, I. Aguirrezabalaga, N. Allende, A. F. Braña, C. Méndez and J. A. Salas, *Chem. Biol.*, 2002, **9**, 721.
37. F. Lombó, M. Gibson, L. Greenwell, A. F. Braña, J. Rohr, J. A. Salas and C. Méndez, *Chem. Biol.*, 2004, **11**, 1709.
38. M. Pérez, F. Lombó, L. Zhu, M. Gibson, A. F. Braña, J. Rohr, J. A. Salas and C. Méndez, *Chem. Commun.*, 2005, **12**, 1604.
39. C. Sánchez, A. P. Salas, A. F. Braña, M. Palomino, A. Pineda-Lucena, R. J. Carbajo, C. Méndez, F. Moris and J. A. Salas, *Chem. Commun.*, 2009, **27**, 4118.
40. C. Zhang, C. Albermann, X. Fu, N. R. Peters, J. D. Chisholm, G. Zhang, E. J. Gilbert, P. G. Wang, D. L. Van Vranken and J. S. Thorson, *ChemBioChem*, 2006, **7**, 795.
41. C. Zhang, R. L. Weller, J. S. Thorson and S. R. Rajski, *J. Am. Chem. Soc.*, 2006, **128**, 2760.
42. S. P. Davies, H. Reddy, M. Caivano and P. Cohen, *Biochem. J.*, 2000, **351**, 95.
43. J. Bain, L. Plater, M. Elliott, N. Shapiro, C. J. Hastie, H. McLauchlan, I. Klevernic, J. S. Arthur, D. R. Alessi and P. Cohen, *Biochem. J.*, 2007, **408**, 297.
44. J. Bain, H. McLauchlan, M. Elliott and P. Cohen, *Biochem. J.*, 2003, **371**, 199.

CHAPTER 7

New Lantibiotics from Natural and Engineered Strains

SONIA I. MAFFIOLI[a], PAOLO MONCIARDINI[a],
MARGHERITA SOSIO[a] AND STEFANO DONADIO*[a,b]

[a] NAICONS (New Anti-Infective Consortium Scrl), Via Fantoli 16/15, 20138 Milano, Italy; [b] KtedoGen Srl, Via Fantoli 16/15, 20138 Milano, Italy
*E-mail: sdonadio@naicons.com and stefano.donadio@ktedogen.com

7.1 Introduction

During the last three decades, the incidence of drug resistance among major bacterial pathogens has greatly outpaced the number of new antibacterial agents introduced into clinical practice, leading to the so-called 'antibiotic crisis', which has been exacerbated by the approval of just two new chemical classes of antibacterial agents during this period. The antibiotic crisis has prompted a renewed interest in poorly explored classes of compounds that, despite interesting antibacterial activities, did not make it into the clinic at a time of plenty. One such class is represented by the lantibiotics, ribosomally and post-translationally modified peptides of which nisin is the best-known example. As these compounds interfere with bacterial metabolism by a mechanism not exploited by the antibiotics currently in clinical use, they usually overcome existing resistance mechanisms and thus offer promise as new therapeutic agents.

Several factors have probably contributed to a renewed interest in this class of compounds. Many of them are active against the major Gram-positive pathogens, irrespective of their mechanism of resistance. At the same time,

RSC Drug Discovery Series No. 25
Drug Discovery from Natural Products
Edited by Olga Genilloud and Francisca Vicente
© The Royal Society of Chemistry 2012
Published by the Royal Society of Chemistry, www.rsc.org

advances in analytical technologies make it now feasible to identify and characterize these high molecular weight compounds in a relatively short time. Perhaps, the largest impetus has come from the possibility of discovering lantibiotics from microbial genomic sequences and/or from the possibility of manipulating their structures by *in vitro* or *in vivo* engineering approaches, thus further expanding the potentially accessible chemical space.

In this chapter, we will first summarize the chemical and biological features of lantibiotics, while providing the understanding of their biosynthesis necessary for the following sections. We will then provide the landscape of lantibiotics known up to around 2006, against which newly discovered compounds must be compared. In the last sections, we will introduce the different approaches used to obtain new compounds, emphasizing especially the molecules that exhibited improved bioactivities. The reader is referred to several recent reviews for a deeper understanding of selected topics related to lantibiotics.[1–5] The lantipeptides described in the text are summarized in Table 7.1.

7.2 Chemistry, Biosynthesis and Biological Activities of Lantipeptides

The term 'lantibiotic', short for 'lanthionine-containing antibiotic', was introduced in 1988 to designate small (19–39 amino acid), gene-encoded peptides that contain one or more lanthionine (Lan) or methyllanthionines (MeLan) residues.[6] The name 'lanthionine' is taken from the Latin word *lana* (wool), from which lanthionine was first extracted.[7] In recent years, as more (Me)Lan-containing peptides were characterized, some devoid of antibacterial activity, the term 'lantipeptides' has been proposed to designate any (Me)Lan-containing peptide, irrespective of its bioactivity.[5] A lantibiotic would then be a lantipeptide endowed with antimicrobial activity.

The modification common to all lantipeptides involves the dehydration of serine and threonine residues to form 2,3-didehydroalanine (Dha) and (Z)-2,3-didehydrobutyrine (Dhb), respectively. This is followed by the stereospecific intramolecular addition of a cysteine thiol onto Dha or Dhb to generate the cognate thioether linkages (Figure 7.1a). In all lantipeptides, the (Me)Lan residues are believed to have the *meso*-stereochemistry, although this has been established only for a subset of compounds.[8] In 2010, the unprecedented carbacyclic amino acid labionin (Lab) was discovered in a newly characterized lantipeptide termed labyrinthopeptin **1**, produced by *Actinomadura namibiensis*.[9] Lab is actually a triamino acid with a $2S,4S,8R$ configuration, consisting of a quaternary central carbon atom joined through a thioether and a methylene bridge to the C- and N-terminal residues, respectively (Figure 7.1a).

In addition to carrying (Me)Lan, Dha, Dhb or Lab residues, lantipeptides may undergo other post-translational modifications, resulting in additional non-proteinogenic amino acids or deaminated residues. Some of these residues are reported in Figure 7.1b. However, compared to the non-ribosomally

Table 7.1 Summary of lantipeptides described in the text

Compound	Class	Structure	Producer strain	Variants[a]	Ref.
Nisin	I	2	*L. lactis*	E, V	15, 16, 21, 22, 78–84
Nisin F	I	6	*L. lactis*		23
Subtilin	I	7	*Bacillus subtilis*	V	69
Epidermin	I	8	*S. epidermidis*		29
Gallidermin	I	9	*Staphylococcus gallinarum*		28, 91, 92
Mutacin B-Ny266	I	10	*Streptococcus mutans*		27, 30
Pep5	I	11	*S. epidermidis*		36, 37, 29
Epicidin 280	I	[b]	*S. epidermidis*		29, 39
Epilancin K7	I	12	*S. epidermidis*		38, 39
97518	I	20	*Planomonospora* sp	C	61–63
NAI-107	I	21	*Microbispora* sp.		64–68
Streptin	I	[c]	*S. pyogenes*		31
Bovicin HC5	I	[c]	*S. bovis*		32–35
Mersacidin	II	3	*Bacillus* sp.	V	17, 41–43, 89
Cinnamycin	II	4	*S. griseoverticillutum*		18
Ancovenin	II	5	*Streptomyces* sp.		19
Actagardine	II	13	*A. garbadinensis/A. liguriae*	C, V	44, 60, 87, 90
Michiganin	II	14	*C. michiganensis*		40
Lacticin 481	II	15	*L. lactis*	E	45, 56–59
Variacin	II	[d]	*Kocuria varians*		45
A-FF22	II	[d]	*Streptococcus* sp.		45
Nukacin ISK1	II	16	*Staphylococcus warneri*	E	46, 88
Plantaricin C	II	17	*Lactobacillus plantarum*		47, 49
Pediocin PD-1	II	[e]	*Pediococcus damnosus*		48
Lactocin S	II	18	*L. sakei*		50
Lacticin 3147	II	19	*L. lactis*	C	51–55, 85, 86
Haloduracin	II	22	*B. halodurans*		70, 71
Lichenicidin	II	23	*B. licheniformis*	*	73, 75
Prochlorosins	II	[f]	*Prochlorococcus* sp.		74
Labyrinthopeptin	III	1	*A. namibiensis*		9, 11

[a] as described in the text, produced by: C, chemical; E, enzymatic;and V, *in vivo* methods; [b] structure partially determined, related to compound **11**; [c] structure partially determined; [d] structure partially determined, related to compound **15**; [e] structure partially determined, related to compound **17**; [f] variable structures.

synthesized family of peptides, the number of non-proteinogenic amino acids encountered in lantipeptides is still relatively small.

Lantipeptides are encoded by a structural gene (generally named *lanA*, with *lan* substituted by a specific three-letter code related to the specific lantipeptide) that encodes an N-terminal leader peptide (up to 59 amino acids in length), followed by a core peptide region that will yield the mature lantipeptide (Figure 7.2). The core peptide region undergoes the required modifications, before the mature product is exported from the cell, coincident

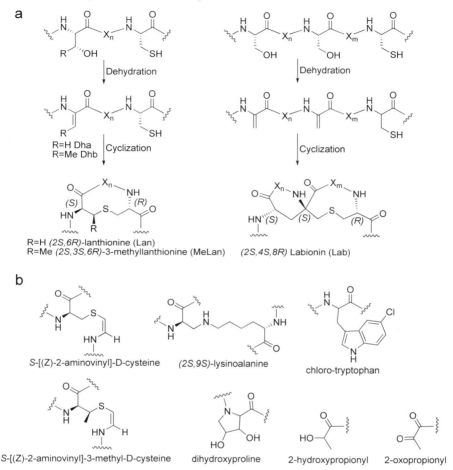

Figure 7.1 (a) Amino acid residues typical of lantipeptides (Dha, Dhb, Lan, MeLan and Lab) and their formation from serine (threonine) and cysteine residues. (b) Some other non-proteinogenic amino acids or deaminated residues found in lantipeptides.

with, or followed by, leader peptide removal. The common modifications to all lantipeptides are the selective dehydration of serine/threonine residues to Dha/ Dhb, followed by regio- and stereospecific addition of the existing cysteines. However, different enzymes are capable of carrying out these reactions (see below). The precise substrates for the other modifications (e.g. decarboxylation of the terminal cysteine, hydroxylation, halogenation, etc.; refer to Figure 7.1b) have not been established in most cases, although they are likely to occur during or after lanthionine formation. In bacterial cells, other proteins participate in the processing and transport of the lantipeptide, in regulating its biosynthesis and in conferring immunity to the producing strain.[1]

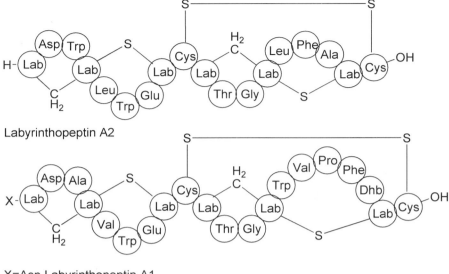

Labyrinthopeptin A2

X=Asp Labyrinthopeptin A1
X=H   Labyrinthopeptin A3

1 Labyrinthopeptin

The recent expansion in the number of known lantipeptides, associated with a deeper understanding of their biogenesis, has resulted in a new classification, which may not be final yet. The 1991 classification, based on the topology of the lanthionine rings,[10] has been superseded by one that divides lantipeptides

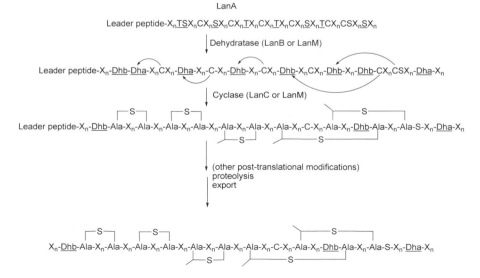

Figure 7.2   General scheme for lantipeptide biosynthesis.

into four classes according to the relatedness of the lanthionine-generating enzymes:[5] in class I, the lanthionine rings are formed by the action of separate dehydratase and cyclase enzymes, termed LanB and LanC, respectively; in the other three classes, they are formed by bifunctional enzymes, termed LanM and LanL for classes II and IV, respectively, which have both been shown to possess kinase activity; class III, originally proposed to accommodate RamC-like enzymes,[1] may actually be involved in forming Lab residues, although this has been demonstrated only in one case.[9,11] Lab formation requires dehydration of two Ser residues followed by a consecutive double Michael addition induced by the nucleophilic attack of a C-terminal cysteine side chain.[11] Thus, except for class III, all enzymes apparently perform the same reactions and we do not yet understand whether a particular lantipeptide requires a specific Lan-generating enzyme. The finding that LanB/C enzymes can process a lantipeptide whose structural gene is adjacent to a *lanM* gene[12] further blurs our understanding.

Up to a few years ago, lantipeptides were believed to be produced mostly by members of the class *Firmicutes* and, to a lesser extent, by the *Actinobacteria*. However, Lan-forming enzymes are also found to be encoded by members of the classes *Proteobacteria* and *Cyanobacteria*. As the number of ribosomally synthesized and post-translationally modified peptides expands, lantipeptides may be found in additional microbial taxa.[5]

As their original name betrays, the majority of known lantipeptides are actually lantibiotics that exert their inhibitory action mainly against Gram-positive bacteria. Lantibiotics are effective against numerous clinically important pathogens, including nosocomial pathogens such as methicillin-resistant *Staphylococcus aureus*, enterococci and *Clostridium difficile*. Some compounds inhibit the outgrowth of bacilli or clostridia spores.[13] Except for selected compounds that show moderate activity against few species only, lantibiotics are generally inactive against Gram-negative bacteria. This is believed to be due to the outer membrane preventing access to the actual target on the outer side of the cytoplasmic membrane. Indeed, when the permeability of the outer membrane is increased by chelators, some Gram-negative bacteria can be sensitized to nisin **2**.[14]

Nisin **2** represents the first and most studied lantibiotic. It was isolated in 1947 from *Lactococcus lactis* and has a 40-year history as a food preservative in many countries. It acts by binding to the pyrophosphate moiety of the cell wall intermediate lipid II, via a cage formed by its two N-terminal rings with five intermolecular hydrogen bonds.[15] This binding is followed by insertion in the cytoplasmic membrane of its C-terminal portion and oligomerization with lipid II and pore formation in the membrane. Thus, in addition to blocking cell wall formation, nisin causes efflux of vital ions and small metabolites leading, within few minutes, to cell death.[16] The N-terminal rings able to form the pyrophosphate cage are conserved in many class I lantibiotics, but many of them lack the C-terminal tail and do not usually form pores or form pores in a strain-specific fashion, dependent on the length of membrane phospholipids.

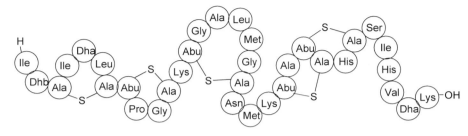

2 Nisin

Mersacidin **3**, as a prototype of class II lantibiotics, also binds lipid II. Binding involves a conformational change and is probably dependent on movement around a hinge region that is located between two lanthionines rings at residues 12 and 13.[17] In contrast to nisin **2**, mersacidin **3** and similar lantibiotics lack the ability to form pores. However, two-component lantibiotics that consist of one mersacidin-like peptide benefit from the presence of a second, synergistically acting peptide that facilitates pore-formation.[2,3]

Some lantipeptides show other bioactivities, such as inhibition of phospholipase A2 by cinnamycin **4** or of angiotensin-converting enzyme by ancovenin **5**,[18,19] compounds produced by *Streptoverticillium griseoverticillutum* and *Streptomyces* sp., respectively. Others are morphogenic agents capable of self-assembly at air/ water interfaces, thus contributing to formation of aerial mycelium in *Streptomyces* strains.[20] Labyrinthopeptin **1** shows *in vitro* activity against herpes simplex virus and efficacy against neuropathic pain in an *in vivo* mouse model.[9]

7.3   The Landscape of Known Lantibiotics

In this section we focus on the different types of lantibiotics reported up to 2006, with a focus on their potential for further clinical development and/or on the drawbacks that have hindered it. It should be noted that the properties of many lantibiotics have been described in qualitative terms only (e.g. presence

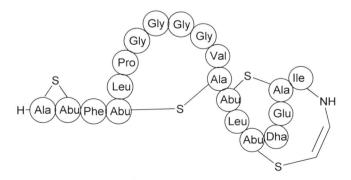

3 Mersacidin

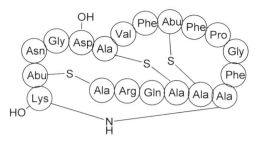

4 Cinnamycin

or absence of activity against selected strains, often relevant to food or veterinary medicine), so that an evaluation of their potential for drug development is not possible.

7.3.1 Class I

Nisin **2** is among the most potent lantibiotics, with minimal inhibitory concentration (MIC) values in the nanomolar range. Despite its potency, *in vivo* efficacy has been reported only for a *Streptococcus pneumoniae* mouse model.[21] While the compound was effective at low doses, its short half-life may have hindered further development. This might be due to nisin's intrinsic instability at physiological pH, which results in total loss of antibacterial activity.[22] More recently, nisin F **6**, a natural variant, was successfully used to control intranasal *S. aureus* infection in a mouse model.[23] Naturally occurring nisin variants, with different amino acids or a slightly shortened C-terminus, have antibacterial activities comparable to those of nisin, and producer strains show cross resistance.[24-26] Other nisin-related antibiotics, such as subtilin **7**, are considerably less active than nisin.

The epidermin group comprises elongated peptides, characterized by the presence of a C-terminal *S*-[(*Z*)-2-aminovinyl]-D-cysteine (Figure 7.1b). In addition to epidermin **8**, the most studied members of this group are gallidermin **9** and mutacin B-Ny266 **10**. All peptides are 22 or 23 amino acids long and their two N-terminal rings are highly related to nisin's. Members of this group are highly active against most Gram-positive strains analyzed.[27]

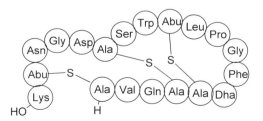

5 Ancovenin

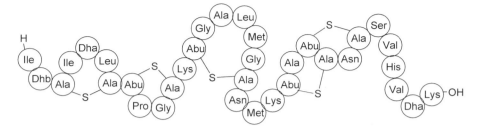

6 Nisin F

Among them, gallidermin **9** has been shown to form pores only in some strains, depending on membrane thickness. However, its killing activity is independent of pore formation, as its MIC values are in the nanomolar range even against strains in which it does not induce pore formation.[28] Thus, its activity is mainly

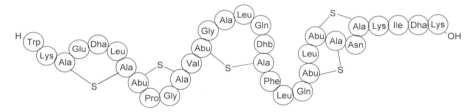

7 Subtilin

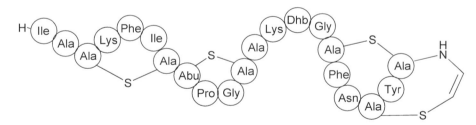

8 Epidermin

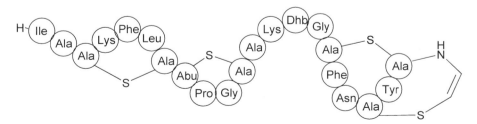

9 Gallidermin

due to inhibition of cell wall biosynthesis. In a study involving multi-resistant *S. aureus* and coagulase-negative staphylococci (CNS) of nosocomial origin, epidermin **8** was active against most *S. aureus* strains, while only 40% of CNS strains were affected.[29] Mutacin B-Ny266 **10** has MIC values in the nanomolar range against several Gram-positive pathogens, lower than those of nisin **2**, and is active also on nisin-resistant strains.[27] This compound demonstrated efficacy comparable to vancomycin against a methicillin-sensitive *S. aureus* strain in an intraperitoneal mouse infection model.[30]

Streptococcus pyogenes produces two lantibiotics, designated streptin 1 and streptin 2, which are 23 and 26 amino acids long, respectively, and derive from a single precursor (Table 7.1).[31] The two N-terminal rings are highly related to nisin rings A and B, while the exact position of streptin's third ring has not been determined. *Streptococcus bovis* produces the related compound bovicin HC5 (Table 7.1), whose structure has also been partially determined.[32] Bovicin HC5 is effective against bacteria involved in food spoilage or of veterinary importance, including nisin-resistant strains.[33,34] Bovicin HC5 induces partial leakage of K^+ ions, but to a different extent than nisin.[35]

Members of the Pep5 group, produced by *Staphylococcus epidermidis* strains, are elongated peptides lacking the nisin-like N-terminal ring system and carrying a net positive charge of +4 to +7. Pep5 **11**, the most active member of the group, has pore-forming activity,[36] but it behaves differently from nisin

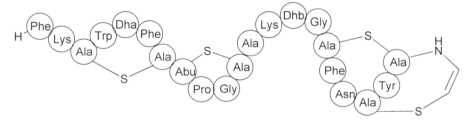

10 Mutacin B-Ny266

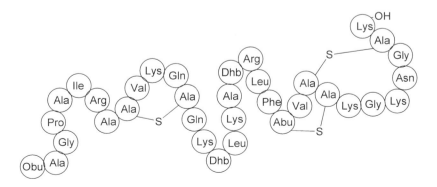

11 Pep5

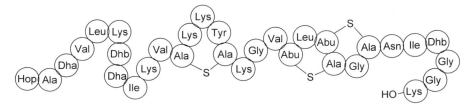

12 Epilancin K7

in having lower affinity for lipid II in liposome models.[37] Interestingly, Pep5 is active against most of the CNS strains of nosocomial origin mentioned above,[29] including those insensitive to epidermin, and has activity comparable to epidermin against *S. aureus*. The related compounds epilancin K7 **12**[38] and epicidin 280 (Table 7.1)[39] were instead active only against a smaller subset of the CNS strains, and inactive against all *S. aureus* strains.[29]

7.3.2   Class II

Mersacidin **3**, produced by *Bacillus* sp., comprises 20 residues forming four rings resulting in a compact structure. Actagardine **13a** and michiganin **14**,[40] produced by the *Actinobacteria Actinoplanes garbadinensis* and *Clavibacter michiganensis*, respectively, are structurally-related lantibiotics, with a similar arrangement of thioether rings. Mersacidin **3** is weakly active against Gram-positive bacteria *in vitro*,[41] but it shows efficacy comparable to vancomycin in an experimental model of infection.[42,43] Similar results were also seen with actagardine **13a**.[44]

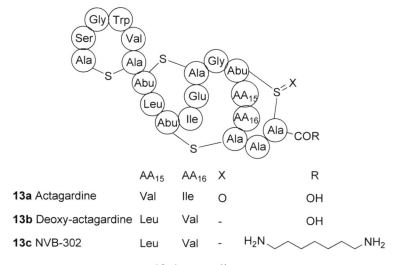

	AA$_{15}$	AA$_{16}$	X	R
13a Actagardine	Val	Ile	O	OH
13b Deoxy-actagardine	Leu	Val	-	OH
13c NVB-302	Leu	Val	-	H$_2$N⌒⌒⌒NH$_2$

13 Actagardines

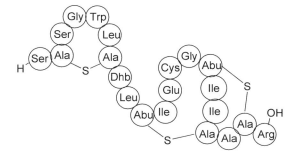

14 Michiganin

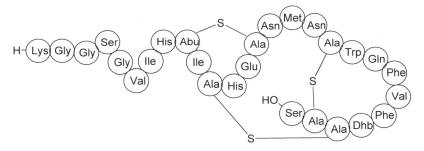

15 lacticin 481

Lacticin 481 **15** and related molecules consist of 22–27 amino acids, have a short N-terminal tail followed by a globular C-terminus with three conserved thioethers including a mersacidin-like ring A.[45] Variability is observed at the N-terminus, where the presence of positive charges might mediate interactions with anionic lipids of target cells. For example, nukacin ISK-1 **16** has three N-terminal Lys residues which are important for activity, lacticin 481 **15** has only one, which is dispensable, and variacin (Table 7.1) does not have any.[4,45] Different modes of actions have been described for different compounds: while

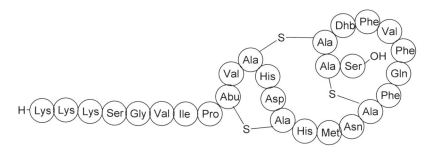

16 Nukacin ISK-1

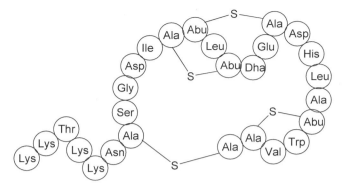

17 Plantaricin C

A-FF22 (Table 7.1) forms transient pores in the membrane, ultimately leading to cell death,[45] nukacin ISK-1 **16** does not, is bacteriostatic and, despite a similar spectrum to nisin **2**, is less potent.[46] The antibacterial activities of other members of the lacticin 481 family are difficult to compare, as they were often determined for possible applications as food preservatives or probiotics.[45]

Plantaricin C **17**[47] and pediocin PD-1 (Table 7.1)[48] have nearly identical structures with a linear, positively charged *N*-terminus and a globular *C*-terminus, containing four thioether bridges including a mersacidin-like lipid II-binding motif. Plantaricin C **17** interacts with lipid II leading to inhibition of cell wall biosynthesis and pore formation, and its MIC values against *Micrococcus flavus* and *L. lactis* are similar to those of nisin.[49] Lactocin S **18**, produced by *Lactobacillus sakei*, was the first lantibiotic reported to contain D-Ala residues, deriving from stereospecific reduction of Dha residues.[50] It probably is the most elongated class II lantibiotic, with only two C-terminal rings. No detailed information is available regarding antimicrobial spectrum and potency.

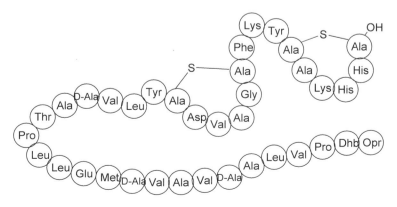

18 Lactocin S

7.3.3 Two-Component Lantibiotics

The activity of two-component lantibiotics depends on the synergistic interaction of two peptides, one carrying a mersacidin-like lipid II-binding motif and the other peptide necessary for pore formation after interaction with the lipid-II bound first peptide, as described for lacticin 3147 **19**.[51] This compound shows high activity against *C. difficile*[52] and clinically relevant mycobacteria.[53] However, despite a good activity against vancomycin-resistant enterococci, it shows variable activity against *S. aureus*, with poor activity on strains with intermediate resistance to vancomycin.[54] It is worth mentioning that the immunity genes of a cluster may confer some cross-protection against other two-component lantibiotics.[55]

7.4 New Lantibiotics by *In Vitro* Methods

The recent development of *in vitro* methods for enzymatic synthesis of lantipeptides offers the opportunity to access additional chemical space. Indeed, issues related to toxicity, processing or transport of the mature lantipeptide can be conveniently bypassed by *in vitro* methodologies. In their pioneering work, van der Donk and co-workers achieved the *in vitro* reconstitution of the LctM enzyme involved in lacticin 481 **15** formation and demonstrated that this enzyme can tolerate changes in the core peptide,

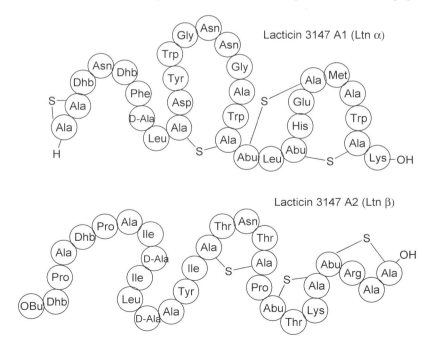

19 Lacticin 3147

provided this is linked to its cognate leader peptide in a LctA prepeptide.[56,57] After developing a method for fusing a triazole-linked leader peptide analog with an azide-modified, synthetically generated structural region,[58] they were able to incorporate non-proteinogenic amino acids into the latter and test the flexibility of LctM in tolerating such substitutions.[59] This approach has resulted into lacticin 481 **15** variants not accessible by *lctA* engineering (see below). Among them, two ring-C analogs, in which Trp19 and Phe23 were replaced with homophenyl- and naphthyl-alanine, respectively, exhibited a two-fold increase in MIC against a *L. lactis* strain.[59] Activity against pathogenic bacteria was not reported.

Approaches to chemical synthesis of lantipeptides (recently reviewed in ref. 4) have provided important answers to the topology of the rings, where substitution of the thioethers with carbons usually results in loss of activity. Activity can be retained in oxa compounds (in which the thioether sulfur is replaced by oxygen) of lacticin 3147 A2 **19**, although interactions with the cognate partner A1 in this two-component lantibiotic was lost.

Chemical modification of lantibiotics has not been actively pursued in recent years, as we are aware of only two reports on this approach. In one case, the class II lantibiotic deoxyactagardine B **13b**, produced by *Actinoplanes liguriae*, was selectively amidated at its C-terminus to generate the basic amide NVB302 **13c**, which showed a general improvement in MIC against several Gram-positive bacteria.[60] The compound is currently a developmental candidate for the treatment of *C. difficile*-associated diarrhea. In the second example, the activity against Gram-positives of the poorly active class I lantibiotic 97518 **20a**, produced by *Planomonospora* sp.,[61,62] was improved several fold by modifying its two carboxyl groups into the corresponding basic amides **20b**.[63]

7.5   New Lantibiotics from Natural Strains

Among the ever-growing number of lantibiotics identified in recent years, one of the most interesting ones, in terms of clinical potential, is NAI-107 **21** (also

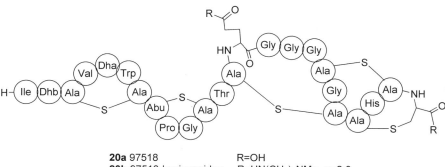

20a 97518   R=OH
20b 97518-basic amides   R=HN(CH$_2$)$_n$NMe$_2$ n=2,3

20 97518

known as 107891 and microbisporicin), produced by *Microbispora* sp.[64,65] This compound, along with the structurally related but less potent lantibiotic 97518 **20a**,[62] was identified as part of a screening strategy campaign involving 120 000 microbial extracts. The screening algorithm consisted in identifying extracts active against an *S. aureus* strain but inactive against its isogenic L-form. Then, those extracts were discarded whose activity was abolished by penicillinases or by excess *N*-caproyl-D-alanyl-D-alanine, presumably containing β-lactams or glycopeptides, respectively.[66] This assay system was originally designed to detect any class of cell wall inhibiting compounds (apart from β-lactams and glycopeptides) but it turned out to be very effective in identifying lantibiotics from actinomycetes and has resulted in the identification of six new compounds, in addition to **21** and **20a** (S.I.M. *et al.*, unpublished results). Interestingly, NAI-107 **21** and 97518 **20a** represent the first class I lantipeptides discovered from *Actinobacteria*. While they share an identical 5-ring system and many amino acids, these two lantibiotics differ substantially in their potency: NAI-107 **21** is active against all Gram-positive pathogens, including clinical isolates carrying drug-resistance phenotypes, often at nanomolar concentrations.[67] Furthermore, the compound is rapidly bactericidal and is also effective in several experimental models of infections.[68] The compound is currently under preclinical development. NAI-107 **21** represents so far the only lantipeptide carrying a chloro-tryptophan residue.

In recent years, alternative routes to bioassay-based screening have been implemented to identify lantipeptides. An interesting method was devised for the rapid identification of microorganisms producing lantipeptides in the 3–3.5 kDa mass range.[69] The method, which involves whole cell matrix-assisted laser desorption/ionization mass spectrometry (MALDI-ToF MS), was validated with a subtilin- and ericin-producing *B. subtilis* strain and used to detect subtilin as the product from several newly isolated *B. subtilis* strains. However, we are unaware of the method being used to discover new lantipeptides. The availability of an increasing number of genomic and metagenomic data and information on lantipeptide biosynthetic pathways has led to the identification of gene clusters specifying the synthesis of previously unknown lantipeptides, taking advantage of the conserved sequences of the LanB/C, LanM and LanA enzymes. A genome-mining strategy, based on searching for mersacidin-like genes in the genome of *Bacillus halodurans*, led to the discovery of the two-

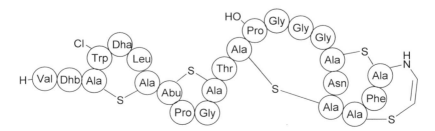

21 NAI-107

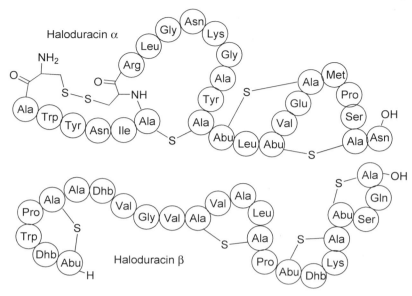

22 Haloduracin

component lantibiotic haloduracin **22**.[70,71] Accurate mass spectrometric and structural characterization data demonstrated that these peptides are part of a two-component system. In parallel, the two *lanA* genes were heterologously expressed and the purified precursor peptides were processed by the modification enzymes, providing an *in vitro* reconstitution of the biosynthesis of a two-component lantibiotic.[70] In a similar approach, the genomes of *S. aureus*, *Bacillus licheniformis* and of the marine cyanobacteria *Prochorococcus* and *Synechococcus* spp. led to the identification of an epidermin-like compound,[72] of lichenicidin **23**[73] and prochlorosins (Table 7.1),[74] respectively. Lichenicidin **23** is another example of a two-component lantibiotic, in which Lchα not only has the typical mersacidin-like lipid II binding motif, but also an N-terminal ring highly related to nisin's A ring, involved in lipid II recognition.[75] Particularly relevant is the genetic potential in the *Prochlorococcus* strain: 29 different lantipeptides, derived from an equal number of *lanA* genes, can be generated by a single promiscuous LanM enzyme.[72]

Among the lantibiotics described above, an evaluation of the antimicrobial activity of compounds **23** showed they were active against many Gram-positive bacteria, although they were less potent against MRSA strains than against methicillin-sensitive *S. aureus* strains.[73] The bioactivity of haloduracin **22** was evaluated only in qualitative terms,[70,71] while no activity against *L. lactis* or *B. subtilis* was observed with four recombinant prochlorosins.[74]

On-line tools and repositories have also been developed to facilitate genome analysis for lantipeptide gene identification.[76,77] A search for class I lantipeptides, screening genomes for the presence of *nisB*- and *nisC*-related genes, has identified putative clusters in bacterial taxa not previously associated with lantipeptide production.[78] However, the actual compounds are yet to be identified.

7.6 New Lantibiotics from Engineered Strains

The ribosomal origin makes lantipeptides attractive candidates for generating variants by mutagenesis of the *lanA* gene, with the aim of improving one or more properties. The simplest systems have involved expression of modified *lanA* genes in the producer strain, either *in cis* (by replacing the wild-type allele) or *in trans* (in a *lanA*-null background).

Among the large number of nisin **2** variants generated by site-directed mutagenesis using *L. lactis* as a host, only a few are worth mentioning because of their improved properties: A mutant carrying the T2S substitution and two ring-A variants (where ITL at position 3–5 is converted to KSI or KFI) showed enhanced activity against at least one Gram-positive target, albeit non-pathogenic;[79,80] while the S5T mutant, with Dhb5 in place of Dha, despite of a lower antibacterial activity, showed considerably greater resistance to acid-catalyzed degradation than wild-type nisin.[81] In another study, the M17N and G18T mutations in ring C led to the production of two congeners, carrying either Thr or Dhb at position 18, both less active than nisin against *Bacillus cereus* and *Streptococcus thermophilus*, but more active against *M. flavus*.[82] Engineering of residues within the hinge region resulted in the formation of two mutants, N20K and M21K, which displayed increased activity against Gram-negatives including *Shigella*, *Pseudomonas* and *Salmonella* spp.[83] These mutants provided information that the central hinge region in nisin **2** plays an important role in providing the conformational flexibility required for antimicrobial activity on the membrane. In recent years, random mutagenesis has been employed to access a larger number of derivatives. In recent years, from a library of approximately 8000 nisin **2** variants, the compound carrying the K22T substitution displayed enhanced activity against *Streptococcus agalactiae*, a human and bovine pathogen, while the M21V replacement afforded a two-fold enhanced activity against medically relevant Gram-positive pathogens.[3,84]

Site-directed mutagenesis has been applied to lacticin 3147 **19**.[85] Two approaches, alanine-scanning and random mutagenesis, were taken to reveal residues or motifs within Ltnα and Ltnβ which may yield active peptides. The work established that conversion to alanine of 36 out of the 59 residues present in the two peptides resulted in compounds that retained some activity, highlighting that many positions are actually tolerant of change, both with respect to the processing enzymes and in terms of bioactivity.[85] This flexibility was confirmed by random mutagenesis that, although extremely valuable for structure–activity relationship studies, failed to generate peptides with enhanced antimicrobial activity.[86]

Alanine-scanning mutagenesis was also undertaken with actagardine **13a**.[87] In this case, only residues not involved in thioether bridges were modified.

Although these mutants demonstrated reduced or lack of biological activity compared with the parent molecule, these results indicated the residues amenable to change that may represent sites for future mutagenesis.

The importance of all 27 amino acids in nukacin ISK-1 **16** was evaluated by saturation mutagenesis, replacing each *lanA* codon with NNK (N = A, C, G or T; K = G or T).[88] Among the resulting 251 variants, only D13E and V22I showed a two-fold increased activity. In addition, variants H15S and H15F were apparently produced in 2–4-fold higher quantities than **16**. However, attempts to combine increased specific activity and increased production in a single peptide were unsuccessful.[88]

Mutagenesis of the mersacidin structural gene *mrsA* was performed. The approach involved trans complementation of a null *mrsA* mutant with a shuttle plasmid expressing the *mrsA* mutant alleles.[89] Two types of mutant libraries were generated: one involved replacing the 12 amino acid residues not involved in thioether linkages in mersacidin **3** with the other possible 19 residues; and the other inserting one amino acid before positions 1 and 12. Variants carrying amino acid replacements were observed in about 1/3 of the cases at each position tested, although positions 6, 14, 17 and 19 were less permissive to change, particularly the latter for which only the variant I19M was produced at appreciable yield. Variants carrying insertions were observed in about 15% of the cases. No insertions were possible into ring A or between rings A and B. While insertions of polar or nonpolar residues in ring B were well tolerated, no mature lantibiotic was detected from alleles specifying a reduction in size of ring B. With respect to bioactivity, insertions yielded compounds with poor activity, while substitution variants were more encouraging. Among them, the F3W variant showed higher activity than mersacidin **3** against one methicillin-resistant *S. aureus* strain, and activity equivalent to **3** against the other tested strains. Other variants showed improved activity against one strain but not another. However, no additive effect was found when double mutants were generated.[89]

Attempts to generate modified lantibiotics have so far exclusively addressed the modification of the *lanA* gene. However, drawing on analogy with other secondary metabolites, the potential exists to generate new structures by adding or removing selected genes encoding post-translational modification enzymes other than those generating the lanthionine bridges.

7.7   Conclusions

In conclusion, recent years have seen a substantial increase in lantipeptide research, which has provided considerable new information on this class of ribosomally synthesized, post-translationally modified peptides. In addition, new technologies and methodologies are now available that can be effectively used to access new lantipeptides or variants thereof. This combination of deeper understanding, new tools and industrial potential is generating increased attention towards this class of antibacterial compounds.

In contrast to many other classes of secondary metabolites, whose synthesis occurs on an enzymatic template, manipulation of which is not straightforward, mutagenesis of *lanA* genes is technically simple and has been used to generate many new variants. However, most of the resulting lantibiotics are less active than the parent compound and, in the few cases where an improvement in bioactivity was observed, this was modest (2–4-fold) and limited to selected strains. This suggests that potency has already been optimized during evolution and there may be little scope for improvement. However, other anthropocentric properties (i.e. stability at physiological pH, solubility, activity in the presence of serum proteins, etc.), which are often critical for efficacy in experimental models of infection, have rarely been investigated in newly discovered lantibiotics, independently of the way they were obtained. Thus, we suspect that the main driver for the many studies performed on lantibiotics was the development of new methods and technologies, not the discovery of new drug candidates.

To our knowledge, only two of the new lantibiotics described here are currently candidates for clinical development. One is NVB302 **13c**, a carboxy amide derivative of deoxy-actagardine B, an actagardine variant produced by *Actinoplanes liguriae* carrying Leu-Val residues at position 15-16 instead of Val-Ile and devoid of the sulfoxide in the C-terminal ring.[90] The other is NAI-107 **21**, produced by *Microbispora* sp., consisting of a nisin-like N-terminal portion and a globular C-terminus.[68] Remarkably, both compounds are produced by members of the *Actinobacteria*, a taxon that not too long ago was considered a modest source of lantipeptides. Compound **13c** is stable in simulated gastric fluid, shows reduced activity against *Bifidobacterium* and *Bacteroides* spp., and is currently in preclinical development by Novacta Therapeutics for the treatment of *C. difficile* associated diarrhea.[60] Because of its rapid bactericidal activity and efficacy in several models of experimental infections,[68] compound **21** is currently under preclinical development for the treatment of multi-drug resistant Gram-positive infections in hospitalized patients. Gallidermin **9** has been reported to have been tested in human volunteers;[89] however, we are unaware of any published results of such studies. Nonetheless, a renewed interest in compound **9** is highlighted by a recent patent application on its use to treat infections of the respiratory tract.[92]

References

1. J. M. Willey and W. A. van der Donk, *Annu. Rev. Microbiol.*, 2007, **61**, 477.
2. G. Bierbaum and H. G. Sahl, *Curr. Pharm. Biotechnol.*, 2009, **10**, 2.
3. D. Field, C. Hill, P. D. Cotter and R. P. Ross, *Mol. Microbiol.*, 2010, **78**, 1077.
4. A. C. Ross and J. C. Vederas, *J. Antibiot.*, 2011, **64**, 27.
5. J. E. Velásquez and W. A. van der Donk, *Curr. Opin. Chem. Biol.*, 2011, DOI 10.1016/j.cbpa.2010.10.027.

6. N. Schnell, K. D. Entian, U. Schneider, F. Götz, H. Zahner, R. Kellner and G. Jung, *Nature*, 1988, **333**, 276.
7. M. J. Horn, D. B. Jones and S. J. Ringel, *J. Biol. Chem.*, 1941, **138**, 141.
8. C. Chatterjee, M. Paul, L. Xie and W. A. van der Donk, *Chem. Rev.*, 2005, **105**, 633.
9. K. Meindl, T. Schmiederer, K. Schneider, A. Reicke, D. Butz, S. Keller, H. Gühring, L. Vértesy, J. Wink, H. Hoffmann, M. Bronstrup, G. M. Sheldrick and R. D. Süssmuth, *Angew. Chem., Int. Ed.*, 2010, **49**, 1151.
10. G. Jung, *Angew. Chem., Int. Ed. Engl.*, 1991, **30**, 1051.
11. W. M. Müller, T. Schmiederer, P. Ensle and R. D. Süssmuth, *Angew. Chem., Int. Ed.*, 2010, **49**, 2436.
12. J. A. Majchrzykiewicz, J. Lubelski, G. N. Moll, A. Kuipers, J. J. Bijlsma, O. P. Kuipers and R. Rink, *Antimicrob. Agents Chemother.*, 2010, **54**, 1498.
13. A. Hurst, *Adv. Appl. Microbiol.*, 1981, **27**, 85.
14. J. Delves-Broughton, P. Blackburn, R. J. Evans and J. Hugenholtz, *Antonie Van Leeuwenhoek*, 1996, **69**, 193.
15. S. T. Hsu, E. Breukink, E. Tischenko, M. A. Lutters, B. de Kruijff, R. Kaptein, A. M. Bonvin and N. A. van Nuland, *Nat. Struct. Mol. Biol.*, 2004, **11**, 963.
16. I. Wiedemann, E. Breukink, C. van Kraaij, O. P. Kuipers, G. Bierbaum, B. de Kruijff and H. G. Sahl, *J. Biol. Chem.*, 2001, **276**, 1772.
17. S. T. Hsu, E. Breukink, G. Bierbaum, H. G. Sahl, B. de Kruijff, R. Kaptein, N. A. van Nuland and A. M. Bonvin, *J. Biol. Chem.*, 2003, **278**, 13110.
18. F. Märki, E. Hanni, A. Fredenhagen and J. van Oostrum, *Biochem. Pharmacol.*, 1991, **42**, 2027.
19. Y. Kido, T. Hamakado, T. Yoshida, M. Anno, Y. Motoki, T. Wakamiya and T. Shiba, *J. Antibiot.*, 1983, **36**, 1295.
20. S. Kodani, M. A. Lodato, M. C. Durrant, F. Picart and J. M. Willey, *Mol. Microbiol.*, 2005, **58**, 1368.
21. B. P. Goldstein, J. Wei, K. Greenberg and R. Novick, *J. Antimicrob. Chemother.*, 1998, **42**, 277.
22. W. C. Chan, B. W. Bycroft, L. Y. Lian and G. C. K. Roberts, *FEBS Lett.*, 1989, **252**, 29.
23. M. De Kwaadsteniet, K. T. Doeschate and L. M. Dicks, *Lett. Appl. Microbiol.*, 2009, **48**, 65.
24. R. E. Wirawan, N. A. Klesse, R. W. Jack and J. R. Tagg, *Appl. Environ. Microbiol.*, 2006, **72**, 1148.
25. G. Le Blay, C. Lacroix, A. Zihler and I. Fliss, *Lett. Appl. Microbiol.*, 2007, **45**, 252.
26. F. Yoneyama, M. Fukao, T. Zendo, J. Nakayama and K. Sonomoto, *J. Appl. Microbiol.*, 2008, **105**, 1982.
27. M. Mota-Meira, G. LaPointe, C. Lacroix and M. C. Lavoie, *Antimicrob. Agents Chemother.*, 2000, **44**, 24.
28. R. R. Bonelli, T. Schneider, H. G. Sahl and I. Wiedemann, *Antimicrob. Agents Chemother.*, 2006, **50**, 1449.

29. J. S. Nascimento, H. Ceotto, S. B. Nascimento, M. Giambiagi-deMarval, K. R. Santos and M. C. Bastos, *Lett. Appl. Microbiol.*, 2006, **42**, 215.
30. M. Mota-Meira, H. Morency and M. C. Lavoie, *J. Antimicrob. Chemother.*, 2005, **56**, 869.
31. P. A. Wescombe and J. R. Tagg, *Appl. Environ. Microbiol.*, 2003, **69**, 2737.
32. H. C. Mantovani, H. Hu, R. W. Worobo and J. B. Russell, *Microbiology*, 2002, **148**, 3347.
33. A. A. de Carvalho, H. C. Mantovani and M. C. Vanetti, *Lett. Appl. Microbiol.*, 2007, **45**, 68.
34. J. R. Lima, A. O. Barros Ribon, J. B. Russell and H. C. Mantovani, *FEMS Microbiol. Lett.*, 2009, **292**, 78.
35. H. C. Mantovani and J. B. Russell, *Antimicrob. Agents Chemother.*, 2008, **52**, 2247.
36. M. Kordel, R. Benz and H. G. Sahl, *J. Bacteriol.*, 1988, **170**, 84.
37. H. Brötz, M. Josten, I. Wiedemann, U. Schneider, F. Götz, G. Bierbaum and H. G. Sahl, *Mol. Microbiol.*, 1998, **30**, 317.
38. M. van de Kamp, H. W. van den Hooven, R. N. H. Konings, G. Bierbaum, H. G. Sahl, O. P. Kuipers, R. J. Siezen, W. M. de Vos, C. W. Hilbers and F. J. M. van de Ven, *Eur. J. Biochem.*, 1995, **230**, 587.
39. C. Heidrich, U. Pag, M. Josten, J. Metzger, R. W. Jack, G. Bierbaum, G. Jung and H. G. Sahl, *Appl. Environ. Microbiol.*, 1998, **64**, 3140.
40. I. Holtsmark, D. Mantzilas, V. G. H. Eijsink and M. B. Brurberg, *Appl. Environ. Microbiol.*, 2006, **72**, 5814.
41. W. W. Niu and H. C. Neu, *Antimicrob. Agents Chemother.*, 1991, **35**, 998.
42. S. Chatterjee, D. K. Chatterjee, R. H. Jani, J. Blumbach, B. N. Ganguli, N. Klesel, M. Limbert and G. Seibert, *J. Antibiot.*, 1992, **45**, 839.
43. D. Kruszewska, H. G. Sahl, G. Bierbaum, U. Pag, S. O. Hynes and A. Ljungh, *J. Antimicrob. Chemother.*, 2004, **54**, 648.
44. A. Malabarba, R. Pallanza, M. Berti and B. Cavalleri, *J. Antibiot.*, 1990, **43**, 1089.
45. A. Dufour, T. Hindré, D. Haras and J. P. Le Pennec, *FEMS Microbiol. Rev.*, 2007, **31**, 134.
46. S. M. Asaduzzaman, J. Nagao, H. Iida, T. Zendo, J. Nakayama and K. Sonomoto, *Antimicrob. Agents Chemother.*, 2009, **53**, 3595.
47. D. L. Turner, L. Brennan, H. E. Meyer, C. Lohaus, C. Siethoff, H. S. Costa, B. Gonzalez, H. Santos and J. E. Suarez, *Eur. J. Biochem.*, 1999, **264**, 833.
48. R. Bauer, M. L. Chikindas and L. M. T. Dicks, *Int. J. Food Microbiol.*, 2005, **101**, 17.
49. I. Wiedemann, T. Böttiger, R. R. Bonelli, T. Schneider, H. G. Sahl and B. Martinez, *Appl. Environ. Microbiol.*, 2006, **72**, 2809.
50. M. Skaugen, J. Nissen-Meyer, G. Jung, S. Stevanovic, K. Sletten, C. I. Mörtved-Abildgaard and I. F. Nes, *J. Biol. Chem.*, 1994, **269**, 27183.
51. I. Wiedemann, T. Böttiger, R. R. Bonelli, A. Wiese, S. O. Hagge, T. Gutsmann, U. Seydel, L. Deegan, C. Hill, P. Ross and H. G. Sahl, *Mol. Microbiol.*, 2006, **61**, 285.

52. M. C. Rea, E. Clayton, P. M. O'Connor, F. Shanahan, B. Kiely, R. P. Ross and C. Hill, *J. Med. Microbiol.*, 2007, **56**, 940.

53. J. Carroll, L. A. Draper, P. M. O'Connor, A. Coffey, C. Hill, R. P. Ross, P. D. Cotter and J. O'Mahony, *Int. J. Antimicrob. Agents*, 2010, **36**, 132.

54. C. Piper, L. A. Draper, P. D. Cotter, R. P. Ross and C. Hill, *J. Antimicrob. Chemother.*, 2009, **64**, 546.

55. L. A. Draper, K. Grainger, L. H. Deegan, P. D. Cotter, C. Hill and R. P. Ross, *Mol. Microbiol.*, 2009, **71**, 1043.

56. L. Xie, L. M. Miller, C. Chatterjee, O. Averin, N. L. Kelleher and W. A. van der Donk, *Science*, 2004, **303**, 679.

57. C. Chatterjee, G. C. Patton, L. Cooper, M. Paul and W. A. van der Donk, *Chem. Biol.*, 2006, **13**, 1109.

58. M. R. Levengood, C. C. Kerwood, C. Chatterjee and W. A. van der Donk, *ChemBioChem*, 2009, **10**, 911.

59. M. R. Levengood, P. J. Knerr, T. J. Oman and W. A. van der Donk, *J. Am. Chem. Soc.*, 2009, **131**, 12024.

60. A. N. Appleyard, T. Ayala, S. Boakes, J. Cortes, M. G. Dawson, S. Choi, A. Lightfoot, D. Read, M. Todd and S. N. Wadman, *49th Intersci Conf. Antimicrob. Agents Chemother.*, 2009, Poster F1-1517.

61. F. Castiglione, L. Cavaletti, D. Losi, A. Lazzarini, L. Carrano, M. Feroggio, I. Ciciliato, E. Corti, G. Candiani, F. Marinelli and E. Selva, *Biochemistry*, 2007, **46**, 5884.

62. S. I. Maffioli, D. Potenza, F. Vasile, M. De Matteo, M. Sosio, B. Marsiglia, V. Rizzo, C. Scolastico and S. Donadio, *J. Nat. Prod.*, 2009, **72**, 605.

63. S. Maffioli, C. Brunati, D. Potenza, F. Vasile and S. Donadio, WO/2010/058238.

64. A. Lazzarini, L. Gastaldo, G. Candiani, I. Ciciliato, D. Losi, F. Marinelli, E. Selva and F. Parenti, PCT/EP2004/007658.

65. F. Castiglione, A. Lazzarini, L. Carrano, E. Corti, I. Ciciliato, L. Gastaldo, P. Candiani, D. Losi, F. Marinelli, E. Selva and F. Parenti, *Chem. Biol.*, 2008, **15**, 22.

66. D. Jabes and S. Donadio, *Methods Mol. Biol.*, 2010, **618**, 31.

67. D. Jabes, C. Brunati, P. Guglierame and S. Donadio, *49th Intersci. Conf. Antimicrob. Agents Chemother.*, 2009, Abs F1-1502.

68. D. Jabes, C. Brunati, G. Candiani, S. Riva, G. Romanò and S. Donadio, *Antimicrob. Agents Chemother.*, 2011, **55**, 1671.

69. T. Stein, *Rapid Commun. Mass Spectrom.*, 2008, **22**, 1146.

70. A. L. McClerren, L. E. Cooper, C. Quan, P. M. Thomas, N. L. Kelleher and W. A. van der Donk, *Proc. Natl. Acad. Sci. USA*, 2006, **103**, 17243.

71. E. M. Lawton, P. D. Cotter, C. Hill and R. P. Ross, *FEMS Microbiol. Lett.*, 2007, **267**, 64.

72. K. M. Daly, M. Upton, S. K. Sandiford, L. A. Draper, P. A. Wescombe, R. W. Jack, P. M. O'Connor, A. Rossney, F. Gotz, C. Hill, C. P. D. Cotter, R. P. Ross and J. R. Tagg, *J. Bacteriol.*, 2010, **192**, 1131.

73. J. Dischinger, M. Josten, C. Szekat, H. G. Sahl and G. Bierbaum, *PLoS One*, 2009, **26**, e6788.
74. B. Li, D. Sher, L. Kelly, Y. Shi, K. Huang, P. J. Knerr, I. Joewono, D. Rusch, S. W. Chisholm and W. A. van der Donk, *Proc. Natl. Acad. Sci. USA*, 2010, **107**, 10430.
75. Z. O. Shenkarev, E. I. Finkina, E. K. Nurmukhamedova, S. V. Balandin, K. S. Mineev, K. D. Nadezhdin, Z. A. Yakimenko, A. A. Tagaev, Y. V. Temirov, A. S. Arseniev and T. V. Ovchinnikova, *Biochemistry*, 2010, **49**, 90.
76. A. De Jong, A. J. van Heel, J. Kok and O. P. Kuipers, *Nucl. Acids Res.*, 2010, **38**, W647.
77. R. Hammami, A. Zouhir, C. Le Lay, J. Ben Hamida and I. Fliss, *BMC Microbiol.*, 2010, **10**, 22.
78. A. J. Marsh, O. O'Sullivan, R. P. Ross, P. D. Cotter and C. Hill, *BMC Genomics*, 2010, **11**, 679.
79. O. P. Kuipers, G. Bierbaum, B. Ottenwalder, H. M. Dodd, N. Horn, J. Metzger, T. Kupke, V. Gnau, R. Bongers, P. van den Bogaard, H. Kosters, H. S. Rollema, W. M. de Vos, R. J. Siezen, G. Jung, F. Götz, H. G. Sahl and M. J. Gasson, *Antonie Van Leeuwenhoek*, 1996, **69**, 161.
80. J. Lubelski, R. Rink, R. Khusainov, G. N. Moll and O. P. Kuipers, *Cell. Mol. Life Sci.*, 2008, **65**, 455.
81. H. S. Rollema, O. P. Kuipers, P. Both, W. M. de Vos and R. J. Siezen, *Appl. Environ. Microbiol.*, 1995, **61**, 2873.
82. O. P. Kuipers, H. S. Rollema, W. M. Yap, H. J. Boot, R. J. Siezen and W. M. de Vos, *J. Biol. Chem.*, 1992, **267**, 24340.
83. J. Yuan, Z. Z. Zhang, X. Z. Chen, W. Yang and L. D. Huan, *Appl. Microbiol. Biotechnol.*, 2004, **64**, 806.
84. D. Field, P. M. O'Connor, P. D. Cotter, C. Hill and R. P. Ross, *Mol. Microbiol.*, 2008, **69**, 218.
85. P. D. Cotter, L. H. Deegan, E. M. Lawton, L. A. Draper, P. M. O'Connor, C. Hill and R. P. Ross, *Mol. Microbiol.*, 2006, **62**, 735.
86. D. Field, B. Collins, P. D. Cotter, C. Hill and R. P. Ross, *J. Mol. Microbiol. Biotechnol.*, 2007, **13**, 226.
87. S. Boakes, J. Cortés, A. N. Appleyard, B. A. Rudd and M. J. Dawson, *Mol. Microbiol.*, 2009, **72**, 1126.
88. M. R. Islam, K. Shioya, J. Nagao, M. Nishie, H. Jikuya, T. Zendo, J. Nakayama and K. Sonomoto, *Mol. Microbiol.*, 2009, **72**, 1438.
89. A. N. Appleyard, S. Choi, D. M. Read, A. Lightfoot, S. Boakes, A. Hoffmann, I. Chopra, G. Bierbaum, B. A. Rudd, M. J. Dawson and J. Cortes, *Chem. Biol.*, 2009, **16**, 490.
90. S. N. Wadman, M. J. Dawson and J. Cortes Bargallo, WO/2009/010763.
91. A. Manosroi, P. Khanrin, W. Lohcharoenkal, R. G. Werner, F. Götz, W. Manosroi and J. Manosroi, *Int. J. Pharmacol.*, 2010, **392**, 304.
92. H. Muellner, T. Bouyssou, J. Daemmgen and B. Disse, WO/2009/141264.

CHAPTER 8

Mining Microbial Genomes for Metabolic Products of Cryptic Pathways

D. OVES-COSTALES*[a] AND G. L. CHALLIS[b]

[a] Fundación MEDINA, Edificio Centro Desarrollo Farmacéutico, Avda Conocimiento n3, Parque Ciencias de la Salud, Armilla, Granada, 18100, Spain; [b] Department of Chemistry, University of Warwick, Coventry CV4 7AL, UK
*E-mail: daniel.oves@medinaandalucia.es

8.1 Introduction

As a consequence of the ever-increasing pace of genome sequencing, the amount of microbial DNA sequence data available in public databases is growing exponentially. As of December 2010, there are 1377 complete microbial genomes and a further 3666 microbial genome projects in progress.[1]

Once microbial genome sequence data began to accumulate, it became apparent that in many microbes there are more gene clusters encoding biosynthetic enzymes likely to be involved in secondary metabolite biosynthesis than natural products known to be produced by the organism. This was first recognized during sequencing of the complete genome of the model actinomycete *Streptomyces coelicolor* A3(2),[2] and since then the same phenomenon has been observed in many other bacteria, e.g. *Streptomyces avermitilis*, *Pseudomonas fluorescens* Pf-5, *Saccharopolyspora erythraea* and *Salinispora arenicola*,[3–6] to name but a few, as well as filamentous fungi.

RSC Drug Discovery Series No. 25
Drug Discovery from Natural Products
Edited by Olga Genilloud and Francisca Vicente
© The Royal Society of Chemistry 2012
Published by the Royal Society of Chemistry, www.rsc.org

'Genome mining' is a term used in several fields to describe the exploitation of genomic data for different purposes. In recent years it has become synonymous, in particular, with the exploitation of genome sequence data to discover new natural products. Many biologically active microbial natural products are biosynthesized by modular multienzymes, such as type I modular polyketide synthases (PKSs) and non-ribosomal peptide synthetases (NRPSs).[7] Both modular PKSs and NRPSs can be conceptualized as biological 'assembly lines' that consist of functional units called modules. Each module is composed of task-specific domains that can select and chemically modify a variety of building blocks from which the metabolic product is assembled.[7] For many PKSs and NRPSs there is a direct correlation between (i) the number and order of modules and (ii) the number and order of building blocks incorporated into the metabolic product. This correlation has been termed the 'co-linearity rule'. An example of the co-linearity rule is given by the 6-deoxyerythronolide B synthase, a type I modular PKS that contains seven modules, each of which is responsible for the incorporation of one of seven propionate-derived building blocks into the erythromycin macrolide core (Figure 8.1).[8] In the 6-deoxyerythronolide B synthase, and for many other modular PKS and NRPS systems, each module is used only once in the assembly of a metabolite molecule.

Our increasing understanding of the biochemical and structural aspects of modular PKS and NRPS systems has proved invaluable for exploiting genomic information. Thus, the alignment of multiple acyl-transferase (AT) and

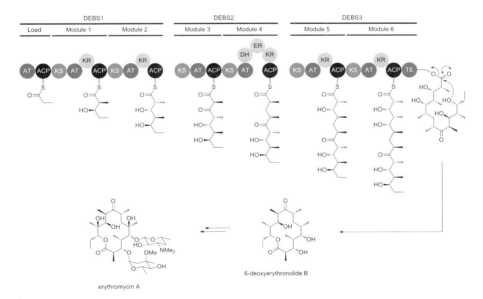

Figure 8.1 Molecular mechanism of 6-deoxyerythronolide B biosynthesis by a type I modular PKS, illustrating the principles of modular natural product assembly lines.

adenylation (A) domains of known substrate specificity has allowed the identification of conserved motifs within these domains that are involved in recognizing substrates.[9–12] This has proved to be useful for predicting the substrates selected by cryptic PKSs and NRPSs systems, identified by microbial genome sequencing, that direct the biosynthesis of uncharacterized natural products. The presence within modules of domains with tailoring functions can indicate the kind of chemical modifications that will take place to each incorporated building block during the metabolite assembly process. Prediction of the stereochemical outcome for some of these tailoring reactions (e.g. ketoreduction, enoylreduction and epimerization) is also possible.[13–15] All this information, together with the high frequency at which the co-linearity rule applies in NRPS and PKS systems, makes possible the tentative proposal of structural features of uncharacterized metabolic products of cryptic biosynthetic pathways.

While in principle this means that predictions of non-ribosomal peptide and polyketide structure can be made from genome sequence data, the approach is not without limitations. For example, sometimes the substrate specificity of the A/AT domains cannot be predicted, especially if they recognize an unusual substrate. Processes such as module skipping and iterative module use cause breakdown of the co-linearity rule, leading to misprediction of the number of building blocks incorporated into the metabolic product of a cryptic pathway.[16] Enzymes that modify the structures of modular PKS/NRPS products, such as glycosyl transferases or cytochrome P450s, for which the precise point of modification is hard to predict, are another complicating factor. For biosynthetic pathways other than those involving type I modular PKSs and NRPSs, structural predictions for the metabolic products of cryptic pathways remains challenging. However, for certain systems, such as type II iterative PKSs and terpene synthases, it is often possible to infer how many building blocks will be incorporated into the metabolic product, although how the intermediate assembled from these building blocks undergoes cyclization and subsequent modification to form the metabolic product is usually not predictable.

Despite the limitations discussed above, genome mining represents a powerful emerging approach for the discovery of novel bioactive natural products. In the following sections, we discuss several recently developed strategies for the discovery of new microbial secondary metabolites by exploitation of genome sequence data. Space constraints prevent us from providing an exhaustive survey of every approach employed to date. Thus we have focused on highlighting key concepts using important recent examples.

8.2   Use of Predicted Physicochemical Properties to Guide the Discovery of Metabolic Products of Cryptic Pathways

As discussed above, analysis of NRPSs and PKSs encoded by cryptic biosynthetic gene clusters can provide considerable insight into structural features of their hitherto unidentified metabolic products. The structural information thus obtained can be used to predict physicochemical properties of the putative metabolite which can be exploited to guide analysis of culture supernatants and extracts (Figure 8.2). This can lead to rapid identification of the putative metabolic product of a cryptic biosynthetic pathway.

A pioneering application of this approach was the discovery of coelichelin as the metabolic product of a cryptic NRPS gene cluster identified by analysis of the *S. coelicolor* genome sequence (see Section 8.5).[17] Recently this approach has been used to guide the identification of the aureusimines, two novel non-ribosomal peptide secondary metabolites of *Staphylococcus aureus* (Figure 8.2).[18] Comparisons of more than 50 *S. aureus* genome sequences led to the identification of a universally conserved, uncharacterized NRPS gene cluster. Sequence analysis of the adenylation domains of the dimodular NRPS encoded by this gene cluster indicated that the amino acids likely to be incorporated into the metabolic product of the NRPS were valine and tyrosine. The presence of a Re (reductase) domain at the C-terminus of the NRPS suggested a reductive chain release mechanism, indicating that the putative Val-Tyr dipeptide-derived product of the NRPS contains an aldehyde group that can react with the amino group of the valine residue to form a cyclic imine. These structural hypotheses led to a tentative prediction of the molecular mass of the metabolic product of the cryptic NRPS. Organic extracts of *S. aureus* cultures were subjected to liquid chromatography-mass spectrometry (LC-MS) analysis using m/z-filtering software. This simplified the analytical process of identifying the metabolic products of the cryptic NRPS by allowing it to focus on metabolites yielding m/z values close to those predicted for the Val-Tyr-derived cyclic imine. Two potential candidate metabolic products of the cryptic NRPS, differing in m/z by 16 units and with m/z 2 and 18 units less than the predicted value, were identified. Purification and structural characterization by high resolution mass spectrometry and NMR spectroscopy led to the proposed structures for aureusimines shown in Figure 8.2. Aureusimines A and B (tyrvalin/phevalin) together with a third congener leuvalin (derived from a Val-Leu dipeptidyl thioester intermediate) were independently identified as metabolic products of the same cryptic NRPS in several *Staphylococcus* species using an approach combining inactivation of the NRPS gene with comparative metabolic profiling (see Section 8.5).[19]

Although, in principle, the use of this approach for the discovery of new natural products does not require the genetic manipulation of the producing microorganism, additional experiments need to be done to confirm that the identified compound(s) are indeed the metabolic product(s) of the cryptic gene cluster (usually either demonstration that inactivation of a biosynthetic gene abolishes metabolite production, or reconstitution of the biosynthetic pathway using purified recombinant enzymes *in vitro*). In the case of aureusimines A and B, this was accomplished by showing that allelic replacement of the NRPS gene abolished aureusimine production.

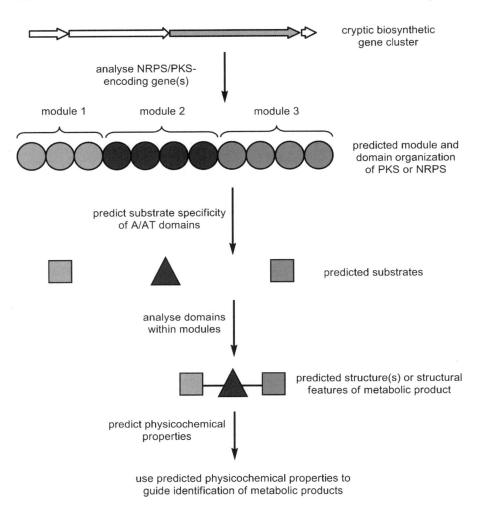

Figure 8.2 Overview of the strategy for discovery of novel natural products by exploiting predicted physicochemical properties and the structures of aureusimines A and B, novel metabolites of *Staphylococcus* species discovered using this approach.

The obvious advantage of this approach is the utility of using the predicted structural information to focus the search for the metabolic products of a cryptic biosynthetic gene cluster, which simplifies the analytical challenges associated with the process. Because the analyses are simplified, large numbers of culture extracts, different fermentation conditions, and a variety of separation methods can be investigated with comparatively little effort. On the other hand, the approach requires a high degree of confidence in the predicted structural features of the cryptic metabolic product; otherwise the corresponding predicted physicochemical properties used to focus the analyses have the potential to mislead. It is also important to remember that the predicted physicochemical properties might be very similar to those of other compounds present in the culture supernatants/extracts, thus increasing the likelihood of incorrectly correlating metabolite(s) with a cryptic biosynthetic gene cluster.

8.3 Tracking the Incorporation of Labeled Predicted Precursors

This approach is based on the combination of substrate specificity predictions from bioinformatics analyses of cryptic biosynthetic enzymes with classical incorporation experiments employing isotope-labeled precursors.

Such methodology was first applied in 2007 by Gerwick and co-workers to the identification of orfamide A, a novel metabolite of *Pseudomonas fluorescens* Pf-5.[20] Analysis of the genome sequence of this microorganism led to the identification of a cryptic biosynthetic gene cluster for which the metabolic product was unknown. A detailed analysis of the gene cluster, in particular the adenylation domains of the NRPSs encoded by the cluster, indicated that the putative metabolic product was likely to be a novel lipopeptide containing four leucine residues. This prediction prompted feeding experiments using ^{15}N-labeled leucine. ^{1}H-^{15}N HMBC NMR analyses of fractionated culture extracts enabled isolation of metabolites containing the labeled leucine, leading to the identification of orfamide A as the major product of the cryptic biosynthetic gene cluster (Figure 8.3).

More recently, Marahiel and co-workers have employed a similar approach for the discovery of erythrochelin, a new siderophore from *Saccharopolyspora erythraea*.[21] The sequencing and annotation of the genome of this industrially important erythromycin producer has been recently published.[5] Twenty-five cryptic biosynthetic gene clusters were identified within the genome.[5] One such cluster contains a gene encoding a tetramodular NRPS that was predicted to assemble a hydroxamate-type siderophore. Analysis of the adenylation domains of the NRPS suggested that one of the building blocks of the tetrapeptide was likely to be a *N*5-hydroxyornithine. This prediction was exploited by carrying out feeding experiments with ^{14}C-labeled ornithine, which should be incorporated into the metabolite. Culture extracts were analyzed by radio-LC-MS, which facilitated rapid identification of metabolites

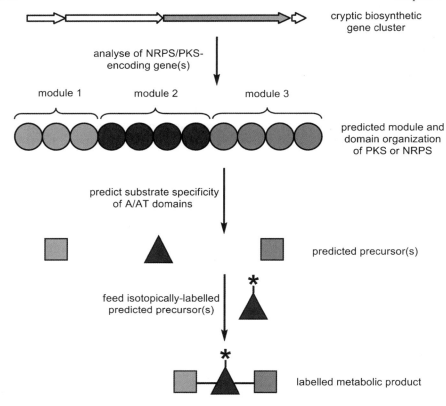

cryptic biosynthetic
gene cluster

analyse of NRPS/PKS-
encoding gene(s)

module 1   module 2   module 3

predicted module and
domain organization
of PKS or NRPS

predict substrate specificity
of A/AT domains

predicted precursor(s)

feed isotopically-labelled
predicted precursor(s)

labelled metabolic product

exploit isotope label to identify metabolite(s)
containing predicted precursors

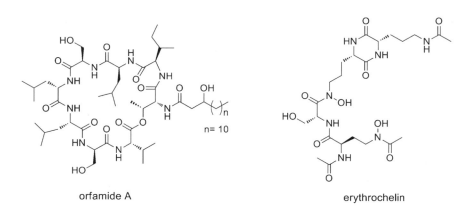

orfamide A   erythrochelin

Figure 8.3   Outline of how tracking the incorporation of labeled predicted precursors can be applied to the identification of metabolic products of cryptic biosynthetic pathways and the structures of orfamide A (left) and erythrochelin (right), novel bacterial natural products discovered using the approach.

containing the labeled precursor. Isolation, purification and structure elucidation led to the proposed structure for erythrochelin shown in Figure 8.3. Erythrochelin was independently identified as the metabolic product of the cryptic NRPS cluster by Leadlay and co-workers.[22]

The two examples above highlight the value of tracking the incorporation of labeled predicted precursors as a strategy for the discovery of new natural products. The strength of this approach lies in the capacity to quickly identify potential products of cryptic biosynthetic pathways from complex fermentation mixtures, as well as the ability to exploit very limited substrate specificity predictions. In principle only a single predicted precursor of the putative metabolic product is required, as was the case for erythrochelin where a reliable substrate-specificity prediction for only one of the adenylation domains of the tetramodular NRPS was sufficient to successfully employ the strategy. However, this approach is not without important limitations. If substrate specificity is hard to predict it might not be possible to choose with confidence a labeled precursor to utilize in the incorporation experiments. On the other hand, there might be cases where the substrate-specificity prediction hypothesizes only very common building blocks, likely to be present in many metabolites. In such cases the analytical chemistry is unlikely to be significantly simplified using this approach. Other limitations are the availability of the requisite precursor in labeled form, and the possible need to optimize the feeding experiments to ensure sufficient levels of incorporation of the labeled precursor into the metabolite. Finally, and analogously to the physicochemical properties prediction approach, this methodology provides only a circumstantial link between the isolated metabolite and the cryptic biosynthetic gene cluster. Thus further genetic or biochemical experiments are also needed, in this case to link production of the metabolite to the gene cluster.

8.4 *In Vitro* Reconstitution of Cryptic Biosynthetic Pathways

This approach is based on the incubation of purified recombinant biosynthetic enzymes encoded by the cryptic gene cluster with their putative substrate(s), followed by isolation and structure elucidation of the resulting product(s) (Figure 8.4).

Such a strategy has recently been used by Cane and co-workers for the identification of avermitilol as a new sesquiterpene metabolite of *Streptomyces avermitilis*.[23] Analysis of the genome sequence of *S. avermitilis* indicated that the *sav76* gene encodes a cryptic sesquiterpene synthase. A synthetic *sav76* gene with codons optimized for expression in *Escherichia coli*, was cloned and overexpressed. Incubation of the resulting purified recombinant Sav76 with the predicted substrate farnesyl diphosphate in the presence of $MgCl_2$ afforded a sesquiterpene alcohol as the major product of the reaction, as determined by capillary GC-MS analysis. Isolation of this alcohol and structure elucidation

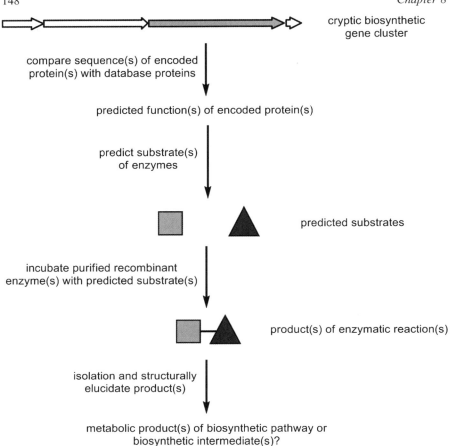

compare sequence(s) of encoded
protein(s) with database proteins

predicted function(s) of encoded protein(s)

predict substrate(s)
of enzymes

predicted substrates

incubate purified recombinant
enzyme(s) with predicted substrate(s)

product(s) of enzymatic reaction(s)

isolation and structurally
elucidate product(s)

metabolic product(s) of biosynthetic pathway or
biosynthetic intermediate(s)?

avermitilol

Figure 8.4   Overview of the *in vitro* reconstitution approach to identifying the
metabolic products of cryptic biosynthetic pathways and the structure of
avermitilol, a novel sesquiterpene metabolite of *S. avermitilis* discovered
using this strategy.

using 1- and 2-D NMR experiments identified it as avermitilol (Figure 8.4).
Although the production of this new metabolite was not observed in wild-type
S. avermitilis, the function of *sav76* was confirmed *in vivo* by ectopic expression

of the *sav76* gene in a genome-minimized derivative of the strain containing the cryptic synthase.

The main advantage of the *in vitro* reconstitution approach is that it offers the opportunity to study the biosynthetic pathway encoded by a cryptic gene cluster free from the constraints imposed by the organism in which it is found. Thus, problems associated with low or no expression of the cryptic gene cluster, which is a frequently encountered problem, can be overcome. Another advantage of the *in vitro* reconstitution approach derives from the fact that the purified biosynthetic enzymes are typically used in well-defined reaction media consisting of only a handful of components. This means that analysis of the reaction mixture and the purification of the putative product(s) are usually straightforward in comparison with the analysis of, and isolation of metabolites from, culture supernatants or extracts. Such simple analyses are particularly beneficial when only the substrate(s) and not the product(s) of the biosynthetic enzymes can be predicted from sequence comparisons, as is the cased for terpene synthases.

On the other hand, this approach requires the cloning and overexpression of the biosynthetic gene(s), purification of the corresponding overproduced protein(s), accurate prediction of the substrate(s) used by the enzyme(s), and careful analysis of the product(s) generated. As a consequence, this strategy is less attractive for large cryptic biosynthetic gene clusters containing several genes, because the likelihood of obtaining all the purified recombinant biosynthetic enzymes in fully functional form is low. The *in vitro* reconstitution approach is much better suited for relatively simple cryptic pathways in which a confident prediction of the substrate(s) used by the biosynthetic enzyme(s) can be made.

8.5 Biosynthetic Gene Inactivation Coupled with Comparative Metabolic Profiling

This approach is based on the inactivation of a putatively essential biosynthetic gene(s) within a cryptic gene cluster, followed by comparison of the profiles of metabolites produced by the wild-type strain and the constructed mutant(s) using an appropriate technique, such as LC-MS. The metabolite(s) missing in the mutant but present in the wild-type strain are likely to be product(s) of the cryptic gene cluster (Figure 8.5).

An example of how this strategy has been combined with the use of predicted physicochemical properties to guide metabolite identification is the discovery of coelichelin as the metabolic product of a cryptic non-ribosomal peptide biosynthetic gene cluster identified in the partial genome sequence of *Streptomyces coelicolor* A3(2).[17] Two tentative tripeptide structures, both of which contain hydroxamic acid functional groups, were proposed for the metabolic product of the trimodular NRPS encoded by this gene cluster. Analysis of intergenic regions within the cryptic gene cluster indicated the presence of known motifs for binding of ferrous iron-dependent repressor

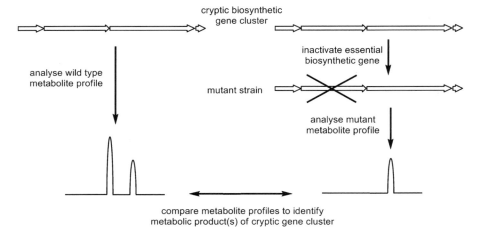

coelichelin

Figure 8.5   The principles underlying biosynthetic gene inactivation coupled with comparative metabolic profiling for the discovery of metabolic products of cryptic biosynthetic gene clusters and the structure of the novel *S. coelicolor* siderophore coelichelin discovered using this approach.

(IdeR) proteins. Taken together, these hypotheses indicated that the metabolic product of this gene cluster is likely to function as a siderophore, suggesting: (i) an iron-deficient culture medium would be needed to maximize production of the metabolite and (ii) selective detection of the ferric complex of this metabolite could be achieved by monitoring absorbance at 435 nm (ferric *tris*-hydroxamate complexes have a characteristic absorbance maximum at this wavelength). An *S. coelicolor* mutant with the NRPS-encoding *cchH* gene inactivated by plasmid insertion was constructed, and the profile of metabolites it produces in iron-deficient medium was compared with that of the wild-type strain. This strategy led to the identification and structure elucidation of a tetrapeptide as the product of the cryptic gene cluster (Figure 8.5). Interestingly, this tetrapeptide appears to be assembled by a trimodular NRPS via a combination of iterative module use *and* module skipping, providing a rare example of the violation of the co-linearity rule.

The main advantage of the gene inactivation/comparative metabolic profiling approach is that, unlike some of the other strategies discussed above, an experimental link between the metabolite and the cryptic biosynthetic pathway is established. A second advantage is that a prediction regarding the structure of the metabolite is not required for this strategy to be effective. However, as the case of coelichelin shows, insights into structural features of the metabolic product of a cryptic biosynthetic pathway from sequence analyses of the enzymes catalyzing its assembly can predict physicochemical properties, which can be targeted to simplify the process of identifying differences in the profile of metabolites produced by wild-type and non-producing mutant strains.

On the other hand, a potential drawback of this approach is the need to construct a mutant lacking one or more essential biosynthetic genes from the cryptic gene cluster. For many microorganisms there are few or no genetic tools available, and, as a consequence, the construction of mutants can be challenging. Another potential problem is that the change in metabolite profile between the mutant and wild-type strains may be too subtle to be easily identified. This could occur when wild-type production of the metabolite occurs in low amounts, or when very complex culture supernatants/extracts are being analyzed.

8.6 Heterologous Expression of the Cryptic Gene Cluster Coupled with Comparative Metabolic Profiling

This strategy is related to the biosynthetic gene inactivation/comparative metabolic profiling approach discussed above, and is based on the introduction of the cryptic biosynthetic gene cluster into a suitable heterologous host. If the gene cluster is expressed, comparative metabolic profiling of the unmodified host and the host containing the cryptic gene cluster can be used to identify the metabolic product of the gene cluster (Figure 8.6).

This approach has recently been used by Clardy and co-workers for the discovery of new formyl-tyrosine metabolites produced by *Pseudoalteromonas tunicata*.[24] A sequence-based search for ATP-grasp-type-like ligases led to the identification of a cryptic gene cluster in the marine bacteria *P. tunicata* that contained a gene encoding an ATP-grasp-like enzyme. The cryptic cluster was PCR amplified and cloned into an expression vector under the control of an inducible promoter. The resulting plasmid was used to transform *E. coli*. Culture supernatants of the transformed *E. coli* were analyzed by LC-MS and compared to culture supernatants of an *E. coli* control carrying an empty vector. This led to the identification of two metabolites present in cultures of the host containing the cryptic gene cluster which were absent in cultures of the control carrying the empty vector. Isolation, purification and structural

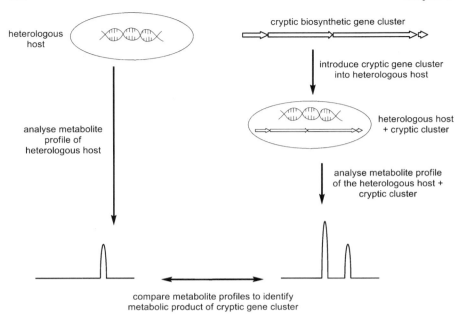

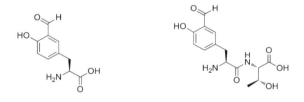

3-formyltyrosine metabolites

Figure 8.6 Outline of how heterologous cryptic biosynthetic gene cluster expression coupled with comparative metabolic profiling can be applied to novel natural product discovery and the structures of two new 3-formyl-tyrosine metabolites discovered using this strategy.

characterization led to the identification of two novel 3-formyl-tyrosine-containing metabolites (Figure 8.6).

There are several advantages associated with the use of this strategy to discover novel natural products. As with the biosynthetic gene inactivation/comparative metabolic profiling approach discussed above, an experimental link between the metabolite and the biosynthetic gene cluster is established, and no prior structural information from bioinformatics analyses is, in principle, required to identify the metabolic product of the cryptic gene cluster. Advantages that derive from using a well-characterized heterologous host for expression of the biosynthetic gene cluster include: growth conditions for high levels of metabolite production are often already known, and a genetically engineered host that lacks endogenous secondary metabolic pathways can be used, thus boosting metabolite titers and reducing the number of metabolites in culture supernatants/extracts. For obvious reasons, this approach is particularly attractive for the expression of cryptic biosynthetic gene clusters identified in metagenomic libraries.

On the other hand, the major drawback of this strategy is the necessity to introduce the entire cryptic biosynthetic gene cluster into the heterologous host; if only part of the gene cluster is introduced, a biosynthetic intermediate or shunt metabolite may be identified rather than the true metabolic product of the cryptic gene cluster. In addition, many natural product biosynthetic gene clusters exceed 100 kb in size. As a result, the use of bacterial artificial chromosomes (BACs) or similar vectors capable of maintaining large inserts is required to introduce the gene cluster into the heterologous host. Several technical challenges are associated with the construction of such vectors containing large inserts.

8.7 Strategies for Inducing Expression of Silent Biosynthetic Gene Clusters

All the approaches discussed so far require the cryptic biosynthetic gene cluster to be expressed under the growth conditions used. It is evident that if the genes are expressed poorly or not at all, the corresponding metabolic product will not be produced, or will be produced in amounts that are too low to detect. Thus, before deciding which of the available strategies to use, it is helpful to ascertain whether the genes within the cryptic gene cluster of interest are being transcribed. This can be quickly achieved using techniques such as reverse transcription-polymerase chain reaction (RT-PCR) or Northern Blot. An alternative to these methods is the employment of DNA microarrays that can simultaneously detect most of the mRNA molecules in a cell at a particular time point. These techniques can also be used to quantify the levels of transcription. This can be used to optimize growth conditions to ensure gene expression is maximized, which should lead to higher metabolite titers that simplify the discovery process. If the genes within the cryptic biosynthetic cluster are not expressed, the gene cluster is often referred to as silent. In such cases, transcription of the gene cluster needs to be induced to initiate the discovery process.

A traditional approach used for the activation of silent gene clusters involves the variation of fermentation conditions, such as media composition, temperature, oxygen supply or pH value. This empirical strategy has been referred to as the one strain–many compounds (OSMAC) approach, and is based on the observation that the conditions under which a microorganism is grown can have a significant impact on the profile of metabolites produced.[25]

In some cases, information gained from bioinformatics analyses of the cryptic biosynthetic gene cluster indicates suitable fermentation conditions for production of the metabolite. This was the case for the siderophore coelichelin, discussed above, where the analysis of intergenic regions within the biosynthetic gene cluster identified known ferrous iron-dependent repressor (IdeR) binding motifs, suggesting that expression of the cluster would be induced under iron-deficient conditions.[17]

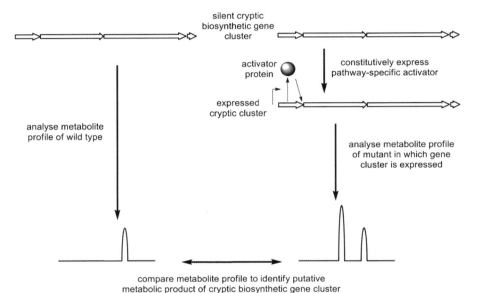

Figure 8.7   The principles of manipulating pathway specific activator genes to induce expression of silent biosynthetic gene clusters and the structures of stambomycins A-D, novel metabolic products of a cryptic biosynthetic gene cluster of *S. ambofaciens* discovered using this approach.

Another approach to inducing transcription of silent biosynthetic gene clusters is based on the manipulation of pathway-specific regulatory genes present within the cryptic gene cluster (Figure 8.7). This strategy has been recently applied to the discovery of stambomycins, a novel family of 51-membered macrolides produced by *Streptomyces ambofaciens*.[26] Analysis of the genome sequence of this actinobacterium revealed a 150 kb cryptic gene cluster encoding a giant type I modular PKS. However, RT-PCR analysis indicated that the PKS-encoding genes within the cluster were poorly expressed under a variety of growth conditions. Examination of the putative regulatory genes within the gene cluster led to the identification of *samR0484*, a gene predicted to encode a large ATP-binding regulator of the LuxR (LAL) regulator. Such regulatory genes have been identified previously in type I modular PKS gene clusters and shown to act as pathway specific activators.[27] Thus, *samR0484* was cloned into an integrative vector under the control of a constitutive promoter, and introduced into *S. ambofaciens*. RT-PCR analysis showed that expression of the cryptic PKS genes was induced in the resulting strain, whereas no increase in transcription was observed in a control strain constructed by integration of the empty vector into *S. ambofaciens*. The profile of metabolites produced by strains with and without the activator under the control of a constitutive promoter was compared using LC-MS, thus allowing the identification of compounds in the LAL-expressing mutant that were absent in the control strain. Isolation and structure elucidation of the compounds identified them as stambomycins A-D, the metabolic products of the cryptic gene cluster (Figure 8.7). The role of the cryptic gene cluster in biosynthesis of the stambomycins was confirmed by inactivating one of the PKS-encoding genes, which abolished stambomycin production.

This example shows that the constitutive expression of genes encoding pathway-specific activators is a useful strategy for the activation of silent biosynthetic gene clusters. Analogously, inactivation of genes encoding pathway-specific repressors is a viable approach for the activation of silent pathways, although to our knowledge this approach has yet to be applied to the discovery of novel metabolites. A disadvantage of these strategies is that genetic manipulation of the organism harboring the silent cluster is required. This can be problematic if genetic tools are not available for the organism.

An alternative to the manipulation of pathway-specific regulators is the induction of silent gene transcription by manipulation of pleiotropic regulators. Three examples of this approach have been reported recently.[28-30]

DasR is a DNA-binding protein in *Streptomyces* species that represses both morphological differentiation and secondary metabolite production. Deletion of the *dasR* gene, or addition of *N*-acetylglucosamine to minimal media results in the overproduction of actinorhodin and prodiginine antibiotics in *S. coelicolor*.[28] Additionally, an increase in extracellular antibiotic production was observed using a bioassay for several other streptomycetes grown in minimal medium supplemented with *N*-acetylglucosamine, suggesting that this could be a general strategy for the discovery of novel natural products.[28]

A related approach has been employed in *Aspergillus nidulans*, and is based on the importance of chromatin-modifying proteins in the pleiotropic regulation of secondary metabolite gene cluster expression. Deletion of *cclA*, a gene encoding a protein involved in the methylation of Lys4 of histone H3 in *A. nidulans*, resulted in a dramatic change in the metabolite profile.[29] The production of a total of eight metabolites was induced using this strategy.

Ochi and co-workers have reported an approach that is based on selection for resistance to certain antibiotics.[30] Growth of actinomycetes on streptomycin and rifampicin creates mutations in the *rpsL* (encoding the ribosomal protein S12) and *rpoB* (encoding the beta subunit of the RNA polymerase) genes, respectively. Some of these mutations can pleiotropically alter the expression of secondary metabolism genes. A study of over a thousand soil actinomycetes showed that 43% of streptomycetes and 6% of non-streptomycetes that do not excrete antibiotics acquired extracellular antibiotic activity after selection for spontaneous streptomycin or rifampicin resistance.[30] One of the strains was selected for further investigation, and it was shown that the antibiotic activity was due, in this case, to the production of a family of novel non-ribosomal peptides named piperidamycin A-H.[30]

Manipulation of pleiotropic regulators represents a promising approach for the discovery of new natural products. However, a significant drawback associated with the global changes in the transcriptome brought about by the application of such approaches is that a global change in the metabolite profile is also induced, which can present a complex and challenging analytical problem.

8.8 Concluding Remarks

Genetic information accumulated in publicly accessible databases has revealed the existence of hundreds of cryptic biosynthetic gene clusters that possess the potential to direct the production of novel natural products. The ever-increasing pace at which genome sequence data is being acquired means that this reservoir of cryptic secondary metabolite biosynthetic gene clusters is likely to increase exponentially in the coming years. Our current understanding of the biochemistry and genetics of natural product biosynthesis has provided the opportunity to apply genomic and post-genomic tools to natural product discovery. This represents an attractive alternative to traditional methods for natural product discovery. Particular advantages of genomics-led approaches to natural product discovery include: (i) the ability to apply bioinformatics-based structural predictions for metabolic products of cryptic biosynthetic gene clusters to the problem of dereplication, by comparing the predicted structures with the structures of known natural products; (ii) significant simplification of the analytical challenge associated with identifying novel natural products by applying predictions of physicochemical properties, gene inactivation or heterologous gene cluster expression coupled with comparative metabolic profiling, or tracking of the incorporation of labeled predicted precursors; (iii) access to a greatly expanded range of novel metabolites compared with traditional natural product discovery approaches via rational manipulation of regulatory networks to uncouple secondary metabolite production from (often unknown) environmental triggers.

While it is possible to infer structural features of metabolic products of certain types of biosynthetic system, such as type I modular PKSs and NRPSs, the site and nature of many modification reactions catalyzed by tailoring enzymes in these biosynthetic pathways remains difficult to predict. For other kinds of biosynthetic system, such as iterative PKSs and terpene synthases, structural predictions from sequence analyses are still challenging. Thus, much work is still required to better understand the molecular mechanisms of natural product biosynthesis. Advances in this area will continue to improve our ability to predict the structure of the likely metabolic product(s) of cryptic biosynthetic pathways. Despite these unsolved problems, there can be no doubt that genomics-based approaches will become an important component of future industrial and academic natural product discovery programs.

References

1. http://www.ncbi.nlm.nih.gov/genomes/lproks.cgi (last accessed December 2010).
2. S. D. Bentley, K. F. Chater, A.-M. Cerdeño-Tárraga, G. L. Challis, N. R. Thomson, K. D. James, D. E. Harris, M. A. Quail, H. Kieser, D. Harper, A. Bateman, S. Brown, G. Chandra, C. W. Chen, M. Collins, A. Cronin, A. Fraser, A. Goble, J. Hidalgo, T. Hornsby, S. Howarth, C.-H. Huang, T. Kieser, L. Larke, L. Murphy, K. Oliver, S. O'Neil, E. Rabbinowitsch, M.-A. Rajandream, K. Rutherford, S. Rutter, K. Seeger, D. Saunders, S. Sharp, R. Squares, S. Squares, K. Taylor, T. Warren, A. Wietzorrek, J. Woodward, B. G. Barrell, J. Parkhill and D. A. Hopwood, *Nature*, 2002, **417**, 141.
3. S. Omura, H. Ikeda, J. Ishikawa, A. Hanamoto, C. Takahashi, M. Shinose, Y. Takahashi, H. Orikawa, H. Nakazawa, T. Osonoe, H. Kikuchi, T. Shiba, Y. Sakaki and M. Hattori, *Proc. Natl. Acad. Sci. USA*, 2001, **98**, 12215.
4. I. T. Paulsen, C. M. Press, J. Ravel, D. Y. Kobayashi, G. S. A. Myers, D. V. Mavrodi, R. T. DeBoy, R. Seshadri, Q. Ren, R. Madupu, R. J. Dodson, A. S. Durkin, L. M. Brinkac, S. C. Daugherty, S. A. Sullivan, M. J. Rosovitz, M. L. Gwinn, L. Zhou, D. J. Schneider, S. W. Cartinhour, W. C. Nelson, J. Weidman, K. Watkins, K. Tran, H. Khouri, E. A. Pierson, L. S. Pierson, L. S. Thomashow and J. E. Loper, *Nat. Biotechnol.*, 2005, **23**, 873.
5. M. Oliynyk, M. Samborskyy, J. B. Lester, T. Mironenko, N. Scott, S. Dickens, S. F. Haydock and P. F. Leadlay, *Nat. Biotechnol.*, 2007, **25**, 447.

6. K. Penn, C. Jenkins, M. Nett, D. W. Udwary, E. A. Gontang, R. P. McGlinchey, B. Foster, A. Lapidus, S. Podell, E. Allen, B. S. Moore and P. R. Jensen, *ISME J.*, 2009, **3**, 1193.
7. M. A. Fischbach and C. T. Walsh, *Chem. Rev.*, 2006, **106**, 3468.
8. J. Staunton and B. Wilkinson, *Chem. Rev.*, 1997, **97**, 2611.
9. S. F. Haydock, J. F. Aparicio, I. Molnar, T. Schwecke, L. E. Khaw, A. Konig, A. F. Marsden, I. S. Galloway, J. Staunton and P. F. Leadlay, *FEBS Lett.*, 1995, **374**, 246.
10. H. Ikeda, T. Nonomiya, M. Usami, T. Ohta and S. Omura, *Proc. Natl. Acad. Sci. USA*, 1999, **96**, 9509.
11. T. Stachelhaus, H. D. Mootz and M. A. Marahiel, *Chem. Biol.*, 1999, **6**, 493.
12. G. L. Challis, J. Ravel and C. A. Townsend, *Chem. Biol.*, 2000, **7**, 211.
13. P. Caffrey, *ChemBioChem*, 2003, **4**, 649.
14. A. T. Keatinge-Clay, *Chem. Biol.*, 2007, **14**, 898.
15. D. H. Kwan, Y. Sun, F. Schulz, H. Hong, B. Popovic, J. C. Sim-Stark, S. F. Haydock and P. Leadlay, *Chem. Biol.*, 2008, **15**, 1231.
16. S. W. Haynes and G. L. Challis, *Curr. Opin. Drug Discovery Dev.*, 2007, **10**, 203.
17. S. Lautru, R. J. Deeth, L. M. Bailey and G. L. Challis, *Nat. Chem. Biol.*, 2005, **5**, 265.
18. M. A. Wyatt, W. Wang, C. M. Roux, F. C. Beasley, D. E. Heinrichs, P. M. Dunman and N. A. Magarvey, *Science*, 2010, **329**, 294.
19. M. Zimmermann and M. A. Fischbach, *Chem. Biol.*, 2010, **17**, 925.
20. H. Gross, V. O. Stockwell, M. D. Henkels, B. Nowak-Thompson, J. E. Loper and W. H. Gerwick, *Chem. Biol.*, 2007, **14**, 53.
21. L. Robbel, T. A. Knappe, U. Linne, X. Xie and M. A. Marahiel, *FEBS J.*, 2010, **277**, 663.
22. O. Lazos, M. Tosin, A. L. Slusarczyk, S. Boakes, J. Cortés, P. J. Sidebottom and P. F. Leadlay, *Chem. Biol.*, 2010, **17**, 160.
23. W. K. W. Chou, I. Fanizza, T. Uchiyama, M. Komatsu, H. Ikeda and D. E. Cane, *J. Am. Chem. Soc.*, 2010, **132**, 8850.
24. L. C. Blasiak and J. Clardy, *J. Am. Chem. Soc.*, 2010, **132**, 926.
25. H. B. Bode, B. Bethe, R. Hofs and A. Zeeck, *ChemBioChem*, 2002, **3**, 619.
26. L. Laureti, L. Song, S. Huang, C. Corre, P. Leblond, G. L. Challis and B. Aigle, *Proc. Natl. Acad. Sci. USA*, submitte for publication.
27. D. J. Wilson, Y. Xue, K. A. Reynolds and D. H. Sherman, *J. Bacteriol.*, 2001, **183**, 3468.
28. S. Rigali, F. Titgemeyer, S. Barends, S. Mulder, A. W. Thomae, D. A. Hopwood and G. P. van Wezel, *EMBO Reports*, 2008, **9**, 670.
29. J. W. Bok, Y.-M. Chiang, E. Szewczyk, Y. Reyes-Dominguez, A. D. Davidson, J. F. Sanchez, H.-C. Lo, K. Watanabe, J. Strauss, B. R. Oakley, C. C. C. Wang and N. P. Keller, *Nat. Chem. Biol.*, 2009, **5**, 462.
30. T. Hosaka, M. Ohnishi-Kameyama, H. Muramatsu, K. Murakami, Y. Tsurumi, S. Kodani, M. Yoshida, A. Fujie and K. Ochi, *Nat. Biotechnol.*, 2009, **27**, 462.

CHAPTER 9

Mining Cyanobacterial Genomes for Drug-Like and Bioactive Natural Products

JOHN A. KALAITZIS AND BRETT A. NEILAN*

School of Biotechnology and Biomolecular Sciences, The University of New South Wales, Sydney, NSW, Australia, 2052
*E-mail: b.neilan@unsw.edu.au

9.1 Introduction

9.1.1 Genome Mining for Natural Product Discovery

Natural products are often described as small to mid-sized, non-polymeric organic molecules derived from biological or natural sources such as plants, fungi, bacteria and marine organisms. Many of these molecules possess pharmacological or biological activity and are thus pursued for a range of applications, most notably as drugs.[1–3] Their ecological roles may be of some consequence in predicting or determining their commercial potential, if any exists; however, their specific roles in nature do not necessarily render them unsuitable or non-desirable as drug candidates.

Over the last 30 years approximately 50% of the new drugs introduced into the clinic are classed as natural products or natural product-derived, e.g. synthetic analogues.[4] Libraries of structurally diverse and unique natural products (molecules) provide the platform for drug-lead discovery.[1] Drug discovery requires large numbers of molecules for screening, and natural molecular diversity is underpinned by a rich biodiversity.[5] One challenge

RSC Drug Discovery Series No. 25
Drug Discovery from Natural Products
Edited by Olga Genilloud and Francisca Vicente
© The Royal Society of Chemistry 2012
Published by the Royal Society of Chemistry, www.rsc.org

confronting researchers is how to best access and harness taxonomically diverse organisms for the purposes of drug discovery. Microorganisms as a group are not only considered taxonomically diverse, but are also largely untapped as a source of novel and/or biologically active natural products.[6–8] While microbes have provided many natural products which have been developed into drugs for use in the clinic, their potential as a source of further candidates is immense, particularly for the treatment of infection and various cancers.[6–8]

In relation to natural products research 'genome mining' is a broad term used to describe several processes that exploit information which is genetically encoded within biosynthesis gene clusters in microbial genomes with the ultimate aim of isolating a novel compound or discovering a novel biosynthetic pathway or enzyme.[9,10] Genome mining encompasses a number of techniques which provide leads or clues for the researcher working at the interface of chemistry and biology to pursue, and is not limited purely to the investigation of organisms whose genomes are sequenced.[9] The field of genome mining for natural products discovery is presented in more detail elsewhere in this book. Throughout this chapter we will present examples of genome mining with an emphasis on those dealing specifically with the natural products of cyanobacteria.

Our cyanobacterial genome mining research has focused upon identifying genes, from sequenced organisms, involved in the biosynthesis of polyketides and non-ribosomal peptides using a bioinformatics-based approach.[9,11] Studies such as these come about through advances in DNA sequencing technologies that have resulted in excess of one thousand microbial genomes being completed to date. The large majority of this sequence data has been deposited into freely available online databases, accessible by the greater research community through websites such as the NCBI Microbial Genomes Page (www.ncbi.nlm.nih.gov/genomes/MICROBES/microbial_taxtree.html). An intimate knowledge of microbial natural product biosynthesis at the genetic level has been aided by development of finely tuned molecular tools, so much so that clusters of biosynthesis genes coding for the assembly of complex natural products can be readily identified on a microorganism's genome.[11] Furthermore, a working knowledge of natural product biosynthesis at the genetics level can allow for structure prediction of the molecule or at very least prediction of the compound type or class and hence provide clues for associated bioactivity.[12] Before proceeding, it is pertinent to overview microbial natural product assembly with an emphasis on those biosynthesis pathways that give rise to drug-like molecules and the linking of these pathways to gene clusters.

9.1.2 The Genetic Basis of Macrocyclic Natural Product Biosynthesis in Cyanobacteria

The structural classes of natural products most commonly associated with microbial sources include non-ribosomal peptides (e.g. penicillin G and vancomycin), polyketides (e.g. erythromycin and tetracycline), and hybrid peptide-polyketides (e.g. cyclosporin A and rapamycin). As the examples cited here illustrate, these particular structure classes give rise to many clinically important antibiotic, anticancer and immunosuppressive agents.[13–15]

Reports of novel natural products from cyanobacteria are also dominated by these structure classes, although terpenes and meroterpenoids, alkaloids and ribosomal peptides also contribute a sizeable portion of the known secondary metabolite complement.[16] Much of this chapter focuses upon non-ribosomal peptides, polyketides and their hybrids because, as we will show, it is these natural products that have been the subject of our biosynthesis and genome mining studies in cyanobacteria.[11,17,18] A lot is now known regarding the biosynthesis of cyanobacteria-derived natural products, particularly polyketides and non-ribosomal peptides, at the genetic level.[19–21] The key to any genome mining study is that in the great majority of cases, genes involved in the biosynthesis of a particular molecule are clustered on the organism's genome (hence the now common term 'biosynthesis gene cluster'). The simple gene and protein architecture of non-ribosomal peptide synthetases (NRPSs) and polyketide synthases (PKSs) makes them identifiable using molecular approaches[13] and allows for chemical structures to be predicted.[20,21] Both NRPSs and type I PKSs possess modules containing a minimum of three enzyme domains. NRPS modules contain an ATP-dependent adenylation (A) domain which activates a specific or preferred amino acid, a peptidyl carrier protein (PCP) for tethering substrates during the assembly, and a condensation domain (C) which catalyses the formation of amide bonds between PCP-bound amino acids. Likewise, the PKS modules contain an acyl transferase (AT), which selects a preferred acyl-CoA thioester substrate, an acyl carrier protein (ACP) and a ketosynthase (KS) which catalyses the condensation of two ACP-bound substrates.[13,20]

Our genome mining approach (Section 9.3.4) involved targeting genes coding for proteins with one or more of the essential catalytic domains. This task was made easier by the ever-growing body of literature describing these systems and the highly conserved nature of gene sequences coding for PKSs and NRPSs across taxonomically distinct organisms. A recent review by Walsh and co-workers should be consulted for an in-depth discussion of the assembly of NRPS and PKS derived products.[20]

Cloning and sequencing biosynthesis gene clusters for the purpose of elucidating a natural product's biosynthesis pathway was popularised in the early to mid 1990s.[13,22] Nowadays the great majority of natural product biosynthesis studies revolve around a molecule's gene cluster and the specific biosynthesis proteins involved in its assembly.[23] Advances in molecular techniques and DNA sequencing technologies have provided the natural products chemist with new tools to assist in deciphering cryptic pathways. Coupling of gene cluster sequence analysis with traditional biosynthetic approaches, such as feeding studies with isotope labelled precursors and analysis of products by NMR, has led to the recent proposal of many more biosynthesis pathways.[24] Of course, efforts such as these were made easier as growing amounts of homologous sequence data was deposited into databases by research groups worldwide. With a cloned and sequenced biosynthesis gene cluster at hand, ultimate proof of its role in a natural product's assembly has been achieved via gene knockouts in the wild-type strain. Complete abolition of the natural product's biosynthesis in the knockout confirmed the gene cluster's product, while partial abolition suggested that the biosynthesis genes may not be essential for a compound's assembly or that the pathway was complemented by other processes in operation within an organism. Expression of biosynthesis gene clusters, or parts thereof, in heterologous hosts and careful analysis of the products either lent support to a proposed pathway or in some cases ruled out a previously plausible proposal.[23,24]

The manipulation of biosynthesis gene clusters in the wild-type or heterologous hosts also led to the production of novel unnatural natural products. Small libraries of new molecules have been constructed using techniques, such as combinatorial biosynthesis, mutasynthesis and precursor-directed biosynthesis, the majority of which are unattainable using synthetic organic methods.[25–27] These techniques rely strongly upon information gained from the characterisation of gene clusters and indeed the biosynthesis proteins encoded by cluster. Harnessing the biosynthetic capability of natural product biosynthesis gene clusters provides researchers with new tools for discovering and producing new chemical entities. Furthermore, discrete biosynthesis proteins coded for by a gene cluster functioned to perform specific chemical reactions not achievable via synthetic methods.[28] Biosynthesis enzymes, and their assorted co-factors, have been used in combination with synthetic methods to derivatise synthetic molecules or precursors and provide further analogues.[29,30] Recent efforts have resulted in the total enzymatic synthesis of some synthetically intractable molecules.[28] Such is the specificity and efficiency of these enzymes, very little in the way of synthetic by-products are observed in the reaction mixtures.[28] For these reasons the discovery and complete characterisation of biosynthesis genes clusters are keenly pursued, particularly if the molecule itself has notable bioactivity. In summary, characterisation of a biosynthesis gene cluster can provide information regarding a natural product's biosynthesis, and genetic manipulation of the cluster can facilitate the production of unnatural natural products, molecules unattainable using synthetic organic methods. Expression, purification, characterisation and harnessing of discrete biosynthesis proteins encoded by the cluster may represent the only means of accomplishing for example, a specific chemical transformation of a synthetic product, and thus these enzymes are of significant commercial interest. Before discussing genome mining of cyano-bacteria (Section 9.3) it is pertinent to provide some background about these organisms, their natural products and relevance to drug discovery (Section

9.2). Selected biosynthesis gene clusters from cyanobacteria will also be highlighted in Section 9.2.

9.2   Cyanobacteria

9.2.1   Introduction to Cyanobacteria

Cyanobacteria are commonly associated with the toxic blooms encountered in many eutrophic fresh and brackish waters, and are widely known for their potential to produce a range of neurotoxic, hepatotoxic and tumour-promoting secondary metabolites.[31] The growth of cyanobacteria, which are broadly described as oxygenic photosynthetic bacteria, is typically seen in conditions of neutral to alkaline pH and moderate levels of light and warmth.[32] A long evolutionary history has allowed cyanobacteria to adapt to and inhabit many extreme and diverse environments[33] such as arid desert soils, thermal springs, rocks, marine, brackish and fresh waters, ice, plants and animals. The immense diversity within this group of microorganisms, apart from the variability of morphology and range of habitats, is also reflected in the extent of their natural production. Cyanobacteria have evolved to produce a diverse array of secondary metabolites that have probably influenced their survival in these varied and highly competitive ecological niches. The reasons why these organisms have evolved to produce such natural products are not well understood nor are the specific roles that these compounds play in the environment, though it has been suggested that they are produced as a means of chemical defence against diverse macrograzers.[34] What is known is that toxicity is increased under favourable (blooming) growth conditions and studies have revealed that stress factors such as high or low cellular pH and iron limitation can stimulate toxin production.[35,36]

9.2.2   Natural Products from Cyanobacteria

Investigations into cyanobacteria as a potential source of novel natural products were pioneered by Professor R. E. Moore and co-workers at the University of Hawaii in the 1970s.[37,38] Since then a startling array of molecules, predominantly peptides and alkaloids, have been isolated from these organisms, some of which have been the focus of intense biosynthesis studies. Only examples pertinent to the topic at hand will be discussed here. For an in-depth discussion of cyanobacterial natural products, see the comprehensive review by Wright and co-authors entitled 'The Biogenetic Diversity of Cyanobacterial Metabolites'.[16] The review presents examples of cyanobacterial natural products which are grouped according to their biosynthetic origins (polyketides, cyanopeptides, alkaloids, isoprenoids and other metabolites), and provides an overview of the major biosynthetic routes in operation in the cyanobacteria.

Cyanobacteria are best known as producers of highly toxic compounds (cyanotoxins), and the majority are commonly grouped according to their physiological effects as either cytotoxins (e.g. cryptophycin 1 **1**, dolastatin 11 **2**, symplostatin **3**), neurotoxins (e.g. anatoxin-a **4**, saxitoxin **5**), hepatotoxins (e.g. microcystin L.R. **6**, nodularin **7**), or as irritants and gastrointestinal toxins (e.g. aplysiatoxin **8** and lyngbyatoxin A **9**). Some of these are discussed in more detail in later sections.[31,39]

Approximately three-quarters of the reported natural products from marine cyanobacteria are from group I and III (of five taxonomic groups), corresponding to the orders *Oscillatoriales* and *Nostocales*, respectively.[40] Due largely to the efforts of the Gerwick Laboratory (Oregon State/Scripps Institution of Oceanography, UCSD), investigations into *Lyngbya* spp. have yielded almost 300 novel compounds, contributing to the prolific number of compounds from the *Oscillatoriales*. Aside from *Lyngbya* spp., the order *Oscillatoriales* includes species from the genera *Oscillatoria*, *Symploca* and *Spirulina* which are also notable secondary metabolite producers.[40] Species of the order *Nostocales* are often associated with toxin production and include *Anabaena*, *Aphanizomenon*, *Cylindrospermopsis*, *Nodularia* and *Nostoc* spp. *Anabaena* spp. are prolific producers of toxic metabolites, and are known to biosynthesise the neurotoxins anatoxin-a **4**, anatoxin-a(s) **10**, and saxitoxin **5**, as well as hepatotoxic microcystins, e.g. **6** and cylindrospermopsin **11**.[41] Due to their unique and diverse structures, as well as their range of biological and pharmacological activities, many cyanobacterial natural products have been the focus of in-depth biosynthesis studies, using traditional feeding type experiments, modern molecular methods, or combinations of both. This has led to a better understanding of cyanobacterial natural product biosynthesis, thus aiding genome mining approaches for the discovery of new molecules.

9.2.3 Bioactive Macrocyclic Natural Products as Drugs

Natural products, derivatives and mimics thereof, constitute a large portion of the new chemical entities that have finalised, or are in the midst of, clinical trials. Newman and Cragg reported that in the period 1981–2006 natural products and derivatives have contributed the large majority of antibacterials, and greater than 25% of all anticancer, anti-inflammatory and anti-ulcer agents.[4] It is beyond the scope of this chapter to discuss all natural product-derived drugs in detail; interested readers should consult reviews dealing with varying aspects of natural product drug discovery, such as those by Newman and Cragg,[4] Butler,[42] Molinski[43] and Fischbach[14] for a more detailed treatise of this subject.

A 2008 review by Terrett and co-authors presented the case for natural product macrocycles as drug candidates.[44] The authors highlighted the fact that the macrocyclic drugs currently on the market (e.g. the antibiotics vancomycin and erythromycin, the immunosuppressants cyclosporin A and FK506, the potent antifungal agent amphotericin B, and the anticancer agent epothilone B), are almost exclusively derived from microbial natural products. They also noted that natural product macrocycles are often ignored as drug candidates because they don't obey Lipinski's rules – the guidelines used to predict whether or not a chemical structure is a suitable drug candidate.[44] The authors provide strong evidence that macrocycles as drug candidates have been underexplored so far, and the discovery of new molecules from this family should become an area of great growth in the future. A review by Zhang and Wilkinson also dispels the applicability of the Lipinski's rule-of-five to natural products.[45]

9.2.4 Macrocyclic Natural Products from Cyanobacteria

Cyanobacteria are prolific producers of complex macrocyclic molecules. In the following sections the biological activity of some of these, as well as some biosynthesis gene clusters encoding their assembly, will be discussed. In-depth discussions of bioactive cyanobacterial natural products belonging to other chemical classes can be found in reviews published elsewhere.[16,46,47] Amongst the plethora of bioactive macrocycles from the cyanobacteria, the dolastatins have garnered a lot of attention. It is pertinent to note that not all of the dolastatins are macrocyclic, but are of non-ribosomal peptide origin. Dolastatin 11 **2** is considered highly cytotoxic and has displayed significant activity against various cancer cell lines.[48] The molecule has not been pursued as a drug due to its deleterious toxic effects; however, the dolastatin 11 **2** analogue, DMMC (desmethoxymajusculamide C) **12**, from a Fijian *Lyngbya* sp. which, like **2**, disrupts actin, has shown promise in preclinical anticancer trials.[49,50]

The NRPS/PKS-derived iejimalides A–D **13–16** were first isolated from two different samples of tunicates suggesting a symbiotic microbe as the true producer.[51,52] Clarification of this was provided by the isolation of **13** from a *Lyngbya* sp. isolated from Papua New Guinea.[50] Of interest is the apparent mode of action displayed by the iejimalides, which on screening against a human cell line panel revealed a unique cytotoxicity profile.[53,54]

The marked anti-proliferative activity shown against transformed fibro-blastic osteosarcoma U2OS cells has attracted attention to the *Symploca* sp. product largazole **17**.[55] The molecule has been the focus of many synthetic and structure–activity studies, culminating in testing which has revealed its potency as a class I histone deacetylase (HDAC) inhibitor.[56,57]

Cytotoxic agents showing serine protease inhibition include the cyanobacterial products symplocamide A **18** and lyngbyastatin 4 **19**.[47] These molecules are only some of the many which possess a 3-amino-6-hydroxypiperidone moiety, the structural motif common to many serine protease inhibitors from cyanobacteria is also indicative of NRPS involvement in their biosynthesis. Further NRPS/PKS products from cyanobacterial origins showing potent cytotoxicities include palau'amide **20** and apratoxins A–C **21–23**. Cytotoxicity testing of **21–23** against various cell lines revealed that those analogues incorporating *N*-methylisoleucine (**21** and **23**), as opposed to isoleucine (**22**), were more potent.[47] Palau'amide **20** exhibited significant (nM) cytotoxicity against KB cells but due to the lack of material available it was not able to be tested further.[58,59] For drug leads such as these to be pursued an alternative method of production needs to be developed. A molecule such as this could potentially be the focus of genetics experiments aimed at harnessing nature's biosynthetic potential. In the next section we overview natural product

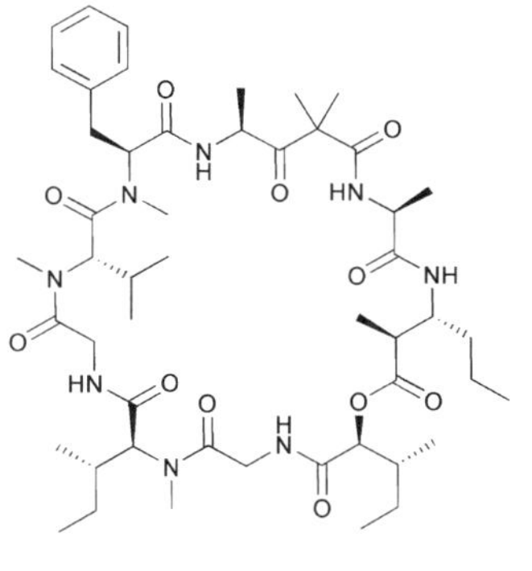

12

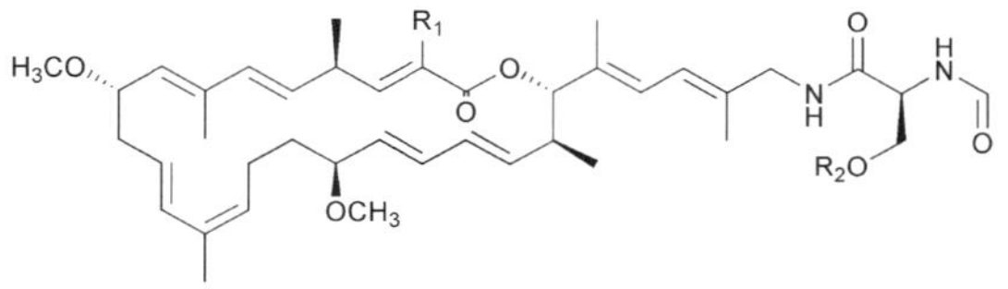

13 R$_1$=R$_2$=H
14 R$_1$=H R$_2$=SO$_3$Na
15 R$_1$=CH$_3$ R$_2$=H
16 R$_1$=CH$_3$ R$_2$=SO$_3$Na

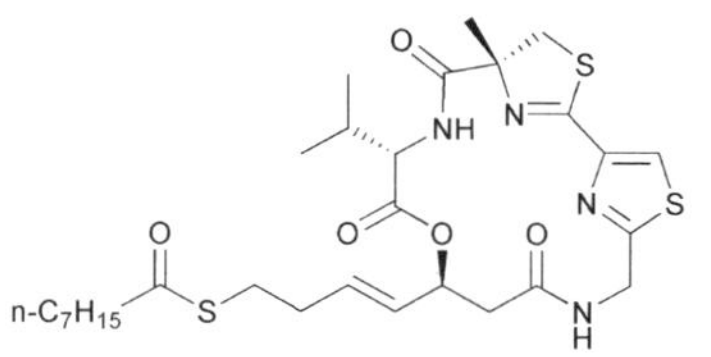

17

biosynthesis gene clusters from cyanobacteria, an understanding of which is required in order to develop expression systems as a stable source of those compounds that are scarce in nature or otherwise difficult to replicate and produce in the laboratory.

9.2.5 Cyanobacterial Natural Product Biosynthesis Gene Clusters

The number of characterised biosynthesis gene clusters from cyanobacteria is limited compared with the numbers reported from other microbes. This is a

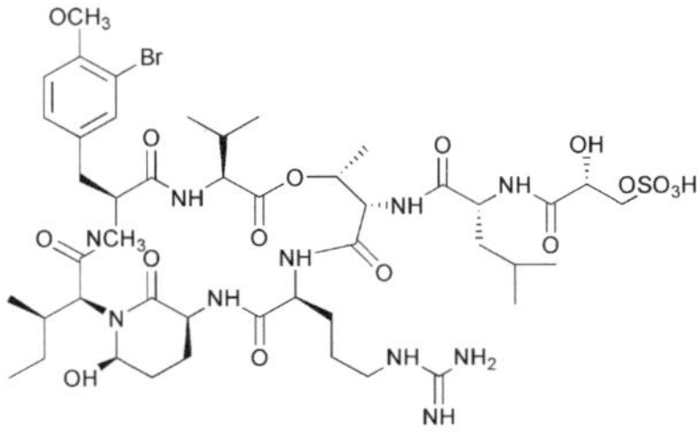

18

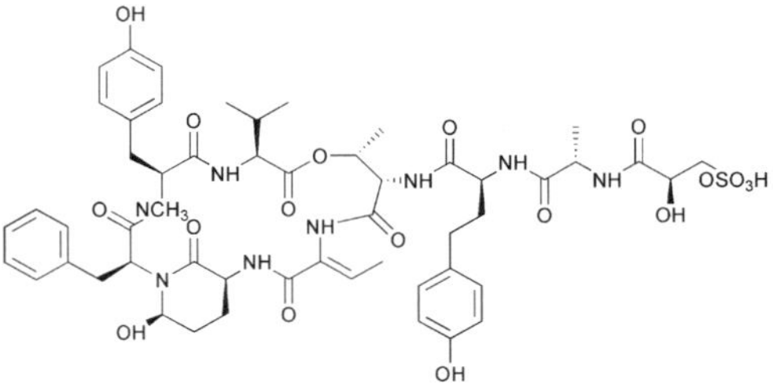

19

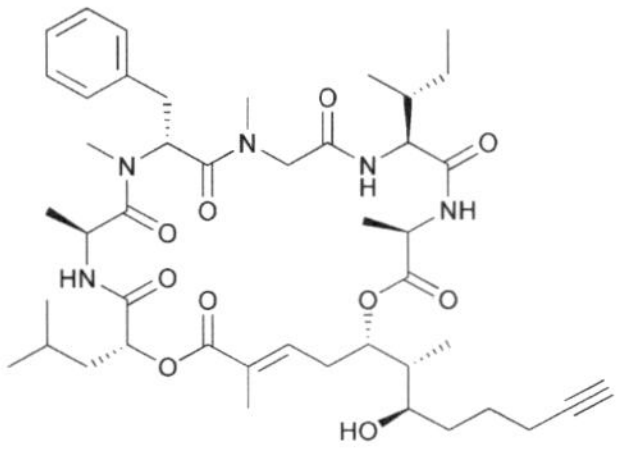

20

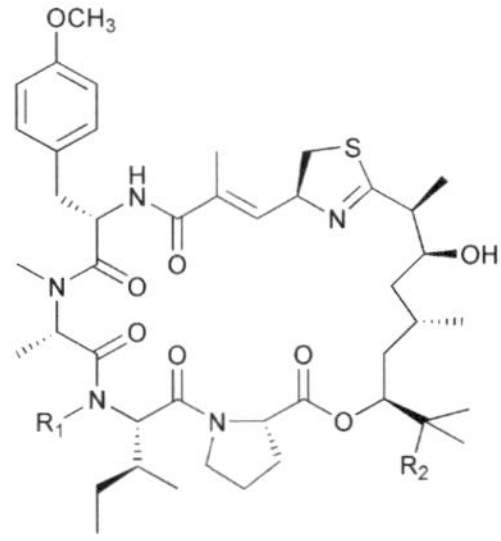

21 R₁=R₂=CH₃
22 R₁=H, R₂=CH₃
23 R₁=CH₃, R₂=H

reflection of the number of researchers pursuing such clusters rather than a lack of interest in the natural products derived from them. Here, some cyanobacteria-derived clusters and the unique biosynthesis pathways in operation will be highlighted. We and others have recently reviewed biosynthesis gene clusters from cyanobacteria in depth so here we highlight gene organisation and architecture rather than provide an in-depth discussion of natural product biosynthesis and enzyme-specific functionality.[11,60] The selected examples will provide the basis for the later discussion of genome mining. Tailoring enzymes encoded by different gene clusters have also been reviewed elsewhere but it is pertinent to note that genes encoding these proteins could serve as potential targets for genome mining studies.[11] Characterised gene clusters from cyanobacteria are listed in Table 9.1; those not mentioned elsewhere in this review include the anabaenapeptilide (*apd*),[61] trunkamide (*tru*)[62] and cyanophycin (*cph*)[63] biosynthesis gene clusters.

9.2.5.1 Gene Clusters Encoding Macrocycle Assembly in Cyanobacteria

As macrocyclic molecules are a focus of this chapter, we present notable biosynthesis gene clusters (with an emphasis on their architectures) from cyanobacteria coding for these complex molecules. Macrocycles serve

Table 9.1   Selected cyanobacteria-derived biosynthesis gene clusters

Gene cluster	Source organism	~ Size (kb)	Biosynthetic origin
barbamide (*bar*)	*Lyngbya majuscula* 19L	26	polyketide/ peptide
cryptophycin (*crp*)	*Nostoc* sp. ATCC53789	40	polyketide/ peptide
cylindrospermopsin (cyr)	*Cylindrospermopsis raciborskii* AWT205	43	polyketide/ peptide
curacin (*cur*)	*Lyngbya majuscula* 19L	64	polyketide/ peptide
hectochlorin (*hct*)	*Lyngbya majuscula* JHB	38	polyketide/ peptide
jamaicamide (*jam*)	*Lyngbya majuscula* JHB	58	polyketide/ peptide
microcystin (*mcy*)	*Microcystis aeruginosa* PCC 7806	55	polyketide/ peptide
microginin (*mic*)	*Planktothrix rubescens* NIVA CYA 98	21	polyketde/peptide
nodularin (*nda*)	*Nodularia spumigena* NSOR10	48	polyketide/ peptide
nostopeptolide (*nos*)	*Nostoc* sp. GSV224	40	polyketide/ peptide
aeruginosin (*aer*)	*Planktothrix agardhii* CYA126/8	34	peptide
anabaenopeptilide (*apd*)	*Anabaena* sp 90	28	peptide
anabaenopeptin (*ana*)	*Planktothrix rubescens* NIVA CYA 98	25	peptide
anatoxin A (*ana*)	*Oscillatoria* PCC 6506	29	polyketide
cyanopeptolin (*mcn*)	*Microcystis* N-C 172/5	30	peptide
cyanophycin	*Synechocystis* sp. PCC 6803	3	peptide
nostocyclopeptide (*ncp*)	*Nostoc* sp. ATCC53789	33	peptide
lyngbyatoxin (*ltx*)	*Lyngbya majuscula*	11	peptide/terpenoid
saxitoxin (*sxt*)	*Cylindrospermopsis raciborskii* T3	35	polyketide/amino acid
scytonemin	*Nostoc punctiforme* ATCC29133	28	shikimate
shinorine	*Anabaena variabilis* ATCC29413	6	peptide derivative
microcyclamide (*mca*)	*Microcystis aeruginosa* NIES298 and PCC 7806	11	ribosomal peptide
microviridin (*mdn/ mvd*)	*Microcystis aeruginosa* NIES298, *P. agardhii* CYA126/8	7	ribosomal peptide
oscillatorin (*osc*)	*Planktothrix rubescens* NIVA CYA 98	3+	ribosomal peptide
patellamide (*pat*)	*Prochloron didemni*	11	ribosomal peptide
trichamide (*tri*)	*Trichodesmium erythraeum* ISM101	13	ribosomal peptide
trunkamide (*tru*)	*Prochloron* sp.	11	ribosomal peptide/ terpenoid

as elegant examples of natural combinatorial biosynthesis. The gene clusters encoding their biosynthesis are associated with prototypical thio-templated modular systems (TMS), and provide the basis for our genome mining study.

9.2.5.1.1 *Cryptophycins Biosynthesis Gene Cluster*

The potent tubulin destabilising agent[64] cryptophycin 1 **1** is the major product of the lichen symbionts[65] *Nostoc* sp. ATCC 53789 and *Nostoc* sp. GSV 224 and approximately 30 structural variants are known from *Nostoc* spp.[66,67] The cryptophycin biosynthesis gene cluster (*crp*) was first identified in *Nostoc* sp. ATCC 53789 using a bioinformatics approach. Initial screening of a cosmid library proved difficult due to the abundance non-overlapping clones containing PKS and NRPS genes.[68] Comparison of adenylation and ketosynthase domain sequences from the non-cryptophycin producing *N. punctiforme* revealed PKS and NRPS pathways unique to the symbiont. Subsequent cloning and analysis of candidate gene sequences revealed a 40 kb NRPS/PKS gene cluster organised collinearly with the proposed order of cryptophycin assembly. The cluster contains four genes *crpA–D* coding for biosynthesis proteins involved in the assembly and cyclisation of the carbon backbone which undergoes subsequent tailoring reactions to produce the final products. The *crp* cluster codes for a loading domain, two PKS modules, three NRPS modules and a terminal thioesterase. The relaxed substrate specificity of several modules within the cluster was exploited to generate a library of unnatural cryptophycins through precursor-directed biosynthesis, further exemplifying the utility of characterising biosynthesis gene clusters and harnessing nature's synthetic capabilities.[68]

9.2.5.1.2 *Microcystins Biosynthesis Gene Cluster*

The microcystins, e.g. **6**, are potent inhibitors of eukaryotic protein phosphatases 1 and 2A. The biosynthetic gene cluster of these cyclic heptapeptides was first characterised in *Microcystis aeruginosa* PCC 7806.[17,69,70] The microcystin biosynthesis gene cluster (*mcy*) spans 55 kb and is composed of ten bidirectionally transcribed open reading frames (ORFs). The PKS McyG contains an N-terminal adenylation-peptidyl carrier protein loading di-domain[71] which is predicted to load the starter unit which then undergoes subsequent malonate and amino acid extensions (encoded by *mcyGJDEABCL*). The C-terminal TE is suspected to facilitate hydrolysis and cyclisation to yield the final product. Tailoring enzymes associated with the cluster include an *O*-methyltransferase (McyJ) and a stand-alone dehydratase (McyI).[17,69,70]

9.2.5.1.3 Nodularin Biosynthesis Gene Cluster

The cyclic pentapeptide nodularin **7** is assembled in a manner similar to the structurally related heptapeptide microcystin L.R **6** and this is clearly reflected in the genetic organisation of the two clusters. The nodularin biosynthesis gene cluster in *Nodularia spumigena* NSOR10 (*nda*) spans 48 kb and consists of nine ORFs. It encodes proteins more or less collinearity with their respective catalytic functions in the assembly of nodularin.[18] Interestingly, the *nda* cluster encodes two NRPS domains modules which, as predicted by structural homology, activate amino acids (D-Ala and L-Leu) that are not present in the final compound. Cases such as these should be considered when predicting compound structures from gene sequence data in any genome mining project. The deletion of two NRPS modules, compared to microcystin synthetase, also raises many questions regarding recombination and shuffling of genes, as well as the transposition of gene clusters between organisms. The cluster also encodes a protein most similar to the ABC transporter proteins associated with microcystin and nostopeptolide biosynthesis.[18] Conserved sequences such as these could also be targeted with the aim of identifying biosynthesis gene clusters of other potentially toxic molecules.

9.2.5.1.4 Nostopeptolides Biosynthesis Gene Cluster

The cryptophycin[66]-producing strain *Nostoc* sp. GSV224 also produces other cyclic peptide polyketide hybrid natural products known as nostopeptolide A_1 **24** and A_2 **25**. The *nos* gene cluster includes eight ORFs, spans 40 kb, and contains most genes required for biosynthesis and transport. The domain organisation is colinear with the proposed order of biosynthetic assembly, with *nosA* encoding a tetramodular NRPS, *nosC* a trimodular NRPS and *nosD* a dimodular NRPS. Located between *nosA* and *nosC* is *nosB*, which encodes a

24 R=CH$_3$
25 R=H

single PKS module.[72] A putative thioesterase is located at the C-terminus of NosD. NosA3, within the third NRPS module encoded by *nosA*, is proposed to adenylate the rare non-proteinogenic amino acid residue L-4-methylproline.[73] Also of interest is *nosG* which encodes an ATP binding cassette (ABC) transporter.[72]

9.2.5.1.5   Nostocyclopeptides Biosynthesis Gene Cluster

The 33 kb nostocyclopeptide (*ncp*) biosynthesis gene cluster in *Nostoc* sp. ATCC53789 has been sequenced and characterised.[74] Like many cyanobacterial NRPS-derived natural products, the cluster is colinear with the proposed order of nostocyclopeptide A₁ **26**, A₂ **27**, and A₃ **28** assembly. The cluster encodes two proteins, NcpA and NcpB, containing three and four NRPS modules, respectively. Like the *nos* cluster, genes encoding for L-4-methylproline biosynthesis and transport enzymes are also present. The cluster architecture mirrors that of *nos* synthetase and also encodes a 265 amino acid protein, NcpC, of unknown function. In both the *ncp* and *nos* clusters, this gene, *ncpC* and *orf5*, respectively, is located between the NRPS modules and a L-4-methylproline biosynthesis gene.[74] A recent BLAST search has shed no further light on its role, if any, in peptide biosynthesis. However, the most striking feature of this cluster is the encoded reductase domain at the C-terminal end of NcpB, which is responsible for the reductive release of the peptide.[75] This reductase also facilitates the cyclisation of the linear peptide to the unusual imine-linked macrocyclic peptide product. NcpB contains a putative NAD(P)H binding domain which suggests that the off-loading of the peptide is reductive in nature therefore generating a linear peptide with a terminal aldehyde. Identification of similar reductase domains through genome mining could ultimately lead to the characterisation of other clusters encoding for imino-linked macrocycles, thus providing further tools to assist in the complete characterisation of this unusual biosynthetic pathway. Kopp and

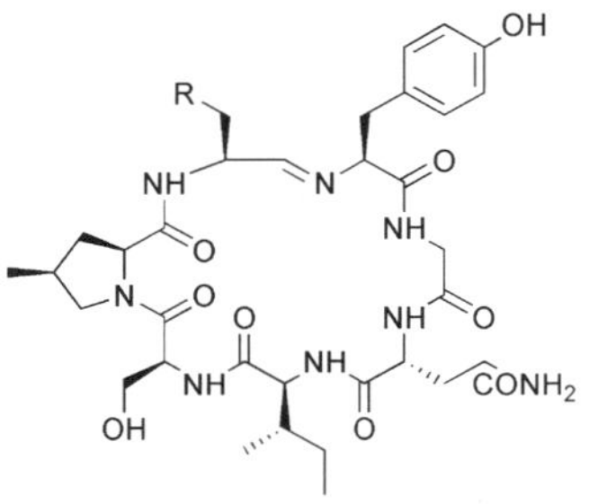

26 R=isopropyl
27 R=phenyl
28 R=p-methylphenyl

Marahiel have recently reviewed 'Macrocylization strategies in polyketide and non-ribosomal peptide biosynthesis' in great depth.[76]

9.2.5.1.6 Hectochlorin Biosynthesis Gene Cluster

The cytotoxic and antifungal agent hectochlorin **29** is a *Lyngyba majuscula* JHB metabolite whose biosynthesis gene cluster is remarkably colinear with its product.[77] The *hct* gene cluster consists of eight ORFs spanning 38 kb, encoding NRPS, PKS, cytochrome P450 and halogenase enzymes. The *hct* gene cluster shares significant sequence similarities with NRPS and PKS elements from other *Lyngbya*-derived biosynthesis gene clusters. *HctB* codes for a putative halogenase (as well as an ACP) that is similar at its N-terminus to BarB1 and BarB2 from the barbamide (*bar*) cluster[78] and which catalyses an unprecedented chlorination reaction.[79] The authors speculated that HctB is involved in the formation of hectochlorin's gem-dichloro group. Another rare feature of the *hct* gene cluster is the C-methyl transferase encoded within the PKS module HctD. Also uncommon are the NRPS-embedded ketoreductases located adjacent to two 2-oxo-isovaleric acid adenylation domains. These unusual dual function KRs[80] reduce 2-oxo-isovaleric acid derived moieties to their corresponding 2-hydroxyisovaleric acid moieties. These are then further oxidised to generate the 2,3-dihydroxyisovaleric derived moieties in hectochlorin. The sequence of the putative transposase encoding gene *hctT* is similar to insertion sequence (IS) elements from other bacteria and is thought to play a role in the plasticity of bacterial genomes.[77] IS elements are of interest in terms of genome mining as they are often involved in the assembly of gene clusters with specialised functions for adaptation to specific environments.

9.2.5.2 Gene Clusters Encoding Non-Macrocyclic NRPS Products in Cyanobacteria

Aside from the cyclic compounds, biosynthesis gene clusters have been characterised for a number of important cyanobacterial natural products, including a small group of toxic alkaloids and a series of structurally diverse molecules from species of *Lyngbya*. Careful analyses of these gene clusters helped unravel the unique and somewhat cryptic biosynthesis pathways in

29

operation in cyanobacteria. While these clusters do not code for macrocycles, they provide excellent examples of the natural product structural diversity derived from complex biosynthesis pathways associated with cyanobacteria. Relevant to the discussion here is that all of the biosynthesis clusters highlighted below possess NRPS coding genes.

9.2.5.2.1 Toxic Alkaloid Biosynthesis Gene Clusters

The biosynthesis gene clusters of alkaloids often code for the incorporation of unusual substrates, rare pathways and numerous tailoring reactions. Also of interest in genome mining terms are the clustered genes encoding toxin efflux and transport proteins. These could prove useful targets for locating biosynthesis clusters encoding such toxins. Gene clusters have been characterised for the toxins saxitoxin **5** (*sxt*) in the cyanobacterium *Cylindrospermopsis raciborskii* T3, *Anabaena circinalis*, *Aphanizomenon* and *Lyngbya*, cylindrospermopsin **11** (*cyr*) in *C. raciborskii* AWT205 and *Aphanizomenon ovalisporum*, and anatoxin-A **4** (*ana*) in *Oscillatoria* sp PCC 6506.[81–83]

9.2.5.2.2 Biosynthesis Gene Clusters of Bioactive Molecules from Marine Lyngbya spp.

Strains of *Lyngbya majuscula* are notable producers of chemically diverse and highly unique metabolites possessing a broad range of biological activities.[60] The biosyntheses of *L. majuscula* natural products are keenly pursued due to their rarely encountered chemical features. To date the biosynthetic gene clusters of several *Lyngbya*-derived metabolites have been functionally characterised including barbamide **30** (*bar*), curacin A **31** (*cur*), jamaicamides A-C **32–34** (*jam*) and lyngbyatoxins A-C **9**, **35–36** (*ltx*).[77,78,84,85] The molecules themselves possess significant bioactivities; **31** is a potent antitubulin natural product,[86,87] while **30** was isolated via a bioassay-guided fractionation protocol using a mollusicidal activity assay.[78] The jamaicamides **32–34** have shown both cytotoxic and neurotoxic effects and the lyngbyatoxins **9**, **35–36** are notorious skin irritants.[88,89] These examples illustrate the diversity of biosynthesis pathways in operation in different strains of the same species; common to all these are the presence of NRPS coding genes and in most cases PKS coding genes. All of these gene clusters are more or less colinear with the assembly of the natural product.

The importance of the colinearity rule with respect to gene cluster characterisation and natural product structure prediction will become more apparent in the following sections. Next we will explore the notion of connecting genes to natural products and vice versa, the theoretical basis of genome mining. We will also introduce the concept of 'orphan clusters' – gene clusters whose natural product is not known.

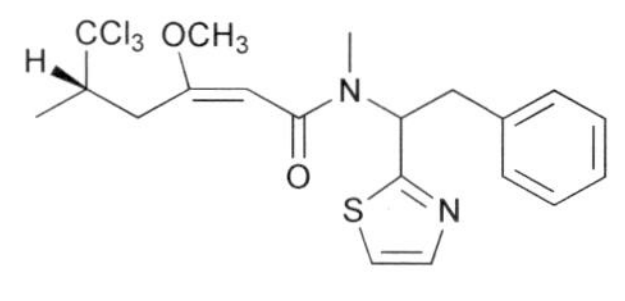

30

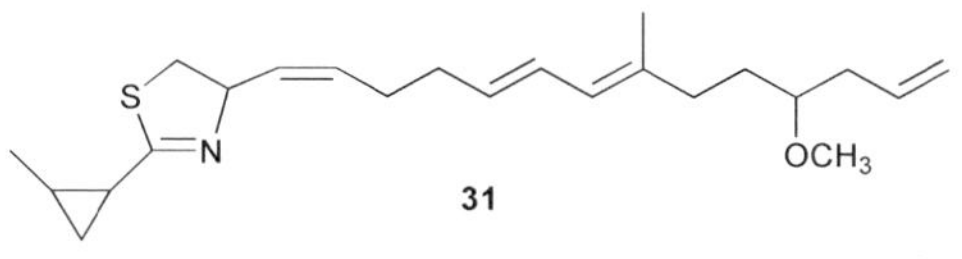

31

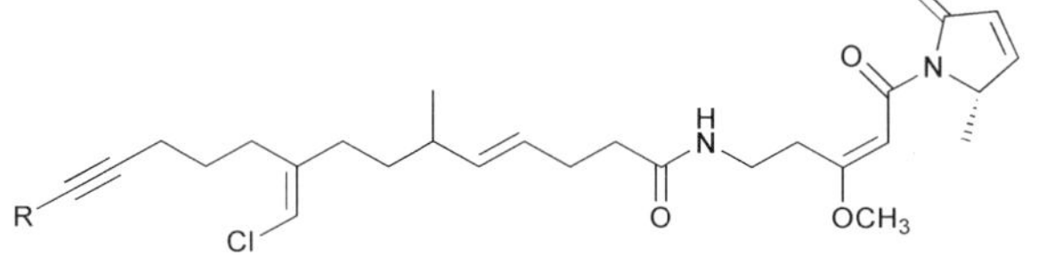

32 R = Br
33 R = H

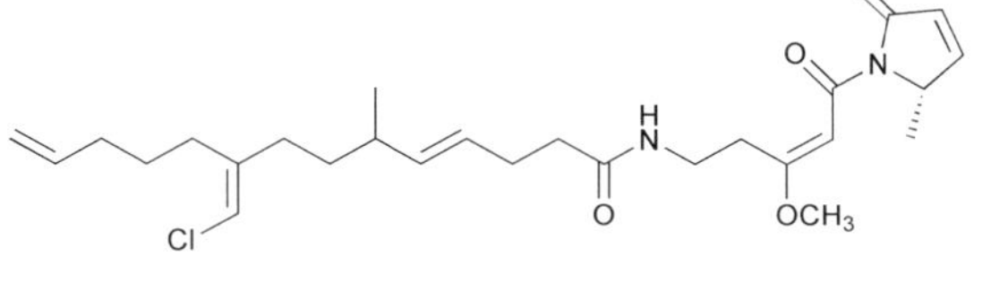

34

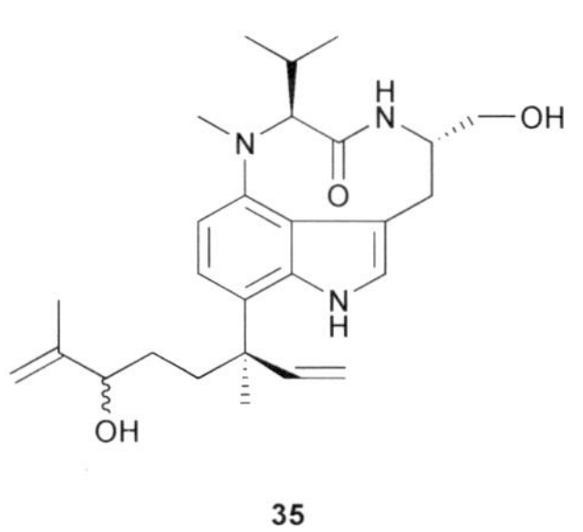

35

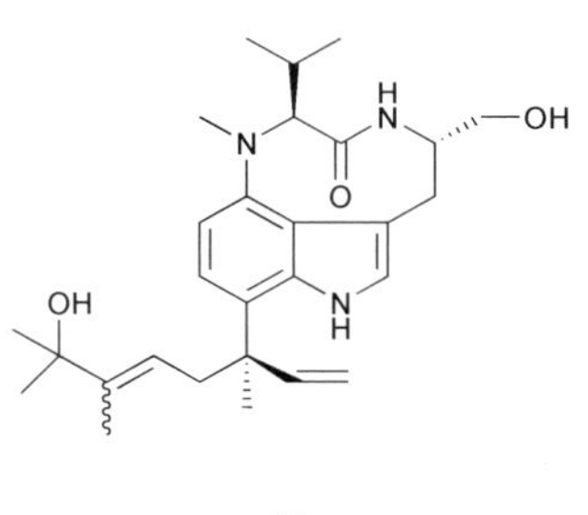

36

9.3   Mining Cyanobacterial Genomes: A New Avenue to Natural Product Discovery

9.3.1   Linking Biosynthesis Genes to Natural Products: The Basis for Genome Mining

For our purposes, research coupling natural product isolation, biosynthetic feeding studies, DNA sequence analysis and genome manipulation with the aim of connecting genes to compounds, discovering new compounds and/or elucidating biosynthesis pathways, can be considered genome mining. We constantly need new drug candidates and therefore novel and unexplored sources of natural products are highly desirable. Genome mining and the complementary techniques it encompasses, is geared toward examining and even providing some of these potentially novel sources.

Aside from those biosynthesis gene clusters characterised in the cyanobacteria, many more have been described from other families of microorganisms.[90] Early discoveries that biosynthesis genes were clustered on an organism's genome and that these clusters encoded sets of enzymes functioning in an assembly line-like manner, provided a platform for future studies.[91,92] The search for gene clusters coding the biosynthesis of structurally interesting natural products began in earnest in order to help decipher cryptic biosynthesis pathways, and with that, better understand unique mechanisms underpinning molecular diversity.

While PKSs use predominantly malonyl-CoA or methylmalonyl-CoA as carbon backbone building blocks, the number of building blocks utilised by NRPSs are vast and include a massive array of non-proteinogenic (often D-) amino and hydroxy acids. Furthermore, analysis of protein sequence data of specific amino acid adenylation domains has revealed a code useful for predicting which amino acid is incorporated into the final structure. Assuming the colinearity rule of assembly is observed, which is not always the case, complete backbones of NRPS, PKS or hybrid molecules could be predicted with some degree of certainty, while products of tailoring and other accessory enzymes could also be predicted, albeit with less confidence. Such is our knowledge base, the corollary of this, that is, predicting the modular architecture of a gene cluster based on the structure of the natural product is also possible.[12]

Numerous reviews dealing with genome mining in general have appeared in the recent literature[11,90,93] and so rather than repeating this information, only key features with an emphasis on mining NRPSs and PKSs are highlighted below.

Modular NRPS and PKS, or TMS (thio-templated modular systems) genes are often the largest open reading frames on an organism's genome, thus making them relatively easy to identify.[90] Genes coding for a molecule's biosynthesis are clustered and their genetic architecture in general mirrors that of an assembly line.[22] A working knowledge of the these modular systems can allow a natural product's structure, or at very least structure type, to be predicted based on gene sequence data alone.[12] Of course, direct evidence that a gene cluster codes for a particular natural product requires further experimentation, and ultimate proof is given by isolation of the compound and confirmation using one or more molecular genetic and/or biochemical techniques.[94]

Mining complete microbial genomes using automated bioinformatics-based search protocols has revealed a great number of orphan biosynthesis gene clusters.[11,94–96] Typically, multiple biosynthesis gene clusters are identified on an organism's genome, and these clusters code for natural products known, suspected or not previously known to be produced by the sequenced strain. Gene clusters coding for molecules not previously known to be produced by the organism may show significant sequence similarity to other fully characterised gene clusters, indicating assembly of a known compound first isolated from another source. Those clusters whose product is not known are termed 'orphans'.[97,98] They may indeed code for a known compound (or structure type) but interrogation of the sequence alone does not allow for the product to be easily identified. In cases such as these rigorous chemical analysis is required to identify the target molecule, the caveat being that the biosynthetic pathway is expressed under laboratory culture conditions. Herein lies one of the main shortcomings of mining genomes for new natural products – the 'silent' gene clusters.[99] Efforts to express silent gene clusters and switch on such pathways in the host are becoming more common; however, these should really only be pursued if the gene cluster or parts of it, appear to be of great significance and likely to result in important findings, otherwise these pursuits may be extremely time consuming for little or no scientific advance.[100]

Examples of genome mining techniques suitable for use with the cyanobacteria have been employed to drive the discovery of novel natural products. Through identifying NRPS genes from sequence data, the siderophore coelichelin was isolated from *Streptomyces coelicolor* and the orfamides from *Pseudomonas fluorescens* Pf-5.[95,97,101] 'The Genomisotopic Approach' as used to discover the orfamides appears to be highly suitable for mining cyanobacteria and relies upon amino acid recognition by NRPS adenylation domains.[97] Models used to predict amino acid recognition by NRPSs have been developed based on critical binding-pocket residues.[102,103] These models were not developed using many cyanobacterial sequences and should thus be used with caution when mining cyanobacterial genomes. The relaxed substrate specificity of adenylation domains encoded by cyanobacterial genomes means that such tools should be used only as a guide, and not considered definitive. Predictions can be verified using the traditional ATP-[^{32}P]pyrophosphate (PPi) exchange assay or the recently described non-radioactive colorimetric assay which quantifies orthophosphate (Pi) derived from degraded PPi as a means of determining activity.[104]

9.3.2 Illustrative Examples of Mining Cyanobacterial Genomes

To date there have been relatively few examples of genome mining in cyanobacteria. The examples that do exist serve to highlight the utility of genome mining, not only for new natural products but also for the identification and characterisation of elusive biosynthesis gene clusters. In Sections 9.3.2.1.1–9.3.2.1.3 the highlighted gene clusters all code for the biosynthesis of cyclic peptides containing heterocyclised residues. Collectively, these ribosomal peptides are known as cyanobactins.[62] Recent genome mining for the biosynthesis clusters coding for the assembly of the natural sunscreens scytonemin and shinorine from cyanobacteria will be highlighted in Section 9.3.2.2, while Sections 9.3.3–9.3.5 will detail genome mining studies focused on the discovery of NRPS and PKS coding genes, the basis of macrocyclic natural product biosynthesis. Genome mining studies focused upon lantibiotic biosynthesis in cyanobacteria are presented in Section 9.3.6.

9.3.2.1 *Mining for Cyanobactins and their Biosynthesis Gene Clusters*

9.3.2.1.1 *The Patellamides Biosynthesis Gene Cluster*

The patellamides are cyclic peptides often isolated from didemnid ascidians and are thought to be biosynthesised by obligate cyanobacterial symbionts of the genus *Prochloron*. Attempts to culture *Prochloron* spp. have so far been unsuccessful and thus biosynthetic studies have proven difficult. In an effort to sustain metabolite production, a shotgun cloning and heterologous expression of *Prochloron* sp. DNA in *E. coli* approach was used to confirm *Prochloron* sp. as a patellamide producer.[105] Independently, and as part of the *Prochloron didemni* genome sequencing project, the draft genome sequence was mined for patellamide biosynthesis genes.[106] The patellamide biosynthesis gene cluster (*pat*) was identified; however, rather than being assembled by an NRPS as first anticipated,[105,107] the patellamides were found to be assembled ribosomally.[62,106] To unequivocally prove that the 11 kbp, seven ORF (*patA–patG*) gene cluster was responsible for patellamide biosynthesis, the cluster was heterologously expressed in *E. coli*. Analysis of the resulting extract by LC-MS revealed production of patellamide A **37** thus providing conclusive evidence for its ribosomal assembly.[62,106]

9.3.2.1.2 *Discovery of Trichamide through Genome Mining*

Homology analysis of patellamide biosynthesis genes in the genome of *Trichodesmium erythraeum* ISM101 revealed the strikingly similar trichamide biosynthesis gene cluster (*tri*) encompassing eleven ORFs and spanning 12.5 kbp. Structure prediction followed by rigorous chemical analysis of the culture extract revealed trichamide **38**, a novel natural product and the first reported metabolite from *Trichodesmium erythraeum*, further illustrating the power of genome mining for natural product discovery.[108]

9.3.2.1.3 Discovery of the Microcyclamides through Genome Mining

The discovery of the *pat* biosynthesis gene cluster, using a PCR directed approach employing degenerate primers based *tri* gene sequences, led to the identification of genes in *Microcystis aeruginosa* NIES 298 responsible for the ribosomal biosynthesis of microcyclamide. Scanning of the *M. aeruginosa* PCC 7806 genome for similar clusters ultimately led to the isolation and structure elucidation of two new microcyclamides, 7806A **39** and 7806B **40**.[109,110] A recent review comparing biosynthesis gene clusters yielding similar peptides in other cyanobacteria revealed a global assembly line responsible for cyanobactins. After PKSs and NRPSs, the cyanobactin biosynthesis assembly line represents another major route to small molecules in cyanobacteria.[111]

9.3.2.2 Mining for Sunscreen Biosynthesis Gene Clusters

9.3.2.2.1 The Scytonemin Biosynthesis Gene Cluster

Scytonemin **41** is an ultraviolet (UV) radiation absorbing pigment which plays an important role in protecting cyanobacteria from harmful exposure.[112] Transposon mutagenesis of the scytonemin-producing strain *Nostoc punctiforme* ATCC 29133 resulted in the generation of a mutant strain, SCY 59, unable to produce scytonemin.[113] Genomic analyses of the mutated region supported a biosynthetic role for the mutated gene, and allowed the putative assignment of the 18 ORF, 28 kbp scytonemin biosynthesis gene cluster. A biosynthetic route to scytonemin, however, was not proposed.[113] Subsequent gene expression studies confirmed that the putative cluster was indeed involved in scytonemin biosynthesis, and the proposed initial steps of its biosynthesis from L-tryptophan and prephenate have since been validated.[114,115] Comparison of scytonemin biosynthesis gene clusters in six cyanobacterial genomes revealed two major architectures, and these appear to have evolved through genetic rearrangement and insertions.[116]

9.3.2.2.2 The Shinorine Biosynthesis Gene Cluster

Other cyanobacterial sunscreens such as the small molecule mycosporine and mycosporine-like amino acids (MAAs) have also been the focus of genome mining studies. A combined analytical chemistry and bioinformatics based approach was used by Häder and co-workers to mine four sequenced cyanobacterial genomes for MAAs (shinorine) and their biosynthesis genes.[117] Analysis of organic extracts derived from *Anabaena variabilis* PCC 7937, *Anabaena* sp. PCC 7120, *Synechocystis* sp. PCC 6803 and *Synechococcus* sp.

37

38

39

40

41

PCC 6301 revealed that only *A. variabilis* PCC 7937 was able to produce the MAA shinorine **42** while the other strains did not produce any MAAs. Bioinformatics revealed that the producing strain possessed clustered homologues of genes encoding a suspected 3-dehydroquinate synthase (DHQS) and an *O*-methyltransferase (OMT), proposed to be involved in MAA biosynthesis, while those that did not produce MAAs lacked this cluster. The authors also reported another DHQS homologue in *A. variabilis* PCC 7937 which they suspected to be involved in shikimate biosynthesis but not MAA assembly. Mining of further sequenced genomes for the homologues of the genes encoding the proposed DHQS (YP_324358) and the OMT (YP_324357) revealed that when they were present, they were often on the same genomic locus and that many cyanobacteria have the genetic potential to assemble MAAs.[117]

Work by Balskus and Walsh expanded upon these findings.[118] Previous feeding studies in cyanobacteria did not provide any direct evidence for the assembly of the mycosporine core from the shikimate pathway intermediate 3-dehydroquinate which had been proposed based on structural features. Identification of biosynthesis genes encoding a DHQS along with an adjacent OMT in the genome sequencing project of a sea anemone (*Nematostella vectensis*) provided the candidate genes for a BLAST-based genome data search, targeting conserved domains of DHQS and OMT protein sequences. Not only were the anticipated ORFs revealed but also a third conserved ORF annotated as a hypothetical protein. Examination of these conserved domains revealed homologies to proteins involved in peptide bond formation and thus the authors suspected a link to amino acid incorporation. Comparison of relevant biosynthesis gene clusters in other cyanobacteria and fungi also indicated differences consistent with the variable natural product diversity associated with fungi (mycosporines) and cyanobacteria (MAAs). With the proposed genetic basis for the UV-blocking MAA biosynthesis in cyanobacteria determined as a result of genome mining, heterologous expression in *E. coli* of the putative 6.5 kb shinorine gene cluster from *Anabaena variabilis* ATCC 29413, along with two truncated versions, confirmed its role in shinorine assembly and, ultimately, allowed its biosynthetic pathway to be proposed.[118] The study revealed an unexpected finding with respect to the biosynthesis pathway.[118] *In vitro* expression of gene products of the DHQS and OMT homologues from the sequenced *Nostoc punctiforme* ATCC 29133 revealed that shinorine was actually derived from sedoheptulose 7-phosphate

42

(SH 7-P), the substrate for 2-epi-5-epi-valiolone synthase (EVS) and not 3-dehydroquinate, the product of DHQS as proposed earlier. The result contradicted the assumption that MMA synthesis proceeds via a shikimate pathway. The final steps of shinorine assembly were elucidated *in vitro* using purified enzymes from *A. variabilis* and isotope-labelled precursors. These experiments not only revealed two distinct and unprecedented strategies for ATP-dependent enzymatic imine formation but also resulted in the total enzymatic biosynthesis of the molecule.[118]

These few examples display some of the innovative techniques that have been used to solve natural product puzzles and serve to demonstrate that a combined chemical and biological approach can be quite effective. Without this dual discipline approach it is probable that some of these findings may never have been realised.

9.3.3   Mining Cyanobacterial Genomes for TMS Genes

In a genome mining study by Wright and co-workers, cyanobacterial strains were analysed for their biosynthetic potential using PCR by amplifying NRPS and PKS genes with degenerate primers. Of the 24 test strains, both NRPS and PKS genes were amplified from 17, while only three showed negative results for both PKS and NRPS genes.[119]

A bioinformatics-based mining study by Donadio and co-workers examined 223 bacterial genome sequences deposited before 2005, including eight cyanobacteria. Again, the focus was to identify TMS genes encoding PKSs and/or NRPSs.[90] TMS genes were detected in only two of the cyanobacterial strains, *Anabaena* sp. PCC 7120 and *Gloeobacter violaceus* PCC 7421, constituting ~1.5 and 1% of their respective genomes. In comparison, the same study revealed that TMS genes constituted ~3.9 and 1.5% of the 9.0 Mb *S. avermitilis* MA-4680 and 8.7 Mb *S. coelicolor* A3(2) genomes, respectively.[90] These prolific producers of secondary metabolites dedicate ~6.6 (*S. avermitilis*) and 8.0% (*S. coelicolor* A3(2)) of their genomes to secondary metabolism.[95,96] The marine obligate actinobacteria *Salinispora tropica* dedicates ~9% of its 5.2 Mb genome to natural product assembly, with at least 6% dedicated to gene clusters (not only TMS genes) encoding the biosynthesis of type I PKS, NRPS, or hybrid PKS/NRPS derived products, known and unknown.[94] Sequencing and analysis of the related *Salinispora arenicola* genome revealed an even greater proportion (~11%) of its 5.8 Mb genome was dedicated to secondary metabolite biosynthesis.[120] It is notable that the biosynthesis of secondary metabolites in cyanobacteria is generally limited to those strains whose genomes are larger than 4 Mb and thus the production of such metabolites can be considered a physiological luxury. To date, the majority of published cyanobacterial genome sequences are *Synechococcus* and *Prochlorococcus* spp., all of which possess small genomes (<4 Mb), and appear to lack PKS/NRPS biosynthesis genes.[90] This observation is also primarily true of most bacterial taxa.

9.3.4   Mining Cyanobacterial Genomes for Orphan Gene Clusters Using Domain Teams

As of December 2010, over 1500 microbial genome projects were completed or were close to completion, including 64 cyanobacteria (Figure 9.1). Up until August 2008, the time of our genome mining study, over 1000 microbial genomes had been completely sequenced, 34 of which were cyanobacteria.[11] Many of these cyanobacterial genomes had never been investigated for their biosynthetic potential. We surveyed these 34 completed cyanobacterial genomes using a purely bioinformatics-based approach employing what could be considered 'selective mining'.[11] We specifically targeted NRPS and PKS (TMS) genes in the search for orphan biosynthesis gene clusters – those whose product is not known. In doing so we were able to assess the biosynthetic potential of sequenced cyanobacteria and make comparisons with other groups of microorganisms. This study, along with some others reported prior to this, revealed that, like many other sequenced microorganisms, cyanobacteria possess the genetic potential to biosynthesise many more natural products than those that are currently known. Our specific approach using purpose-built software uncovered a greater number of TMS genes than were previously reported from these very same organisms. These genome mining results supported the generally held notion that larger genomes give rise to more biosynthesis gene clusters.

9.3.4.1   The Domain Teams Approach

Rather than searching for genes and gene clusters, our approach involved identifying replicated genomic features across the 34 sequenced genomes in our dataset. We specifically targeted common catalytic protein domains essential

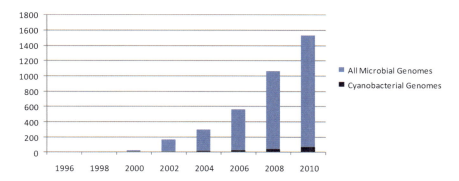

Figure 9.1   Sequenced cyanobacterial genomes as a proportion of total microbial genomes sequenced for the period 1996–2010.

to NRPSs and PKSs using a method which extended the work of Pasek and co-workers.[121,122] The technique was designed to target protein domains such as those found in the Protein Family database (Pfam),[123] and allows for the detection of strings of domains (termed 'domain teams') that are conserved in their content but not necessarily their order. In designating domain teams, we essentially deconstructed NRPS and PKS genes into the catalytic domains of the proteins they code for. So rather than searching for genes *per se*, we mined sequenced genomes for conserved protein domains, as domain teams, in order to detect natural product biosynthesis gene clusters. As seen from the gene clusters presented earlier, the highly ordered and repetitive nature of such NRPS and PKS systems makes them ideal candidates for such analyses.

Pfam (version 22.0) is a collection of protein families (9318 in total) and domains including those involved in natural product assembly, and covers 73% of sequences found in UniProtKB version 9.7.[123] In generating our data, Pfam identifiers PF00109/PF02801 (N- and C-terminal domains of β-ketoacyl synthase) were used to identify ketosynthase modules while NRPS modules were detected using the PF00501 (AMP-binding enzyme), PF00608 (condensation domain), and PF00550 (phosphopantetheine attachment site). The final search dataset comprised 59 chromosomes (genomes and genomic plasmids) and a total of 154 279 domains. A summary of our findings can be found elsewhere.[11] A subsection of our Domain Teams analysis is shown in Table 9.2. It is important to note that the data included in the table are the

Table 9.2 Numbers of domain teams identified in selected cyanobacterial strains.

Organism	Genome size[b]	PKS	NRPS	Hybrid
Acaryochloris marina MBIC11017	8.3	3	2	1
Anabaena variabilis ATCC 29413	7.1	5	1	4
Gloeobacter violaceus PCC 7421	4.7	4	3	2
Microcystis aeruginosa NIES-843	5.8	4	3	3
Nostoc punctiforme PCC 73102	9.1	7	6	8
Nostoc sp. PCC 7120 [a]	7.2	6	1	5
Prochlorococcus marinus AS9601	1.7	2	1	1
Synechococcus sp. CC9311	2.6	3	1	2
Thermosynechococcus elongatus BP-1	2.6	3	1	1
Trichodesmium erythraeum IMS101	7.8	3	2	1

[a] Also known as *Anabaena* sp. PCC 7120; [b] size in Mbp and plasmid inclusive.

raw, de-replicated, output of our genome mining analysis and refers to domain teams not gene clusters.

In analysing the data generated, the logical first step is to determine whether the output can be truly considered a *bona fide* biosynthesis cluster, a task simplified by the highly interactive graphical output of the 'Domain Teams' software. An example is shown in Figure 9.2. Another consideration is one which still causes some consternation, and that is determining gene cluster boundaries and linker regions between genes. While our Domain Teams-based approach doesn't allow us to unequivocally determine these, automated bioinformatics tools such as UMA (Udwary-Merski algorithm)[124] and CLUSEAN (CLUster SEquence ANalyzer)[125] can also be used for determining the structure of thio-templated systems.

9.3.4.2 The Domain Teams Approach: Mining Outcomes

Genome mining using Domain Teams revealed a rich array of potential non-ribosomal peptide and polyketide biosynthesis gene clusters in cyanobacteria. We determined that TMS genes account for at least 127 kb (or 2.2%) of the *Microcystis aeruginosa* NIES-843 genome, 119 kb (1.7%) of the *Nostoc* sp. PCC 7120 genome, 56 kb (1.2%) of the *Gloeobacter violaceus* PCC 7421 genome, 348 kb (3.8%) of the *Nostoc punctiforme* PCC 73102 genome and 160 kb (2.3%) of the *Anabaena variabilis* ATCC 29413 genome. While these figures are lower than those reported for other families of organisms such as the actinobacteria, they are higher than what was anticipated at the time of the study. The following section highlights some of our findings and compares and contrasts genome mining studies predating our own. A complete overview has been published elsewhere.[11]

9.3.4.2.1 The Nostoc sp. PCC 7120 Genome

The Domain Teams analysis supported Donadio's findings in *Nostoc* sp. PCC 7120.[90] Of note is a large gene cluster encoding two PKSs and ten NRPSs adjacent to a gene encoding a putative ABC transporter. ABC transporters represent one of the largest, most highly conserved protein superfamilies in existence and these proteins, found in bacteria, eukaryotes and archaea, are responsible for the ATP-dependent transport of allocrites across intracellular and cell surface membranes.[126–128] Few of the bacterial secondary metabolite transporters have been functionally characterised; however, they have been shown to confer self-resistance to the producing organisms.[129,130] Zhang and co-workers have proposed that this gene cluster is involved in the biosynthesis of siderophores; however, a product was not identified as part of their study.[131]

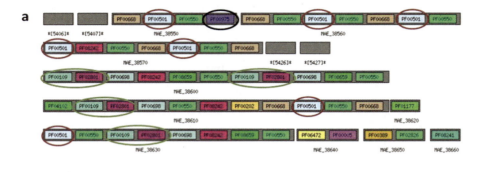

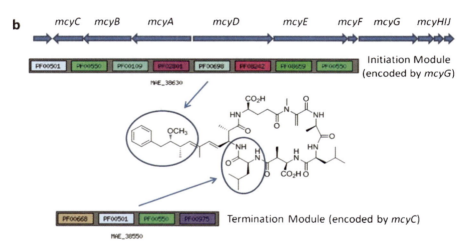

Figure 9.2   An example of Domain Teams output of the *mcy* cluster in *M. aeruginosa* NIES-843. Circled in Figure 9.2a are seven adenylation domains (Pfam identifier PF00501), four ketosynthases (PF00109/PF02801) and a thioesterase (PF00975) indicative of the *mcy* cluster. Figure 9.2b shows the genetic arrangement of the *mcy* cluster and highlights the Domain Teams output corresponding to the McyG (initiation module) and McyC (termination module). *McyG* codes for functional domains involved in priming (PF00501), polyketide extension (PF00109/PF02801) and tailoring reactions (PF08242 methyltransferase/PF08659 ketoreduction). *McyC* codes for functional domains involved the adenylation of leucine (PF00501) and the cyclisation and release (PF00975) of the product from the PKS/NRPS complex. Other PFAM identifiers correspond to condensation (PF00668), phosphopantetheine attachment site, i.e. ACPs/PCPs, (PF00550/PF00698) and acyl transferase (PF00698) domain families.

9.3.4.2.2 The Nostoc punctiforme PCC 73102 Genome

A bioinformatics based overview of the ∼9 Mb *Nostoc punctiforme* PCC 73102 (also known as ATCC 29133) revealed 62 ORFs encoding proteins involved in TMS-derived secondary metabolite biosynthesis.[132] Two clusters, consisting of 12 and 14 biosynthesis genes spanning 47 and 49 kb, respectively, were identified along with a host of smaller gene sets. A subsequent BLAST search based, genomic survey of *N. punctiforme* ATCC 29133 revealed 17 genes encoding NRPSs (possessing 42 A-domains) and 10 genes encoding PKSs (possessing 22 KS-domains).[133] These data suggest that *N. punctiforme*'s biosynthetic potential is greater than any other sequenced cyanobacterium and supports the notion that organisms with larger genomes encode a greater number of secondary metabolite pathways. Our analyses revealed that seven major biosynthesis clusters, including the characterised nostopeptolide cluster, when combined encode 45 A-domains and 24 KS-domains. One orphan NRPS/PKS hybrid cluster spanning at least 57 kb includes one gene 16 kb in length encoding three NRPSs and one PKS.[72] Of the seven major clusters, six are hybrids, and the other encodes an NRPS with six adenylation domains.

9.3.4.2.3 The Anabaena variabilis ATCC 29413 Genome

The 7.1 Mb *Anabaena variabilis* ATCC 29413 genome codes for several small to mid-sized hybrid NRPS/PKSs including a 36 kb cluster encoding seven adenylation domains and a single PKS which we determined to be the largest cluster from this organism. The remainder of this strain's 130 kb of TMS sequence is accounted for by a 27 kb locus coding for four ketosynthase domains and two adenylation domains, a small PKS/NRPS (16 kb) containing four adenylation domains and one ketosynthase, a trimodular NRPS (13 kb), a small PKS (16 kb) and two 11 kb fragments in close proximity to each other each encoding two NRPSs.[90]

9.3.4.2.4 The Microcystis aeruginosa NIES-843 Genome

The original report and annotation of the *Microcystis aeruginosa* NIES-843 genome revealed, as expected, gene clusters for the biosynthesis of the microcystins (*mcyA–J*) and the cyanopeptolins (A **43**) (*mcnA–C* and *mcnE–G*), as well as one other orphan NRPS spanning 17 kb.[134] Using the Domain Teams approach we also uncovered an additional hybrid NRPS/PKS spanning at least 17 kb over five ORFs.

9.3.5 Mining *Planktothrix rubescens* NIVA CYA 98 for Biosynthesis Gene Clusters

A genome wide analysis of *Planktothrix rubescens* NIVA CYA 98, while not technically a mining study, revealed an impressive array of NRPSs, including two novel gene clusters.[135] This strain's genome codes for clusters involved in the biosynthesis of the microcystins (PKS/NRPS), cyanopeptolins, anabaeno-peptins, aeruginosins and the ACE (angiotensin-converting enzyme) inhibitor microginin (PKS/NRPS) analogue oscillaginin B **44**. NRPS encoding gene clusters accounted for 4.1% (226 kb) of the 5.5 Mb *Planktothrix* genome, which makes it potentially one of the most prolific cyanobacterial NRP producers.[135] Also detected were clusters coding for the microviridin and oscillatorin families of ribosomal peptides. Assembly of the oscillatorins is not completely understood and the cluster has not been fully characterised. Analysis of culture extracts by mass spectrometry allowed the identification of known compounds produced by this strain, and these were in turn matched to their respective biosynthesis gene clusters.[135] The two novel gene clusters mentioned above remain orphans.

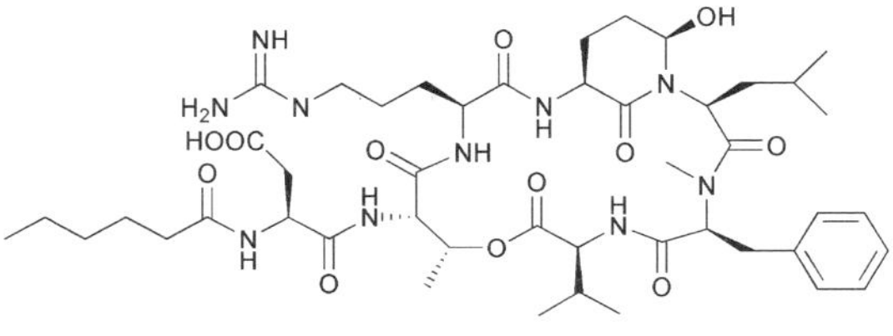

43

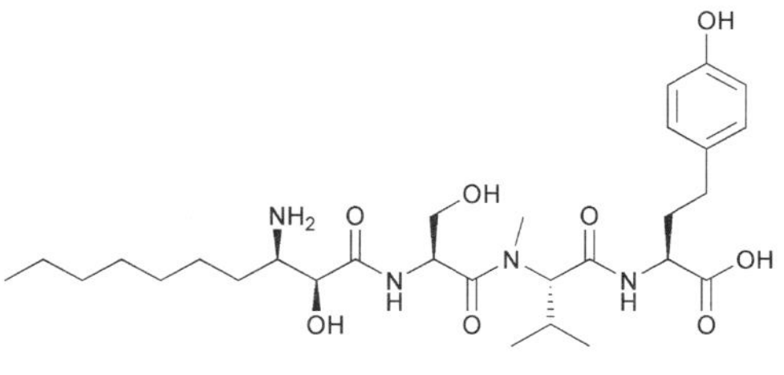

44

9.3.6   Mining Cyanobacterial Genomes for Lantibiotic Biosynthesis Gene Clusters

Lantibiotics are lanthionine-containing ribosomally synthesised peptides which have antibiotic activity.[136] They possess intramolecular thioether cross-linkages formed between a cysteine and either a dehydrated serine or threonine residue. Like other ribosomal peptides, the lantibiotics result from a leader peptide encoded by genes generally termed *lanA*s. Translated LanA peptides undergo several biosynthetic processes including dehydration and cyclisation to yield the final product. These final processes are catalysed by a bifunctional synthetase broadly termed LanM. In nature, LanM enzymes generally modify a single LanA peptide; however, under laboratory conditions they exhibit low substrate specificity and thus the potential to generate a variety of lantibiotics from different peptide precursors. A genome mining study of three sequenced cyanobacteria, two strains of *Prochlorococcus* (MIT 9313 and MIT 9303) and *Synechococcus* RS9916, for genes involved in lantibiotic biosynthesis, revealed 29, 15 and 10 putative *lanA*s, respectively. Further examination of the *Prochlorococcus* MIT 9313 genome revealed that a single *lanM* homologue (*procM*) was clustered with seven (of 29) *lanA* (*procA*) homologues while the remaining 22 *procA* genes were found elsewhere on the genome.[136]

Combinatorial *in vitro* assays with purified ProcM and 17 linear peptides derived from the expression of 17 *procA* homologues (ProcAs) resulted in the production of 17 unique lantibiotics.[136] Based on the high sequence conservation of leader peptides, the authors concluded it is likely that all 29 ProcA peptides are substrates for ProcM. Given the extreme diversity of *Prochlorococcus* spp. in the oceans the inherent structural diversity of this family of molecules, termed the prochlorosins (Pcn), is potentially immense. In terms of genome mining, the search for new molecules from those cyanobacteria with smaller genomes probably needs to be heightened. *Prochlorococcus* MIT 9313 has the genetic capacity to biosynthesise at least as many secondary metabolites as other strains considered to be prolific producers, and in doing so it dedicates a much smaller proportion of its genome for that task.[136]

9.3.7   Sequenced Cyanobacterial Genomes: Future Targets of Opportunity

A recent survey of the NCBI Microbial Genomes webpage revealed that in the period since our 2008 genome mining study up until the end of 2010, a further 9 cyanobacterial genomes had been completely sequenced and assembled. An additional 21 organisms whose genomes had been sequenced at that time were at the draft assembly stage (Figure 9.3). While strains from the orders *Prochlorales* and *Chroococales* continue to dominate these sequencing projects, there is a marked increase in the relative number of *Nostocales* and *Oscillatoriales* strains being sequenced. Cyanobacteria from the *Nostocales* and *Oscillatoriales* orders are rich sources of bioactive natural products and thus in terms of mining, these genomes represent future targets of opportunity for the discovery of new chemical entities and drug leads.

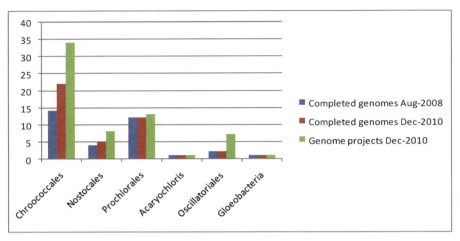

Figure 9.3 Cyanobacterial genomes completed and in draft status as of December 2010. Genomes are grouped according to their taxonomic order. Data sets reflect relative increase in cyanobacterial genome mining projects between August 2008 and December 2010.

9.4 Concluding Remarks

Genome mining of microorganisms for novel natural products, a field of endeavour still in its infancy, has already reaped substantial rewards. Studies continue to reveal further natural products from microbes whose secondary metabolite complement had previously been examined. With new molecules come new biosynthesis pathways, novel enzymes, a greater understanding of microbial natural product assembly at the genetic level and new drug candidates. Natural product screening programs for drug discovery rely on structurally diverse molecules, and chemodiversity is underpinned by a rich biodiversity. Microorganisms are inherently diverse and continue to provide drug candidates; one group of microbes underexplored or often ignored as a source of drug-like molecules are the cyanobacteria. Genome mining of cyanobacteria has revealed that they are a richer than anticipated source of new molecules, particularly those organisms with larger genomes. In general cyanobacterial natural products display varied bioactivities which are either extremely harmful (cyanotoxins) or beneficial to humans, and in some cases compounds from both classes are produced by the same organism. Cyanobacteria are renowned for their ability to assemble macrocyclic natural products, a class of compounds shown to be excellent drug candidates – many of which have advanced through phase clinical trials. So why aren't cyanobacteria providing more drug candidates? Perhaps difficulties associated with culturing and accessing secondary metabolites as compared with other microbes makes them less attractive subjects and thus fewer research groups are pursuing their natural products. Or could it be that initial biological screening of cyanobacterial extracts is focused upon cytotoxicity at the expense of other treatment areas because of their noted roles as toxin producers in the environment? Based on genome mining we would argue that cyanobacteria are a good source of NRPS/PKS products and hence a good source of drug candidates in the form of macrocyclic compounds.

With DNA sequencing technologies rapidly advancing it is not unreasonable to expect the costs associated with sequencing a microorganism will be negligible within the next 5 years. With the continued development of bioinformatics tools to assemble and interrogate genomes, it is feasible that in the near future, sequencing and genome mining for biosynthesis clusters could routinely become the first step in isolating a novel natural product. Our intimate knowledge of biosynthesis gene clusters is such that genome mining also provides an efficient means of de-replicating natural product isolations, further emphasising its utility for those dedicated to natural products discovery. These are exciting times for those of us working at the interface of chemistry and biology.

Acknowledgements

B.A.N. is a Federation Fellow of the Australian Research Council. J.A.K. is supported by the UNSW Environmental Microbiology Initiative through a Post-Doctoral Research Fellowship.

References

1. F. E. Koehn and G. T. Carter, *Nat. Rev. Drug Disc.*, 2005, **4**, 206.
2. J. W. H. Li and J. C. Vederas, *Science*, 2009, **325**, 161.
3. A. L. Harvey, *Curr. Opin. Chem. Biol.*, 2007, **11**, 480.
4. D. J. Newman and G. M. Cragg, *J. Nat. Prod.*, 2007, **70**, 461.
5. G. M. Cragg and D. J. Newman, *Pure Appl. Chem.*, 2005, **77**, 1923.
6. M. A. Fischbach and C. T. Walsh, *Science*, 2009, **325**, 1089.
7. D. J. Newman and G. M. Cragg, *Curr. Opin. Invest. Drug*, 2009, **10**, 1280.
8. X. Y. Liu, E. Ashforth, B. A. Ren, F. H. Song, H. Q. Dai, M. Liu, J. A. Wang, Q. O. Xie and L. X. Zhang, *J. Antibiot.*, 2010, **63**, 415.
9. C. Corre and G. L. Challis, *Nat. Prod. Rep.*, 2009, **26**, 977.
10. H. Gross, *Curr. Opin. Drug Disc.*, 2009, **12**, 207.
11. J. A. Kalaitzis, F. M. Lauro and B. A. Neilan, *Nat. Prod. Rep.*, 2009, **26**, 1447.
12. C. T. Walsh and M. A. Fischbach, *J. Am. Chem. Soc.*, 2010, **132**, 2469.
13. C. E. Salomon, N. A. Magarvey and D. H. Sherman, *Nat. Prod. Rep.*, 2004, **21**, 105.
14. M. A. Fischbach, *Curr. Opin. Microbiol.*, 2009, **12**, 520.
15. A. L. Demain and S. Sanchez, *J. Antibiot.*, 2009, **62**, 5.

16. R. M. Van Wagoner, A. K. Drummond and J. L. C. Wright, *Adv. Appl. Microbiol.*, 2007, **61**, 89.
17. D. Tillett, E. Dittmann, M. Erhard, H. von Dohren, T. Borner and B. A. Neilan, *Chem. Biol.*, 2000, **7**, 753.
18. M. C. Moffitt and B. A. Neilan, *Appl. Environ. Microbiol.*, 2004, **70**, 6353.
19. A. C. Jones, E. A. Monroe, E. B. Eisman, L. Gerwick, D. H. Sherman and W. H. Gerwick, *Nat. Prod. Rep.*, 2010, **27**, 1048.
20. E. S. Sattely, M. A. Fischbach and C. T. Walsh, *Nat. Prod. Rep.*, 2008, **25**, 757.
21. A. Koglin and C. T. Walsh, *Nat. Prod. Rep.*, 2009, **26**, 987.
22. J. Staunton and K. J. Weissman, *Nat. Prod. Rep.*, 2001, **18**, 380.
23. A. L. Lane and B. S. Moore, *Nat. Prod. Rep.*, 2011, **28**, 411.
24. A. A. Roberts, K. S. Ryan, B. S. Moore and T. A. M. Gulder, *Top. Curr. Chem.*, 2010, **297**, 149.
25. R. McDaniel, S. Ebert-Khosla, D. A. Hopwood and C. Khosla, *Science*, 1993, **262**, 1546.
26. J. A. Kalaitzis, M. Izumikawa, L. K. Xiang, C. Hertweck and B. S. Moore, *J. Am. Chem. Soc.*, 2003, **125**, 9290.
27. J. A. Kalaitzis, Q. Cheng, P. M. Thomas, N. L. Kelleher and B. S. Moore, *J. Nat. Prod.*, 2009, **72**, 469.
28. Q. Cheng, L. Xiang, M. Izumikawa, D. Meluzzi and B. S. Moore, *Nat. Chem. Biol.*, 2007, **3**, 557.
29. F. Kopp, J. Grunewald, C. Mahlert and M. A. Marahiel, *Biochemistry*, 2006, **45**, 10474.
30. B. Ostash, X. Yan, V. Fedorenko and A. Bechthold, *Top. Curr. Chem.*, 2010, **297**, 105.
31. G. A. Codd, S. G. Bell, K. Kaya, C. J. Ward, K. A. Beattie and J. S. Metcalf, *Eur. J. Phycol.*, 1999, **34**, 405.
32. R. Castenholz and J. B. Waterbury, in *Bergey's Manual of Systematic Bacteriology*, ed. J. T. Staley, Williams & Wilkins, Sydney, 1989, vol. 3, pp. 1710.
33. H. W. Paerl, J. L. Pinckney and T. F. Steppe, *Environ. Microbiol.*, 2000, **2**, 11.
34. D. G. Nagle and V. J. Paul, *J. Phycol.*, 1999, **35**, 1412.
35. A. J. Vanderwesthuizen and J. N. Eloff, *Z. Pflanzenphysiol.*, 1983, **110**, 157.
36. M. Lukac and R. Aegerter, *Toxicon*, 1993, **31**, 293.
37. F. J. Marner, R. E. Moore, K. Hirotsu and J. Clardy, *J. Org. Chem.*, 1977, **42**, 2815.
38. J. S. Mynderse, R. E. Moore, M. Kashiwagi and T. R. Norton, *Science*, 1977, **196**, 538.
39. K. Harada, *Chem. Pharm. Bull.*, 2004, **52**, 889.
40. W. H. Gerwick, R. C. Coates, N. Engene, L. Gerwick, R. V. Grindberg, A. C. Jones and C. M. Sorrels, *Microbe*, 2008, **3**, 277.

41. K. Gademann and C. Portmann, *Curr. Org. Chem.*, 2008, **12**, 326.
42. M. S. Butler, *Nat. Prod. Rep.*, 2008, **25**, 475.
43. T. F. Molinski, D. S. Dalisay, S. L. Lievens and J. P. Saludes, *Nat. Rev. Drug Disc.*, 2009, **8**, 69.
44. E. M. Driggers, S. P. Hale, J. Lee and N. K. Terrett, *Nat. Rev. Drug Disc.*, 2008, **7**, 608.
45. M. Q. Zhang and B. Wilkinson, *Curr. Opin. Biotech.*, 2007, **18**, 478.
46. A. M. Burja, B. Banaigs, E. Abou-Mansour, J. G. Burgess and P. C. Wright, *Tetrahedron*, 2001, **57**, 9347.
47. L. T. Tan, *Phytochemistry*, 2007, **68**, 954.
48. M. A. Ali, R. B. Bates, Z. D. Crane, C. W. Dicus, M. R. Gramme, E. Hamel, J. Marcischak, D. S. Martinez, K. J. McClure, P. Nakkiew, G. R. Pettit, C. C. Stessman, B. A. Sufi and G. V. Yarick, *Bioorg. Med. Chem.*, 2005, **13**, 4138.
49. T. L. Simmons, L. M. Nogle, J. Media, F. A. Valeriote, S. L. Mooberry and W. H. Gerwick, *J. Nat. Prod.*, 2009, **72**, 1011.
50. T. L. Simmons and W. H. Gerwick, in *Oceans and Human Health*ed. P. J. Walsh, S. L. Smith, L. E. Fleming, H. M. Solo-Gabriele and W. H. Gerwick, Elsevier, San Diego, 2008, pp. 431.
51. J. Kobayashi, J.-F. Cheng, T. Ohta, H. Nakamura, S. Nozoe, Y. Hirata, Y. Ohizumi and T. Sasaki, *J. Org. Chem.*, 1988, **53**, 6147.
52. Y. Kikuchi, M. Ishibashi, T. Sasaki and J. Kobayashi, *Tetrahedron Lett.*, 1991, **32**, 797.
53. K. Nozawa, M. Tsuda, H. Ishiyama, T. Sasaki, T. Tsuruo and R. Kobayashi, *Bioorg. Med. Chem.*, 2006, **14**, 1063.
54. A. Furstner, C. Nevado, M. Waser, M. Tremblay, C. Chevrier, F. Teply, C. Aissa, E. Moulin and O. Muller, *J. Am. Chem. Soc.*, 2007, **129**, 9150.
55. K. Taori, V. J. Paul and H. Luesch, *J. Am. Chem. Soc.*, 2008, **130**, 1806.
56. Y. C. Ying, K. Taori, H. Kim, J. Y. Hong and H. Luesch, *J. Am. Chem. Soc.*, 2008, **130**, 8455.
57. Y. X. Liu, L. A. Salvador, S. Byeon, Y. C. Ying, J. C. Kwan, B. K. Law, J. Y. Hong and H. Luesch, *J. Pharmacol. Exp. Ther.*, 2010, **335**, 351.
58. P. G. Williams, W. Y. Yoshida, M. K. Quon, R. E. Moore and V. J. Paul, *J. Nat. Prod.*, 2003, **66**, 1545.
59. B. Zou, K. Long and D. W. Ma, *Org. Lett.*, 2005, **7**, 4237.
60. A. C. Jones, L. C. Gu, C. M. Sorrels, D. H. Sherman and W. H. Gerwick, *Curr. Opin. Chem. Biol.*, 2009, **13**, 216.
61. L. Rouhiainen, L. Paulin, S. Suomalainen, H. Hyytiainen, W. Buikema, R. Haselkorn and K. Sivonen, *Mol. Microbiol.*, 2000, **37**, 156.
62. M. S. Donia, J. Ravel and E. W. Schmidt, *Nat. Chem. Biol.*, 2008, **4**, 341.
63. S. Picossi, A. Valladares, E. Flores and A. Herrero, *J. Biol. Chem.*, 2004, **279**, 11582.
64. K. Kerksiek, M. R. Mejillano, R. E. Schwartz, G. I. George and R. H. Himes, *FEBS Lett.*, 1995, **377**, 59.

65. N. Biondi, R. Piccardi, M. C. Margheri, L. Rodolfi, G. D. Smith and M. R. Tredici, *Appl. Environ. Microbiol.*, 2004, **70**, 3313.
66. G. Trimurtulu, I. Ohtani, G. M. L. Patterson, R. E. Moore, T. H. Corbett, F. A. Valeriote and L. Demchik, *J. Am. Chem. Soc.*, 1994, **116**, 4729.
67. C. D. Smith, X. Q. Zhang, S. L. Mooberry, G. M. L. Patterson and R. E. Moore, *Cancer Res.*, 1994, **54**, 3779.
68. N. A. Magarvey, Z. Q. Beck, T. Golakoti, Y. S. Ding, U. Huber, T. K. Hemscheidt, D. Abelson, R. E. Moore and D. H. Sherman, *ACS Chem. Biol.*, 2006, **1**, 766.
69. T. Nishizawa, M. Asayama, K. Fujii, K. Harada and M. Shirai, *J. Biochem.*, 1999, **126**, 520.
70. T. Nishizawa, A. Ueda, M. Asayama, K. Fujii, K. Harada, K. Ochi and M. Shirai, *J. Biochem.*, 2000, **127**, 779.
71. L. M. Hicks, M. C. Moffitt, L. L. Beer, B. S. Moore and N. L. Kelleher, *ACS Chem. Biol.*, 2006, **1**, 93.
72. D. Hoffmann, J. M. Hevel, R. E. Moore and B. S. Moore, *Gene*, 2003, **311**, 171.
73. H. Luesch, D. Hoffmann, J. M. Hevel, J. E. Becker, T. Golakoti and R. E. Moore, *J. Org. Chem.*, 2003, **68**, 83.
74. J. E. Becker, R. E. Moore and B. S. Moore, *Gene*, 2004, **325**, 35.
75. F. Kopp, C. Mahlert, J. Grunewald and M. A. Marahiel, *J. Am. Chem. Soc.*, 2006, **128**, 16478.
76. F. Kopp and M. A. Marahiel, *Nat. Prod. Rep.*, 2007, **24**, 735.
77. A. V. Ramaswamy, C. M. Sorrels and W. H. Gerwick, *J. Nat. Prod.*, 2007, **70**, 1977.
78. Z. X. Chang, P. Flatt, W. H. Gerwick, V. A. Nguyen, C. L. Willis and D. H. Sherman, *Gene*, 2002, **296**, 235.
79. N. Sitachitta, B. L. Marquez, R. T. Williamson, J. Rossi, M. A. Roberts, W. H. Gerwick, V. A. Nguyen and C. L. Willis, *Tetrahedron*, 2000, **56**, 9103.
80. C. T. Calderone, S. B. Bumpus, N. L. Kelleher, C. T. Walsh and N. A. Magarvey, *Proc. Natl. Acad. Sci. USA*, 2008, **105**, 12809.
81. R. Kellmann, T. K. Michali and B. A. Neilan, *J. Mol. Evol.*, 2008, **67**, 526.
82. T. K. Mihali, R. Kellmann, J. Muenchhoff, K. D. Barrow and B. A. Neilan, *Appl. Env. Microbiol.*, 2008, **74**, 716.
83. A. Mejean, S. Mann, T. Maldiney, G. Vassiliadis, O. Lequin and O. Ploux, *J. Am. Chem. Soc.*, 2009, **131**, 7512.
84. Z. X. Chang, N. Sitachitta, J. V. Rossi, M. A. Roberts, P. M. Flatt, J. Y. Jia, D. H. Sherman and W. H. Gerwick, *J. Nat. Prod.*, 2004, **67**, 1356.
85. D. J. Edwards and W. H. Gerwick, *J. Am. Chem. Soc.*, 2004, **126**, 11432.
86. A. V. Blokhin, H. D. Yoo, R. S. Geralds, D. G. Nagle, W. H. Gerwick and E. Hamel, *Mol. Pharmacol.*, 1995, **48**, 523.

87. P. Verdier-Pinard, J. Y. Lai, H. D. Yoo, J. R. Yu, B. Marquez, D. G. Nagle, M. Nambu, J. D. White, J. R. Falck, W. H. Gerwick, B. W. Day and E. Hamel, *Mol. Pharmacol.*, 1998, **53**, 62.
88. D. J. Edwards, B. L. Marquez, L. M. Nogle, K. McPhail, D. E. Goeger, M. A. Roberts and W. H. Gerwick, *Chem. Biol.*, 2004, **11**, 817.
89. R. E. Moore, *J. Ind. Microbiol.*, 1996, **16**, 134.
90. S. Donadio, P. Monciardini and M. Sosio, *Nat. Prod. Rep.*, 2007, **24**, 1073.
91. D. H. Sherman, M. J. Bibb, T. J. Simpson, D. Johnson, F. Malpartida, M. Fernandez Moreno, E. Martinez, C. R. Hutchinson and D. A. Hopwood, *Tetrahedron*, 1991, **47**, 6029.
92. Z. H. Hu, K. Bao, X. F. Zhou, Q. Zhou, D. A. Hopwood, T. Kieser and Z. X. Deng, *Mol. Microbiol.*, 1994, **14**, 163.
93. M. Zerikly and G. L. Challis, *ChemBioChem*, 2009, **10**, 625.
94. D. W. Udwary, L. Zeigler, R. N. Asolkar, V. Singan, A. Lapidus, W. Fenical, P. R. Jensen and B. S. Moore, *Proc. Natl. Acad. Sci. USA*, 2007, **104**, 10376.
95. S. D. Bentley, K. F. Chater, A. M. Cerdeno-Tarraga, G. L. Challis, N. R. Thomson, K. D. James, D. E. Harris, M. A. Quail, H. Kieser, D. Harper, A. Bateman, S. Brown, G. Chandra, C. W. Chen, M. Collins, A. Cronin, A. Fraser, A. Goble, J. Hidalgo, T. Hornsby, S. Howarth, C. H. Huang, T. Kieser, L. Larke, L. Murphy, K. Oliver, S. O'Neil, E. Rabbinowitsch, M. A. Rajandream, K. Rutherford, S. Rutter, K. Seeger, D. Saunders, S. Sharp, R. Squares, S. Squares, K. Taylor, T. Warren, A. Wietzorrek, J. Woodward, B. G. Barrell, J. Parkhill and D. A. Hopwood, *Nature*, 2002, **417**, 14.
96. S. Omura, H. Ikeda, J. Ishikawa, A. Hanamoto, C. Takahashi, M. Shinose, Y. Takahashi, H. Horikawa, H. Nakazawa, T. Osonoe, H. Kikuchi, T. Shiba, Y. Sakaki and M. Hattori, *Proc. Natl. Acad. Sci. USA*, 2001, **98**, 12215.
97. H. Gross, V. O. Stockwell, M. D. Henkels, B. Nowak-Thompson, J. E. Loper and W. H. Gerwick, *Chem. Biol.*, 2007, **14**, 53.
98. J. M. Winter, S. Behnken and C. Hertweck, *Curr. Opin. Chem. Biol.*, 2011, **15**, 22.
99. Y. M. Chiang, S. L. Chang, B. R. Oakley and C. C. Wang, *Curr. Opin. Chem. Biol.*, 2011, **15**, 137.
100. S. Bergmann, A. N. Funk, K. Scherlach, V. Schroeckh, E. Shelest, U. Horn, C. Hertweck and A. A. Brakhage, *Appl. Environ. Microbiol.*, 2010, **76**, 8143.
101. S. Lautru, R. J. Deeth, L. M. Bailey and G. L. Challis, *Nat. Chem. Biol.*, 2005, **1**, 265.
102. T. Stachelhaus, H. D. Mootz and M. A. Marahiel, *Chem. Biol.*, 1999, **6**, 493.
103. G. L. Challis, J. Ravel and C. A. Townsend, *Chem. Biol.*, 2000, **7**, 211.

104. T. J. McQuade, A. D. Shallop, A. Sheoran, J. E. DelProposto, O. V. Tsodikov and S. Garneau-Tsodikova, *Anal. Biochem.*, 2009, **386**, 244.

105. P. F. Long, W. C. Dunlap, C. N. Battershill and M. Jaspars, *ChemBioChem*, 2005, **6**, 1760.

106. E. W. Schmidt, J. T. Nelson, D. A. Rasko, S. Sudek, J. A. Eisen, M. G. Haygood and J. Ravel, *Proc. Natl. Acad. Sci. USA*, 2005, **102**, 7315.

107. E. W. Schmidt, S. Sudek and M. G. Haygood, *J. Nat. Prod.*, 2004, **67**, 1341.

108. S. Sudek, M. G. Haygood, D. T. A. Youssef and E. W. Schmidt, *Appl. Environ. Microbiol.*, 2006, **72**, 4382.

109. N. Ziemert, K. Ishida, A. Liaimer, C. Hertweck and E. Dittmann, *Angew. Chem., Int. Ed.*, 2008, **47**, 7756.

110. N. Ziemert, K. Ishida, P. Quillardet, C. Bouchier, C. Hertweck, N. T. de Marsac and E. Dittmann, *Appl. Environ. Microbiol.*, 2008, **74**, 1791.

111. J. A. McIntosh, M. S. Donia and E. W. Schmidt, *Nat. Prod. Rep.*, 2009, **26**, 537.

112. P. J. Proteau, W. H. Gerwick, F. Garcia-Pichel and R. Castenholz, *Experientia*, 1993, **49**, 825.

113. T. Soule, V. Stout, W. D. Swingley, J. C. Meeks and F. Garcia-Pichel, *J. Bacteriol.*, 2007, **189**, 4465.

114. T. Soule, F. Garcia-Pichel and V. Stout, *J. Bacteriol.*, 2009, **191**, 4639.

115. E. P. Balskus and C. T. Walsh, *J. Am. Chem. Soc.*, 2008, **130**, 15260.

116. C. M. Sorrels, P. J. Proteau and W. H. Gerwick, *Appl. Environ. Microbiol.*, 2009, **75**, 4861.

117. S. P. Singh, M. Klisch, R. P. Sinha, D.-P. Häder, *Genomics*, 2010, **95**, 120.

118. E. P. Balskus and C. T. Walsh, *Science*, 2010, **329**, 1653.

119. M. E. Barrios-Llerena, A. M. Burja and P. C. Wright, *J. Ind. Microbiol. Biotech.*, 2007, **34**, 443.

120. K. Penn, C. Jenkins, M. Nett, D. W. Udwary, E. A. Gontang, R. P. McGlinchey, B. Foster, A. Lapidus, S. Podell, E. E. Allen, B. S. Moore and P. R. Jensen, *ISME J.*, 2009, **3**, 1193.

121. S. Pasek, A. Bergeron, J. L. Risler, A. Louis, E. Ollivier and M. Raffinot, *Genome Res.*, 2005, **15**, 867.

122. S. Pasek, in *Methods in Molecular Biology*, ed. N. H. Bergman, Humana Press, Totowa, 2007, vol. 396, pp. 17.

123. R. D. Finn, J. Tate, J. Mistry, P. C. Coggill, S. J. Sammut, H. R. Hotz, G. Ceric, K. Forslund, S. R. Eddy, E. L. L. Sonnhammer and A. Bateman, *Nucleic Acids Res.*, 2008, **36**, 281.

124. D. W. Udwary, M. Merski and C. A. Townsend, *J. Mol. Biol.*, 2002, **323**, 585.

125. T. Weber, C. Rausch, P. Lopez, I. Hoof, V. Gaykova, D. H. Huson and W. Wohlleben, *J. Biotech.*, 2009, **140**, 13.

126. W. Saurin, M. Hofnung and E. Dassa, *J. Mol. Evol.*, 1999, **48**, 22.

127. R. J. Jovell, A. J. L. Macario and E. C. deMacario, *Gene*, 1996, **174**, 281.

128. P. M. Jones and A. M. George, *FEMS Microbiol. Lett.*, 1999, **179**, 187.

129. D. M. Gardiner, R. S. Jarvis and B. J. Howlett, *Fungal Gen. Biol.*, 2005, **42**, 257.
130. R. H. Proctor, D. W. Brown, R. D. Plattner and A. E. Desjardins, *Fungal Gen. Biol.*, 2003, **38**, 237.
131. R. Jeanjean, E. Talla, A. Latifi, M. Havaux, A. Janicki and C. C. Zhang, *Environ. Microbiol.*, 2008, **10**, 2574.
132. J. C. Meeks, J. Elhai, T. Thiel, M. Potts, F. Larimer, J. Lamerdin, P. Predki and R. Atlas, *Photosynth. Res.*, 2001, **70**, 85.
133. I. M. Ehrenreich, J. B. Waterbury and E. A. Webb, *Appl. Environ. Microbiol.*, 2005, **71**, 7401.
134. T. Kaneko, N. Nakajima, S. Okamoto, I. Suzuki, Y. Tanabe, M. Tamaoki, Y. Nakamura, F. Kasai, A. Watanabe, K. Kawashima, Y. Kishida, A. Ono, Y. Shimizu, C. Takahashi, C. Minami, T. Fujishiro, M. Kohara, M. Katoh, N. Nakazaki, S. Nakayama, M. Yamada, S. Tabata and M. M. Watanabe, *DNA Res.*, 2007, **14**, 247.
135. T. B. Rounge, T. Rohrlack, A. J. Nederbragt, T. Kristensen and K. S. Jakobsen, *BMC Genomics*, 2009, **10**, 396.
136. B. Li, D. Sher, L. Kelly, Y. Shi, K. Huang, P. J. Knerr, I. Joewono, D. Rusch, S. W.Chisholm and W. A. Van Der Donk, *Proc. Natl. Acad. Sci. USA*, 2010, **107**, 10430.

CHAPTER 10

Epigenetic Approaches to Natural Product Synthesis in Fungi

ALEXANDRA A. SOUKUP[a] AND NANCY P. KELLER*[b,c]

[a] Department of Genetics, University of Wisconsin – Madison, 3455 Microbial Sciences, 1550 Linden Drive Madison, WI 53706, USA; [b] Department of Medical Microbiology and Immunology, University of Wisconsin – Madison, 3455 Microbial Sciences, 1550 Linden Drive Madison, WI 53706, USA; [c] Department of Bacteriology, University of Wisconsin – Madison, 3455 Microbial Sciences, 1550 Linden Drive Madison, WI 53706, USA
*E-mail: npkeller@wisc.edu

10.1 Introduction to Chromatin Structure

The basic unit of DNA compaction in eukaryotes is the nucleosome. This structure consists of approximately 147 bp of DNA wrapped around a heterooctamer of histone particles. Within each nucleosome there are two molecules each of histones H2A, H2B, H3, and H4. Each of the core histone particles consists of a globular domain known as the histone fold, as well as unstructured N- and C-terminal tails. Further condensation of nucleosome particles may then be mediated by addition of linker histones, such as H1. However, linker histones are not required in several fungal genera, including *Ascobolus*, *Aspergillus*, *Neurospora*, and *Saccharomyces*.[1]

Chromatin structure can be modified through a number of different mechanisms. Nucleosome displacement can be mediated through ATP-dependent chromatin remodeling, such as that mediated by the SWI/SNF complex in yeast.[2] Histones can be post-translationally modified by addition and removal of small molecules and peptides.[3] These modifications are most

RSC Drug Discovery Series No. 25
Drug Discovery from Natural Products
Edited by Olga Genilloud and Francisca Vicente

commonly added to the N-terminal tail, and can include acetylation,[4] methylation,[5] phosphorylation,[6] ubiquitination,[7] ADP-ribosylation,[8] SUMOlation,[9] glycosylation,[10] carbonylation,[11] propionylation, butyrylation,[12] and biotinylation.[13] The combinations of modifications have been termed the histone code, with different combinations stimulating different responses.[14] In addition to decoration of the histone proteins, DNA itself can be subjected to cytosine methylation.[15] The combinations of different histone and DNA modifications result in different states of chromatin compaction and transcription levels, with open structures at high transcriptional levels being known as euchromatin, and more highly condensed, less transcriptionally active areas known as heterochromatin.[16]

The body of research regarding chromatin regulation of secondary metabolism in filamentous fungi has focused on histone modifications by acetylation and methylation, as well as DNA methylation, and these modifications will be discussed in further detail below.[17] Many of these remodeling events can occur in concert to activate gene expression, although the exact mechanistic details are not known at this point. As discussed later in this chapter, these events can be utilized in order to increase expression of genes required for secondary metabolite biosynthesis (Figure 10.1).

10.1.1 Acetylation

Histone acetylation is a reversible process mediated by histone acetyltransferases (HATs) and histone deacetylases (HDACs). Although the common generalization is that increased histone acetylation results in gene activation, histone deacetylation can also be responsible for increased activation of a gene.[18] This was shown to be true in *S. cerevisiae*, where deletion of the HDAC RPD3 leads to increased acetylation and impaired induction of genes required for responding to osmotic stress, such as *HSP12* and *CTT1*.[19] There are a number of different classes of HDACs and HATs shown to be conserved in fungi, each differing slightly in their target specificity.[20] These proteins are similar in identity to mammalian orthologs, although recently it has been

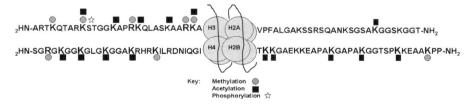

Figure 10.1 Known histone modifications in the filamentous fungi. This figure depicts the histone modifications identified in the filamentous fungus *Neurospora crassa*.[17] Modifications displayed here are not unique to the filamentous fungi, but are conserved throughout eukaryotes. Based on this conservation, many other modifications can be predicted to occur based on identified decorations on histone tails of other organisms, but they have not yet been identified in these fungi.

found that the RPD3-type HDAC contains a novel motif required for catalytic activity in filamentous fungi.[21] While there is not an optimal method of increasing histone acetylation by increasing HAT activity, there are a number of nonspecific HDAC inhibitors commonly used, including trichostatin A (TSA) and suberoylanilide hydroxamic acid (SAHA), which indirectly result in increased acetylation.[22]

10.1.2   Methylation

Numerous residues of histones are subject to reversible methylation, including many arginines and lysines. The two best-studied residues include histone 3 lysine 4 (H3K4) and histone 3 lysine 9 (H3K9). These modifications can vary in their effects. In mammalian cells, H3K4 trimethylation is a common mark of transcriptionally active genes, while enhancer elements are typically monomethylated on this residue.[23] In contrast, budding yeast also employs this mark, but can use it to both activate or silence transcription as observed at the mating type loci and sub-telomeric regions.[24,25] H3K9 is seen as a mark of transcriptionally repressed genes, while this same residue may be acetylated in transcriptionally active genes.[26] Filamentous fungi contain putative orthologs to those identified in other eukaryotes, with several studies focused on H3K9 and arginine methyltransferases.[27,28]

10.1.3   Phosphorylation

Phosphorylation of histones was originally studied in the context of cell division, where an increase in phosphorylated histones, especially the linker histone H1, signals entry into mitosis.[29] However, phosphorylation can occur independently of mitosis in order to direct gene expression at promoters throughout the genome. Phosphorylation of histone 3 serine 10 has been shown to block H3K9 methylation and prevent its spread through removal of heterochromatin protein 1 (HP1) homologs. Phosphorylation of this residue also activates gene expression by stimulating acetylation of H3K9 and H3K14.[30] The *Aspergillus nidulans* H3S10 kinase NimA has served as a model for studies of orthologs in mammalian systems.[31,32]

10.1.4   DNA Methylation

In *Neurospora*, histone methylation of H3K9 acts as a signal to initiate DNA methylation.[33] This has also been shown to be the case for *Ascobolus immersus*.[34] In these and other organisms where methylation of cytosine residues occurs, the result is often altered initiation and elongation of transcription. This methylation of DNA is associated with a genome defense process in both *Neurospora* and *Ascobolus*.[35] Treatment of these species with the methyltransferase inhibitor 5-azacytidine (5-AC) can abrogate this silencing effect.[36] However, a number of fungi, including the Aspergilli –

despite containing putative DNA methyltransferases – have barely detectable levels of DNA methylation that do not significantly affect chromatin structure or development.[37]

10.2   Secondary Metabolism of Filamentous Fungi

10.2.1   Clustering

Early experiments designed at elucidating the genetic pathways of secondary metabolite (SM) production discovered that genes required for the biosynthesis of a particular secondary metabolite were often located adjacent to one another in a cluster. These clusters typically contain all necessary genes required for the biosynthesis of a given product. This usually includes a backbone gene required for synthesizing the body of the metabolite and additional enzymes capable of decorating the backbone with alternative or modified groups. Major classes of biosynthetic enzymes include the polyketide synthases (PKS), non-ribosomal peptide synthases (NRPS), prenyltransferases (DMATSs), and terpene cyclases. In addition to the backbone gene, modifying enzymes are present that can include ketoreductases, monooxygenases, *O*-methyltransferases, esterases, and acyl transferases. Also commonly located within these clusters are cluster-specific transcription factors and genes used for self-protection against toxicity of the compound. Several recent reviews on this topic are available for the reader to peruse.[38,39]

Prior to 2005, examples of known clusters included those responsible for producing aflatoxins,[40,41] cephalosporin,[42] compactin,[43,44] ergot alkaloids,[45] fumonisin,[46] gibberellins,[47] HC toxin,[48] lovastatin,[49] melanin,[50,51] paxillin,[52] penicillin,[53] sterigmatocystin,[54] sirodesmin,[55] and several tricothecenes.[56,57] With the development of advance sequencing technologies and increased genomic knowledge, new SM gene clusters are being discovered at an astonishing rate. Algorithms have recently been developed to search entire genomes for these clusters, predicting approximately 800 SM clusters within only 27 publicly published genomes.[58] Proteome analyses are also currently being used in attempts to identify clusters.[59] The conserved domain structures of backbone genes have also been used extensively to predict new clusters.

It has been predicted that more than 70% of secondary metabolite gene clusters are not expressed under standard laboratory conditions.[60] In fact, once bioinformatics approaches were used on the published *Neurospora crassa* genome, many putative SM clusters were found, including seven PKS, four NRPS, and several terpene biosynthetic gene clusters; this is in a fungus not found to produce secondary metabolites.[61] The fact that a well-studied laboratory model could harbor so many silent clusters emphasizes the need for stimulating activity of these cryptic clusters and determining their mechanisms of regulation (Figure 10.2).

There is evidence that some clusters, such as that required for penicillin biosynthesis, originated by horizontal transfer from prokaryotes.[62] However,

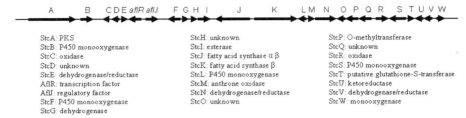

StcA: PKS
StcB: P450 monooxygenase
StcC: oxidase
StcD: unknown
StcE: dehydrogenase/reductase
AflR: transcription factor
AflJ: regulatory factor
StcF: P450 monooxygenase
StcG: dehydrogenase

StcH: unknown
StcI: esterase
StcJ: fatty acid synthase α β
StcK: fatty acid synthase β
StcL: P450 monooxygenase
StcM: anthrone oxidase
StcN: dehydrogenase/reductase
StcO: unknown

StcP: O-methyltransferase
StcQ: unknown
StcR: oxidase
StcS: P450 monooxygenase
StcT: putative glutathione-S-transferase
StcU: ketoreductase
StcV: dehydrogenase/reductase
StcW: monooxygenase

Figure 10.2   The sterigmatocystin cluster in *A. nidulans*. This figure depicts the sterigmatocystin cluster from *Aspergillus nidulans*. This cluster contains typical marks of a fungal SM cluster, including the backbone PKS (StcA), a cluster-specific transcription factor (AflR), and numerous modifying enzymes.[62]

many metabolites produced by fungi have no precedent in the bacterial kingdom. Another possibility of SM cluster evolution is that evolution led to the grouping of the enzymes. Examination of several *Fusarium* species provided evidence for loss, non-functionalization, and rearrangement of genes required for tricothecene biosynthesis.[63] Horizontal transfer among different fungi may also play a role in diversification, or propagation, of clusters. *A. clavatus* contains a six-gene *ACE1* cluster that appears to have originated from a relative of *Magnaporthe grisea*, although the *M. grisea* cluster currently contains 15 genes.[64] *Podospora anserina* contains a cluster near identical to the *A. nidulans* sterigmatocystin cluster.[65] These mechanisms all likely contribute to the diversity of SMs that may be produced.

10.2.2   Regulation

A number of different factors may contribute to the coordinated expression or repression of these gene clusters. It has previously been shown that environmental cues can have a large effect on pathway expression. These effects are often mediated through global response transcription factors. For example, in *Aspergillus nidulans*, SM clusters are known to contain binding sites for transcription factors in response to changes in carbon source (*creA*),[66] nitrogen source (*areA*),[67] and pH (*pacC*).[68] *Fusarium fujikuroi* also contains a well-characterized SM response to oxidative stress through the TOR pathway.[69] In addition to these global transcription factors, many SM clusters contain a cluster-specific transcription factor that more specifically directs expression of the cluster.

The SM-specific transcription factors are subject to regulation with the remainder of the cluster, but may also be manipulated to either increase or decrease expression of the genes in their native cluster. These pathway-specific transcription factors are required for synthesis of many SMs, including: gliotoxin (*gliZ*),[70] tricothecene (*tri6*),[71] aflatoxin/sterigmatocystin (*aflR*),[40,72] citrinin (*ctnA*),[73] aspyridone (*apdR*),[74,75] asperfuranone (*afoA*),[76] and ustilagic acid (*rua1*).[77] Characterization of the sterigmatocystin cluster transcription factor *aflR* led to the identification of a sequence-specific binding motif present in several copies within the promoter region of most sterigmatocystin cluster genes.[78]

Removal of these cluster-specific transcription factors typically results in decreased expression of the entire cluster.[79] Additionally, moving the transcription factor out of its native chromatin environment to an ectopic location in the genome can either stimulate or repress other genes in the cluster. For example, ectopic expression of the *A. nidulans* gene *aflR*, a transcription factor for the sterigmatocystin cluster, was able to restore expression of the repressed genes in the cluster caused by deletion of the global activator *laeA*.[79] This suggests that the location of genes within the cluster is important for their proper regulation. Supporting this theory, decreased expression upon ectopic insertion was also shown for other members of SM cluster genes, such as the *A. parasticus* ketoreductase *nor-1*.[80] Similarly, expression of *A. parasticus* aflatoxin cluster biosynthetic gene, *ver-1*, in its native location resulted in approximately 500-fold increased expression relative to an ectopically localized copy.[81]

The proximity of genes within an SM cluster led to the theory of co-regulation by a physical mechanism. In accordance with this, numerous papers have begun to link chromatin state to activation and repression of SM clusters. An early study tracked levels of histone 4 acetylation and transcript levels during activation of the aflatoxin cluster in *A. parasiticus*. The authors of this study showed a temporal link between levels of H4 acetylation and gene activation. The genes first to show transcription were also the first to become acetylated, and the same pattern was true for both middle and late genes.[82] Since acetylation patterns follow those predicted from studies of other systems, it was a logical extension to examine orthologs of other chromatin modifiers.

Increasing acetylation levels by deletion of HDACs in *A. nidulans* has been shown to increase levels of known metabolites. Deletion of the deacetylase *hdaA* led to greatly increased levels of the SMs penicillin and sterigmatocystin.[83] Similarly, removal of ClrD and HepA in this organism, enzymes proposed to act in stabilizing heterochromatin formation through spreading of H3K9 trimethylation, led to increases in both penicillin and terraquinone production.[84] Both studies demonstrated that increased metabolite production was via increased transcription of the biosynthetic genes which, in the latter study, was shown to be associated with loss of heterochromatin marks.

Altering other modifications also results in differing SM profiles. Decreasing H3K4 methylation by deletion of the COMPASS complex component *cclA* led to decreases in sterigmatocystin production, while orsellinic acid and monodictyphenone derivatives were increased in production.[85] Results from this study implicated a role of H3K4 crosstalk in regulating H3K9 methylation, as decreased levels of methylated lysine 4 in the *cclA* mutant led to a decrease in the heterochromatic H3K9me mark.[85] This suggests that the increase in some metabolites in the *cclA* mutant may be due to inhibited heterochromatin formation at these loci. Prevention of SUMOlation by deletion of the *sumO* gene also led to altered profiles, including increased levels of aspterthecin and decreased levels of austinol, dehydroaustinol, and sterigmatocystin.[86] Like H3K4me3, SUMOlation was previously reported to be a repressive mark in yeast, preventing the binding of activating marks such as acetylation.[87] Again, as in the H3 modification studies in the prior paragraph, increased SM production was associated in increased transcription of cluster genes in both *cclA* and *sumO* works.

10.2.3   Boundary Elements

Most SM clusters appear to have clearly defined borders which limit the extent of gene activation or repression. As indicated earlier, several studies examining the sterigmatocystin cluster have found that activation of the cluster does not result in increased transcription of those genes directly either proximal or distal to the cluster.[83,84,88] Even more convincingly, these genes fail to exhibit histone modifications and localization of chromatin modifiers typical of those genes in the cluster.[84] Transcriptional profiling of mutants of the global SM regulator *laeA* also indicated clear boundaries of the terraquinone cluster.[89]

It has recently been shown that transposon relics may play a role in modulating these border elements. Deleting portions of repetitive DNA surrounding the penicillin cluster in *A. nidulans* resulted in decreased gene expression and deregulation of the cluster.[90] This was true even in a sensitized genetic background lacking a histone deacetylase, where the genes were constitutively overexpressed. This suggests that the facultative heterochromatin consisting of repetitive elements surrounding gene clusters may serve as a guidance system for recruitment of machinery required for proper activation and transcription. Examination of genomes of several other fungi show many SM clusters are either bordered or embedded with transposons, transposon relics, and/or repeated DNA.[91,92]

10.3   Stimulation of Secondary Metabolites

Using *A. niger* as an experimental model, it has been predicted that over 70% of this organism's gene clusters are not expressed under standard laboratory conditions.[60] This led to the development of new techniques targeted at stimulating gene expression, with the hope of identifying novel compounds. Techniques fall into a number of different categories, exploiting environmental cues, genetic modifications, and chemical treatments. In some cases, separate approaches through several of these categories led to the identification of the same previously unknown metabolites. This emphasizes the epigenetic levels of control in activating these clusters, with different upstream signals and events leading to gene cluster expression (Table 10.1).

Table 10.1 Metabolites identified by cluster stimulation

Compound	Species	Method	Figure
Aspoquinolones A–D	*A. nidulans*	Culture conditions[96]	3A
Cytosporones F–I	*Paraphaeosphaeria quadriseptata*	Culture conditions[95]	
Quadriseptin A	*Paraphaeosphaeria quadriseptata*	Culture conditions[95]	3B
5′-Hydroxymonocillin	*Paraphaeosphaeria quadriseptata*	Culture conditions[95]	
Emericellimides A & B	*Emericella* sp.	Co-culture[99]	
Pestalone	*Pestalotia* sp.	Co-culture[101]	
F-9775 A and B	*A. nidulans*	Co-culture[97] Reverse Genetics[85] Culture conditions[98]	10.3C
Aspyridone A & B	*A. nidulans*	TF[74]	
Asperfuranone	*A. nidulans*	TF (afoA)[76] TF (scpR)[75]	10.4A
Tuberculariols A–C	*Tubercularia* sp.	Genome shuffling[105]	10.4B
7-Chloroscytalone	*Tubercularia* sp.	Genome shuffling[105]	
7-Chloro-4-hydroxymellein	*Tubercularia* sp.	Genome shuffling[105]	10.4C
Nygerone A	*A. niger*	SAHA[125]	10.5A
Lunalides A & B	*Diatrype* sp.	5-AC[121]	
Sclerotonin	*P. citreonigrum*	5-AC[129]	
Ochrephilone	*P. citreonigrum*	5-AC[129]	10.5B
Dechloroisochromophilone III & IV	*P. citreonigrum*	5-AC[129]	
6-((3*E*,5*E*)-5,7-dimethyl-2-methylenenona-3,5-dienyl)-2,4-dihydroxy-3-methylbenzaldehyde	*P. citreonigrum*	5-AC[129]	10.5C
Atlantinones A and B	*P. citreonigrum*	5-AC[129]	
(9*Z*,12*Z*-)11-hydroxyoctadeca-9,12-dienoic acid	*Cladosporium cladosporoides*	5-AC[121]	

In this table is described the compounds identified by the stimulation of fungal SM gene clusters and the type of method used for stimulation. The wide range of compounds produced and method of stimulation used emphasizes the variety of fungal SMs.

10.3.1 Environmental Cues

10.3.1.1 Culture Conditions

One approach to activate new fungal gene clusters is to vary growth conditions, including media composition, temperature, time, and other parameters. This approach, coined OSMAC ('one strain many compounds'), utilizes the natural fungal response to varied environmental conditions to stimulate gene expression of compounds that would not otherwise be produced.[93,94] This takes advantage of global regulatory elements responding to changes in pH, carbon source, nitrogen source, or environmental factors.[39] Variations of this approach have successfully been used to identify several novel compounds. Interestingly, the simple shift from tap water to purified water stimulated the production and identification of cytosporones F-I, quadriseptin A (Figure 10.3A), and 5′-hydroxymonocillin from *Paraphaeosphaeria quadriseptata* and *Chaetomium chiversii*.[95] A more varied set of fermentation parameters were used with *A. nidulans* in order to identify aspoquinolones A–D[96] (Figure 10.3B).

10.3.1.2   Co-culture

Similar to the OSMAC approach, it is predicted that exposing a given strain to interactions with another organism can increase metabolite expression as a likely protective mechanism. In an attempt to simulate fungal/bacterial cohabitation of soil environments, co-culture of *A. nidulans* with the actinomycete *Streptomyces hygroscopicus* yielded compounds previously not known to be produced by *A. nidulans*, including the orsellinic acid derivatives F-9775A and B[97] (Figure 10.3C). These same compounds were found to be expressed in a strain defective in H3K4 methylation,[85] and later detected in pure culture of a wild-type strain through alterations in culture conditions.[98] The identification of these compounds through three different approaches confirms the idea that regulation of SM gene cluster expression can be mediated by alterations in chromatin structure, with different upstream effectors signaling activation.

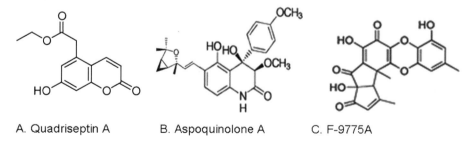

A. Quadriseptin A   B. Aspoquinolone A   C. F-9775A

Figure 10.3   Compounds isolated after varying culture conditions. This figure depicts three compounds isolated after altering culture conditions. (A) Quadriseptin A[95] was isolated from *Paraphaeosphaeria quadriseptata* after altering water sources for the culture. (B) Aspoquinolone A[96] was isolated from *A. nidulans* after varying a series of culture conditions. (C) F-9775A[85,97,98] was isolated from *A. nidulans* in three different approaches, including altered culture conditions, co-culture with *Streptomyces hygroscopicus*, and deletion of the H3K4 methyltransferase component *cclA*.

Co-culture has successfully been used in probing several marine fungi for novel metabolites. In one example, a marine derived *Emericella* sp. and the bacterium *Salinispora arenicola* were cultured together, yielding the identification of emericellimides A and B.[99] The entire gene cluster was subsequently identified and characterized in *A. nidulans*.[100] Earlier analysis of cultures of a *Pestalotia* sp. and an unknown marine bacterium produced the novel compound pestalone.[101]

In a more general approach, 10 different strains of Aspergilli were subjected to co-culture under various conditions. Although not deeply investigated, the authors noted that co-culture led to altered metabolite profiles for all strains examined.[102] This suggests that competition is likely to be an effective activator of SM clusters, even among species within a single genus.

10.3.2 Genetic Modifications

10.3.2.1 Forward Genetics

Traditional genetic tools have proven valuable in large-scale screens of mutants, although the screening process can often be burdensome. One study, which examined mutagenized strains for loss of norsolorinic acid (NOR) production, examined 100 000 colonies in order to identify 83 mutants defective in production of this metabolite.[103] Although most of the implicated genes outside of the sterigmatocystin cluster were not identified, complementation tests identified one as the global regulator *laeA*, discussed in further detail below.[104]

In another large-scale screen, a *Tubercularia* sp. was subjected to protoplast mutation by UV light and chemical treatment, followed by genome shuffling by protoplast fusion to rearrange the context of the modifications.[105] Although the goal of the authors was to isolate strains with restored taxol production, they isolated two strains with greatly altered SM profiles, including 18 novel compounds from only one of these strains (Figures 10.4B and 10.4C).

10.3.2.2 Reverse Genetics

More targeted approaches deal with deletion or overexpression of specific genes in order to activate metabolites. This can be done by targeting specific gene clusters, or by modulating global effectors.

Commonly targeted global regulators include genes responsible for histone modifications and chromatin structure. As explained above, shifting the balance of the cell towards a more euchromatic state commonly activates many SM clusters. This has been shown to occur when the histone deacetylase *hdaA* is deleted in *A. nidulans*.[83] Alternatively, mutations in genes that produce modifications of a less clearly defined role can lead to both gene activation and repression. Deletion of *cclA*, a component of the COMPASS complex responsible for H3K4 methylation in *A. nidulans*, leads to decreased

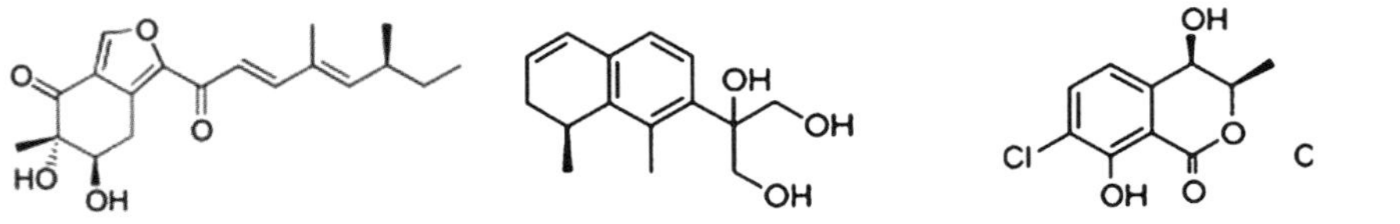

A. asperfuranone
B. tuberculariol A
C. 7-chloro-4-hydroxymellein

Figure 10.4   A selection of metabolites identified by genetic modifications. This figure shows a selection of compounds isolated after employing genetic approaches. (A) Asperfuranone[76,75] was isolated from *A. nidulans* by increasing levels of either the cluster-specific transcription factor *afoA*, or the nonspecific transcription factor *scpR*. (B) Tuberculariol A[105] was isolated from a *Tubercularia* species after protoplast mutation and genome shuffling. (C) 7-Chloro-4-hydroxymellein[105] was also was isolated from a *Tubercularia* species after protoplast mutation and genome shuffling.

production of sterigmatocystin, but increased production of a number of compounds previously not thought to be produced by this fungus, such as orsellinic acid and monodictyphenone derivatives, including F-9775 A and B.[85] Deletion of the *sumO* gene in this fungus, which encodes the peptide responsible for sumolation, resulted in increased production of the sexual spore pigment asperthecin.[86]

Although less clear in its mechanism of action, the global regulator *laeA* may also be manipulated to increase SM production. The protein encoded by this gene has no homology to those of known function, although it contains a putative *S*-adenosylmethionine-binding domain typical of methyltransferases. Homologs of *laeA* have been found in the Aspergilli,[104] *Penicillium citrum*,[106] *Fusarium fujkuroi*,[107] and *Penicillium chrysogenum*,[108] and in all cases have been found to positively regulate secondary metabolism. Deletion of this gene severely decreases metabolite production from many different clusters, and its overexpression stimulates SM production.[104,109]

Examination of *laeA* null and overexpression mutants led to the identification of several *laeA*-regulated gene clusters in many species.[88,92,107,110] Functional analysis of the terraquinone cluster later confirmed the boundaries of this cluster based on the *laeA* mutant data set.[89] Examination of the sterigmatocystin cluster in *A. nidulans* confirmed the specificity of this cluster-specific activation.[84]

Further experiments examining the pathway through which LaeA acts have supported its role in altering the chromatin landscape of metabolite clusters. In *A. nidulans*, deletion of either *clrD* or *hepA*, proposed to aid in H3K9 spreading, partially restored SM production to a *laeA* null mutant.[84] Similarly, deletion of the HDAC *hdaA* also partially complemented the SM defect of this mutant.[83] Together, these results indicate that *laeA* aids in opening chromatin structure to increase gene expression, although its exact mechanism is still unknown at this time. Recently, a novel bZip protein, RsmA, has been found to partially remediate *laeA* loss through transcriptional up-regulation of secondary metabolite cluster genes in *A. nidulans*.[111] The identification of another bZip protein, MeaB, involved in regulation of SMs in *Fusarium fujikuroi* suggests that this class of transcription factors may also be involved in global regulation of SM clusters.[112]

As alluded to above, specific clusters may also be induced through targeted overexpression of their components. This is most commonly done through up-regulation of the pathway-specific transcription factor, and can be applied to either known clusters in order to facilitate increased expression, or to currently uncharacterized clusters in hopes of identifying the metabolite produced. This approach has successfully been used to identify aspyridones A and B by introduction of ectopic copies of the regulator *apdR*.[74] Additionally, the compound asperfuranone was identified after induction of the locus-specific transcription factor *afoA*, as well as a regulator for another cluster (*scpR*) which undergoes cross-talk with the asperfuranone cluster[76,75] (Figure 10.4A).

10.3.3 Chemical Epigenetics

Two classes of inhibitors that can lead to increased gene expression of SM clusters are commonly used, the HDAC inhibitors, such as TSA or synthetic derivatives thereof, or DNA methyltransferase inhibitors such as 5-AC. Although stimulating activities of the HATs may also seem a logical path for increasing cluster-specific gene expression, this approach is less technically feasible, as activities of this class of enzymes have already been optimized through evolution.[22]

Even before they were known as inhibitors of histone and DNA methyltransferases, agents such as trichostatin A (TSA) and 5-azacytidine (5-AC) were initially shown to have a role in fungal development and other processes in several different species of fungi, including *C. albicans*, *Aspergillus* spp., *F. oxysporum*, and *N. crassa*.[36,113–120] Their application in development applies to metazoans as well, as many HDAC and DNMT inhibitors were discovered in a screen for anticancer agents, and have since progressed to clinical testing.[20,22]

Combinations of nine different HDAC and DNMT inhibitors against twelve different fungi, including *A. flavus*, *A. westerdijkiae*, *Cladosporium cladosporoides*, *Clonostachys* sp., *Diatrype* sp., *Penicillium chrysogenum*, *Penicillium citrinum*, *Rhizopus* sp., *Verticillium psallotae*, and three unknown filamentous fungi resulted in altered metabolite profiles under at least one condition.[121] The interplay among these compounds may help to dissect the pathways coordinating expression of SM clusters. For example, in *Neurospora crassa*, treatment with the HDAC inhibitor TSA mimics treatment with the DNMT inhibitor 5-AC, in that decreased DNA methylation and increased gene expression occur.[122]

10.3.3.1 HDAC Inhibitors

Commonly used HDAC inhibitors including trichostatin A (TSA) and the synthetically derived suberoylanilide hydroxamic acid (SAHA). These compounds nonselectively inhibit traditional histone deacetylases through coordination of the zinc ion in the active site of these enzymes.[123,124]

The use of HDAC inhibitors has proved valuable in both examining the transcriptional landscape of SM clusters and identifying novel compounds. Treatment of *Alternaria alternata* and *Penicillium expansum* with TSA led to increased production of several unidentified secondary metabolites.[83] As above, treatment of *A. niger* with SAHA led to identifying increased transcription of 14 of the 33 known clusters. Interestingly, SAHA treatment resulted in an even larger bias for increased transcription of telomeric gene clusters, with 13 of the 14 up-regulated clusters being located within 1.5 kb of the chromosome termini.[60] Further investigation of treated *A. niger* cultures led to the identification of nygerone A[125] (Figure 10.5A). Treatment of *A. fumigatus* with TSA resulted in the up-regulation of at least one PKS and one NRPS.[126]

10.3.3.2   5-Azacytidine

The compounds 5-azacytidine (5-AC) and 5-aza-2′-deoxycytidine have commonly been referred to as DNA methyltransferase inhibitors. However, these compounds have also been shown to have activities against histone methyltransferases.[127,128] Action of this drug in activation gene expression may therefore occur through a combination of inhibiting silencing DNA and histone methylation. This may be especially effective in activating clusters located in specific regions of the genomes, such as telomeres.

In a study of the effects of 5-AC treatment against *A. niger*, Fisch *et al.* found that treatment with this reagent resulted in 9 of the 33 selected SM clusters having increased transcription as detected by qPCR. Interestingly, there was enrichment for strongly regulated clusters lying within 1.5 kB of the

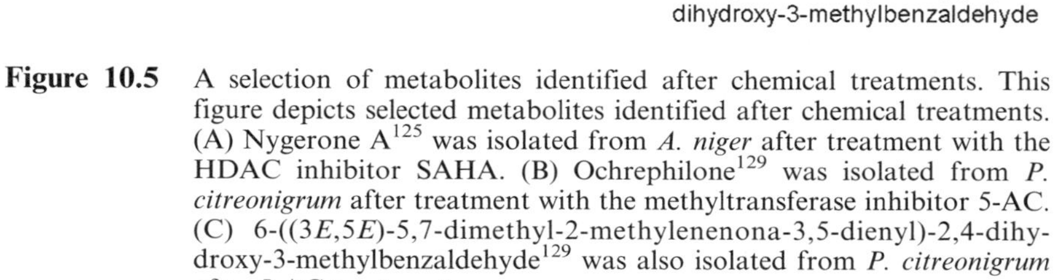

A. nygerone A   B. ochrephilone   C. 6-((3E,5E)-5,7-dimethyl-2-
                                                  methylenenona-3,5-dienyl)-2,4
                                                  dihydroxy-3-methylbenzaldehyde

Figure 10.5   A selection of metabolites identified after chemical treatments. This figure depicts selected metabolites identified after chemical treatments. (A) Nygerone A[125] was isolated from *A. niger* after treatment with the HDAC inhibitor SAHA. (B) Ochrephilone[129] was isolated from *P. citreonigrum* after treatment with the methyltransferase inhibitor 5-AC. (C) 6-((3E,5E)-5,7-dimethyl-2-methylenenona-3,5-dienyl)-2,4-dihydroxy-3-methylbenzaldehyde[129] was also isolated from *P. citreonigrum* after 5-AC treatment.

telomeres, as 7 or the 9 up-regulated clusters were located within these regions.[60]

In attempts to isolate novel compounds from 5-AC treated cultures, analysis of treated cultures of *Cladosporium cladosporoides* by NMR and MS identified novel production of several oxylipins, including (9Z,12Z-)11-hydroxyocta-deca-9,12-dienoic acid, its methyl ester, and a glycerol conjugate.[121] Similarly, treatment of *Penicillium citreonigrum* yielded six azaphilones, pencolide, and two new meroterpenes[129] (Figures 10.5A and 10.5B). *Diatrype* species treated with 5-AC resulted in the identification of lunalides A and B.[121] Interestingly, treatment of *A. fumigatus* with 5-AC resulted in high rates of conversion to a developmental variant which showed altered transcriptional responses and increased production of aspergillopepsin F[130] (Table 10.1).

10.4 Conclusions

The large number of clusters predicted by genome mining far outnumbers the known metabolites being produced, emphasizing the potential bounty of metabolites that may be produced. In order to address this, many methods are currently being used to stimulate gene clusters in fungi in order to identify new compounds from biosynthetic gene clusters. Many of these methods are still in their infancy, although their use has been confirmed by the identification of new compounds, as summarized in this review. Novel techniques derived from other methods are also currently being developed. As a derivation of using co-culture, the cyclic depsipeptide jasplakinolide derived from marine sponges was used to stimulate the production of new chaetoglobosins from *Phomospis asparagi*.[131] We predict co-culturing will be a favored method in future studies. Alternatively, examining extracts of spores of *Metarhizium anisopliae* allowed for the identification of tyrosine betaine that could not be detected from whole cell extracts,[132] suggesting advances if efforts are directed towards specific organelles. There is no doubt that with the advent of continued fungal genome sequencing coupled with technological advances in gene manipulation and metabolite detection methodology, the characterization of novel fungal secondary metabolites will soar in the coming decade.

Acknowledgement

This was funded in part by NIH grant GM084077 to N.P.K.

References

1. R. H. Cichewicz, *Nat. Prod. Rep.*, 2010, **1**, 11.
2. A. Eberharter and P. B. Becker, *J. Cell Sci.*, 2004, **117**, 3707.
3. B. M. Turner, *Nat. Cell. Biol.*, 2007, **9**, 2.
4. D. E. Sterner and S. L. Berger, *Microbiol. Mol. Biol. Rev.*, 2000, **64**, 435.

5. Y. Zhang and D. Reinberg, *Genes Dev.*, 2001, **15**, 2343.
6. S. J. Nowak and V. G. Corces, *Trends Genet.*, 2004, **20**, 214.
7. J. R. Davie and L. C. Murphy, *Biochemistry*, 1990, **29**, 4752.
8. P. Adamietz and A. Rudolph, *Biol. Chem.*, 1984, **259**, 6841.
9. D. Nathan, D. E. Sterner and S. L. Berger, *Proc. Natl. Acad. Sci. USA*, 2003, **100**, 13118.
10. H. M. Liebich, E. Gesele, C. Wirth, J. Wöll, K. Jobst and A. Lakatos, *Biol. Mass. Spectrom.*, 1993, **22**, 121.
11. J. Hymes, K. Fleischhauer and B. Wolf, *Biochem. Mol. Med.*, 1995, **56**, 76.
12. Y. Chen, R. Sprung, Y. Tang, H. Ball, B. Sangras, S. C. Kim, J. R. Falck, J. Peng, W. Gu and Y. Zhao, *Mol. Cell. Proteomics*, 2007, **6**, 812.
13. G. T. Wondrak, D. Cervantes-Laurean, E. L. Jacobson and M. K. Jacobson, *Biochem. J.*, 2000, **351**, 769.
14. T. Jenuwein and C. D. Allis, *Science*, 2001, **293**, 1074.
15. J. R. Dobosy and E. U. Selker, *Cell. Mol. Life Sci.*, 2001, **58**, 721.
16. S. I. Grewal and S. Jia, *Nat. Rev. Genet.*, 2007, **8**, 35.
17. L. Xiong, K. K. Adhvaryu, E. U. Selker and Y. Wang, *Biochemistry*, 2010, **49**, 5236.
18. A. Wang, S. K. Kurdistani and M. Grunstein, *Science*, 2002, **298**, 1412.
19. E. De Nadal, M. Zapater, P. M. Alepuz, L. Sumoy, G. Mas and F. Posas, *Nature*, 2004, **427**, 370.
20. M. Biel, V. Wascholowski and A. Giannis. *Angew. Chem., Int. Ed.*, 2005, **44**, 3186.
21. M. Tribus, I. Bauer, J. Galehr, G. Rieser, P. Trojer, G. Brosch, P. Loidl, H. Haas and S. Graessle, *Mol. Biol. Cell*, 2010, **21**, 345.
22. P. A. Cole, *Nat. Chem. Biol.*, 2008, **4**, 590.
23. N. D. Heintzman, R. K. Stuart, G. Hon, Y. Fu, C. W. Ching, R. D. Hawkins, L. O. Barrera, S. Van Calcar, C. Qu, K. A. Ching, W. Wang, Z. Weng, R. D. Green, G. E. Crawford and B. Ren, *Nat. Genet.*, 2007, **39**, 311.
24. M. Bryk, S. D. Briggs, B. D. Strahl, M. J. Curcio, C. D. Allis and F. Winston, *Curr. Biol.*, 2002, **12**, 165.
25. J. E. Mueller, M. Canze and M. Bryk, *Genetics*, 2006, **173**, 557.
26. V. E. MacDonald and L. J. Howe, *Epigenetics*, 2009, **4**, 139.
27. Z. A. Lewis, K. K. Adhvaryu, S. Honda, A. L. Shiver, M. Knip, R. Sack and E. U. Selker, *PLoS Genet.*, 2010, **6**, e1001196.
28. P. Trojer, M. Dangl, I. Bauer, S. Graessle, P. Loidl and G. Brosch, *Biochemistry*, 2004, **43**, 10834.
29. E. M. Bradbury, R. J. Inglis and H. R. Matthews, *Nature*, 1974, **247**, 257.
30. A. Soloaga, S. Thomson, G. R. Wiggin, N. Rampersaud, M. H. Dyson, C. A. Hazzalin, L. C. Mahadevan and J. S. C. Arthur, *EMBO J.*, 2003, **22**, 2788.
31. C. P. De Souza, A. H. Osmani, L. P. Wu, J. L. Spotts and S. A. Osmani, *Cell*, 2000, **102**, 293.

32. E. Feige, O. Shalom, S. Tsuriel, N. Yissachar and B. Motro, *Biochim. Biophys. Acta*, 2006, **1763**, 272.
33. H. Tamaru, X. Zhang, D. McMillen, P. B. Singh, J. Nakayama, S. I. Grewal, C. D. Allis, X. Cheng and E. U. Selker, *Nat. Genet.*, 2003, **34**, 75.
34. J. L. Barra, A. M. Holmes, A. Gregoire, J. L. Rossignol and G. Faugeron, *Mol. Microbiol.*, 2005, **57**, 180.
35. E. U. Selker, *Trends Genet.*, 1997, **13**, 296.
36. M. Tamame, F. Antequera, J. R. Villanueva and T. Santos, *Mol. Cell. Biol.*, 1983, **3**, 2287.
37. D. W. Lee, M. Freitag, E. U. Selker and R. Aramayo. *PLoS One*, 2008, **3**, e2531.
38. A. A. Brakhage and V. Schroeckh, *Fungal Genet. Biol.*, 2010, **48**, 15.
39. D. Hoffmeister and N. P. Keller, *Nat. Prod. Rep.*, 2007, **24**, 393.
40. J. H. Yu, R. A. Butchko, M. Fernandes, N. P. Keller, T. J. Leonard and T. H. Adams, *Curr. Genet.*, 1996, **29**, 549.
41. J. H. Yu, P. K. Chang, K. C. Ehrlich, J. W. Cary, D. Bhatnagar, T. E. Cleveland, G. A. Payne, J. E. Linz, C. P. Woloshuk and J. W. Bennett, *Appl. Environ. Microbiol.*, 2004, **70**, 1253.
42. S. Gutiérrez, J. Velasco, F. J. Fernandez and J. F. Martín, *J. Bacteriol.*, 1992, **174**, 3056.
43. Y. Abe, C. Ono, M. Hosobuchi and H. Yoshikawa, *Mol. Genet. Genomics*, 2002, **268**, 352.
44. Y. Abe, T. Suzuki, T. Mizuno, C. Ono, K. Iwamoto, M. Hosobuchi and H. Yoshikawa, *Mol. Genet. Genomics*, 2002, **268**, 130.
45. P. Tudzynski, K. Hölter, T. Correia, C. Arntz, N. Grammel and U. Keller, *Mol. Gen. Genet.*, 1999, **261**, 133.
46. H. Proctor, D. W. Brown, R. D. Plattner and A. E. Desjardins, *Fungal Genet. Biol.*, 2003, **38**, 237.
47. B. Tudzynski, *Appl. Microbiol. Biotechnol.*, 1999, **52**, 298.
48. J. H. Ahn, Y. Q. Cheng and J. D. Walton. *Fungal Genet. Biol.*, 2002, **35**, 31.
49. J. Kennedy, K. Auclair, S. G. Kendrew, C. Park, J. C. Vederas and C. R. Hutchinson, *Science*, 1999, **284**, 1368.
50. N. Kimura and T. Tsuge, *J. Bacteriol.*, 1993, **175**, 4427.
51. H. F. Tsai, M. H. Wheeler, Y. C. Chang and K. J. Kwon-Chung, *J. Bacteriol.*, 1999, **181**, 6469.
52. C. Young, L. McMillan, E. Telfer and B. Scott, *Mol. Microbiol.*, 2001, **39**, 754.
53. D. J. Smith, M. K. Burnham, J. H. Bull, J. E. Hodgson, J. M. Ward, P. Browne, J. Brown, B. Barton, A. J. Earl and G. Turner, *EMBO J.*, 1990, **9**, 741.
54. D. W. Brown, J. H. Yu, H. S. Kelkar, M. Fernandes, T. C. Nesbitt, N. P. Keller, T. H. Adams and T. J. Leonard, *Proc. Natl. Acad. Sci. USA*, 1996, **93**, 1418.

55. D. M. Gardiner, A. J. Cozijnsen, L. M. Wilson, M. S. Pedras and B. J. Howlett, *Mol. Microbiol.*, 2004, **53**, 1307.
56. S. C. Trapp, T. M. Hohn, S. McCormick and B. B. Jarvis, *Mol. Gen. Genet.*, 1998, **257**, 421.
57. D. W. Brown, S. P. McCormick, N. J. Alexander, R. H. Proctor and A. E. Desjardins, *Fungal Genet. Biol.*, 2001, **32**, 121.
58. N. Khaldi, F. T. Seifuddin, G. Turner, D. Haft, W. C. Nierman, K. H. Wolfe and N. D. Fedorova, *Fungal Genet. Biol.*, 2010, **47**, 736.
59. S. B. Bumpus, B. S. Evans, P. M. Thomas, I. Ntai and N. L. Kelleher, *Nat. Biotechnol.*, 2009, **27**, 951.
60. K. M. Fisch, A. F. Gillaspy, M. Gipson, J. C. Henrikson, A. R. Hoover, L. Jackson, F. Z. Najar, H. Wägele and R. H. Cichewicz, *J. Ind. Microbiol. Biotechnol.*, 2009, **36**, 1199.
61. J. E. Galagan, S. E. Calvo, K. A. Borkovich, E. U. Selker, N. D. Read, D. JaVe, W. FitzHugh, L.-J. Ma, S. Smirnov, S. Purcell, B. Rehman, T. Elkins, R. Engels, S. Wang, C. B. Nielsen, J. Butler, M. Endrizzi, D. Qui, P. Ianakiev, D. Bell-Pedersen, M. A. Nelson, M. Werner-Washburne, V. C. P. Selitrenniko, J. A. Kinsey, E. L. Braun, A. Zelter, U. Schulte, G. O. Kothe, G. Jedd, W. Mewes, C. Staben, E. Marcotte, D. Greenberg, A. Roy, K. Foley, J. Naylor, N. Stange-Thomann, R. Barrett, S. Gnerre, M. Kamal, M. Kamvysselis, E. Mauceli, C. Bielke, S. Rudd, D. Frishman, S. Krystofova, C. Rasmussen, R. L. Metzenberg, D. D. Perkins, S. Kroken, C. Cogoni, G. Macino, D. Catcheside, W. Li, R. J. Pratt, S. A. Osmani, C. P. C. DeSouza, L. Glass, M. J. Orbach, J. A. Berglund, R. Voelker, O. Yarden, M. Plamann, S. Seiler, J. Dunlap, A. Radford, R. Aramayo, D. O. Natvig, L. A. Alex, G. Mannhaupt, D. J. Ebbole, M. Freitag, I. Paulsen, M. S. Sachs, E. S. Lander, C. Nusbaum and B. Birren, *Nature*, 2003, **422**, 859.
62. N. P. Keller and T. M. Hohn, *Fungal Genet. Biol.*, 1997, **1**, 17.
63. H. Proctor, S. P. McCormick, N. J. Alexander and A. E. Desjardins, *Mol. Microbiol.*, 2009, **74**, 1128.
64. N. Khaldi, J. Collemare, M. H. Lebrun and K. H. Wolfe, *Genome Biol.*, 2008, **9**, R18.
65. J. C. Slot and A. Rokas, *Curr. Biol.*, 2011, **21**, 1.
66. C. E. Dowzer and J. M. Kelly, *Curr. Genet.*, 1989, **15**, 457.
67. M. J. Hynes, *Aust. J. Biol. Sci.*, 1975, **28**, 301.
68. J. Tilburn, S. Sarkar, D. A. Widdick, E. A. Espeso, M. Orejas, J. Mungroo, M. A. Peñalva and H. N. Arst Jr, *EMBO J.*, 1995, **14**, 779.
69. S. Teichert, M. Wottawa, B. Schönig and B. Tudzynski, *Eukaryot. Cell*, 2006, **5**, 1807.
70. J. W. Bok, D. Chung, S. A. Balajee, K. A. Marr, D. Andes, K. F. Nielsen, J. C. Frisvad, K. A. Kirby and N. P. Keller, *Infect. Immun.*, 2006, **74**, 6761.
71. H. Proctor, T. M. Hohn, S. P. McCormick and A. E. Desjardins, *Appl. Environ. Microbiol.*, 1995, **61**, 1923.

72. C. P. Woloshuk, K. R. Foutz, J. F. Brewer, D. Bhatnagar, T. E. Cleveland and G. A. Payne, *Appl. Environ. Microbiol.*, 1994, **60**, 2408.

73. T. Shimizu, H. Kinoshita and T. Nihira, *Appl. Environ. Microbiol.*, 2007, **73**, 5097.

74. S. Bergmann, J. Schümann, K. Scherlach, C. Lange, A. A. Brakhage and C. Hertweck, *Nat. Chem. Biol.*, 2007, **3**, 213.

75. S. Bergmann, A. N. Funk, K. Scherlach, V. Schroeckh, E. Shelest, U. Horn, C. Hertweck and A. A. Brakhage, *Appl. Environ. Microbiol.*, 2010, **24**, 8143.

76. Y. M. Chiang, E. Szewczyk, A. D. Davidson, N. P. Keller, B. R. Oakley and C. C. Wang, *J. Am. Chem. Soc.*, 2009, **131**, 2965.

77. B. Teichmann, L. Liu, K. O. Schink and M. Bölker, *Appl. Environ. Microbiol.*, 2010, **76**, 2633.

78. M. Fernandes, N. P. Keller and T. H. Adams, *Mol. Microbiol.*, 1998, **28**, 1355.

79. J. W. Bok, D. Noordermeer, S. P. Kale and N. P. Keller, *Mol. Microbiol.*, 2006, **61**, 1636.

80. C. H. Chiou, M. Miller, D. L. Wilson, F. Trail and J. E. Linz, *Appl. Environ. Microbiol.*, 2002, **68**, 306.

81. S. H. Liang, T. S. Wu, R. Lee, F. S. Chu and J. E. Linz, *Appl. Environ. Microbiol.*, 1997, **63**, 1058.

82. L. V. Roze, A. E. Arthur, S. Hong, A. Chanda and J. E. Linz, *Mol. Microbiol.*, 2007, **66**, 713.

83. E. K. Shwab, J. Bok, M. Tribus, J. Galehr, S. Graessle and N. P. Keller, *Eukaryot. Cell*, 2007, **6**, 1656.

84. Y. Reyes-Dominguez, J. Bok, H. Berger, E. K. Shwab, A. Basheer, A. Gallmetzer, C. Scazzocchio, N. P. Keller and J. Strauss, *Mol. Microbiol.*, 2010, **76**, 1376.

85. J. W. Bok, Y. Chiang, E. Szewczyk, Y. Reyes-Domingez, A. D. Davidson, J. F. Sanchez, H. Lo, K. Watanabe, J. Strauss, B. R. Oakley, C. C. C. Wang and N. P. Keller, *Nat. Chem. Biol.*, 2009, **5**, 462.

86. E. Szewczyk, Y. M. Chiang, C. E. Oakley, A. D. Davidson, C. C. Wang and B. R. Oakley, *Appl. Environ. Microbiol.*, 2008, **74**, 7607.

87. D. Nathan, K. Ingvarsdottir, D. E. Sterner, G. R. Bylebyl, M. Dokmanovic, J. A. Dorsey, K. A. Whelan, M. Krsmanovic, W. S. Lane, P. B. Meluh, E. S. Johnson and S. L. Berger, *Genes Dev.*, 2006, **20**, 966.

88. J. W. Bok, D. Hoffmeister, L. Maggio-Hall, R. Murillo, J. D. Glasner and N. P. Keller, *Chem. Biol.*, 2006, **13**, 31.

89. S. Bouhired, M. Weber, A. Kempf-Sontag, N. P. Keller and D. Hoffmeister, *Fungal Genet. Biol.*, 2007, **44**, 1134.

90. M. Shaaban, J. M. Palmer, W. A. El-Naggar, M. E. El-Sokkary, S. E. El-Habib and N. P. Keller, *Fungal Genet. Biol.*, 2010, **47**, 423.

91. R. A. Dean, N. J. Talbot, D. J. Ebbole, M. L. Farman, T. K. Mitchell, M. J. Orbach, M. Thon, R. Kulkarni, J. R. Xu, H. Pan, N. D. Read,

Y. H. Lee, I. Carbone, D. Brown, Y. Y. Oh, N. Donofrio, J. S. Jeong, D. M. Soanes, S. Djonovic, E. Kolomiets, C. Rehmeyer, W. Li, M. Harding, S. Kim, M. H. Lebrun, H. Bohnert, S. Coughlan, J. Butler, S. Calvo, L. J. Ma, R. Nicol, S. Purcell, C. Nusbaum, J. E. Galagan and B. W. Birren, *Nature*, 2005, **434**, 980.

92. R. M. Perrin, N. D. Fedorova, J. W. Bok, R. A. Cramer, J. R. Wortman, H. S. Kim, W. C. Nierman and N. P. Keller, *PLoS Pathol.*, 2007, **4**, e50.
93. H. B. Bode, B. Bethe, R. Hofs and A. Zeeck. *ChemBioChem*, 2002, **3**, 619.
94. H. Gross, *App. Microbiol. Biotechnol.*, 2007, **75**, 267.
95. A. Paranagama, E. M. Wijeratne and A. A. Gunatilaka. *J. Nat. Prod.*, 2007, **70**, 1939.
96. K. Scherlach and C. Hertweck, *Org. Biomol. Chem.*, 2006, **4**, 3517.
97. V. Schroeckh, K. Scherlach, H. W. Nützmann, E. Shelest, W. Schmidt-Heck, J. Schuemann, K. Martin, C. Hertweck and A. A. Brakhage, *Proc. Natl. Acad. Sci. USA*, 2009, **106**, 14558.
98. J. F. Sanchez, Y. Chiang, E. Szewczyk, A. D. Davidson, M. Ahuja, C. E. Oakley, J. W. Bok, N. Keller, B. R. Oakley and C. C. C. Wang, *Mol. BioSyst.*, 2010, **6**, 587.
99. D. C. Oh, C. A. Kauffman, P. R. Jensen and W. Fenical, *J. Nat. Prod.*, 2007, **70**, 515.
100. Y. M. Chiang, E. Szewczyk, T. Nayak, A. D. Davidson, J. F. Sanchez, H. C. Lo, W. Y. Ho, H. Simityan, E. Kuo, A. Praseuth, K. Watanabe, B. R. Oakley and C. C. Wang, *Chem. Biol.*, 2008, **15**, 527.
101. M. Cueto, P. R. Jensen, C. Kauffman, W. Fenical, E. Lobkovsky and J. Clardy, *J. Nat. Prod.*, 2001, **64**, 1444.
102. L. Losada, O. Ajayi, J. C. Frisvad, J. Yu and W. C. Nierman., *Med. Mycol.*, 2009, **47**, S88.
103. R. A. Butchko, T. H. Adams and N. P. Keller, *Genetics*, 1999, **153**, 715.
104. J. W. Bok and N. P. Keller, *Eukaryot. Cell*, 2004, **3**, 527.
105. M. Wang, S. Liu, Y. Li, R. Xu, C. Lu and Y. Shen, *Curr. Microbiol.*, 2010, **61**, 254.
106. W. Xing, C. Deng and C. H. Hu, *Biotechnol. Lett.*, 2010, **32**, 1733.
107. P. Wiemann, D. W. Brown, K. Kleigrewe, J. W. Bok, N. P. Keller, H. U. Humpf and B. Tudzynski, *Mol. Microbiol.*, 2010, **77**, 972.
108. K. Kosalková, C. García-Estrada, R. V. Ullán, R. P. Godio, R. Feltrer, F. Teijeira, E. Mauriz and J. F. Martín, *Biochimie*, 2009, **91**, 214.
109. S. Amaike and N. P. Keller, *Eukaryot. Cell*, 2009, **8**, 1051.
110. D. R. Georgianna, N. D. Fedorova, J. L. Burroughs, A. L. Dolezal, J. W. Bok, S. Horowitz-Brown, C. P. Woloshuk, J. Yu, N. P. Keller and G. A. Payne, *Mol. Plant Pathol.*, 2010, **11**, 213.
111. M. I. Shaaban, J. W. Bok, C. Lauer and N. P. Keller, *Eukaryot. Cell*, 2010, **9**, 1816.
112. D. Wagner, A. Schmeinck, M. Mos, I. Y. Morozov, M. X. Caddick and B. Tudzynski, *Eukaryot. Cell*, 2010, **9**, 1588.

113. S. Pancaldi, L. Del Senno, M. P. Fasulo, F. Poli and G. L. Vannini, *Cell Biol. Int. Rep.*, 1988, **12**, 35.
114. M. Tamame, F. Antequera, J. R. Villanueva and T. Santos, *J. Gen. Microbiol.*, 1983, **129**, 2585.
115. M. Tamame, F. Antequera and E. Santos, *Mol. Cell. Biol.*, 1988, **8**, 3043.
116. K. Akiyama, T. Ohguchi and K. Takata, *Nippon Shokubutsu Byori Gakkaiho*, 1997, **63**, 385.
117. K. Akiyama, H. Katakami and R. Takata, *Plant Biotechnol.*, 2007, **24**, 345.
118. M. S. Kritsky, S. Y. Filippovich, T. P. Afanasieva, G. P. Bachurina and V. E. A. Russo, *Appl. Biochem. Microbiol.*, 2001, **37**, 243.
119. M. S. Kritsky, V. E. A. Russo, S. Y. Filippovich, T. P. Afanasieva and G. P. Bachurina, *Photochem. Photobiol.*, 2002, **75**, 79.
120. S. Y. Filippovich, G. P. Bachurina and M. S. Kritsky, *Appl. Biochem. Microbiol.*, 2004, **40**, 398.
121. R. B. Williams, J. C. Henrikson, A. R. Hoover, A. E. Lee and R. H. Cichewicz, *Org. Biomol. Chem.*, 2008, **6**, 1895.
122. E. U. Selker, *Proc. Natl. Acad. Sci. USA*, 1998, **95**, 9430.
123. S. Finnin, J. R. Donigian, A. Cohen, V. M. Richon, R. A. Rifkind, P. A. Marks, R. Breslow and N. P. Pavletich, *Nature*, 1999, **401**, 188.
124. A. Vannini, C. Volpari, G. Filocamo, E. C. Casavola, M. Brunetti, D. Renzoni, P. Chakravarty, C. Paolini, R. De Francesco, P. Gallinari, C. Steinkühler and S. Di Marco, *Proc. Natl. Acad. Sci. USA*, 2004, **101**, 15064.
125. J. C. Henrikson, A. R. Hoover, P. M. Joyner and R. H. Cichewicz, *Org. Biomol. Chem.*, 2009, **7**, 435.
126. H. Nishida, T. Motoyama, Y. Suzuki, S. Yamamoto, H. Aburatani and H. Osada, *PLoS One*, 2010, **5**, e9916.
127. H. Wada, M. Kagoshima, K. Ito, P. J. Barnes and I. M. Adcock, *Biochem. Biophys. Res. Commun.*, 2005, **331**, 93.
128. R. J. Wozniak, W. T. Klimecki, S. S. Lau, Y. Feinstein and B. W. Futscher, *Oncogene*, 2006, **26**, 77.
129. X. Wang, J. G. Sena Filho, A. R. Hoover, J. B. King, T. K. Ellis, D. R. Powell and R. H. Cichewicz, *J. Nat. Prod.*, 2010, **73**, 942.
130. R. Ben-Ami, V. Varga, R. E. Lewis, G. S. May, W. C. Nierman and D. P. Kontoyiannis, *Virulence*, 2010, **1**, 164.
131. O. E. Christian, J. Compton, K. R. Christian, S. L. Mooberry, F. A. Valeriote and P. Crews, *J. Nat. Prod.*, 2005, **68**, 1592.
132. C. A. Carollo, A. L. Calil, L. A. Schiave, T. Guaratini, D. W. Roberts, N. P. Lopes and G. U. Braga, *Fungal Biol.*, 2010, **114**, 473.

Section 2
New Methodologies and Screening Technologies for the Exploitation of Microbial Resources

CHAPTER 11
Novel Approaches to Exploit Natural Products from Microbial Resources

OLGA GENILLOUD* AND FRANCISCA VICENTE*

Fundación MEDINA, Centro de Excelencia en Investigación de Medicamentos Innovadores en Andalucía, Parque Tecnológico de Ciencias de la Salud, Avda. de Conocimiento 3, E-18100 Armilla, Granada, Spain
*E-mail: olga.genilloud@medinaandalucia.es and francisca.vicente@medinaandalucia.es

11.1 Introduction

Microbial natural products have played a major role as one of the most important sources for the discovery of novel drugs, especially antibacterials, with a huge number of drugs and derivatives successfully introduced in the market. The discovery of penicillin revealed that microorganisms could be a source, not only for antibiotics, but also of novel molecules with therapeutic potential and fostered the development of natural products-based research operations both in academia and industry. Worldwide concerted efforts for the next 20 years contributed to the discovery of most of the natural product-derived chemical scaffolds that are today in clinical practice, but no major classes of antibiotics were introduced between 1962 and 2000.[1–5] All classes of antibiotics have seen the emergence of resistance at various degrees, compromising their use and utility, with antibiotics of same class exhibiting rapid cross resistance to that class of drugs. Resistance is inevitable and resistance management is part of the process for life cycle management of a

RSC Drug Discovery Series No. 25
Drug Discovery from Natural Products
Edited by Olga Genilloud and Francisca Vicente
© The Royal Society of Chemistry 2012
Published by the Royal Society of Chemistry, www.rsc.org

new antibiotic immediately after launch. Chemical modifications of existing chemical scaffolds have been the sources of a large number of highly potent broad-spectrum antibiotics and continue to deliver newer drugs that provide temporary coverage for resistant organisms.[3,4,6,7] However, further modifications of existing scaffolds have become increasingly challenging, and the discovery of new antibiotics scaffolds with known modes of action or, more importantly, novel scaffolds with completely novel mechanisms of action are urgently needed.

In spite of this enormous success of the past in delivering novel structures, and their privileged nature as chemical structures evolved from their interaction with molecular targets, large pharmaceutical companies have been abandoning the use of natural products in the discovery programs. The rediscovery problem of old known molecules has become a hurdle, and the lower outputs, the isolation limitations, and the difficulties in developing better candidates from the original molecule have faced strong competition of very promising synthetic and combinatorial libraries prone to high throughput screening (HTS) scenarios.[5,8] Furthermore, natural products screening programs had to compete for resources and for targets and the crude extract collections were not amenable to be used in centralized HTS facilities. Nevertheless, natural products, their semi-synthetic derivatives and natural products inspired compounds still represent one of the most important sources of chemical diversity and bioactive novel structures ever described.[9–12] From a therapeutic perspective, and more specially in the case of antibiotics, there is still an urgent need for new molecules, with a major threat of resistant emerging pathogens for which the demand has not been fulfilled yet.[7,8]

11.2 New High Throughput Screening Technologies

Screening has played a critical role in the discovery of chemical leads that are further optimized for their properties eventually leading to clinical candidates and drugs. The tremendous progress made in life sciences has resulted in the definition of many pathological processes and mechanisms of drug action. This advancement has led to the establishment of various molecular and cellular bioassays in conjunction with HTS methods.[13] HTS decreases the amount of compound required in the assay. This is advantageous for certain natural products that are difficult to isolate and purify, and permits compounds that are difficult to synthesize to be assayed.

One of the most critical aspects of drug discovery is the selection of an appropriate screening strategy. Screening programs have relied heavily upon *in vitro* biochemical assays and synthetic libraries to detect activity on defined targets and utilized secondary or counter screen assays to provide evidence of specificity or efficacy. Most of these initiatives failed to deliver new leads that could be developed into the clinic.[14] For obvious reasons, pharmaceutical companies have not been prone to reveal their screening strategies, but, since screening trends have been to focus on mechanism-based approaches, a variety of screening procedures were used in successful discovery programs.[15] Different groups have approached the discovery of novel molecules using resistant or multidrug resistant strains in whole cell assays coupled to secondary assays to define the mechanism of action (MOA).[16–18]

In the last decade, tremendous progress has been made in HTS technology and so a variety of screening strategies have contributed to the discovery of novel natural products, ranging from classical empiric whole cell assays to the more sophisticated target-based whole cell assays.

The development of whole cell target-based tests, more amenable to be used in conjunction with crude extract libraries of natural products, have represented a shift in the paradigm of antibiotic research. Several of these new technologies were developed and implemented to target novel scaffolds with novel modes of action. The basis of these technologies is summarized below and representative natural products discovered using such assays are described in further sections.[19]

11.2.1 The Antisense Approach

A rapid shotgun antisense procedure for the comprehensive identification of *Staphylococcus aureus* genes essential for growth was developed by Elitra Pharmaceuticals.[20] Inhibition of gene expression by antisense RNA has been observed in natural bacterial systems and has been used for silencing gene expression utilizing a *tet* regulatory system in which tetracycline or anhydrotetracycline was used to include antisense RNA production.[21–23] The generation of antisense RNA to selectively decrease the production of intracellular gene products involved a xylose inducible promoter very useful for primary screening.[20] This approach has led to the identification of a comprehensive set of *S. aureus* essential genes that would enable the discovery of new antimicrobial compounds (Table 11.1). In this method, short sequences of a selectively expressed mRNA strand bind to a complementary gene sequence, reducing transcription of that targeted gene and translation of the corresponding essential gene product. While still viable, these weakened strains are sensitized to an inhibitor that targets the depleted gene product. Figure 11.1 describes this type of differential sensitization of strains to drug targets by manipulation of target protein levels through induction of antisense RNA. The mechanism of silencing by inducible antisense RNA in this system is probably due to the degradation of the antisense-targeted mRNA. By expressing the antisense RNA to essential genes, the cell becomes hypersensitive to inhibitors of that target (10–100 times more sensitive to specific inhibitors) (Figure 11.2). The shotgun antisense approach described above is a simple method to rapidly and comprehensively determine the essential genes for the growth of microorganisms. This information was used in the determination of true minimal genome sets and to create conditional lethal strains to be used in the identification of antimicrobial compounds against important pathogens. In this sense, these antisense screening strategies in

Table 11.1   A wide range of bacterial essential genes from different pathways
identified with Antisense technology.

Pathway	Function of the genes	Some of the important S. aureus essential genes*
Cell Wall Synthesis	Cell division Membrane bioenergetics	*fts*H, *fts*Z, gidA, *mur*A, *oppF, opp-1D* *atpF*
Nucleic Acid Synthesis	DNA replication Transcription, RNA processing	*dnaC, dnaX, dnaA, polC, gyrB, parC* *rpoC, rpoA, tmD, SA1885*
Intermediary Metabolism	Energy metabolism	*upp, gap, adk, fruB, gntK, lacA, metK*
Cellular processes	Protein secretion Sporulation	*secA* *obg*
Protein synthesis	Initiation / Elongation / Termination factors Ribosomal proteins Aminoacyl-tRNA synthetases Protein modification	*infA, infB, tufA, prfA, ptsH, ptsI* *rpsD, rpsE, rplA, rplL, rpsJ, rpsP* *metS, lysS, pheT, valS, leuS,* *SA1715* *ptsH, ptsI, SA1063, SA1854*
Other Pathways	Biosynthesis of cofactors, Prosthetic groups,..etc	*clpC, clpB, clpL*
Novel Targets	Unknown function	*SA0437, SA0722, SA1957, SA1966*

* This table shows some selection of *S. aureus* essential genes identified by shotgun antisense.[20]

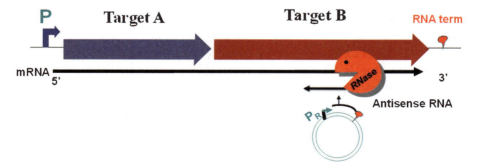

Figure 11.1   Antisense RNA technology: growth inhibition by expression of
antisense RNA to essential gene.

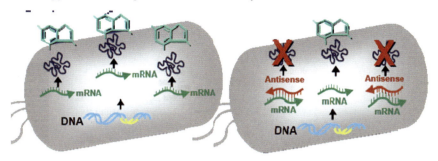

Normal levels of target essential protein   Increase antisense Decrease target essential protein

Drug inhibits growth   Less drug needed to inhibit growth

Figure 11.2   The basic concept for the antisense screens: an induction of the antisense RNA gene decreases the expression of the target protein and sensitizes cells for target inhibitors.

Staphylococcus aureus have shown to have the advantage of being unique discovery platforms, and have recently been used to successfully identify known and novel antibacterials.[24–29]

11.2.2   *Staphylococcus aureus* Antisense Target-based Whole Cell Screens

The hypersensitive strains obtained through the antisense technology described in the above section have been used in drug screens to find new classes of inhibitors for novel bacterial targets. The *Staphylococcus aureus* antisense target-based whole cell screens reported by Elitra were developed in liquid format.[20,30] Although they served well for testing compounds, this method required titration of each and every compound and extract for both antisense and control strains. Therefore, this methodology was neither cost effective nor practical for high throughput primary screening on an industrial scale. These *Staphylococcus aureus* antisense target-based assays were transferred at Merck to an agar diffusion whole cell screening assay format, which would be used as a standard primary screening model.[26] The two-plate antisense assay was optimized as a differential sensitivity assay in which one plate contained *S. aureus* cells expressing AS-RNA (AS plate) and the other one *S. aureus* cells lacking AS-RNA expression (control plate).[31] Upon induction of the antisense, a decrease in the target protein enhances the sensitivity for target-specific inhibitors. The effect of the xylose inducible promoter on the growth/viability of the antisense deleted strain was studied versus the effect on a strain containing the vector with no antisense. This technique presents several advantages, as it can be applied as a differential whole cell screen targeting essential ORFs of unknown function. Different agar diffusion two-plate assays

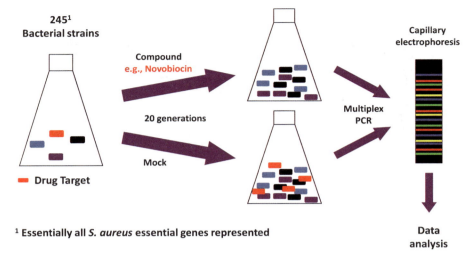

Figure 11.3 The genome-wide target-based whole cell screening in *S. aureus*.

were set up by the Merck team using this methodology covering diverse genes from several pathways, such as those described in Table 11.2.

11.2.3 *Staphylococcus aureus* Genome-wide Fitness Test

The genome-wide fitness test is a mode-of-action screening approach that was developed by Merck in collaboration with Elitra Pharmaceuticals on the basis of the antisense technology. This method was validated by mechanistically profiling a diverse set of 59 antibacterial compounds.[32] This technology relies on the use of regulated expression of target-specific antisense RNA to make bacterial strains hypersensitive to target-specific inhibitors of growth. The *S. aureus* fitness test consists of a collection of 245 inducible *S. aureus* antisense RNA strains engineered for reduced expression of a single target that correspond to essential genes for which xylose-inducible antisense RNA

Table 11.2 Different agar diffusion two-plate antisense assays. Samples tested, confirmed hits and hit rate of actives per assay

Antisense pathway assays	*Genes*	*Samples tested*	*Confirmed hits*	*% active*
Cell wall	*murA*	129,152	415	0.3
DNA synthesis	*dnaC*	203,855	275	0.1
Fatty acid synthesis	*fabF*	250,000	750	0.3
Protein synthesis	*rpsD*	138,288	251	0.2
	pheT	224,624	210	0.1
Signal peptidase I	*spsB*	7,354	46	0.6
Secretory pathway	*secA*	114,327	116	0.1

expression imparts a growth phenotype. When pooled, fitness test strains are grown together for approximately 20 population doublings in the presence of test compounds with antibacterial activity. The strains which are rendered sensitive to the biological effects of the compound by their targeted antisense RNA preferentially drop out of the population. Using multiplex PCR, capillary electrophoresis and gene fragment analysis we compared the abundance of the strains at the end of the experiment with mock-treated controls. Antisense induced strain sensitivity (AISS) profiles reflected the mechanistic selectivity of a structurally diverse set of reference antibiotics for a variety of known targets including the cell wall, the protein synthesis, nucleic acid biosynthesis and several metabolic pathways (Figure 11.3). This type of competitive growth assay is called a 'fitness test' and provides profile of strain sensitivities that are specific for the mechanism of action (MOA) of the compound being tested. This semi-automated system is capable of analyzing the compound MOA either by direct inference from the AISS profile or by comparison to a database of AISS profiles of known inhibitors.[32] The platform has broad applications in the field of antibacterial drug discovery, including MOA determination for newly discovered antibacterial agents, the identification of novel drug targets, and the detection of reporter strains whose chemically induced hypersensitivity correlates with specific antibacterial modes of action.[36]

11.3 Drug Discovery: High Throughput Screening from Natural Products Extracts

The evolution of discovery screening using natural products extracts and synthetic compounds has covered a large number of approaches in the last decade.[19,34,35] The most commonly used methods for natural products are summarized in Figure 11.4 and were extensively covered by recent reviews.[19,35] In addition, the success of these new screening technologies have required the integration of multidisciplinary teams to exploit the microbial chemical arsenal. Some of them are described in detail in the following sections.

11.3.1 Strategies for the Generation of Natural Products Extracts Collection

The exploration of the microbial secondary metabolism of bacteria and fungi was intensively pursued in industrial screening programs, and the extremely prolific production of novel molecules by some groups of microorganisms, especially some taxa of actinomycetes and fungi, did not require the use of an unlimited number of cultivation conditions to ensure that novel molecules were produced. In fact for decades it has been well accepted that the application of a number of three to four production media at a time were sufficient to exploit the production of new bioactive molecules.[37] In recent years, expanding

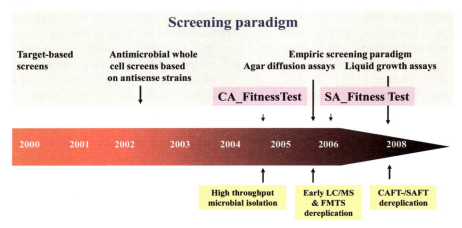

Figure 11.4 Evolution of natural products screening paradigm.

numbers of complete genome annotations have been confirming the presence of a huge biosynthetic potential in bacteria and in fungi, in many cases only detected as cryptic pathways from genome mining of biosynthetic pathways.[38–41] We still ignore the most important factors conditioning the nutritional requirements and secondary metabolism regulation factors of most of the species screened that can be producing molecules quite below the detection threshold. On another side, the difficulty of discovering novel molecules and the recurrent re-discovery problem of well-known old molecules has required to move away from traditional initiatives and challenge the secondary metabolism of these species from quite different perspectives. The recent history of antimicrobial screening from natural products extract libraries shows that these questions were approached from quite different perspectives, always combining the objectives of maximizing the diversity of the secondary metabolites in the extracts and the possibility of identifying novel compounds, and minimizing the impact on resources required to generate this large collections of extracts. At the same time, these microbial resources were better exploited with the introduction of different screening strategies. The challenge of finding new antibiotics classes from libraries of secondary metabolites produced by these microorganisms required a change of paradigm with a shift in the numbers of new extracts tested, and an improvement in the strain selection conditions and the nutritional conditions required for the production of novel molecules. To respond to these needs, we introduced small-scale fermentations for bacteria and fungi in tubes, instead of the traditional flasks, that could be readily adapted to the automated liquid handling equipment for further extraction. By reducing the fermentation volumes from 40 ml to 10 ml, and selecting the best three fermentation media covering the largest metabolic space of the producing strains, we managed to ensure an increase in the numbers and diversity of the strains tested.[42] In the course of four years a large collection of more than 250,000 microbial extracts was generated from the largest diversity of fungi and actinomycetes. These collections were systematically screened against target-based antisense-sensitized *S. aureus* assays developed for essential *S. aureus* genes.[43]

The OSMAC approach (one strain, many active compounds), has been described for the possibility of one strain to produce many different activities, and the term has been used to describe the effect of the manipulation of the production conditions of the strains to further exploit their microbial biosynthetic potential.[44–47] To expand the possibilities of triggering the production of new secondary metabolites from a larger number of strains of actinomycetes and fungi, we introduced the use of multiple nutritional conditions with miniaturized parallel fermentation devices. High throughput miniaturized formats for cultivation of bacteria was previously optimized by other groups for unicellular bacteria.[48–50] This high throughput cultivation of bacteria offered by System Duetz allowed the fermentation in as many production media as desired, with volumes as low as 750 μl in 96-well plates; a format that could be easily integrated in later stages in the highly automated extraction and screening platforms in place.[48,49] Without increasing our resources we managed to increase by five-fold the number of extracts ready to be evaluated for the production of antibiotic activities. The activities against the Gram-positive pathogen *Staphylococcus aureus* were scored as zones of inhibition and as a representative measure of the production rate. This progressive increase in the number of nutritional conditions, up to 8–12 different media, was translated in all the microbial groups tested in an increase in the number of antibiotic activities produced.[45,46] This approach was developed for almost three years and generated more than 450,000 screening samples that were used to feed the empiric screening strategy targeting the identification of potent and broad spectrum antibacterial inhibitors on a subset of test strains (*S. aureus*, *Acinetobacter baumannii*, *Pseudomonas aeruginosa*, *E. coli* and *C. albicans*). These activities were de-replicated from known compounds by low-resolution LC-MS against a proprietary library of antibacterial molecules, to be selected to be further characterized for a mode of action against the antisense fitness test.[32,36]

11.3.2 Natural Products Screening Campaigns

11.3.2.1 Antisense Whole Cell Screening

Conditional expression of antisense RNA of essential genes in *S. aureus* cells has been found to inhibit cell growth and alter cell sensitivity to selective inhibitors, as was previously described, leading to the possibility of developing new, hypersensitive target-based whole-cell assays.[20,23] This approach appears to be general in nature and is applicable to all essential bacterial targets in *S. aureus*. These technologies were applied with different essential genes selected from diverse pathways and carried out several screening campaigns over several years with the microbial natural products extracts libraries

(Table 11.2). In each case, the two-plate assay specific for each gene was validated using a large number of reference antibiotics with known modes of action. The antibiotics that inhibited the specific target in each assay showed a differential inhibition zone in the assay whereas the antibiotics that inhibited other targets or pathways did not show this effect (Figure 11.5).

These antisense assays determined the hypersensitivity of the antisense strains to the target-based inhibitors, permitting the detection of active compounds even when they were present in relatively low concentrations or low titers – even as mixtures.

Young *et al.* reported for the first time the combined use of the antisense and the agar-diffusion technologies, an agar-diffusion two-plate differential sensitivity assay (Figure 11.5, two-plate assay) using the fatty acid synthesis (FAS) pathway as the target, which is essential for the viability of bacteria.[26] Table 11.2 shows the global numbers involved in the different antisense screening campaigns that were performed with natural products extracts. It includes in addition to the total number of samples tested in the different antisense assays at a single concentration, the numbers of confirmed hits which display target-specific activities in each of the assays, and the hit rates per assay. All these assays were implemented in the HTS format as a novel way to approach the antisense screening with diverse targets to identify in parallel potential hits from natural products extract libraries. These screening efforts led to the identification and isolation of a series of known and novel antibacterial compounds, which will be described in detail in the next section.

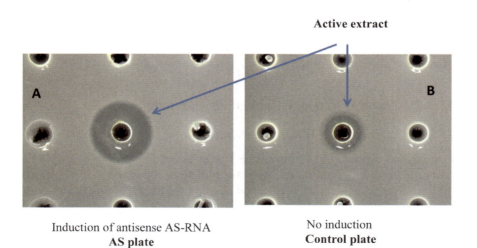

Figure 11.5   A differential two-plate sensitivity assay.

11.3.2.2   Whole-Cell Empiric and Genome-Wide Fitness Test Screens

The most productive of these approaches was the empiric whole-cell screening in which antibacterial activities were identified *a priori* followed by the determination of the targets or modes of action. The obvious advantage of the empiric whole-cell screening is that it ensures inherent antibacterial activity. However, it does not allow for the differentiation of antibacterial activity from general toxicity. This disadvantage was potentially circumvented when the whole-cell screens were designed to be target based. In the past, many target-based whole-cell screening methods have been used to find target-specific inhibitors. However, most of these approaches suffer from generality and are not applicable to all targets. An alternative strategy would be the use of empiric screening to look for activity against a target microorganism, very much in the way it was done in the early days of antibiotics discovery, followed by the examination of the mode of action using, for instance, a simultaneous antisense screen for all essential bacterial gene targets, which has been described in the above paragraphs.[32]

The microbial world is the ideal starting place for exploiting genome-wide approaches for antibacterial drug discovery. Since 2005 the Merck team actively pursued a paradigm to rapidly differentiate crude natural products extracts that contained novel activities from those containing known compounds/activities. The strategy used was focused on identifying preliminary mechanistic information and on obtaining comparative information of high value in de-replication of natural products. As was mentioned before, the front end of this program procured new bacterial and fungal crude acetone extracts from a large diversity of strains cultivated under 8–12 different conditions that were tested for antibacterial activity against *S. aureus*. The screening of more than 450 000 acetone extracts followed by mass spectrometry de-replication of an average of 18 000 hits against proprietary LCMS-UV databases of known antibiotics permitted the early identification of almost all major known classes of antibiotics distributed across more than 4500 active extracts with characteristic depletion profiles (Figure 11.6).[46]

Only extracts that showed a low minimal inhibitory concentration (MIC) against *S. aureus* were subjected to the *S. aureus* fitness test for AISS profiling[47] to discover new antibacterial agents working through target-based MOAs and prioritization for natural product chemistry.

Extracts producing AISS profiles that are similar to profiles produced by known drugs with attractive target-based MOA's may contain the same known compounds, novel analogs of these compounds, or potentially new structural classes of inhibitors working through established MOA (Figure 11.7). After this triaging process, extracts exhibiting a known AISS profile were further analyzed with high resolution mass spectral analysis to either confirm or rule out the presence of known inhibitors before any isolation chemistry steps were undertaken. Continued screening efforts led to the identification of a large number of extracts exhibiting known AISS profiles that matched according to

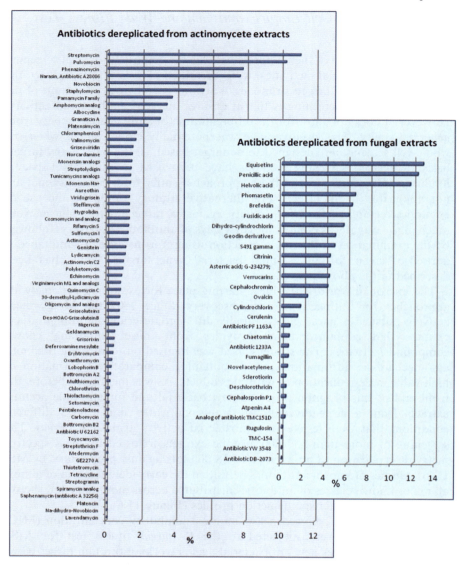

Figure 11.6 Early de-replication of major known antibiotics from crude extracts by low resolution mass spectrometry.

their depletion profile with known antibacterials (i.e. affecting DNA replication, cell wall synthesis, fatty acid synthesis, or translation patterns) that were confirmed by HRMS. Figure 11.8 shows, as an example, the AISS profiles for novobiocin. This antibiotic is characterized by strong depletions in the *parE* and *parC*-antisense (AS) strains (the two subunits of topoIV) and the *gyrB*-AS and *gyrA*-AS strains (the two subunits of DNA gyrase).[32] The

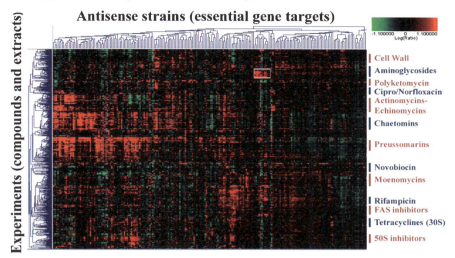

Figure 11.7 Two-dimensional cluster of antisense-induced strain sensitivity profiles of different classes of compounds with defined mechanisms. The cluster contains 637 experiments performed with 245 antisense strains. The compounds and extracts cluster into branches grouping different families of molecules according to their mode of action, as shown on the right-hand side of the figure. Strain depletions are depicted in red while strain resistances are depicted in green.

compounds which exhibited a novel AISS profile and were likely to contain chemical inhibitors not seen before in the assay were assigned to isolation chemistry. Screening of natural products extracts and AISS profiling led to selection of reasonable number of extracts with a potentially novel MOA. It is within this framework that the screening led to the discovery of several unknown compounds among which are included coelomycin and kibdelomycin.[19,36,47]

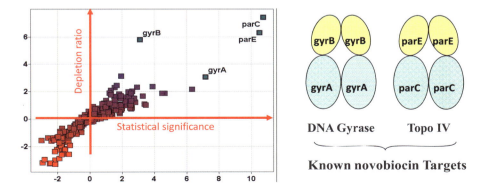

Figure 11.8 Novobiocin depletion profile.

In a similar way to the AISS profiling test, the *Candida albicans* fitness test assay was extremely productive and contributed to the discovery of parnafungin and other novel natural products as antifungal agents.[51–54]

11.4 Discovery of New Antibacterial Compounds

11.4.1 Antibacterial Compounds Derived from the Target-based Antisense-Sensitized *S. aureus* Screening

The target-based antisense-sensitized *S. aureus* screening was applied intensively on a group of essential target genes, many of them with known inhibitors. The screening approach was rapidly validated with the identification of known compounds acting on the selected targets and more importantly with the discovery of important novel compounds with specific activity against these targets. Some targets used were especially prolific at selecting novel compounds with selective antibacterial activity.

11.4.1.1 Fatty Acid Synthesis: FabF/H

Fatty acid synthesis has been shown to be essential for survival in most bacterial species and different condensing enzymes involved in the synthesis had been previously validated as targets.[56] FabH and FabF are two essential enzymes in type II fatty acid synthesis that were used as targets for antibacterial drug discovery. The screening of 250 000 natural products extracts with the two-plate FabF/H antisense assay allowed the detection of the previously known inhibitors of condensing enzymes cerulenin as well as the *Streptomyces*-derived thiolactomycin and thiotetromycin (Figure 11.9).[26] Both

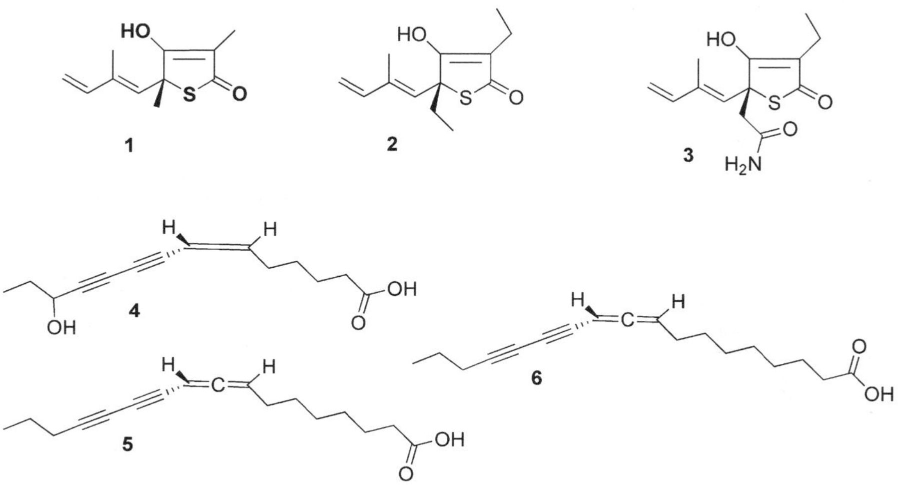

Figure 11.9   Inhibitors of fatty acid condensing enzymes.

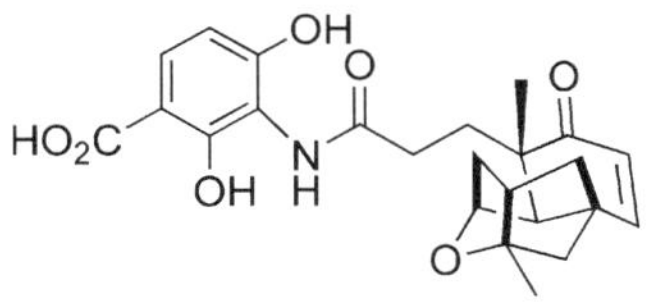

Figure 11.10   Structure of platensimycin and platencin.

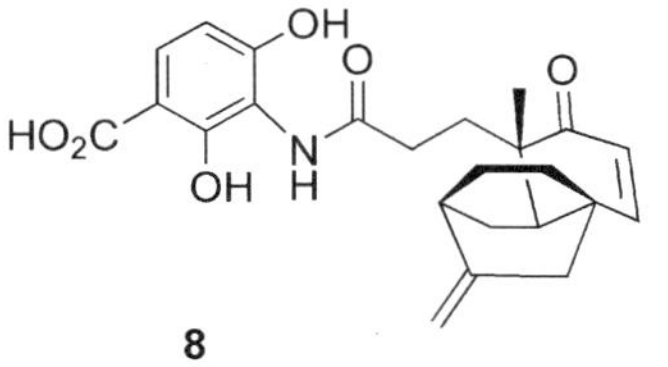

compounds showed very little activity against *S. aureus* with MICs between 64 and 128 µg/mL, in contrast to the analog of thiotetromycin Tu3010 (3) that showed higher MICs of 4 µg/mL, in spite of exhibiting similar IC_{50} values for FASII inhibition than thiolactomycin and its analog thiotetromycin (11–15 µg/mL) (Figure 11.9). Phomallenic acids A–C (4–6) are acetylenic acid versions of fatty acids produced by a fungus of *Phoma* sp. also found in this assay (Figure 11.9).[55] They exhibit activities that differ by 64 fold (MICs against *S. aureus* are 250, 12.5 and 3.9 µg/mL, respectively), but they also show antibacterial and FASII activities against the Gram-negative *Haemophilus influenzae* and *E. coli*. But most importantly, this screening approach led to the discovery of the novel antibiotics platensimycin (7) and platencin (8), two specific inhibitors of condensing enzymes that were shown to be produced by several strains of *Streptomyces platensis* from our collection that had been isolated from different geographic origins (Figure 11.10).[46] Both compounds show potent broad-spectrum Gram-positive antibiotic activities with low MIC values (0.06–2 µg/mL) against major drug-resistant bacteria, such as MRSA and vancomycin-resistant enterococci VRE.[28,29] Platensimycin and platencin have a completely novel mode of action, by inhibition of the acyl enzyme intermediate of the fatty acid condensing enzyme, but they differ in their specificity as determined in the cell-free enzyme FASII assay.[24] Platensimycin presents 800-fold specificity for the elongation condensing enzyme FabF (IC_{50} 48 nM) whereas it has almost no activity against the initiation-condensing enzyme (FabH; IC_{50} 67 µM). Both compounds exhibited excellent *in vivo* efficacy in an *S. aureus* mouse model of infection when infused intravenously. Further details of the discovery and development of these novel leads have been covered extensively in Chapter 12 of this book by Dr. S. Singh.

11.4.1.2   Protein Secretion: SecA ATPase

Bacterial protein secretion is highly conserved among a wide range of species and it has been considered by many as an attractive target for antibacterial agents given the differences with the eukaryotic machinery and the potential little toxicity. The sec-dependent pathway is responsible for protein translocation across the cytoplasmic membrane and a multicomponent system of at least seven proteins, five of which are essential.[57] Homologs of these proteins have been identified in both Gram-negative and Gram-positive bacteria.

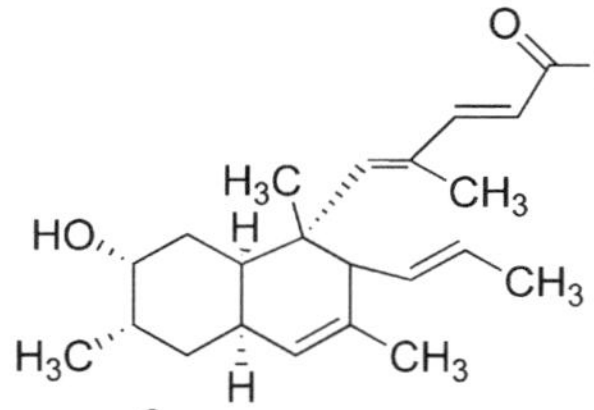

Figure 11.11   Structure of pannomycin.

Whereas the SecYEG heterotrimer forms an integral membrane channel, SecA is an ATPase with high affinity for SecYEG with a central role played in the protein translocase pathway, making it an attractive target for the inhibition of secretion.

The two-plate assay based on the sensitization of the cell by a secA antisense was used to screen a collection of 115,000 natural products extracts. From this campaign we could identify from the broth extract of a strain of the fungus *Geomyces pannorum* the novel natural product pannomycin (Figure 11.11). This cis-decalin natural product is structurally similar to known inhibitors of SecA cissetin and CJ-21,058.[58,59] When the compound was tested on a panel of microorganisms, only weak antimicrobial activity was observed with MICs of 1.4 mM against *S. aureus* Smith and *Enterococcus faecalis*, and activity was observed at this concentration against *Streptococcus pneumoniae*, *H. influenzae* and *Escherichia coli*.[58]

11.4.1.3   Bacterial Protein Synthesis: RpsD

Protein synthesis is one of the well-validated targets with clinically useful antibiotic classes in the market. The bacterial ribosome protein S4 is encoded by the rpsD gene and is an essential component of the 30S ribosomal subunit, as shown in the antisense genome-wide strategy used to identify essential genes in *S. aureus*.[20] Its structure appears to be conserved across bacterial species and it was selected as a broad spectrum target in our screening. To discover novel natural products that interact with RpsD, a two-plate assay with a reduced expression of the rpsD gene by antisense and leading to hypersensitivity against RpsdD inhibitors was implemented for the screening of natural products libraries. The screening of about 138 000 microbial extracts in a rpsD two-plate antisense whole-cell assay has led to the isolation of different molecules such as the lucensimycins (10–16), coniothyrione, pleosperone, phaenosphenone, okilactomycin and glabramycins.

Lucensimycins A–B, C and D–G: the screening of the sensitized rpsD strain led to the discovery of the lucensimycins, seven related natural products polyketides produced by the strain of *Streptomyces lucensis* MA7349. Lucensimycins A–B (10) (11) display weak antibiotic activity and may weakly bind to or inhibit the S4 protein (Figure 11.12).[60–62] Further chemical analysis

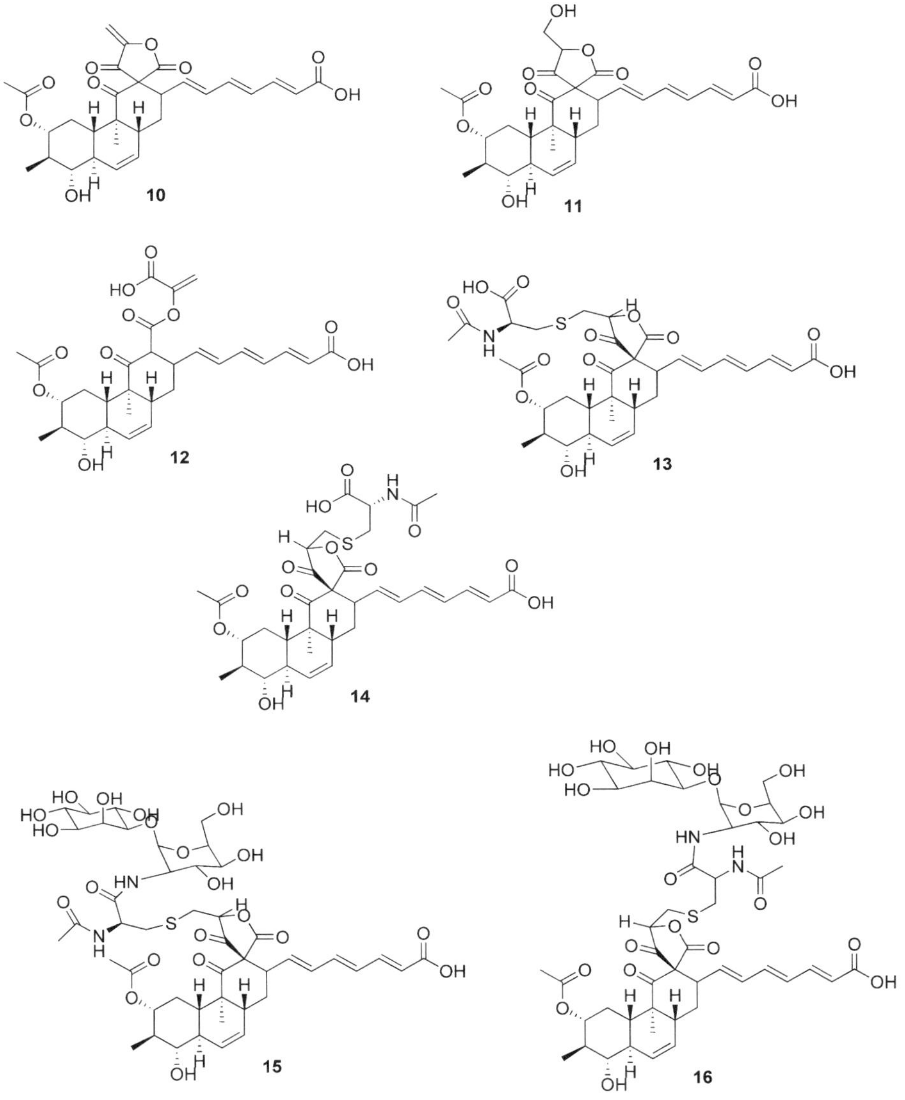

Figure 11.12 Structure of lucensimycins.

of the extracts led to the isolation of novel compound lucensimycin C, an ultimate intermediate of lucensimycin A (Figure 11.12). Lucensimycin C is 80-fold less active than lucensimycin A. As lucensimycin A, it did not show any selectivity against rpsD versus other antisense strains. These compounds are distantly related to antibiotic cytotoxic delaminomycins A–C.[63] Lucensimycin E (13) showed the best activity with MICs of 38 µg/mL against *S. aureus* and 8 µg/mL against *Streptococcus pneumoniae*.

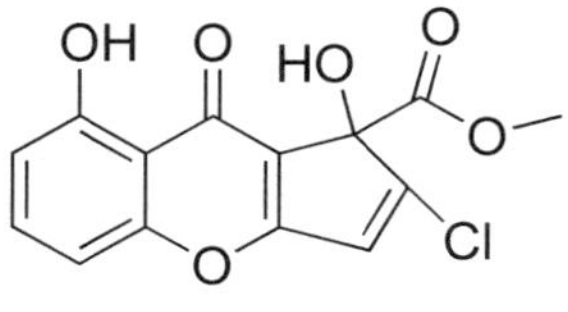

17

Figure 11.13 Structure of coniothyrione and related compounds.

Coniothyrione is another new natural product isolated from a fermentation extract of *Coniothyrium cerealis* MF7209 derived from the rpsD antisense screening. The compound is related to coniochaetone B and remisporines A and B (Figure 11.13), two molecules with weak antibacterial activity.[64,65] The molecules exhibited moderate broad spectrum antibacterial activity, with liquid MICs of 16–32 µg/mL against *Staphylococcus aureus*, *Bacillus subtilis*, *Haemophilus influenzae*, *Streptococcus pneumoniae* and *Enterococcus faecalis* and >64 µg/mL against *Escherichia coli*. The compound showed a differential zone of inhibition in the dual-plate screening assay, suggesting an interaction with the S4 protein, and inhibition of protein synthesis in *S. aureus* with IC_{50} of 5 µg/mL. Nevertheless, inhibition of macromolecular synthesis indicated that this action was not selective as the compound also inhibited DNA synthesis with an IC_{50} of 3 µg/mL.[67]

The tetrahydro-tetrahydroxy anthraquinone pleosperone was identified from an extract produced by a fungal strain of *Pleosporales* (Figure 11.14). The new molecule showed differential sensitivity against rpsD, and was selective for RNA synthesis inhibition. When tested against a panel of strains, it showed modest antibacterial activity against Gram-positive and Gram-negative strains.[67] Pleosperone shows MIC values of 64 µg/mL against wild-type *S. aureus* strains and shows better activities against *Streptococcus pneumoniae* regardless of the medium used (MIC value of 4 µg/mL). The activity against the other key pathogenic Gram-positive bacteria *Enterococcus faecalis* was similar to the activity against *S. aureus* (MIC 64 µg/mL) and it inhibited the growth of *Haemophilus influenzae* with a MIC value of 1 µg/mL. The lack of activity in *E. coli* has been shown to be related to poor penetration.[67] Pleosperone has more preference to inhibit RNA synthesis

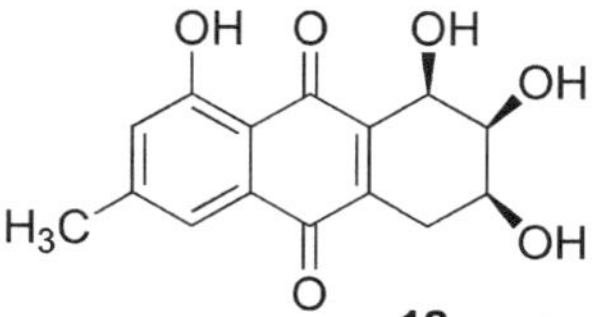

18 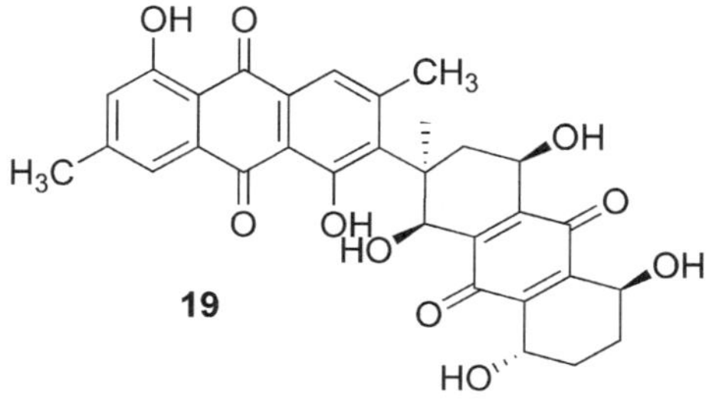   **19**

Figure 11.14 Structure of pleosperone (18) and phaeosphenone (19).

(IC_{50} 1.3 µg/mL) over DNA (IC_{50} 8.4 µg/mL) and protein synthesis (IC_{50} 15.4 µg/mL). This preferential inhibition of RNA synthesis suggests another mode of action in addition to the weak interaction with RpsD protein.

Phaeosphenone is a new dimeric compound from an unidentified species of the ascomycete *Phaeosphaeria* consisting of an anthraquinone and an octahydro anthraquinone (Figure 11.14). This inhibitor was identified from antisense dual-plate screening with reduced expression of rpsD.[68] The compound shows broad spectrum Gram-positive antibacterial activity similar to the anthraquinone pleosperone. It inhibited the growth of wild-type *S. aureus* strain and *S. pneumoniae* (MIC 32–64 µg/mL) and similar MICs were observed with *E. faecalis* and *B. subtilis*. Unlike pleosperone, the compound did not inhibit the growth of Gram-negative bacteria *Haemophilus influenzae* and *Escherichia coli*, but inhibited the growth of *C. albicans* with MIC of 8 µg/mL with no selectivity over fungal strains. With regard to the MOA, phaeosphenone preferentially inhibited *S. aureus* RNA synthesis (IC_{50} 6 µg/mL) over protein and DNA synthesis, but it is likely that another unknown mode of action exists in addition to this weak interaction with RpsD.[68]

Okilactomycin (20) is an old known molecule described as produced by a strain of *Streptomyces griseoflavus* subsp. *zamamiensis* that exhibited weak antimicrobial activity against Gram-positive bacteria with MICs ranging from 12.5 to 50 µg/mL (Figure 11.15). The compound was also described as cytotoxic against some leukemia cell lines with IC_{50} ranging from 0.09 to 0.037 µg/mL.[69] Okilactomycin and four congeners okilactomycin A, B, C and D (21–24) were recently described by our group as produced by a strain of *Streptomyces scabrisporus*.[70] These compounds were identified from the two-plate antisense RpsD screening where okilactomycin remained the most active compound inhibiting *Streptococcus pneumoniae* (MIC of 8–16 µg/mL), *Enterococcus faecalis* (MIC of 16 µg/mL) and *Bacillus subtilis* (MIC of 4–16 µg/mL). The four congeners were significantly less active. Okilactomycin as well as okilactomycin A and C showed preferential inhibition for RNA

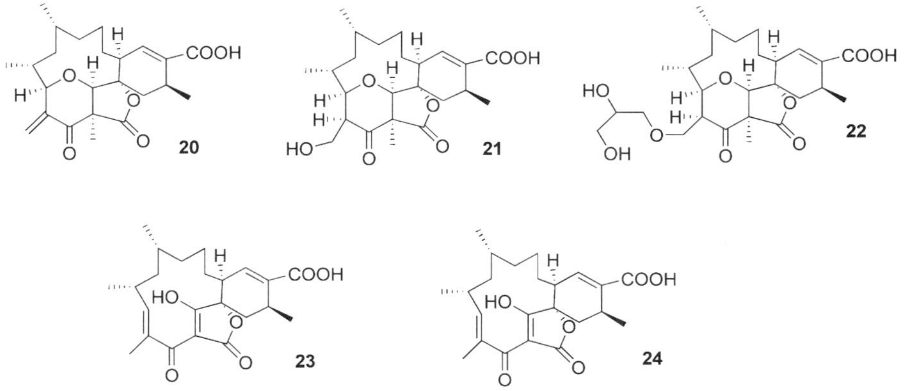

Figure 11.15 Structure of okilactomycins.

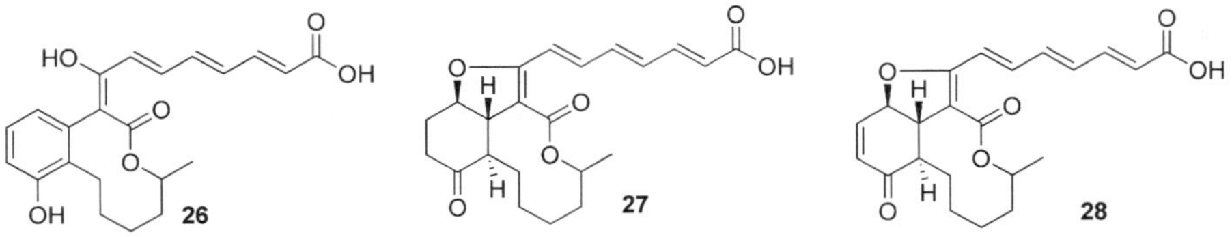

Figure 11.16   Structure of glabramycins A–C.

synthesis rather than protein synthesis, as observed with pleosporone and phaeosphenone.

Glabramycins A–C are new bicyclic macrolactones produced by *Neosartorya glabra*, being glabramycin A the most abundant of the three compounds produced by submerged fermentation of the fungus (Figure 11.16).[71] Glabramycin C was shown to have the most potent activity in the *S. aureus* antisense rpsD-sensitized two-plate assays, with minimum detection concentration (MDC) of 62 µg/mL. This compound showed good growth inhibition activity against *Streptococcus pneumoniae* (MIC of 2 µg/mL), but was less active against *S. aureus* and *B. subtilis* (MICs of 16 µg/mL), and even lower against *Enterococcus faecalis* (MIC of 32 µg/mL). Glabramycin A and B were less active against the whole panel of strains. In addition, none of the compounds were active against the Gram-negative bacteria *H. influenzae* and *E. coli*, or the yeast *C. albicans*. It has been shown that glabramycin A exhibited a preferential inhibition of RNA synthesis over DNA and protein synthesis, a mechanism that has been observed with many other inhibitors obtained from this assay and that differs from the expected protein synthesis.

11.4.2   Novel Antibacterial Compounds Derived from Whole-Cell Empiric and Genome-Wide *Fitness* Test

The application of an *S. aureus* whole-cell empiric combined test against the whole genome *S. aureus* fitness test as a new screening technology has represented a major shift in the way that antimicrobial discovery was approached from natural products crude extracts. Not only has the mode of action of many known compounds been validated by means of an antisense gene depletion profiling of crude natural products extracts, but more importantly novel chemical scaffolds have been identified with this chemical genetic approach in the pathogenic bacteria *S. aureus*.

Coelomycin is a highly substituted 2,6-dioxo-pyrazine identified from a fungal extract obtained from an unidentified coelomycete of the Dothileales isolated from *Juniperus phoenicea* from a salt marsh environment in south Spain (Figure 11.17). The diketopiperazine structure of the molecule has an extended conjugation similar to the antiviral flutimide.[72] The same compound was reported at the same time by other authors as a molecule produced from the plant *Menisporopsis theobroma* with poor antimalarial and antimycobacterial activity.[73] The unique AISS profile produced by this extract was

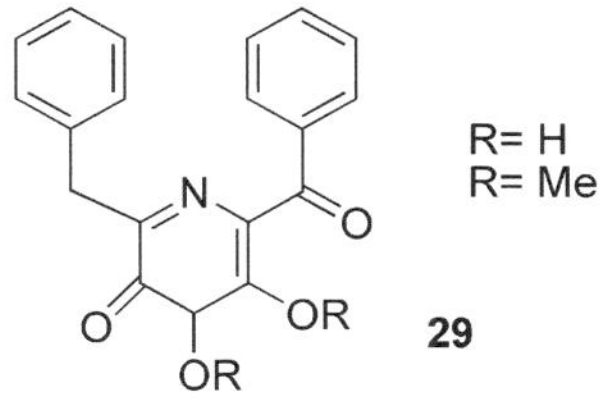

Figure 11.17 Structure of coelomycin.

characterized among others by an important depletion of antisense strains for genes involved in cell wall (*mra*Y, *mur*G, *upp*S and *pbp*A) and fatty acid biosynthesis (*fab*I and *hmr*B), and to a lesser extent in the case of the *map* (methionine aminopeptidase-protein synthesis) and *thy*A (thymidilate synthase-DNA synthesis) genes. This depletion profile could not be correlated with any known inhibitor and no known antibiotic that could account for part of this profile could be detected by HRLC-MS analysis.[47] Further evaluation of the activity of coelomycin against a panel of Gram-positive and Gram-negative bacterial strains showed good broad spectrum activity of the compound with MICs against *S. aureus*, *Streptococcus pneumoniae* or *Haemophilus influenzae* in the range of 4 to 8 μg/mL, an activity that was also shown to be media-dependent. In fact bivalent cations present in the test media have been shown to abolish the activity of this molecule in the whole cell assay. The molecule was not active against wild-type *Escherichia coli* but was effective against a permeable and efflux pump mutant *E. coli* strain (*env*A1, *tol*C)?(MIC 0.5–2 μg/mL). Macromolecular synthesis studies have shown that the compound inhibits all DNA, RNA, protein, peptidoglycan and phospholipid synthesis, with only four-fold selectivity for cell wall synthesis. These results are well in agreement with the complexity of the AISS profile, which supports a broad mode of action of coelomycin. The compound did not show any cytotoxicity to mammalian cell lines (HEK, CHO or Hela cells). When tested in a murine disseminated target organ assay against *S. aureus* the compound did not show *in vivo* activity and showed poor PK and solubility.[47]

Kibdelomycin is a novel class of bacterial gyrase inhibitors with both potent *in vitro* and whole-cell antibacterial activity, and the first one to be discovered from natural products sources since the 1950s.

The structure of kibdelomycin is completely novel compared to other bacterial gyrase inhibitors (novobiocin, chlorobiocin and coumermycin A1) discovered in the 1950s and a recently reported related compound (amycolamycin).[74] Kibdelomycin is comprised of four novel structural motifs, a central hydroxydecalin tetramic acid flanked on each side with highly substituted glycosidic residues terminating with a fourth unit, a dichloro-5-methyl-pyrrole-2-carboxamide (Figure 11.18).[36] Kibdelomycin was identified from an extract obtained from the cultivation of a novel *Kibdelosporangium* strain, isolated from a soil sample collected in a forest in the Central African Republic. This compound generated AISS profiles with strong depletions in

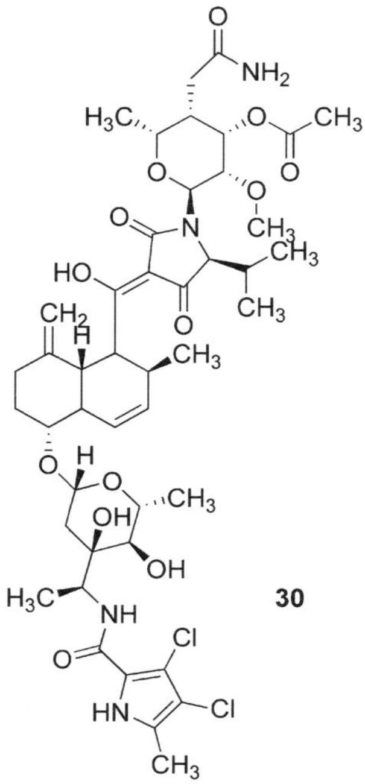

Figure 11.18   Structure of kibdelomycin.

*par*E/C and *gyr*B antisense strains and weak depletions of *gyr*A antisense strains that were highly correlated to those generated in the *S. aureus* fitness test for the natural product coumarin antibiotic novobiocin targeting the two subunits of topoIV and of DNA gyrA respectively.[32,36] The coumarin antibiotics, novobiocin and coumermycin A1, inhibit the catalytic activity of DNA gyrase and topoIV through inhibition of the ATPase activity of these heterodimeric type II topoisomerases and the depletion profiles suggest that kibdelomycin is likely working through an MOA similar to that of the coumarin antibiotics that involves inhibition of the ATPase activity of DNA gyrase and topoIV.[76,77] The major difference between the AISS profiles of novobiocin and kibdelomycin was the lack of significant *gyr*A depletions for kibdelomycin where strong depletions in *gyr*A (GyrA subunit of DNA gyrase) are observed in the novobiocin profiles.

In spite of the predictive fitness test profile pointing to a known coumarin antibiotic, no coumarin antibiotics were identified by high-resolution electrospray ionization Fourier transform mass spectral analysis (HRESIFTMS).

Kibdelomycin exhibits strong antibacterial activity against wild-type *S. aureus* strains (MIC 2 µg/mL) and methicillin-resistant *S. aureus* (MRSA) (MIC 0.5 µg/mL), as well as against the Gram-positive pathogens, *Streptococcus pneumoniae* and *Enterococcus faecalis* (MICs of 1 and 2 µg/mL, respectively) and the Gram-negative pathogen, *Haemophilus influenzae* (MIC 2 μg/mL). Kibdelomycin has no activity against other Gram-negative wild-type pathogens such as *Escherichia coli* (MIC >64 μg/mL), although some inhibitory effect is observed on a permeable and efflux pump mutant *E. coli* (*envA1*, *tolC*) (MIC 32 μg/mL). The activity of kibdelomycin was also drastically reduced to MIC of 64 μg/mL in the presence of 50% human serum. Macromolecular synthesis analysis confirmed that kibdelomycin inhibited DNA synthesis (IC_{50} 0.03 μg/mL) in *S. aureus* with more than 100-fold selectivity over RNA, peptidoglycan, phospholipids and protein synthesis. The inhibition of DNA synthesis is the result of selective inhibition of the ATPase activity of DNA gyrase and topoIV. Kibdelomycin inhibits the *in vitro* DNA gyrase and topoIV activities of both *E. coli* and *S. aureus*, with an activity profile similar to that of the coumarin antibiotic novobiocin. In fact cell-free topoisomerase supercoiling and decatenation assays used to evaluate the inhibitory effect on DNA gyrase and DNA topoIV, have shown that kibdelomycin inhibits *E. coli* gyrase supercoiling and topoIV decatenation activity with IC_{50} values of 60 and 29 000 nM, respectively. Catalytic ATPase assays also show that similarly to novobiocin, kibdelomycin presents more potent inhibition on *E. coli* DNA gyrase ATPase activity (IC_{50} 11 nM) than topoIV ATPase activity (IC_{50} 900 nM). In addition, no cross resistance has been observed with other gyrase inhibitors when tested against novobiocin- and ciprofloxacin-resistant *S. aureus* strains). Kibdelomycin shows very low frequency of resistance ($<5 \times 10^{-10}$), unlike other known GyrB inhibitors (e.g. novobiocin, 10^{-8}) suggesting a different binding than that of known GyrB inhibitors.[26]

In summary, kibdelomycin is the first novel class of potent inhibitor of bacterial type II topoisomerases (DNA gyrase subunit GyrB and topoisomerase IV subunit ParE) and the first in a new class of natural products bacterial gyrase inhibitors with broad spectrum Gram-positive antibacterial activity. Kibdelomycin represents a new chemistry, with apparently a new binding mode, that could help respond to issues of resistance and toxicity shown in the class of the coumarin antibiotics.

11.5 Conclusions and Perspectives

The emergence of bacterial resistance has emphasized the urgent medical need for new antibacterial agents to fight pathogenic bacteria. Microbial natural products have been for decades one of the major sources of clinical antibacterial agents and today all the evidence suggests that novel molecules with potential therapeutic application are still waiting to be discovered from natural sources. Whereas the rediscovery of thousands of known inhibitors, the perceived lack of innovation and the difficulty to identify novel classes of antibiotics has been major limitations for most of the recent approaches developed in the last years in the pharma sector, recent advances in the application of miniaturization and high throughput cultivation techniques, improvements in mass spectrometry and nuclear resonance techniques, combined with new whole-cell target-based screening technologies have shown that novel leads can be delivered from microbial natural products. The discoveries of platensimycin and kibdelomycin are excellent examples that microbial natural products can continue to deliver novel scaffolds when the appropriate technologies are applied in drug discovery. With no doubt, future advances in the field of natural products research will be key to promote a resurgence of natural products as sources of novel discoveries of leads and drugs.

Acknowledgements

In this chapter we would like to acknowledge all our former colleagues from the Basic Research Center of Merck Sharp and Dohme in Spain and the Natural Products Chemistry and Anti-infective Research at Merck in the US who contributed to the development of one of the most productive antibiotic discovery platforms of the last decades.

References

1. S. B. Singh and J. F. Barrett, *Biochem. Pharmacol.*, 2006, **71**, 1006.
2. S. B. Singh and F. Pelaez, *Prog. Drug Res.*, 2008, **65**, 143.
3. C. T. Walsh, *Antibiotics: Actions, Origin, Resistance*, ASM Press, Washington, DC, USA, 2003.
4. C. Walsh, *Nat. Rev. Microbiol.*, 2003, **1**, 65.
5. M. A. Fischbach and C. T. Walsh, *Science*, 2009, **325**, 1089.
6. H. Brötz-Oesterhelt and P. Sass, *Future Microbiol.*, 2010, **5**, 1553.
7. M. S. Butler, *Nat. Prod. Rep.*, 2008, **25**, 475.
8. L. L. Silver, *Expert Opin. Drug Discov.*, 2008, **3**, 487.
9. D. J. Newman and G. M. Cragg, *J. Nat. Prod.*, 2007, **7**, 461.
10. M. S. Butler and A. D. Buss, *Biochem. Pharmacol.*, 2006, **71**, 919.
11. Ganesan, *Curr. Opin. Chem. Biol.*, 2008, **12**, 306.
12. R. Bade, H.-F. Chan and J. Reynisson, *Eur. J. Med. Chem.*, 2010, **45**, 5646.
13. D. Kell, *Trends Biotechnol.*, 1999, **17**, 89.
14. D. J. Payne, M. N. Gwynn, D. J. Holmes and D. L.Pompliano, *Nat. Rev. Drug Discov.*, 2007, **6**, 29.
15. R. L. Monaghan and J. S. Tkacz, *Ann. Rev. Microbiol.*, 1990, **44**, 271.
16. S. Donadio, L. Carrano, L. Brandi, S. Serina, A. Soffientini, E. Raimondi, N. Montanini, M. Sosio and C. O. Gualerzi, *J. Biotechnol.*, 2002, **99**, 175.
17. N. M. Haste, V. Perera, K. N. Maloney, D. N. Tran, P. Jensen, W. Fenical, V. Nizet and M. E. Hensler, *J. Antibiot.*, 2010, **63**, 219.
18. W. Li, J. E. Leet, H. A. Ax, D. R. Gustavson, D. M. Brown, L. Turner, K. Y. Brown, J. Clark, H. Yang, J. Fung-Tomc and K. S. Lam, *J. Antibiot.*, 2003, 56, 226.

19. S. B. Singh, K. Young and L. Miesel, *Expert Rev. Anti Infect. Ther.*, 2011, **9**, 589.
20. R. A. Forsyth, R. J. Haselbeck, K. L. Ohlsen, R. T. Yamamoto, H. Xu, J. D. Trawick, D. Wall, L. Wang, V. Brown-Driver, J. M. Froelich, G. C. Kedar, P. King, M. McCarthy, C. Malone, B. Misiner, D. Robbins, Z. Tan, Z. Zhu, G. Carr, D. A. Mosca, C. Zamudio, J. G. Foulkes and J. W. Zyskind, *Mol. Microbiol.*, 2002, **43**, 1387.
21. S. Altuvia and E. G. Wagner, *Proc. Natl. Acad. Sci. USA*, 2000, **97**, 9824.
22. Y. Ji, A. Marra, M. Roenberg and G. Woodnutt, *J. Bacteriol.*, 1999, **181**, 6585.
23. Y. Ji, B. Zhang, S. F. Van Horn, P. Warren, G. Woodnutt, M. K. R. Burnham and M. Rosenberg, *Science*, 2001, **293**, 2266.
24. H. Jayasuriya, K. B. Herath, C. Zhang, D. L. Zink, A. Basilio, O. Genilloud, M. T. Diez, F. Vicente, I. Gonzalez, O. Salazar, F. Pelaez, R. Cummings, S. Ha, J. Wang and S. B. Singh, *Angew Chem., Int. Ed.*, 2007, **46**, 4684.
25. J. G. Ondeyka, D. Zink, A. Basilio, F. Vicente, G. Bills, M. T. Diez, M. Motyl, G. Dezeny, K. Byrne and S. B. Singh, *J. Nat. Prod.*, 2007, **70**, 668.
26. K. Young, J. Hiranthi, J. G. Ondeyka, K. Herath, C. Zhang, D. L. Zink, A. Galgoci, S. Kodali, R. Painter, L. L. Silver, S. Ha, V. Brown-Driver, R. Yamamoto, A. Forsyth, P. Youngman, J. Sigmund, A. Basilio, F. Vicente, J. Tormo, F. Peláez, D. Cully, J. F. Barrett, D. Schmatz, S. B. Singh and J. Wang, *Antimicrob. Agents Chemother.*, 2006, **50**, 519.
27. S. B. Singh, D. L. Zink, J. Huber, O. Genilloud, O. Salazar, M. T. Díez, A. Basilio, F. Vicente and K. M. Byrne, *Org. Lett.*, 2006, **8**, 5449.
28. J. Wang, S. Kodali, S. H. Lee, A. Galgoci, R. Painter, K. Dorso, F. Racine, M. Motyl, L. Hernandez, E. Tinney, S. L. Colletti, K. Herath, R. Cummings, O. Salazar, I. González, A. Basilio, F. Vicente, O. Genilloud, F. Peláez, H. Jayasuriya, K. Young, D. F. Cully and S. B. Singh, *Proc. Natl. Acad. Sci. USA*, 2007, **104**, 7612.
29. J. Wang, S. M. Soisson, K. Young, W. Shoop, S. Kodali, A. Galgoci, R. Painter, G. Parthasarathy, Y. S. Tang, R. Cummings, S. Ha, K. Dorso, M. Motyl, H. Jayasuriya, J. Ondeyka, K. Herath, C. Zhang, L. Hernandez, J. Allocco, A. Basilio, J. R. Tormo, O. Genilloud, F. Vicente, F. Peláez, L. Colwell, S. Ho Lee, B. Michael, T. Felcetto, C. Gill, L. L. Silver, J. D. Hermes, K. Bartizal, J. Barret, D. Schmatz, J. W. Becker, D. Cully and S. B. Singh, *Nature*, 2006, **441**, 358.
30. J. G. Foulkes, *DDT*, 2002, **7**, S12.
31. S. B. Singh, J. W. Phillips and J. Wang, *Curr. Opin. Drug Discovery Dev.*, 2007, **10**, 160.
32. R. G. K. Donald, S. Skwish, R. A. Forsyth, J. W. Anderson, T. Zhong, C. Burns, S. Lee, X. Meng, L. LoCastro, L. W. Jarantow, J. Martin, S. Ho Lee, I. Taylor, D. Robbins, C. Malone, L. Wang, C. S. Zamudio, P. J. Youngman and J. W. Phillips, *Chem. Biol.*, 2009, **16**, 826.

33. D. Xu, B. Jiang, T. Ketela, S. Lemieux, K. Veillette, N. Martel, J. Davison, S. Sillaots, S. Trosok, C. Bachewich and T. Roemer, *Chem. Biol.*, 2007, **3**, 92.
34. M. J. Valler and D. Green, *DDT*, 2000, **5**, 286.
35. L. L. Silver, *Clin. Microbiol. Rev.*, 2011, **24**, 71.
36. J. W. Phillips, M. A. Goetz, S. K. Smith, D. L. Zink, J. Polishook, R. Onishi, S. Salowe, J. Wiltsie, J. Allocco, J. Sigmund, K. Dorso, S. Lee, S. Skwish, M. de la Cruz, J. Martín, F. Vicente, O. Genilloud, J. Lu, R. E. Painter, K. Young, K. Overbye, R. G. K. Donald and S. B. Singh, *Chem. Biol.*, 2011, **18**, 955.
37. G. G. Yarbrough, D. P. Taylor, R. T. Rowlands, M. S. Crawford and L. L. Lasure, *J. Antibiot.*, 1993, **46**, 535.
38. R. H. Baltz, *Microbe*, 2007, **2**, 12.
39. J. Clardy, M. A. Fischbach and C. T. Walsh, *Nature Biotechnol.*, 2006, **24**, 1541.
40. Y. M. Chiang, K. H. Lee, J. F. Sanchez, N. P. Keller and C. C. Wang, *Nat. Prod. Commun.*, 2009, **4**, 1505.
41. G. Challis, *Microbiology*, 2008, **154**, 1555.
42. J. R. Tormo, J. B. García, M. DeAntonio, J. Feliz, A. Mira, M. T. Diez, P. Hernáncez and F. Pélaez, *J. Ind. Microbiol. Biotechnol.*, 2003, **30**, 582.
43. H. B. Bode, B. Bethe, R. Hofs and A. Zeeck, *ChemBioChem*, 2002, **3**, 619.
44. K. Scherlach and C. Hertweck, *Org. Biomol. Chem.*, 2009, **7**, 1753.
45. G. F. Bills, G. Platas, A. Fillola, M. R. Jimenez, J. Collado, F. Vicente, J. Martin, A. Gonzalez, J. Bur-Zimmermann, J. R. Tormo and F. Peláez, *J. Appl. Microbiol.*, 2008, **104**, 1644.
46. O. Genilloud, I. González, O. Salazar, J. Martín, J. R. Tormo and F. Vicente, *J. Ind. Microbiol. Biotechnol.*, 2011, **38**, 375.
47. M. A. Goetz, C. Zhang, D. L. Zink, M. Arocho, F. Vicente, G. F. Bills, J. Polishook, K. Dorso, R. Onishi, C. Gill, E. Hickey, S. Lee, R. Ball, S. Skwish, R. G. K. Donald, J. W. Philips and S. B. Singh, *J. Antibiot.*, 2010, **63**, 512.
48. W. A. Duetz and B. Witholt, *Biochem. Eng. J.*, 2001, **7**, 113.
49. W. A. Duetz, *Trends Microbiol.*, 2007, **15**, 469.
50. H. F. Zimmermann and J. Rieth, *J. Assoc. Lab. Automat.*, 2006, **11**, 134.
51. B. Jiang, D. Xu, J. Allocco, C. Parish, J. Davison, K. Veillette, S. Sillaots, W. Hu, R. Rodriguez-Suarez, S. Trosok, L. Zhang, Y. Li, F. Rahkhoodaee, T. Ransom, N. Martel, H. Wang, D. Gauvin, J. Wiltsie, D. Wisniewski, S. Salowe, J. N. Kahn, M. J. Hsu, R. Giacobbe, G. Abruzzo, A. Flattery, C. Gill, P. Youngman, K. Wilson, G. Bills, G. Platas, F. Pelaez, M. T. Diez, S. Kauffman, J. Becker, G. Harris, P. Liberator and T. Roemer, *Chem. Biol.*, 2008, **15**, 363.
52. C. A. Parish, S. K. Smith, K. Calati, D. Zink, K. Wilson, T. Roemer, B. Jiang, D. Xu, G. Bills, G. Platas, F. Peláez, M. T. Díez, N. Tsou, A. E. McKeown, R. G. Ball, M. A. Powles, L. Yeung, P. Liberator and G. Harris, *J. Am. Chem. Soc.*, 2008, **130**, 7060.

53. K. Herath, G. Harris, H. Jayasuriya, D. Zink, S. Smith, F. Vicente, G. Bills, J. Collado, A. Gonzalez, B. Jiang, J. N. Kahn, S. Galuska, R. Giacobbe, G. Abruzzo, E. Hickey, P. Liberator, D. Xu, T. Roemer and S. B. Singh, *Bioorg. Med. Chem.*, 2009, **17**, 1361.

54. J. Ondeyka, G. Harris, D. Zink, A. Basilio, F. Vicente, G. Bills, G. Platas, J. Collado, A. Gonzalez, M. de la Cruz, J. Martin, J. N. Kahn, S. Galuska, R. Giacobbe, G. Abruzzo, E. Hickey, P. Liberator, B. Jiang, D. Xu, T. Roemer and S. B. Singh, *J. Nat. Prod.*, 2009, **72**, 136.

55. J. G. Ondeyka, D. L. Zink, K. Young, R. Painter, S. Kodali, A. Galgoci, J. Collado, J. R. Tormo, A. Basilio, F. Vicente, J. Wang and S. B. Singh, *J. Nat. Prod*, 2006, **69**, 377.

56. R. J. Heath, S. W. White and C. O. Rock, *Appl. Microbiol. Biotechnol.*, 2002, **58**, 695.

57. H. Mori and K. Ito, *Trends Microbiol.*, 2001, **9**, 494.

58. C. A. Parish, M. de la Cruz, S. K. Smith, D. Zink, J. Baxter, S. Tucker-Samaras, J. Collado, G. Platas, G. Bills, M. T. Díez, F. Vicente, F. Peláez and K. Wilson, *J. Nat. Prod.*, 2009, **72**, 59.

59. Y. Sugie, S. Inagaki, Y. Kato, H. Nishida, C.-H. Pang, T. Saito, S. Sakemi, F. Dib-Hajj, J. P. Mueller, J. Sutcliffe and Y. Kojima, *J. Antibiot.*, 2002, **55**, 25.

60. S. B. Singh, D. L .Zink and J. Huber, *Org. Lett.*, 2006, **8**, 5449.

61. S. B. Singh, D. L. Zink, K. B. Herath, O. Salazar and O. Genilloud, *Tetrahedron Lett.*, 2008, **49**, 2616.

62. S. B. Singh, D. L. Zink and K. Dorso, *J. Nat. Prod.*, 2008, **72**, 345.

63. M. Ueno, M. Amemiya, K. Yamazaki, M. Iijima, M. Osono, T. Someno, H. Iimuma, M. Hamada, M. Ishizuka, T. Takeuchi, *J. Antibiot.*, 1993, **46**, 1156.

64. H. Wang and J. B. Gloer, *Tetrahedron Lett.*, 1995, **36** 5847.

65. F. Kong and G. T. Carter, *Tetrahedron Lett.*, 2003, **44**, 3119.

66. J. G. Ondeyka, D. Zink, A. Basilio, F. Vicente, G. Bills, M. T. Diez, M. Motyl, G. Dezeny, K. Byrne and S. B. Singh, *J. Nat. Prod.*, 2007, **70**, 668.

67. C. Zhang, J. G. Ondeyka, D. L. Zink, A. Basilio, F. Vicente, J. Collado, G. Platas, J. Huber, K. Dorso, M. Motyl, K. Byrne and S. B. Singh, *Bioorg. Med. Chem.*, 2009, **17**, 2162.

68. C. Zhang, J. G. Ondeyka, D. L. Zink, A. Basilio, F. Vicente, J. Collado, G. Platas, J. Huber, K. Dorso, M. Motyl, K. Byrne and S. B. Singh, *J. Nat. Prod.*, 2008, **71**, 130.

69. H. Imai, K.-i. Suzuki, M. Morioka, Y. Numasaki, S. Kadota, K. Nagai, T. Sato, M. Iwanami and T. Saito, *J. Antibiot.*, 1987, **40**, 1475.

70. C. Zhang, J. G. Ondeyka, D. L. Zink, A. Basilio, F. Vicente, O. Salazar, O. Genilloud, K. Dorso, M. Motyl, K. Byrne and S. B. Singh, *J. Antibiot.*, 2009, **62**, 655.

71. H. Jayasuriya, D. Zink, A. Basilio, F. Vicente, J. Collado, G. Platas, J. Huber, K. Dorso, M. Motyl, K. Byrne and S. B. Singh, *J. Antibiot.*, 2009, **62**, 265.

72. S. B. Singh and J. E. Tomassini, *J. Org. Chem.*, 2001, **66**, 5504.
73. M. Chinworrungsee, P. Kittakoop, J. Saenboonrueng, P. Kongsaeree and Y. Thebtaranonth, *J. Nat. Prod.*, 2006, **69**, 1404.
74. S. Tohyama, Y. Takahashi and Y. Akamatsu, *J. Antibiot.*, 2010, **63**, 147.
75. J. W. Phillips, M. A. Goetz, S. K. Smith, D. L. Zink, J. Polishook, R. Onishi, S. Salowe, J. Wiltsie, J. Allocco, J. Sigmund, K. Dorso, S. Lee, S. Skwish, M. de la Cruz, J. Martín, F. Vicente, O. Genilloud, J. Lu, R. E. Painter, K. Young, K. Overbye, R. G. K. Donald and S. B. Singh, *Chem. Biol.*, 2011, **18**, 955.
76. M. Oblak, M. Kotnik and T. Solmajer, *Curr. Med. Chem.*, 2007, **14**, 2033.
77. A. Maxwell and D. M. Lawson, *Curr. Top. Med. Chem.*, 2003, **3**, 283.

CHAPTER 12
Discovery and Development of Platensimycin and Platencin

SHEO SINGH

Merck Research Laboratories, Rahway, NJ 07065, USA
E-mail: sheo.singh@merck.com

12.1 Introduction

Small molecule natural products produced by microorganisms and screening methods that were primarily based on inhibiting bacterial growth have played crucial roles in the discovery and development of antibiotics. This discovery paradigm dates back to the discovery of penicillin in 1928. The latter discovery revolutionized the spirit of collaboration; academic scientists very closely worked with the industrial scientists to produce penicillin for treatment of infected patients during the Second World War. The discovery of penicillin was not only crucial for the discovery and its development as an antibiotic but also was pivotal in launching the field of microbial chemistry that allowed for the discovery of a large number of small molecule natural products from microbial sources, initially as antibiotics and subsequently as drugs for other human health targets such as statins and immunosuppressants. Worldwide concerted efforts, by academia and industry, for the next twenty years (1940–1962) culminated in the discovery of most of the natural product-derived antibacterial chemical scaffolds that are in clinical practice today.[1,2] While these successes were admirable and joyful celebration was very reasonable, speculation and naïve conclusions that these antibiotics or their chemical variants would be sufficient to conquer bacterial infections was untimely. It turns out that bacteria are very resourceful and evade drugs by changing

RSC Drug Discovery Series No. 25
Drug Discovery from Natural Products
Edited by Olga Genilloud and Francisca Vicente
© The Royal Society of Chemistry 2012
Published by the Royal Society of Chemistry, www.rsc.org

themselves. As it stands now, antibiotic resistance pre-dated the introduction of antibiotics, such as penicillins, into the clinic and resistance continues to spread, compromising antibiotic utility.

Of the penicillins developed earlier, methicillin was one of the more important ones and soon after its introduction to the market various hospital strains of *Staphylococcus aureus* started to demonstrate resistance to it, and now methicillin-resistant *Staphylococcus aureus* (MRSA) accounts for 60–89% of nosocomial infections.[3–6] A recent study shows that MRSA infections cause a large number of deaths every year in the United States and elsewhere in the world.[7] Vancomycin, a glycopeptide natural product, was considered the antibiotic of last resort for treatment of Gram-positive infections; however, 20–30% of *Enterococci* are now found to be resistant to this drug.[3,6] Linezolid, first of the synthetic oxazolidinones, and daptomycin, a peptidic natural product, are newer antibiotics that represent options for treatment of Gram-positive infections including MRSA. Resistance is prevalent among other organisms as well, including Gram-negative bacteria which are more difficult to treat. Recent surveillance data from the Centers for Disease Control and Prevention (CDC) suggest that greater than 20% of *Pseudomonas aeruginosa* are imipenem resistant in hospital environments with limited or no treatment options available.[3] Chemical modifications of existing chemical scaffolds have been the sources for a large number of highly potent broad spectrum antibiotics and continue to deliver newer drugs that provide temporary coverage for resistant organisms. However, further modifications of existing scaffolds has become increasingly challenging and discovery of new antibiotic scaffolds with known modes of action or, more importantly, novel scaffolds with completely novel mechanisms of action are urgently needed. To address this massive task a combination of following is needed: (i) identification of new targets, (ii) identification of new sources for antibiotics leads, and (iii) development of new screening techniques/tools.

12.1.1 Identification of New Targets: Fatty Acid Synthase II (FASII) as a Novel Antibiotic Target

Antibiotics that are in clinical practice today inhibit DNA, RNA, protein and cell wall synthesis and inhibit/interact with no more that 25 molecular targets.[2] Bacterial genome analysis and subsequent gene knock-out studies show more than 300 highly conserved essential genes both in Gram-positive and Gram-negative bacteria and represent broad spectrum drug targets.[8–12] Fatty acids are an integral component of phospholipids and are essential for bacterial survival. Bacterial fatty acids are synthesized by type II fatty acid synthases which are comprised of a large number of dissociated enzymes representing a series of molecular targets (Figure 12.1). In contrast, mammalian fatty acid synthesis is catalyzed by type I fatty acid synthases, a single large protein. Enough differences in FASI and FASII enzymes exist, which allows for the rationalization that inhibitors of FASII versus FASI targets should show

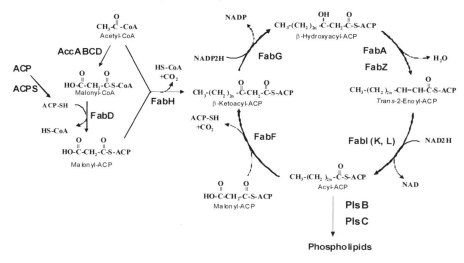

Figure 12.1 Biosynthetic pathway of bacterial fatty acid synthesis.

selectivity. Fatty acid synthesis begins with acetyl CoA which gets converted to malonyl CoA by acetyl carboxylases. FabD converts malonyl CoA to malonyl-ACP (malonyl acyl carrier protein). FabH catalyzes acetylation to form FabH acyl enzyme intermediate followed by a two-carbon transfer from malonyl-ACP to form a four-carbon β-ketoacyl-ACP, which enters in the catalytic cycle of reduction by FabG, dehydration by FabA/FabZ, and reduction by enoyl reductase to produce acyl-ACP. The $n + 2$ elongated acyl-ACP becomes the substrate of the elongation condensing enzyme FabF which adds two more carbons to produce two-carbon elongated β-acyl($n + 2$C)-ACP. Each ($n + 2$C) elongated acyl-ACP goes through the reduction, dehydration, and reduction cycle until the appropriate acyl chain length is achieved. Once the appropriate chain length is achieved it becomes a substrate for PlsB/C to acylate glycerol to form phospholipids (Figure 12.1). Of the FASII enzymes, the elongation condensing enzyme (FabF) is the most attractive of the targets due to its much higher conservation across most bacterial species. Triclosan and isoniazid are clinical agents that inhibit the enoyl reducing enzyme (FabI) which is one of the fatty acid synthesis pathway enzymes, thus providing clinical validation of the pathway. Despite these clinical validations, the essentiality of fatty acid synthesis as a target in certain bacterial species has been challenged recently with supportive evidence provided only for *Streptococcus agalactiae*.[13] More recently, additional evidence has been provided which reinforces the validity of the FASII targets in *S. aureus*.[14]

12.1.2 Sources of New Antibiotics

Natural products, particularly those derived from the microbial sources, have played a major role directly or indirectly in the discovery and development of antibiotics. Essentially all antibiotics that occupy today's clinical practice are unmodified natural products, modified by semi-synthesis or synthesized by total synthesis.[2] Oxazolidinones (e.g. linezolid), quinolones (e.g. ciprofloxacin), and sulfonamides are the rare exceptions. While compounds from synthetic/ chemical collections have served well as starting points (leads) for non-antibacterial biological targets, these sources have not been very useful in delivering leads for antibacterial targets, particularly those that can be advanced to the clinic. This is not a surprise since most of the corporate chemical collections are the extension of the medicinal chemistry programs of other disease targets which often require dialing out antibacterial activity. Therefore, quality antibacterial leads have not been discovered by screening of synthetic collections.[2,15] Unfortunately, natural products screening efforts have not fared well in recent years and have not afforded significant new antibiotic leads. Rediscovery of known compounds predominate natural products antibacterial discoveries despite significant advances in the analytical methods of detection and characterization of natural products. Rediscovery of known compounds leads to significantly increased cost of discovery of each new natural product. This has led to a significant negative impact for the natural product programs, leading to the abandonment of all but a few programs in the large pharmaceutical companies. Differentiation of a known from unknown natural products when screening by whole-cell empiric assays remains the biggest challenge, even with the availability of highly sensitive chemical de-replication methods, especially when compounds are highly potent and are present in very low concentrations. While that challenge is real the opportunities for new discoveries are also real since only a small portion of the bacteria and fungi have been explored for their potential to produce natural products. Therefore, natural products remain a powerful source for discovery of novel antibiotics. The keys to new discoveries are the discovery of new sources capable of producing new natural products – such as deep marine environments, thermophilic environments, unstudied terrestrial habitats, recombinant genetic manipulation of biosynthetic genes of cultivatable and silent genetic machinery, etc. – and couple that with innovative new target-specific whole-cell screening technologies with emphasis on the new targets.

12.1.3   Novel Screening Technologies: Target-Based Whole-Cell Antisense Screening

Conditional expression of antisense RNA of essential genes in *S. aureus* cells has been known to inhibit cell growth and alter cell sensitivity to selective inhibitors,[10,12] leading to the development of new hypersensitive target-based whole-cell screening assays. In the last decade, a number of inducible antisense expression systems for *S. aureus* were reported.[10,12,16] One of the most useful systems was reported by Forsyth *et al.*,[12] which involved a xylose inducible promoter. Xylose does not possess inherent antibacterial activity, which allowed this method to be very useful for the primary screening of antibacterial agents. A recent review describes its application to drug discovery[17] and so it is only very briefly described here. Xylose induction of target-specific antisense RNA leads to proportional decrease of the target mRNA, and hence the target protein, leading to increased sensitivity for the target-specific inhibitors. In order to determine target specificity, the activity of a given compound and/or extract is tested for its ability to inhibit the growth of an antisense sensitized *S. aureus* strain and then compared with its ability to inhibit growth of the control *S. aureus* strain under identical conditions. If the compound and/or extract has target-specific activity it will show hypersensitivity for the antisense strain versus the control strain and show a left shift of the growth inhibition curve (Figure 12.2). The hypersensitivity of target-based inhibitors allows for the discovery of compounds even when they are present in relatively lower concentrations or low titers or even when present as mixtures.

The liquid assay (Figure 12.2) was modified to an agar-diffusion two-plate differential sensitivity assay (two-plate assay)[18] in which one plate was seeded with *S. aureus* cells expressing *fabF* AS-RNA (the AS plate) and the other with cells lacking AS-RNA expression (the control plate). When equal amounts of true FabH and/or FabF inhibitor is applied to both plates, it should selectively inhibit the growth of the *fabF* AS strain and form a larger inhibition zone on the AS plate compared to the control plate. The assay was validated by application of cerulenin, 5 µg, a known FabF selective inhibitor, spotted onto each plate. As expected, cerulenin produced a large zone of growth inhibition on the *fabF* AS plate (18 mm) and no zone of inhibition on the control plate (Figure 12.3). In contrast, the compounds that do not work by inhibition of FabF/H enzymes and work by other modes of action show no zone differentials between both plates. Results of the screening of natural products extracts was reported by Singh *et al.*[17] This screening method was amenable for large-scale primary screening and provided a higher level of sensitivity, which allowed for the detection of antibiotics that are present in lower concentrations.

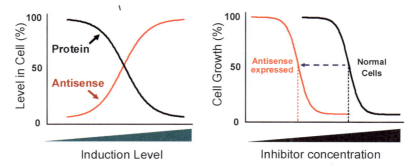

Figure 12.2   Schematic representation of how induction of antisense RNA leads to reduction of the protein levels of the targeted mRNA, leading to differential growth sensitivities between target depleted cells and control cells without antisense induction.

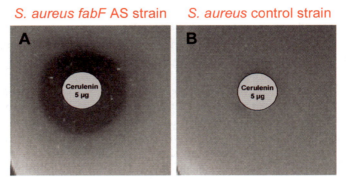

Induction of FabF antisense RNA   *no induction*

Figure 12.3   Two-plate differential sensitivity assay. (A) A plate seeded with *S. aureus* fabF AS-RNA strain where antisense expression was induced with 50 mM xylose. (B) A control plate seeded with *S. aureus* without expression of fabF antisense RNA. A filter disc containing 5 µg cerulenin was placed on the top of agar of each plate.

12.2   Screening of Inhibitors of Condensing Enzymes (FabH/FabF)

To complement the two-plate FabF/H antisense assay, a robust cell-free enzyme assay (the FASII assay) representing enzymes (FabD, FabF/B, FabG, FabA/Z, and FabI) of the fatty acid synthesis pathway was constituted using extracts from wild-type *E. coli*, *S. aureus*, and *B. subtilis*.[19] This assay nicely complemented the whole-cell assay for the confirmation of activities. Screening of over 250 000 natural product microbial extracts using the two-plate FabH/F assays at a single concentration produced a 0.3% hit rate which was reduced to 0.1% after confirmation in FASII biochemical assay.[18,20] The bioassay-guided fractionation allowed rapid discovery of all known inhibitors of the condensing enzymes such as cerulenin, thiolactomycin, and thiotetromycin. These compounds are rather poor antibiotics (MIC > 64 µg/mL) but were easily detected by the antisense-based two-plate FabF assay even when their titer (production levels) in fermentation broths was very poor and did not show any zone of inhibition on the control plate and, therefore, would have not been detected by wild-type strain screening alone. These findings provided validation of the screening process and continued screening allowed for the identification of Tü3010,[21] an amide analog of thiotetromycin. While Tü3010 showed no difference in activity in the cell-free FASII assay[19] compared to thiolactomycin and thiotetromycin, it showed >15-fold better potency in antibacterial assays.[18] Similarly, the screening efforts led to the isolation of a series of novel acetylenic allenes, phomallenic acids A–C (Figure 12.4).[18,22]

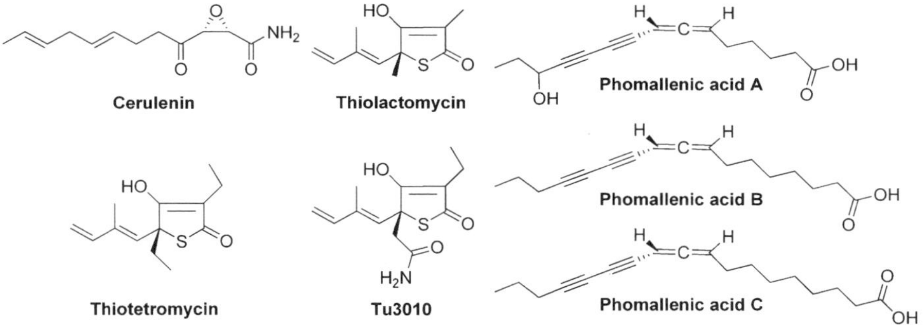

Figure 12.4 Chemical structures of natural product inhibitors of bacterial condensing enzymes (FabH/F).

12.3 Platensimycin and Platencin

Rapid discovery of the both known and unknown inhibitors of FabH/FabF (Figure 12.4) validated the antisense-based screening strategy and allowed for implementation of the strategy on a wider scale for routine natural products screening. Screening of the samples from freshly isolated strains of soil bacteria led to the discovery of platensimycin and platencin as selective inhibitors of condensing enzymes, FabF/H (Figure 12.5).[20,23–25]

Platensimycin (PTM) was discovered from an extract of *Streptomyces platensis* MA7327 (Figure 12.6), which was first isolated from a soil sample that was collected in South Africa. Subsequent routine screening led to the discovery of several strains of *S. platensis* from soil samples collected from various parts of Spain that produced platensimycin. Platencin (PTN) was isolated from another strain of *S. platensis* MA7339 isolated from a soil sample collected in Spain. Originally MA7327 produced only platensimycin and MA7339 produced only platencin. However, during biosynthetic studies of platensimycin it was noted that MA7327 also produced minor amounts of platencin. This co-production of PTM and PTN was recently confirmed by the Scripps group.[26] Structures of platensimycin and platencin were primarily elucidated by nuclear magnetic resonance (NMR) and mass spectral analysis.[20,23–25] The absolute configuration of platensimycin was established

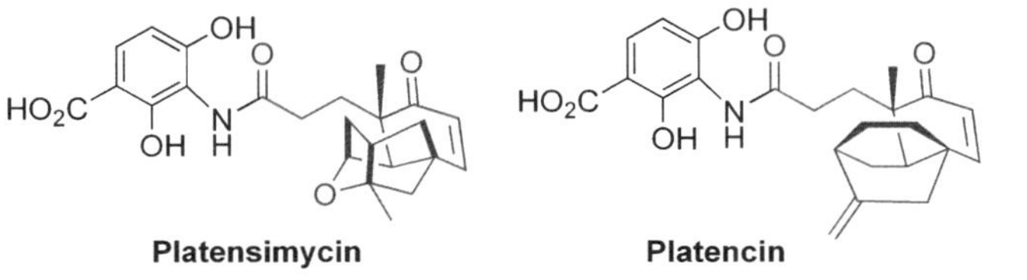

Figure 12.5 Structures of platensimycin and platencin. Structural moiety: 3-amino-2,4-dihydroxy benzoic acid (black drawing) and platensic and platencinic acid (blue drawing).

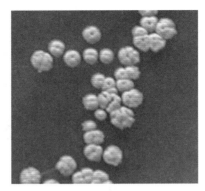

Figure 12.6 Photograph of platensimycin producing culture of *Streptomyces platensis* MA7327.

by a single X-ray crystal diffraction analysis of a 6'-bromo derivative.[23] The structure of platensimycin is comprised of two novel moieties, an amine (3-amino-2,4-dihydroxybenzoic acid) and a C-17 tetracyclic enone acid (platensic acid) joined by an amide bond. The structure of platencin consists of two moieties as well: an amino benzoic acid moiety which is the same as present in platensimycin and a C-17 tricyclic enone acid (platencinic acid).

12.3.1 Antibacterial Activity and *in vivo* Efficacy

Platensimycin and platencin showed potent broad spectrum Gram-positive antibiotic activities, exhibiting potent MIC values (0.06–2 µg/mL) against clinically relevant bacteria including major drug-resistant pathogens such as MRSA, VRE, and VISA.[20,24] Platensimycin and platencin exhibited excellent *in vivo* efficacy in an *S. aureus* mouse infection model when infused intravenously, without exhibiting any overt toxicity. Both compounds exhibited 4 to 5 log reductions of colony-forming units (CFU) of *Staphylococcus aureus* burden in kidney at 100–150 µg/hr infusion. These compounds did not show significant efficacy when dosed by conventional routes of administration such as subcutaneous, intraperitoneal, and oral.

12.3.2 Fatty Acid Synthesis Inhibition and Target Specificity

Platensimycin showed potent inhibition (IC_{50} 0.1 µg/mL) of phospholipid synthesis and demonstrated over 10 000-fold selectivity over the inhibition of other macromolecules such as DNA, protein, cell wall and RNA synthesis (Figure 12.7) in *S. aureus* and *Streptococcus pneumoniae*. Platencin similarly inhibited phospholipid synthesis with similar specificity against the other macromolecular targets. The exquisite selectivity displayed by these compounds for the inhibition of fatty acid synthesis was further confirmed by the analysis of their AISS (antisense induced strain sensitivity) profile in the SaFT

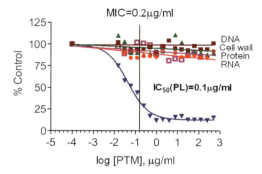

Figure 12.7 Inhibition of macromolecular synthesis by platensimycin.

(*S. aureus* fitness test) assay.[27] SaFT is a chemical genetic assay which consists of 245 antisense strains with reduced expression of over 245 essential target genes.[27] Platensimycin showed significant depletions of the target gene *fabF*-AS, and upstream genes *accA*-AS, *accC*-AS and *accB*-AS (Figure 12.8). Platencin showed identical gene depletions except for some what stronger gene depletion of *accB*-AS (Figure 12.8). Since condensing enzyme genes *fabH* and *fabF* are members of the same operon, *fabF*-AS represents the effect of both genes.[18,28] The AISS profile of platensimycin was unaltered whether it was tested as a purified compound or tested as crude extracts of *Streptomyces platensis* (Figure 12.9) containing various titers of platensimycin.

12.3.3 Inhibition of Condensing Enzymes, FabF and FabH

Platensimycin and platencin impart their antibacterial activity by selectively inhibiting biosynthesis of fatty acid in bacteria; however, they differ in their precise modes of action. Platensimycin selectively inhibited the *S. aureus* elongation condensing enzyme FabF (IC_{50} 48 nM) and was essentially inactive against initiation condensing enzyme (FabH) (IC_{50} 67 μM) thus showing over 800-fold specificity for FabF. Platensimycin showed a binding affinity of 19 nM in a FabF direct binding assay.[20] Platencin showed essentially identical potency as platensimycin in all the cellular assays including inhibition of *S. aureus* fatty acid synthesis (IC_{50} 0.23 μg/mL) in macromolecular synthesis inhibition assays. However, it was six-fold less potent in the FabF direct binding assay (IC_{50} 0.11 μM). It did show a four fold better activity against *S. aureus* FabH (IC_{50} 16.2 μM) compared to PTM. This apparent difference in the cellular and cell-free activities could be better explained by comparison of activities of the two compounds in a gel elongation assay using *S. aureus* cell lysate in which FabH and FabF inhibitory activities were measured simultaneously in the same assay.[24] In this assay, platensimycin exhibited IC_{50} values against FabF (0.29 μM) and FabH (247 μM) and platencin showed IC_{50} values of FabF (4.58 μM) and FabH (9.17 μM). Comparison of the FabF and FabH activities of platensimycin indicated 850-fold selectivity for FabF

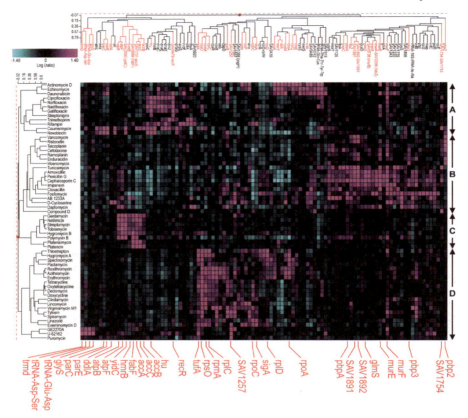

Figure 12.8   Two-dimensional cluster analysis of AISS profiles of various compound classes belonging to defined mechanisms and 245 strains showing depletion (magenta) and resistance (cyan). The compounds representing a specific mechanism cluster together. (A) A cluster of 12 nucleic acid synthesis/replication inhibitors; (B) a cluster of 16 cell wall biosynthesis inhibitors along with isoprenoid biosynthesis inhibitor AB1233A and membrane depolarizer daptomycin; (C) a cluster of five aminoglycosides along with FAS inhibitors, platensimycin, platencin and polymyxin B; and (D) a cluster of 20 protein synthesis inhibitors along with U-62162, a natural product inhibitor with an unknown mode of action. For detailed explanation and more precise profiling data, see for example Donald *et al.*[27]

whereas platencin showed only two-fold selectivity for FabF (Table 12.1). Cross comparison of the activity between the two compounds revealed that platensimycin had a 15-fold better *S. aureus* FabF activity and platencin had a 30-fold better *S. aureus* FabH activity. The balanced, albeit lower, FabF/H potencies of platencin appear to be responsible, and account, for the potent fatty acid synthesis inhibition which is essentially equal to the fatty acid synthesis inhibition of platensimycin.

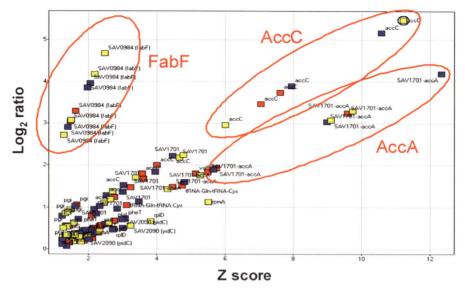

Figure 12.9 Two-dimensional AISS profiles of *Streptomyces platensis* extracts containing platensimycin (3 mg/L), the natural titer at screening extract. Data points represent independent doses across a minimum of two profiling experiments. All the data is represented. FabF, the enzyme target for platensimycin, is significantly depleted and is clustered away from the baseline noise. The acetyl CoA carboxylases (ACC) A and C, the enzymes early in the pathway, are also depleted perhaps due to a pathway effect.

The condensing enzymes, FabH and FabF, are bi-functional enzymes that catalyze two discrete steps of the biochemical reactions. The first biochemical reaction that is catalyzed by both of these enzymes is the acylation of active site cysteine followed by the binding of the second substrate malonyl-ACP. Claisen condensation followed by decarboxylation leads to the addition of two carbons to the growing acyl chain (Figure 12.10). Iteration of this process continues until appropriate chain lengths of fatty acids are achieved, at which point they get cleaved from the enzyme and ligated to glycerol to form triglyceride. It has

Table 12.1 Summary of key biochemical data of platensimycin and platencin

Enzyme	Platensimycin (IC_{50}, μM)	Platencin (IC_{50}, μM)	Fold selectivity
FabF (S. aureus)	0.048	ND	—
FabF binding (E. coli)	0.019	0.113	6× for PTM
FabH (S. aureus)	67	16.2	4× for PTN
Gel elongation cell lysate			
FabF (S. aureus)	0.29	4.58	15× for PTM
FabH (S. aureus)	247	9.17	30× for PTN

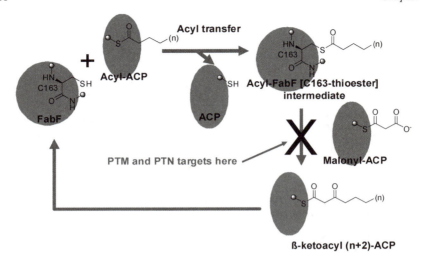

Figure 12.10   Enzymatic reactions catalyzed by FabF and target of platensimycin.

been shown by binding experiments that platensimycin does not bind to the enzyme in the absence of the fatty acid substrate and does bind in the presence of the fatty acid substrate, confirming that platensimycin binds only to the acyl-enzyme intermediate and thus inhibits only acylated enzyme.[20] This phenomenon was confirmed by X-ray co-crystallographic studies. A co-crystal structure of platensimycin bound to C163Q ecFabF provides insight of binding and mechanism of inhibition (Figure 12.11).[20] The cysteine to glutamine

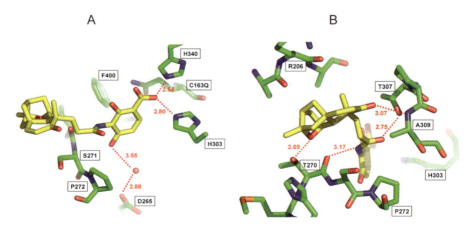

Figure 12.11   X-ray co-crystal structure of platensimycin bound to C163Q ecFabF. (A) Interactions of aminobenzoic acid and (B) interaction of the C-17 tetracyclic enone acid. Dotted lines show polar interactions with amino acids numbered in the black box. The numbers represent the distance between the polar groups in Å.

mutation is known to mimic acylated enzyme and opens up the binding site by 45° movement of gatekeeper residue Phe400.[20] Once in the active site the benzoic acid moiety binds to the catalytic triad and the phenyl ring provides an edge to face interaction with the gatekeeper residue phenylalanine (Phe400). The carboxyl group of the benzoic acid unit forms a salt bridge interaction with the two active site histidines (His303 and 340) and the C-4 phenolic OH group engages in a water mediated H-bond interaction with the aspartic acid (D265). Most of the tetracyclic enone acid portion of the molecule binds to the mouth of the enzyme and is solvent exposed but still provides significant binding energy through van der Waals interactions and polar interactions with the backbone residues. The amide group that connects the two structural moieties shows two strong polar interactions. The NH of the amide group shows an H-bond with the keto group of the backbone threonine (Thr270) and the amide keto group shows an equally strong H-bond with threonine (Thr307). The two oxygen atoms in the tetracyclic cage structure show H-bond interactions with the OH group of threonine (Thr270) and NH of the alanine (Ala309).

X-Ray crystal structure of *Escherichia coli* FabF (C163A) bound to platensimycin (2.75 Å) shows identical polar interactions to that observed for saFabF (C163Q) (Figure 12.12). The crystal structure of *Escherichia coli* FabF (C163A) bound to platencin also showed similar polar interactions except for the missing H-bond with Thr270 due to loss of the ether oxygen in platencin (Figure 12.12). The loss of this polar interaction is likely enough to account for the loss of binding affinity by six-fold (see Table 12.1).

It has been established thoroughly that platensimycin is a selective inhibitor of elongation condensing enzyme (FabF) and platencin is a balanced inhibitor of both condensing enzymes, H and F (Figure 12.13). It has been postulated that compounds that inhibit/interact with more than one protein target will have a higher barrier of resistance. Whether that is the case for platencin compared to platensimycin is unknown as yet. The cellular activity of PTM and PTN is essentially identical but the cell-free single enzyme potencies are very different. It is clear that the potent cellular activity of platencin is due to inhibition of both enzymes. Whether FabH inhibition leads to synergistic antibacterial activity is not reported.

12.3.4 Antidiabetic and Antisteatotic Activity

The lack of antibiotic activity of platensimycin by conventional routes of administration was attributed to the lack of systemic exposure. In a drug disposition study using [³H]-platensimycin in bile duct cannulated (BDC) rats it was determined that the platensimycin was highly concentrated in bile, suggesting that it must be extracted by the liver and therefore never goes back for recirculation in the plasma. This insight provided rationale and impetus to the Merck team to study its effect on the inhibition of mammalian fatty acids, which are mostly synthesized in the liver.[29] Platensimycin was found to be a

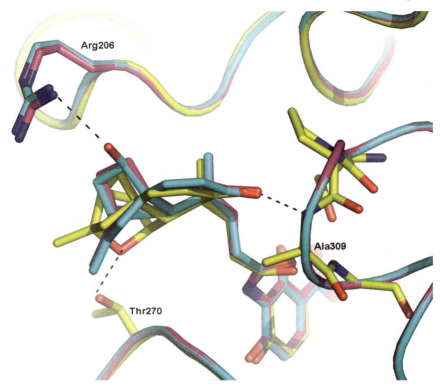

Figure 12.12   Superposition of the structures of platensimycin (yellow), platencin (magenta) and platencin A₁.

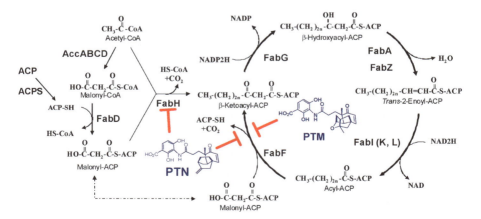

Figure 12.13   Fatty acid synthesis II pathway, showing steps inhibited by platensimycin and platencin.

potent and selective inhibitor of the human and rat fatty acid syntheses in the cell-free system with IC_{50} values of 0.3 μM and 0.2 μM, respectively. It inhibited fatty acid synthesis in rat hepatocytes with an IC_{50} of 0.1 μM.[29] It inhibited *de novo* lipogenesis (DNL) in the mouse liver but not in other tissues such as adipose tissues. The chronic treatment of diabetic models of mice with platensimycin showed significant lowering of the plasma glucose, liver triglyceride and increased insulin sensitivity at 30 mg/kg. The inhibition of fatty acid synthesis leads to increased levels of malonyl CoA, which is a negative regulator of carnithine palmatoyl transferase 1 (CPT1), the enzyme that transports fatty acids to mitochondria for the β-oxidation. Clearly the lower levels of liver fatty acids and increased levels of malonyl CoA leads to lower activity of CPT1 and decreased inhibition of β-oxidation. Lower β-oxidation levels results in concomitant increase of glycolysis and thus lowering of glucose. These data validate that FAS inhibition by platensimycin leads to antidiabetic activity in rodent model of diabetes. The compelling scientific rationale for such an effect exhibited by this compound provides impetus for its potential development as an antidiabetic agent for treatment of type II diabetes mellitus.[29]

12.3.5   De-replication of Platensimycin and Platencin

Successful natural products screening programs often continue in order to find better leads after discovery of potent and desirable inhibitors. The FabF screening program was no exception. After the discovery of platensimycin and platencin, the screening campaign continued to discover new fatty acid synthesis inhibitors and it became necessary to develop an efficient and quantitative method for rapid de-replication of known FAS inhibitors including PTM and PTN. This task was easily accomplished by employing HPLC with an integrated photo-diode array detector and using standard curve for quantification. The quantity of each compound present in the broth was predicted by quantification of the biological activity. When the predicted amount of each compound was matched with the amount of platensimycin and/or platencin present in the broth, those extracts were readily eliminated from further chemistry work. This procedure allowed us to discover about a dozen additional producers of platensimycin and platencin. When necessary, the structure verification of these compounds was easily accomplished by mass spectral fragmentation of platensimycin which exhibited a characteristic fragment ion at *m/z* 273 due to the loss of the anilide unit (3-amino-2,4-dihydroxy benzoic acid).[30–32] The corresponding mass spectral fragment ion from platencin appeared at *m/z* 257 but was relatively weaker. When the platencin tricyclic terpenoid unit is substituted with a hydroxy group the corresponding fragment ion appeared at *m/z* 273, which was isomeric with the fragment ion with platensimycin, but the HPLC retention time would be different and differentiation between the compounds would be trivial. This differentiation only became necessary during search of congeners during the scale-up production of platensimycin (see below) and was never an issue during de-replication of extracts from new screening samples. The SaFT AISS profiling data (Figures 12.8 and 12.9) was not available until after the screening campaign had ended; had it been available, it could have served as an extremely powerful de-replication tool for these compounds due to specific depletion of only three strains.

12.3.6   Large-Scale Production of Platensimycin

The original production of platensimycin from the natural *S. platensis* strain was approximately 3 mg/L. Several rounds of seed and production media optimization of the producing organism led to a production of 180 mg/L allowing for the purification of over 500 g of platensimycin.[30,32] The fermentation and purification process was readily scalable from the shake flasks to 20–600 L fermentation tanks. Initial purification of platensimycin was accomplished by a three-step chromatographic process involving solid-phase extraction on a reversed phase resin (Amberchrome), size exclusion chromatography on Sephadex LH20, and reversed phase HPLC.[23] With the knowledge of the structural features and increased production, the purification of platensimycin was improved and was readily accomplished in over 60% yield using a solid-phase extraction followed by an acid/base switch directly on the solid-phase extraction column and subsequent precipitation.[30]

12.4   Discovery of Congeners

12.4.1   Production from Native Organism

Congeners play a significant role in the understanding of the structure–activity relationship and provide rationale for the biosynthesis of natural products. Often many congeners are isolated during the discovery of original natural products and are present in relatively higher abundance. However, in the case of platensimycin, no congeners were isolated during the original purification of platensimycin or platencin from the early shake flask or small-scale fermentation studies in fermentation tanks using native producer. During the large-scale fermentation studies for the isolation of 10s to 100s grams of platensimycin using native producer and modified media a concerted effort was undertaken to isolate congeners of platensimycin. LCMS with photo-diode array detector was used to identify peaks that corresponded to PTM and PTN class of compounds. Gel filtration on Sephadex LH20 played a pivotal role in the concentration of these compounds in a series of fractions allowing for very efficient subsequent purifications. This led to the purification and characterization of over 32 congeners of platensimycin and platencin. Most of the congeners were hydroxylated analogs of platensimycin and platencin and were isolated in yields of less than 0.8 mg/L in the fermentation mixtures where the titer of platensimycin was 50–180 mg/L. Surprisingly, during this effort platencin was not isolated from the platensimycin producers but its hydroxy analogs were abundantly produced. The congeners were named based on their substitution patterns: series A congeners are those with hydroxylation in terpenoid moieties and series B congeners are those with substitutions in the benzoic acid moiety. The molecules with other type of modifications were given independent names. Often congeners were isolated as methyl esters due to the extraction by acidic methanol; as a result the formation of methyl esters is likely to be an artifact of isolation. The congeners are shown in Figure 12.14: platensimycin A_1 (14-hydroxyplatensimycin),[33] platensimycin A_2 (13-hydroxyplatensimycin),[32] platensimycin A_3 (12-hydroxyplatensimycin),[32] platensimycin A_4 (17-hydroxyplatensimycin),[32] platensimycin A_5 (10,16-seco-15,17-dihydroxyplatensimycin),[32] platensimycin A_6 (7-hydroxy-6,7-dihydroplatensimycin),[32] platensimycin B_1 (platensimycin-1′-carboxymide),[34] platensimycin B_2 (platensimycin-1′-cyclic carbamate),[34] platensimycin B_1 (descarboxyplatensimycin),[34] platensimycin B_4 (platensimycin-4′-glucoside),[31] platensimide A,[35] homoplatensimide A[36], platensic acid,[35] 14-hydroxyplatensic acid,[33] platencin A_1 (13-hydroxyplatencin),[37] platencin A_2 (12-hydroxyplatencin),[30] platencin A_3 (14-hydroxyplatencin),[30] platencin A_4 (15-hydroxyplatencin),[30] platencin A_{11} (15,16-dihydroxyplatencin),[32] 12-hydroxy-platencinic acid, 13-hydroxy-platencinic acid, and platencin glycerol ester.[32]

12.4.2 Overproduction of Platensimycin and Platencin and Production of Congeners by Gene Deletions

Genetic suppression of biosynthetic regulation genes and various other genetic manipulations of biosynthetic machinery is an extremely powerful modern technique for overproduction of natural products. Recently, Shen and his associates described 100-fold increase in the production (323 mg/L) of platensimycin from the original *Streptomyces platensis* MA7327 by deletion of gene *ptmR1*, a gene which encodes a putative *GntR*-like transcriptional regulator.[26] Shen's group produced two strains (SB12001 and SB12002) by *ptmR1* deletions.[26] Both of these strains not only overproduced platensimycin but also remarkably overproduced platencin (255 mg/L) in the same culture broth. While 180 mg/L production of platensimycin was demonstrated in fermentation tanks from wild-type producing organism,[30,32] similar overproduction of platensimycin and platencin by *ptmR1* deleted organism has been only demonstrated in shake flasks and not in fermentation tanks. However 255 mg/L production of platencin is highly noteworthy since its production by wild-type strains has remained very low (<3 mg/L). Originally it was noted that *S. platensis* MA7327 only produced platensimycin and *S. platensis* MA7339 only produced platencin.[23,25] However, during biosynthetic studies MA7327 produced platensimycin[38] along with platencin,[39] which was subsequently confirmed by Shen's group.[26] The Shen group also reported deletion of a similar *GntR*-like transcriptional regulator, *ptnR1*, in the original platencin-only producer *S. platensis* MA7339 and generated strain SB12600

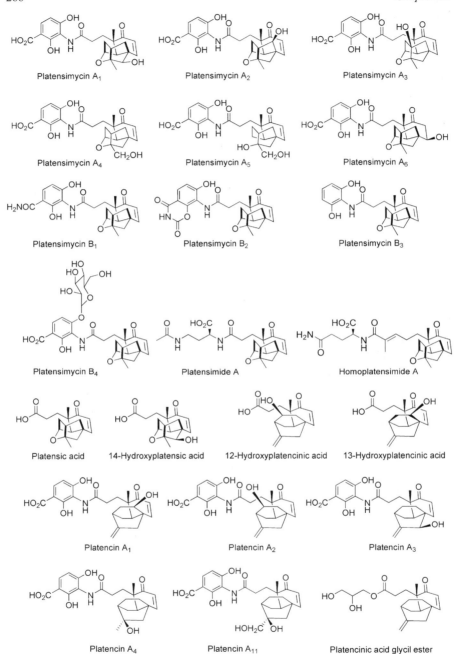

Figure 12.14   Chemical structures of platensimycin and platencin congeners produced by the native *S. platensis*.

which overproduced platencin (22 mg/L). From the latter genetically modified strain, the production of a number of platencin congeners were also reported but no production of platensimycin was reported.[40] The hydroxylated platencins produced by the strain SB12600 listed here did not follow the naming conventions discussed above. These are platencin A_2, A_3, A_5, A_6, A_7, A_8, A_9 (13-hydroxyplatencin-4'-glucoside-1'-thiomethyl ester), and platencin A_{10} (13-hydroxyplatencin-4'-glucoside-1'-methyl ester) (Figure 12.15).[40] They also reported isolation of another compound which they named platencin A_4, a name that was already assigned to 15-hydroxyplatencin. Therefore structure assigned to platencin A_4 was renamed in this article as platencin A_{12}.[40] It is noteworthy that the *ptnR1* deletion led to the production of five out of nine compounds that possess the homoplatencinic acid (C-20) terpenoid moiety,[40] suggesting that the deletion also altered the oxidative step in which three carbons are excised off to produce platencinic acid.

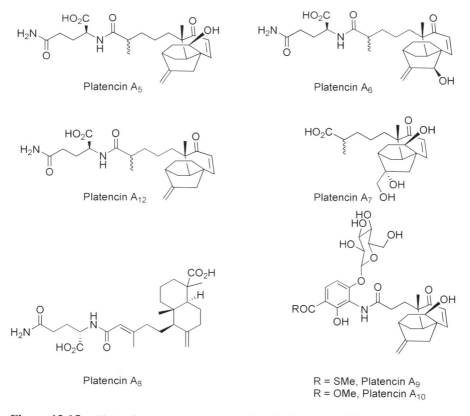

Figure 12.15 Platencin congeners produced by *S. platensis* ptnR1 mutant.

12.5   Biosynthesis of Platensimycin and Platencin

Biosynthesis of platensimycin and platencin was deduced first by the precursor-feeding experiments shown in Figures 12.16–12.18. Precursor-feeding experiments showed that the benzoic acid moiety was derived via a C7N intermediate (Figure 12.16) originating from a C4 intermediate from the Krebs cycle and a C3 intermediate from the glycolytic pathway (Figure 12.17).[38,39] The feeding experiments showed lack of incorporation of the labeled acetate and strong incorporation of the labeled pyruvate in the platensic and platencinic acid moieties of platensimycin and platencin, respectively, suggesting their biosynthetic origin from other than the mevalonate pathway. Examination of the precise labeling pattern indicated that MEP pathway is involved in the biosynthesis of platensic acid moiety and goes through advanced diterpenoid intermediate geranyl-geranyl diphosphate (GGPP), *ent*-copalyl diphosphate (*ent*-CPP), *ent*-kaurene or seco-*ent*-kaurene-A (Figure 12.17).[38] The intermediacy of the latter precursor could be gleaned from the congener, homoplatensimide A, produced by the native producer

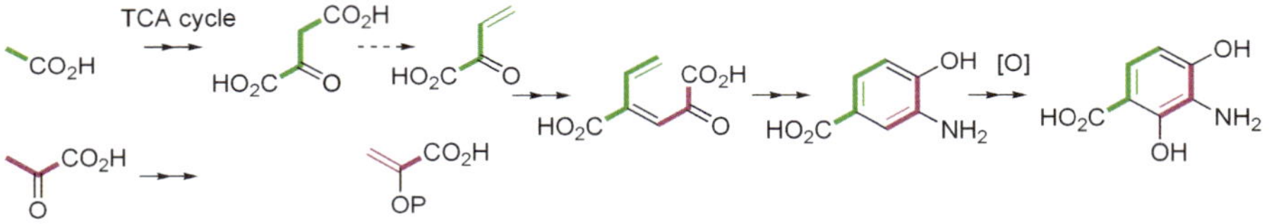

Figure 12.16   Biosynthesis of 3-amino-2,4-dihydroxy-benzoic acid.

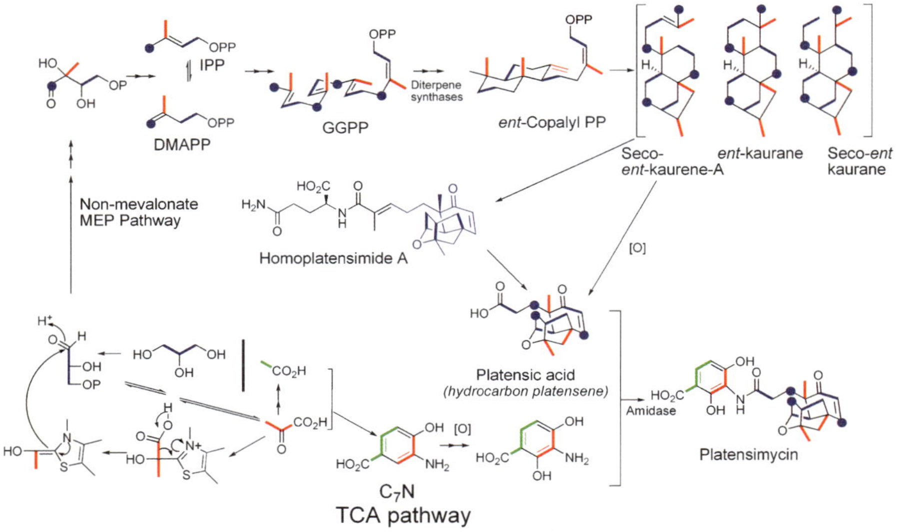

Figure 12.17   Biosynthetic scheme of platensimycin based on precursor-feeding experiments.

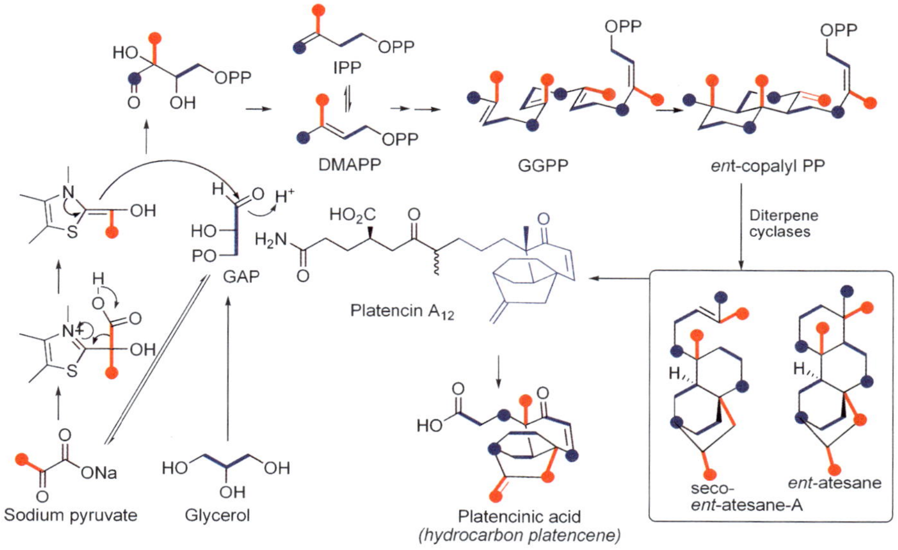

Figure 12.18   Biosynthetic scheme of platencin based on precursor-feeding experiments.

MA7327.[36] An oxidative cleavage of the terminal enoic acid followed by a loss of C3 unit will lead to the formation of platensic acid (via hydrocarbon platensene). The platencinic acid moiety of platencin is similarly biosynthesized by the MEP pathway with divergence at the latter step involving intermediacy of *ent*-atesane or seco-*ent*-atesane-A leading to new putative hydrocarbons homoplatencene and platencene (Figure 12.18).[36] Like platensimycin, purification of platencin A$_{12}$ (originally named platencin A$_4$)[26] from the *ptnR1*-deleted mutant SB12600 may suggest that seco-*ent*-artesane may be a late-stage intermediate of the platencin biosynthesis. Platencin A$_8$, a direct analog of *ent*-copalyl alcohol, produced from SB12600 also provides a direct evidence that *ent*-copalyl diphosphate is involved in the biosynthesis of platensimycin and platencin via *ent*-kaurene and *ent*-atesane,[41] as originally proposed by the Singh group (Figures 12.17 and 12.18).[25,35,39] Shen and colleagues have recently cloned the biosynthetic gene clusters of platensimycin and platencin and shown that they are produced by dedicated *ent*-kaurene and *ent*-artiserene synthases.[41] In the *ptm* and *ptn* gene clusters, they deduced the function of the open reading frames (ORFs) that are consistent with the proposed pathway. Eventual characterization of the individual enzymes will help validate not only the proposed pathway but also specifically delineate the late-stage intermediates. It is noteworthy that the amine partners of all C-20 enoic acid amides (e.g. homoplatensimide A and platencin A$_{12}$) is other than the 3-amino-2,4-dihydroxy benzoic acid moiety, indicating the involvement of a C-17 enone acid specific amidase for the coupling of platensic and platencinic acids with the amino benzoic acid moiety.

12.6   Chemistry of Platensimycin

12.6.1   Role of Enone in the Chemical Stability of Platensimycin

Catalytic hydrogenation of platensimycin produced the dihydroplatensimycin (dihydro PTM) which slowly converted to the enamido product (enamido PTM) in methanol (Figure 12.19). In the presence of catalytic amounts of a mineral acid or an organic acid the dihydro PTM rapidly and quantitatively converted to the enamido PTM. Similarly, the reduction of the C-5 keto group with sodium borohydride led to the formation of a mixture of dihydro and tetrahydro products which rapidly converted to a mixture of four lactones with the release of 3-amino-2,4-dihydroxy-benzoic acid (Figure 12.20). The formation of both enamido and the lactone products could be avoided if the reactions were performed with the sodium salt of the acid or with the methyl ester. The [6,7]-dihydroplatensimycin showed four-fold loss of antibacterial activity whereas all the other compounds shown in Figures 12.19 and 12.20 did not show any antibacterial activity. Most often dihydro type compounds would be co-produced with enone type natural products by producing organisms. Lack of the discovery of such compounds in *S. platensis* is quite remarkable. It appears that the organism protected the structure of the biologically active platensimycin by strategically placing the enone olefin thus preventing it from decomposition to biologically inactive compounds. This observation may be highly noteworthy and may suggest that platensimycin plays a direct role in the protection of *S. platensis* from predatory bacteria surrounding it in the environment and may eventually provide a direct evidence of the role of secondary metabolites in the protection of the

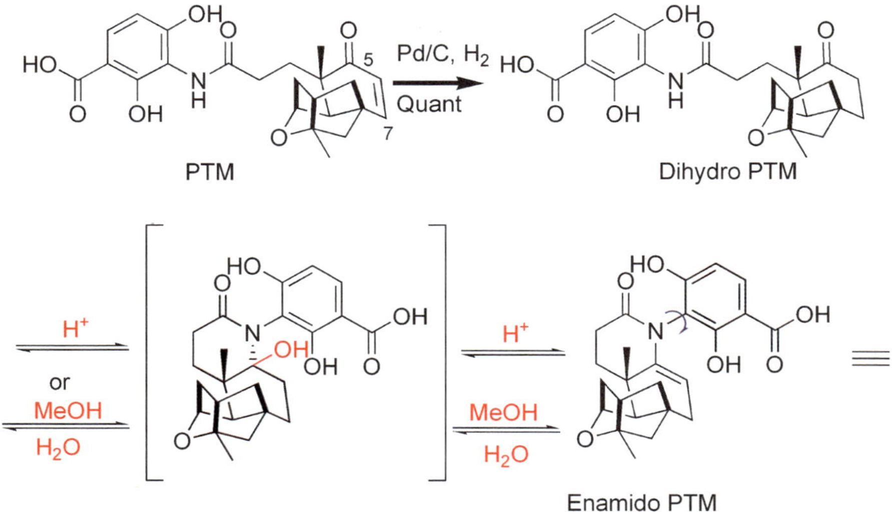

Figure 12.19   Synthesis of [6,7]-dihydroplatensimycin and acid-catalyzed cyclization.

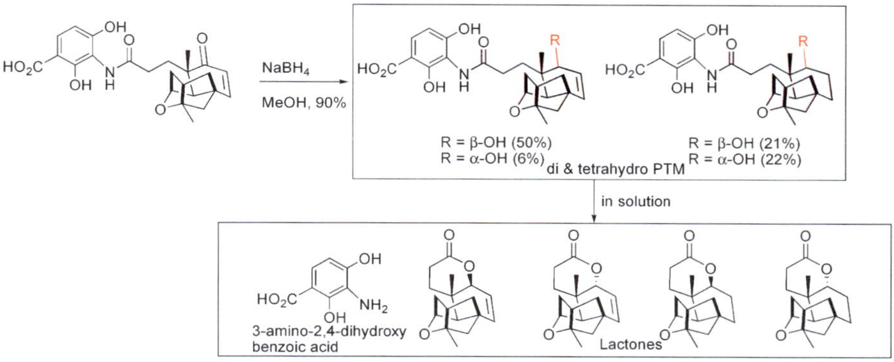

Figure 12.20 Reduction of C-5 keto of platensimycin and resulting products.

producing organism. More broadly it may explain why secondary metabolites are produced by many organisms. The detailed analysis of the genome may help resolve whether the dihydro PTM and related compounds are also produced by *S. platensis* and may answer some of these critical fundamental questions.

12.6.2 Structure–Activity Relationships

It is evident from the work described above that the enone is critical for the activity of platensimycin. Conjugate addition of a methyl group (7-Me dihydro PTM) or phenyl group (7-Ph dihydro PTM) at the C-7 position led to eight-fold loss of the antibacterial activity (Figure 12.21).[42] A similar reduction of the antibacterial activity was also observed by an addition of a 7-methyl group (7-Me PTM) while maintaining the enone functionality. The loss of activity was also observed in the cell-free FASII assay but magnitudes were different. The compound with the enone methyl group (7-Me PTM) showed five-fold loss of the enzyme activity while the 7-Ph dihydro PTM showed eight-fold loss

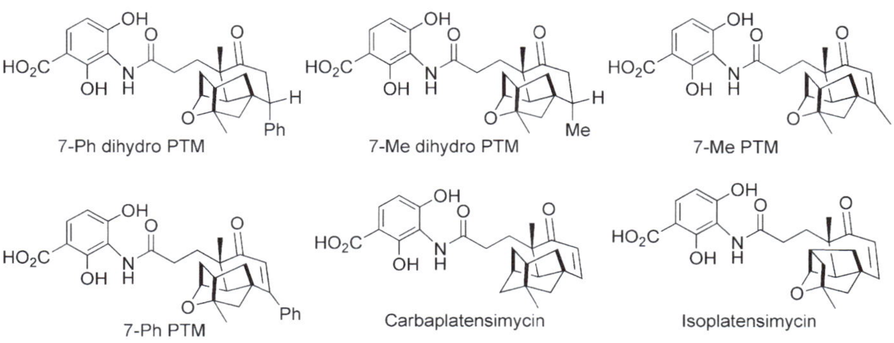

Figure 12.21 Key platensimycin derivatives with modest retention of activity.

and the methyl-substituted compound (7-Me dihydro PTM) showed 26-fold loss of the enzyme activity.[43] 7-Phenyl platensimycin (7-Ph PTM) showed almost equal or perhaps even slightly better activity than platensimycin in the antibacterial cellular assays.[44] No enzymatic activity of the latter compound was reported. Therefore it is not clear whether the equal activity was due to the better enzyme activity or simply slightly better cell penetration. These observations suggest that the unsubstituted enone group is critical for the activity. The structural chemistry and the computer modeling experiments suggest that the saturation of the enone olefin leads to conformational changes in the cyclohexenone ring, affecting the extent of hydrogen bonding of the C-5 keto group. However, it is not clear why a similar loss of activity was observed by the C-7 methyl enone compound (7-Me PTM). The deletion of the ether oxygen leads to the loss of a critical hydrogen bond and leads to 6–8-fold loss of antibacterial activity (e.g. carboplatensimycin).[45] Similar six-fold loss of the FabF binding affinity was also observed with platencin as compared to platensimycin due to loss of the ether hydrogen bond (see Table 12.1). The polar group substitutions (e.g. hydroxy groups in platensimycins A_1–A_6 and platencins A_1–A_{12}) lead to significant reduction or complete loss of antibacterial and FabF enzymatic activities, which is consistent with the mostly hydrophobic pocket surrounding the cage structure.[25,33,37] However, added hydrophobicity of the core structure appears to be beneficial for FabH activity due to even more hydrophobic nature of the binding pocket of the FabH enzyme.[25] Alteration of the bond connectivity (C_{13}–C_{12} to C_{13}–C_{11}, e.g. isoplatensimycin) of the tetracyclic enone core completely abolished the activity (Figure 12.21).[46]

The carboxyl group of the benzoic acid makes a salt bridge interaction with the two histidines at the catalytic site. Breaking of this interaction is catastrophic to the activity (e.g. methyl ester or carboxamide). Although only one of the two phenolic groups is involved in apparent polar interactions, both phenolic groups are equally important for the potency of platensimycin.[42] It is believed that the phenolic group at C-2 of the benzoic acid provides weight balance with the C-4 hydroxy group, allowing the benzoic acid unit to adopt an optimal conformation for maximal binding.

Other structural modifications or substitutions of the structural motifs have led to loss or the complete abolition of activity. The full summary of the SAR is shown in Figure 12.22.

12.7 Total Synthesis

Platensimycin and platencin have been very extensively studied molecules and over 30 total or formal syntheses have been reported. The first total synthesis of platensimycin was reported only four months after the first publication on platensimycin.[47] The total syntheses have been reviewed in a number of very good review articles and the detailed discussion of total synthesis is out of scope of this article.[48–51] Unfortunately, none of the total syntheses is short

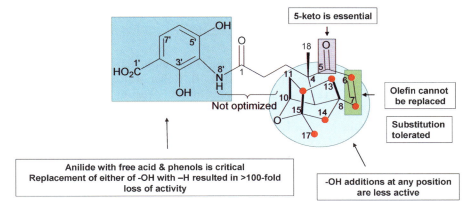

Figure 12.22 Summary of SAR of platensimycin.

and efficient enough to enable preparation of platensimycin or platencin by synthetic routes for a large-scale production.

Summary

Platensimycin and platencin are two novel antibiotics that were discovered by an innovative antisense target-based whole-cell screening approach. These two compounds are potent inhibitors of the condensing enzymes of fatty acid biosynthesis, which leads to their potent activities against Gram-positive bacteria including *in vivo* efficacy in a murine model of *Staphylococcus aureus* infection. Because of their novel structure and novel mode of action, these compounds drew a lot of attention in the worldwide scientific community, resulting in a number of total syntheses, medicinal chemistry, biosynthetic, and biological studies. New studies continue to appear in the literature on a regular basis. Recently, it was reported that platensimycin inhibited mammalian fatty acid synthesis, which led to further studies in various mouse models of diabetes. In these models it showed significant increase in insulin sensitivity, and lowered glucose and liver triglyceride levels. These effects were mechanism dependent and thus provided validation for the FAS mechanism and platensimycin as a potential treatment for diabetes.

References

1. C. T. Walsh, *Antibiotics: Actions, Origin, Resistance*, ASM Press, Washington, DC, 2003.
2. S. B. Singh and J. F. Barrett, *Biochem. Pharmacol.*, 2006, **71**, 1006.
3. A report from NNIS system, *Am. J. Infection Control*, 2004, **32**, 470.
4. G. H. Talbot, J. Bradley, J. E. Edwards, Jr., D. Gilbert, M. Scheld and J. G. Bartlett, *Clin. Infect. Dis.*, 2006, **42**, 657.

5. B. Spellberg, R. Guidos, D. Gilbert, J. Bradley, H. W. Boucher, W. M. Scheld, J. G. Bartlett and J. E. Edwards, Jr., *Clin. Infect. Dis.*, 2008, **46**, 155.

6. B. Spellberg, M. Blaser, R. J. Guidos, H. W. Boucher, J. S. Bradley, B. I. Eisenstein, D. Gerding, R. Lynfield, L. B. Reller, J. Rex, D. Schwartz, E. Septimus, F. C. Tenover and D. N. Gilbert, *Clin. Infect. Dis.*, 2011, 52 Suppl **5**, S397.

7. R. M. Klevens, M. A. Morrison, J. Nadle, S. Petit, K. Gershman, S. Ray, L. H. Harrison, R. Lynfield, G. Dumyati, J. M. Townes, A. S. Craig, E. R. Zell, G. E. Fosheim, L. K. McDougal, R. B. Carey and S. K. Fridkin, *J. Am. Med. Assoc.*, 2007, **298**, 1763.

8. B. J. Akerley, E. J. Rubin, V. L. Novick, K. Amaya, N. Judson and J. J. Mekalanos, *Proc. Natl. Acad. Sci. USA*, 2002, **99**, 966.

9. C. A. Hutchinson III, S. N. Peterson, S. R. Gill, R. T. Cline, O. White, C. M. Fraser, H. O. Smith and J. Craig Venter, *Science*, 1999, **286**, 2165.

10. Y. Ji, B. Zhang, S. F. Van Horn, P. Warren, G. Woodnutt, M. K. Burnham and M. Rosenberg, *Science*, 2001, **293**, 2266.

11. K. Kobayashi, S. D. Ehrlich, A. Albertini, G. Amati, K. K. Andersen, M. Arnaud, K. Asai, S. Ashikaga, S. Aymerich, P. Bessieres, F. Boland, S. C. Brignell, S. Bron, K. Bunai, J. Chapuis, L. C. Christiansen, A. Danchin, M. Debarbouille, E. Dervyn, E. Deuerling, K. Devine, S. K. Devine, O. Dreesen, J. Errington, S. Fillinger, S. J. Foster, Y. Fujita, A. Galizzi, R. Gardan, C. Eschevins, T. Fukushima, K. Haga, C. R. Harwood, Hecker, M. D. Hosoya, M. F. Hullo, H. Kakeshita, D. Karamata, Y. Kasahara, F. Kawamura, K. Koga, P. Koski, R. Kuwana, D. Imamura, M. Ishimaru, S. Ishikawa, I. Ishio, D. Le Coq, A. Masson, C. Mauel, R. Meima, R. P. Mellado, A. Moir, S. Moriya, E. Nagakawa, H. Nanamiya, S. Nakai, P. Nygaard, M. Ogura, T. Ohanan, M. O'Reilly, M. O'Rourke, Z. Pragai, H. M. Pooley, G. Rapoport, J. P. Rawlins, L. A. Rivas, C. Rivolta, A. Sadaie, Y. Sadaie, M. Sarvas, T. Sato, H. H. Saxild, E. Scanlan, W. Schumann, J. F. M. L Seegers, J. Sekiguchi, A. Sekowska, S. J. Seror, M. Simon, P. Stragier, R. Studer, H. Takamatsu, T. Tanaka, M. Takeuchi, H. B. Thomaides, V. Vagner, J. M. van Dijl, K. Watabe, A. Wipat, H. Yamamoto, M. Yamamoto, Y. Yamamoto, K. Yamane, K. Yata, K. Yoshida, H. Yoshikawa, U. Zuber and N. Ogasawara, *Proc. Natl. Acad. Sci. USA*, 2003, **100**, 4678.

12. R. A. Forsyth, R. J. Haselbeck, K. L. Ohlsen, R. T. Yamamoto, H. Xu, J. D. Trawick, D. Wall, L. Wang, V. Brown-Driver, J. M. Froelich, G. C. Kedar, P. King, M. McCarthy, C. Malone, B. Misiner, D. Robbins, Z. Tan, Z.-Y. Zhu, G. Carr, D. A. Mosca, C. Zamudio, J. G. Foulkes and J. W. Zyskind, *Mol. Microbiol.*, 2002, **43**, 1387.

13. S. Brinster, G. Lamberet, B. Staels, P. Trieu-Cuot, A. Gruss and C. Poyart, *Nature*, 2009, **458**, 83.

14. J. B. Parsons, M. W. Frank, C. Subramanian, P. Saenkham and C. O. Rock, *Proc. Natl. Acad. Sci. USA*, 2011, **108**, 15378.

15. D. J. Payne, M. N. Gwynn, D. J. Holmes and D. L. Pompliano, *Nat. Rev. Drug Discovery*, 2007, **6**, 29.
16. Y. Ji, A. Marra, M. Rosenberg and G. Woodnutt, *J. Bacteriol.*, 1999, **181**, 6585.
17. S. B. Singh, J. W. Phillips and J. Wang, *Curr. Opin. Drug Discovery Dev.*, 2007, **10**, 160.
18. K. Young, H. Jayasuriya, J. G. Ondeyka, K. Herath, C. Zhang, S. Kodali, A. Galgoci, R. Painter, V. Brown-Driver, R. Yamamoto, L. L. Silver, Y. Zheng, J. I. Ventura, J. Sigmund, S. Ha, A. Basilio, F. Vicente, J. R. Tormo, F. Pelaez, P. Youngman, D. Cully, J. F. Barrett, D. Schmatz, S. B. Singh and J. Wang, *Antimicrob. Agents Chemother.*, 2006, **50**, 519.
19. S. Kodali, A. Galgoci, K. Young, R. Painter, L. L. Silver, K. B. Herath, S. B. Singh, D. Cully, J. F. Barrett, D. Schmatz and J. Wang, *J. Biol. Chem.*, 2005, **280**, 1669.
20. J. Wang, S. M. Soisson, K. Young, W. Shoop, S. Kodali, A. Galgoci, R. Painter, G. Parthasarathy, Y. Tang, R. Cummings, S. Ha, K. Dorso, M. Motyl, H. Jayasuriya, J. Ondeyka, K. Herath, C. Zhang, L. Hernandez, J. Alloco, Á. Basilio, J. R. Tormo, O. Genilloud, F. Vicente, F. Pelaez, L. Colwell, S. H. Lee, B. Michael, T. Felcetto, C. Gill, L. L. Silver, J. Hermes, K. Bartizal, J. Barrett, D. Schmatz, J. W. Becker, D. Cully and S. B. Singh, *Nature*, 2006, **441**, 358.
21. C. Rapp, G. Jung, C. Isselhorst-Schart and H. Zahner, *Liebigs Ann. Chem.*, 1988, 1043.
22. J. G. Ondeyka, D. L. Zink, K. Young, R. Painter, S. Kodali, A. Galgoci, J. Collado, J. R. Tormo, A. Basilio, F. Vicente, J. Wang and S. B. Singh, *J. Nat. Prod.*, 2006, **69**, 377.
23. S. B. Singh, H. Jayasuriya, J. G. Ondeyka, K. B. Herath, C. Zhang, D. L. Zink, N. N. Tsou, R. G. Ball, A. Basilio, O. Genilloud, M. T. Diez, F. Vicente, F. Pelaez, K. Young and J. Wang, *J. Am. Chem. Soc.*, 2006, **128**, 11916 and 15547.
24. J. Wang, S. Kodali, S. H. Lee, A. Galgoci, R. Painter, K. Dorso, F. Racine, M. Motyl, L. Hernandez, E. Tinney, S. Colletti, K. Herath, R. Cummings, O. Salazar, I. Gonzalez, A. Basilio, F. Vicente, O. Genilloud, F. Pelaez, H. Jayasuriya, K. Young, D. Cully and S. B. Singh, *Proc. Natl. Acad. Sci. USA*, 2007, **104**, 7612.
25. H. Jayasuriya, K. B. Herath, C. Zhang, D. L. Zink, A. Basilio, O. Genilloud, M. T. Diez, F. Vicente, I. Gonzalez, O. Salazar, F. Pelaez, R. Cummings, S. Ha, J. Wang and S. B. Singh, *Angew. Chem., Int. Ed.*, 2007, **46**, 4684.
26. M. J. Smanski, R. M. Peterson, S. R. Rajski and B. Shen, *Antimicrob. Agents Chemother.*, 2009, **53**, 1299.
27. R. G. Donald, S. Skwish, R. A. Forsyth, J. W. Anderson, T. Zhong, C. Burns, S. Lee, X. Meng, L. LoCastro, L. W. Jarantow, J. Martin, S. H. Lee, I. Taylor, D. Robbins, C. Malone, L. Wang, C. S. Zamudio, P. J. Youngman and J. W. Phillips, *Chem. Biol.*, 2009, **16**, 826.

28. S. B. Singh, K. Young and L. Miesel, *Expert Rev. Anti. Infect. Ther.*, 2011, **9**, 589.

29. M. Wu, S. B. Singh, J. Wang, C. C. Chung, G. Salituro, B. V. Karanam, S. H. Lee, M. Powles, K. P. Ellsworth, M. E. Lassman, C. Miller, R. W. Myers, M. R. Tota, B. B. Zhang and C. Li, *Proc. Natl. Acad. Sci. USA*, 2011, **108**, 5378.

30. C. Zhang, J. Ondeyka, L. Dietrich, F. P. Gailliot, M. Hesse, M. Lester, K. Dorso, M. Motyl, S. N. Ha, J. Wang and S. B. Singh, *Bioorg. Med. Chem.*, 2010, **18**, 2602.

31. C. Zhang, J. Ondeyka, Z. Guan, L. Dietrich, B. Burgess, J. Wang and S. B. Singh, *J. Antibiot. (Tokyo)*, 2009, **62**, 699.

32. C. Zhang, J. Ondeyka, K. Herath, H. Jayasuriya, Z. Guan, D. L. Zink, L. Dietrich, B. Burgess, S. N. Ha, J. Wang and S. B. Singh, *J. Nat. Prod.*, 2011, **74**, 329.

33. S. B. Singh, H. Jayasuriya, K. B. Herath, C. Zhang, J. Ondeyka, D. L. Zink, S. Ha, G. Parthasarathy, J. W. Becker, J. Wang and S. M. Soisson, *Tetrahedron Lett.*, 2009, **50**, 5182.

34. C. Zhang, J. Ondeyka, D. L. Zink, B. Burgess, J. Wang and S. B. Singh, *Chem. Commun. (Camb.)*, 2008, 5034.

35. K. B. Herath, C. Zhang, H. Jayasuriya, J. G. Ondeyka, D. L. Zink, B. Burgess, J. Wang and S. B. Singh, *Org. Lett.*, 2008, **10**, 1699.

36. H. Jayasuriya, K. B. Herath, J. G. Ondeyka, D. L. Zink, B. Burgess, J. Wang and S. B. Singh, *Tetrahedron Lett.*, 2008, **49**, 3648.

37. S. B. Singh, J. G. Ondeyka, K. B. Herath, C. Zhang, H. Jayasuriya, D. L. Zink, G. Parthasarathy, J. W. Becker, J. Wang and S. M. Soisson, *Bioorg. Med. Chem. Lett.*, 2009, **19**, 4756.

38. K. B. Herath, A. B. Attygalle and S. B. Singh, *J. Am. Chem. Soc.*, 2007, **129**, 15422.

39. K. Herath, A. B. Attygalle and S. B. Singh, *Tetrahedron Lett.*, 2008, **49**, 5755.

40. Z. Yu, M. J. Smanski, R. M. Peterson, K. Marchillo, D. Andes, S. R. Rajski and B. Shen, *Org. Lett.*, 2010, **12**, 1744.

41. M. J. Smanski, Z. Yu, J. Casper, S. Lin, R. M. Peterson, Y. Chen, E. Wendt-Pienkowski, S. R. Rajski and B. Shen, *Proc. Natl. Acad. Sci. USA*, 2011, **108**, 13498.

42. S. B. Singh, K. B. Herath, J. Wang, N. N. Tsou and R. G. Ball, *Tetrahedron Lett.*, 2007, **48**, 5429.

43. H. C. Shen, F. X. Ding, S. B. Singh, G. Parthasarathy, S. M. Soisson, S. N. Ha, X. Chen, S. Kodali, J. Wang, K. Dorso, J. R. Tata, M. L. Hammond, M. Maccoss and S. L. Colletti, *Bioorg. Med. Chem. Lett.*, 2009, **19**, 1623.

44. K. P. Jang, C. H. Kim, S. W. Na, D. S. Jang, H. Kim, H. Kang and E. Lee, *Bioorg. Med. Chem. Lett.*, 2010, **20**, 2156.

45. K. C. Nicolaou, Y. Tang, J. Wang, A. F. Stepan, A. Li and A. Montero, *J. Am. Chem. Soc.*, 2007, **129**, 14850.

46. K. P. Jang, C. H. Kim, S. W. Na, H. Kim, H. Kang and E. Lee, *Bioorg. Med. Chem. Lett.*, 2009, **19**, 4601.
47. K. C. Nicolaou, A. Li and D. J. Edmonds, *Angew Chem., Int. Ed.*, 2006, **45**, 7086.
48. K. C. Nicolaou, J. S. Chen and S. M. Dalby, *Bioorg. Med. Chem.*, 2009, **17**, 2290.
49. K. Tiefenbacher and J. Mulzer, *Angew Chem., Int. Ed.*, 2008, **47**, 2548.
50. D. T. Manallack, I. T. Crosby, Y. Khakham and B. Capuano, *Curr. Med. Chem.*, 2008, **15**, 705.
51. M. Saleem, H. Hussain, I. Ahmed, T. van Ree and K. Krohn, *Nat. Prod. Rep.*, 2011, **28**, 1534.

CHAPTER 13

Coupling Chemical Genomics and Natural Products: The Discovery of Parnafungins and Novel Antifungal Leads

HAO WANG[a], CRAIG A. PARISH[b], DEMING XU[a,c] AND TERRY ROEMER*[a]

[a] Department of Infectious Diseases, Merck Research Laboratories, 2000 Galloping Hill Road, Kenilworth, NJ 07033, USA; [b] Natural Products Chemistry, Merck Research Laboratories, 126 East Lincoln Ave., Rahway, NJ 07065, USA; [c] Present address: WuXi AppTec (Shanghai) Co. Ltd., Shanghai 200131, China
*E-mail: terry_roemer@merck.com

13.1 Introduction

Natural products (NPs) have provided a valuable source of chemical entities for the development of therapeutic agents to treat infectious diseases. Underscoring this reality, ~80% of antibiotics currently in clinical use are derived from NPs.[1] This remarkable success can be attributed to the unique characteristics of these small molecules – such as immense chemical diversity, intrinsic cell permeability, and specific bioactivity. Accordingly, NPs from microbial sources represent privileged chemical scaffolds. Their target selectivity and intrinsically favorable pharmacological properties serve as critical starting points in drug development. Antibacterial discovery has

RSC Drug Discovery Series No. 25
Drug Discovery from Natural Products
Edited by Olga Genilloud and Francisca Vicente
© The Royal Society of Chemistry 2012
Published by the Royal Society of Chemistry, www.rsc.org

yielded over a dozen distinct and clinically relevant chemical scaffolds,[2] all of which, excluding oxazolidinones, quinolones and sulfonamides, are NP derived. Current clinically relevant antifungal therapeutics, however, are represented by only three structural classes, of which two (polyenes and echinocandins, excluding the azole class) are naturally derived.[3] Exacerbated by their therapeutic limitations, the medical need for more efficacious and broad spectrum antifungal agents is clear. To this end, we exploited the chemical diversity that NPs provide to discover novel antifungal leads.

Compared to synthetic chemical drug discovery, NP-based drug discovery does come with significant limitations, ranging from the high demand of resources and expertise in building and maintaining a high quality library of NPs and producing microbes, to the incompatibility of NP chemistry with the 'blitz screen' mentality of high throughput screening (HTS), and ever-increasing inefficiency caused by the 'rediscovery' of known compounds. These limitations, in particular the rediscovery, have exhausted the conventional 'grind-and-find' approach. Efficient and robust NP de-replication, by comparing the chemical and biological properties of the unknown with those of the known, is needed. Any large-scale, solely chemical de-replication, which has been increasingly resource consuming, is strategically prohibitive. Understanding the mechanism of action (MOA) of an inhibitor upfront could implicate the presence of known compounds and therefore serve as a biological de-replication strategy. However, the conventional single-target based biochemical assay is not amenable to this challenge, since biological de-replication requires more comprehensive information of MOA than is provided by a single-target assay. Chemical genomic strategies, on the other hand, could serve as a means to MOA-based biological de-replication. With comprehensive coverage, up to an entire genome, and concise readouts, chemical genomic assays can faithfully reflect the MOA of a given antifungal compound that exerts its effect in a mechanistically specific manner.[4-6] The material requirements, both in quantity and purity, of this assay are minimal as the starting sample of unfractionated NPs can be analyzed directly, and used immediately for LC/MS-based chemical de-replication, should a desirable biological activity be identified. Thus, biological de-replication coupled with analytical chemistry profiling can predict the presence of a known antifungal in NPs and therefore serve as a de-replication strategy.

The fitness test was first developed in baker's yeast and adapted for the primary human fungal pathogen, *Candida albicans*.[7,8] The assay relies on a phenomenon known as chemically induced haploinsufficiency, where deletion of one allele of a given gene in a diploid fungus renders the mutant hypersensitive to a compound that inhibits the corresponding protein target at sub-lethal concentrations. In fact, when a collection of heterozygous deletion strains covering the entire or significant portion of a fungal genome was tested against a panel of antifungal compounds with known MOAs, hypervariable growth (either hypersensitivity or resistance, hence the term fitness test) was restricted to only a small set of strains that correspond to different aspects of their respective MOAs. In almost all these cases, the concise fitness test profile was reflective of the corresponding MOA.[4,6,8,9]

Applying the *C. albicans* fitness test (CaFT) to screen a large collection of NPs from microbial sources (in the format of fermentation extracts), we identified those extracts with discernible biological activities. By comparing their CaFT profiles to those of known antifungal NPs, we were able to prioritize the extracts whose mechanistic novelty implies chemical novelty. Thus, chemical resources were allocated accordingly. Often, CaFT profiling was also used during chemical fractionation to track the biological activities sought after.

In what follows, we present our biological and chemical de-replication strategies used in screening ~1800 extracts of fungal and bacterial fermentation broths. We summarize the overall outcome of this campaign, and illustrate how biological de-replication by CaFT profiling successfully led to chemical de-replication and eventual identification of novel antifungal small molecules from microbial sources, including the parnafungins – the first significant novel class of NPs with antifungal development potential since the discovery of sordarins and enfumafungins in the 1990s.[10,11]

13.2   *C. albicans* Fitness Test and its Application

13.2.1   Antifungal Target Identification and Validation

Mycoses have emerged as important and prevalent microbial infections in the hospital setting worldwide. Of pathogenic fungi, *Candida* spp. account for 8–9% of all bloodstream infections, with a mortality rate as high as 40%, and remain as the fourth leading cause of nosocomial infections.[12] Antifungal therapeutic drugs are restricted largely to three mechanistic classes that target plasma membrane (polyenes), cell wall (echinocandins) and ergosterol biosynthesis (azoles). They are increasingly challenged by their respective limitations and drawbacks, such as spectrum, toxicity, route of administration, and the emergence of drug resistance. In this context, novel target identification is a needed first step to expand the armamentarium of antifungal drugs.

Unlike the bakers' yeast, *C. albicans*, the leading cause of all mycoses, is an obligatory diploid without a sex lifecycle based on meiosis. Therefore, conventional genetic analyses cannot be applied directly to this opportunistic pathogen. Since sequential chromosomal deletion is also not applicable to identify and characterize essential target genes, we developed a two-step gene r̲eplacement a̲nd c̲onditional e̲xpression (GRACE) strategy. One allele of the gene in question was first deleted to construct a heterozygous deletion mutant, and the native promoter of the remaining allele was replaced with a tetracycline regulable promoter. This enables controlled repression of even essential genes whose expression, under the control of the *TET* promoter, is shut off only in the presence of tetracycline.[13] Conditional shut-off strains were built for genes covering ~35% of the genome and over 850 *C. albicans* genes were directly demonstrated to encode conserved essential functions in the pathogen.[13] More medically significant, this tetracycline-based conditional repression system was extended to an *in vivo* infection setting.[14–16] In total, over 100 new antifungal targets have been genetically validated as essential for establishing and maintaining an acute systemic candidiasis infection in immune competent mice, thereby serving as suitable targets for prophylactic and acute infection intervention.[17]

These studies, together with others performed in *C. albicans*,[18] *Aspergillus fumigatus*,[19] *Cryptococcus neoformans*,[20] and *Neurospora crassa*,[21] provide an immense diversity of validated novel targets potentially suitable for antifungal screening. With such an abundance of possible antifungal targets, the single-target based biochemical assay is inefficient, because the selection of a particular or a small set of target(s) for HTS automatically precludes a larger set of equally attractive targets whose cognate inhibitors may exist in the screening library (Figure 13.1A). Furthermore, the single-target based biochemical assay is somewhat incompatible with the complexity of NPs. A genome-wide hetero-zygous deletion strain set, representing all potential antifungal drug target genes, provides an innovative and comprehensive screening alternative.

13.2.2 *C. albicans* Fitness Test

For most yeast genes, heterozygous deletion confers no or minimal growth defects. However, in the presence of sub-lethal concentrations of antifungal compounds, a small group of heterozygotes display significant growth variations, reflecting altered fitness. We adapted and developed the fitness test approach in a more medically relevant pathogen, *C. albicans*. Heterozygous strains representing genes of greatest potential as antifungal targets were constructed, during which two specific 20 base pair DNA barcode sequences were introduced at the deleted alleles (up- and down-tags, Figure 13.1B). These strains were pooled, and small aliquots were tested under identical growth conditions in the absence (mock) or presence (compound treatment) of an inhibitory compound at different sub-lethal concentrations. The relative abundance of each strain in response to chemical perturbation was determined by competitive hybridization of PCR amplified and labeled tags on DNA microarrays (Figure 13.1B).

Initially, we used a collection of ∼2900 bar coded heterozygous deletion strains and a panel of well-characterized antifungal compounds, including all three classes of therapeutic drugs, to validate the *C. albicans* fitness test (CaFT).[8] In most cases, the CaFT profiles were concise, and the heterozygotes corresponding to the targets were readily identified by their consistent and significant hypersensitivities to the cognate inhibitors. Other aspects of MOAs, including drug import, efflux, and metabolism, were also clearly recognized in the profiles. Occasionally, more complex profiles were obtained. However, the underlying biology remained coherent and consistent with the respective MOAs.[8] To validate the robust ability of linking novel antifungal compounds

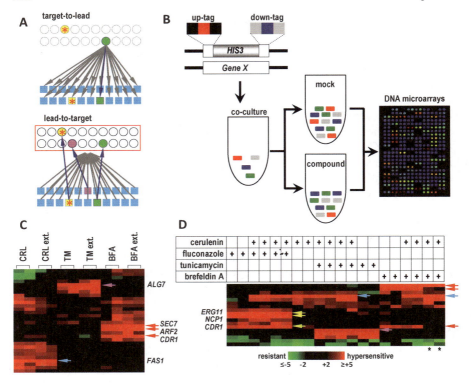

Figure 13.1 The *C. albicans* fitness test and its applications. (A) Schematic comparison of a target-based screening approach (top) and a genomic approach (bottom). Antifungal targets are represented by circles, and compounds as filled squares, with hits indicated by blue arrows. The selection of a target (top) may exclude compounds that inhibit other targets. In some cases, other compound/target pairs (indicated by red asterisks) are better suited for lead development and optimization. In a genomic approach, a set of targets (with genomic coverage) is screened against any given bioactive compound. The biological inclusivity could identify potentially all compound/target pairs available in a defined chemical library. (B) An overview of the *C. albicans* fitness test (CaFT). Each heterozygous deletion strain is double bar-coded with unique, strain-identifying 20 bp DNA sequences or tags (red and blue bars). These strains are pooled, and tested in the absence (mock) and presence (compound) of a sub-inhibitory concentration of a bioactive compound. The relative abundance of each strain is then determined by the hybridization signals of their corresponding tags on DNA microarrays to infer the growth rate (or 'fitness') and drug sensitivity of all heterozygote strains. For illustrative purposes, the strain highlighted in red is specifically hypersensitive to compound treatment, suggesting the corresponding gene directly or indirectly participates in the MOA of the compound. (C) CaFT profiling of crude fermentation extracts (ext.) and pure NPs. CRL, cerulenin; TM, tunicamycin; and BFA, brefeldin A. (The scale of heat is shown in part D.) Red arrows indicate the targets of BFA (*SEC7* and *ARF2*) and its efflux pump (*CDR1*), a pink

to their protein targets, we profiled synthetic compounds from chemical libraries that showed inhibitory activities against wild-type *C. albicans*. Their predicted MOAs, to interdict tubulin (Tub1p), GMP synthesis (Gua1p), and fatty acid biosynthesis (Ole1p), were each validated by genetic and biochemical means.[8,14,16] These results collectively demonstrate the suitability of using the CaFT profiling approach to determine the MOAs of unknown biological activities without prior selection of targets. In contrast with the target-to-lead approach, this represents a new screening paradigm that is compound centric rather than target centric and in which a pool of targets in the form of defined heterozygous deletion strains are tested against any compound with whole-cell antifungal activity. Such an inclusive approach extends biology associated with a given chemical library to those that would have been missed otherwise (Figure 13.1A). The present version of the CaFT contains ∼5400 heterozygous deletion strains, such that any single compound of interest could be screened against ∼90% of the *C. albicans* genome.[22]

13.2.3   Application of the CaFT to NPs

Application of the CaFT screening approach to microbial unfractionated fermentation NP extracts has to address several issues, two of which – chemical impurity and biological complexity – are most challenging. Initially, we applied the CaFT to crude extracts of two fungi and a streptomycete known to produce cerulenin, brefeldin A, and tunicamycin, respectively. The MOAs of the biological activities in these extracts were faithfully reflected in their profiles (Figure 13.1C). The presence of these compounds was confirmed subsequently by LC/MS.[15]

When two equipotent antifungal activities (of pure compounds) were mixed within one extract and analyzed, the resulting CaFT profiles were largely additive, and no additional strains displayed significant growth variations in response to mixed treatment. However, we noted that mixtures consisting of a larger ratio of one active over the other produced CaFT profiles reflective of only the major bioactivity within the mixture[15] (Figure 13.1D). Nevertheless, these early proof-of-concept experiments indicated that CaFT can reliably detect the MOA of bioactive NPs within defined mixtures or crude extracts

arrow the target of TM (*ALG7*); and a blue arrow the target of CRL (*FAS1*). Note the highly related CaFT profiles between pure NP compound and its crude NP extract. (D) CaFT profiling of mixtures of two antifungal compounds. Yellow arrows indicate the target (*ERG11*) and its co-factor (*NCP1*), and the efflux pump (*CDR1*) of fluconazole, and red arrows the targets (*SEC7* and *ARF2*) and the efflux pump (*CDR1*) of brefeldin A (from top to bottom). The blue and pink arrows are as indicated in part C. Note that in the two mixtures marked by asterisks, only partial profile of brefeldin A was observed. This is due to the relatively low amount of brefeldin A versus cerulenin contributing to the bioactivity of the mixture. The experiments presented in parts C and D were described by Jiang *et al.*[15]

and therefore it is possible to use the CaFT to biologically de-replicate prominent NPs within crude extracts.

13.3 Antifungal Discovery by *C. albicans* FT Screening from NP Extracts

13.3.1 General Strategy

As shown in Figure 13.2A, crude fermentation extracts with *in vitro C. albicans* whole-cell activity are first analyzed using the CaFT to identify a putative mechanism of action. Extracts with non-specific MOA (i.e. lacking any strain depletions in the FT or complex FT profiles lacking obvious biological relevance of target/pathway) are excluded from further analysis. Analytical techniques, mainly LC/MS, are employed to assess extracts, yielding informative profiles. Comparative analysis of FT profiles in context with LC/MS data can then be used to de-replicate previously known NPs and prioritize the following chromatography-based compound purification procedures. Once the CaFT profile of the purified component in question is confirmed, it is then fully characterized by 1D and 2D NMR spectroscopic methods along with chemical derivatization when necessary in order to elucidate the structure.[23–25] Target-specific biochemical and/or genetic approaches are then employed to confirm the specific MOA of a purified compound predicted by CaFT profiling.

The robust feature of this screening approach is the ability to rapidly de-replicate known bioactive molecules from those that are novel early in the screening process and therefore allow allocation of the resource- and time-consuming chemical process towards key bioactive extracts. A library of CaFT profiles for a broad reference set of pure, well-characterized NPs was used for comparative analysis with profiles from new and uncharacterized extracts. The extracts with novel MOAs, thus likely to contain novel compounds, were prioritized for chemical procedures. Those with known but desirable MOAs were subject to further chemical de-replication by a standardized LC/MS method. The retention time, UV spectrum, and apparent mass spectrum of the components were obtained and used to compare with the Merck historical database of isolated NPs.[26] As the screening progressed, the libraries of CaFT profiles and Merck NP database were routinely updated with information from internally discovered compounds as well as literature reports of new antifungal structures. To purify the desired bioactivity, a series of chromatographic processes were employed and bioactive NP fractions were tracked by bioassay, normally with wild-type *C. albicans*. At times, when multiple active fractions were identified, CaFT profiling was performed to determine the fraction that rendered the desired MOA. Two examples are provided below to illustrate how our strategy was implemented.

Two fungal extracts generated CaFT profiles corresponding to strains heterozygous for multiple subunits of the translation initiation factor eIF3, including *TIF34*, *NIP1*, *PRT1*, and *RPG1* (Figure 13.2B and data not shown). A literature search revealed that the eukaryotic eIF3 is the target of the structurally related mycotoxins roridin A and verrucarin A. When pure NPs were tested, both the CaFT and the LC/MS analysis confirmed the presence of mycotoxins in the original extracts (Figure 13.2B and data not shown). Therefore, neither extract was pursued further. To flag these ubiquitous mycotoxins during continued screening, their profiles were included in the reference library.

Another fungal extract yielded a CaFT profile indicating that ribosome biogenesis, in particular the U3 snoRNP, was affected by the activity present. As the profile was distinct from that of tubercidin, a known NP targeting ribosome biogenesis (Figure 13.2C), we proceeded with fractionation. Two components with antifungal activity against wild-type *C. albicans* were isolated (Figure 13.2D). A panel of heterozygous mutants displaying significant growth variations in response to the extract were selected based on the original CaFT profile (Figure 13.2C). By analyzing the growth shift of each strain in the presence of each component, we determined that the CaFT profile detected is attributed to one of the activities (Figure 13.2D). Indeed, the CaFT profile of this purified material was largely consistent with that of the original extract (Figure 13.2C). Structure elucidation revealed that the active compound is a homoarginine derivative of 3′-aminoadenosine (data not shown).

13.3.2 Overview of the CaFT-based NP Screening Campaign

In total, 1801 NP extracts (1231 fungal and 570 actinomycete) with whole-cell inhibitory activity against *C. albicans* were profiled in the CaFT.[22] Of these, over 770 were derived from fresh microbial isolates discovered internally with maximized taxonomic and geographic diversity[27] and coupled with nutritional array methodologies to stimulate normally unexpressed secondary metabolic pathways to produce novel bioactive NPs.[28] Approximately 380 extracts (~20%) generated informative CaFT profiles, of which 61 were prioritized for bioactive isolation and structure elucidation. In the end, 16 novel antifungal compounds were identified, including two new inhibitors of the fungal 1,3-glucan synthase, the cell wall biosynthetic enzyme which echinocandins inhibit. In addition, several known compounds that had not been previously characterized by past Merck NP discovery efforts were also identified (Table 13.1).

A survey of novel and known target-specific inhibitors identified through this work validates our approach and reflects (in part) the achievable chemical and mechanistic diversity of bioactivities that exist in NP extracts. Importantly, novel NPs with new MOAs of potential antifungal therapeutic utility were also identified.[22]

In the following two sections, we present a group of novel antifungal NPs discovered in our campaign, in particular parnafungins, a new class of NPs with a novel scaffold and MOA. We describe our efforts in the purification and structure elucidation of parnafungins, their chemical and biological characterization, and their potential development as a new antifungal therapeutic class.

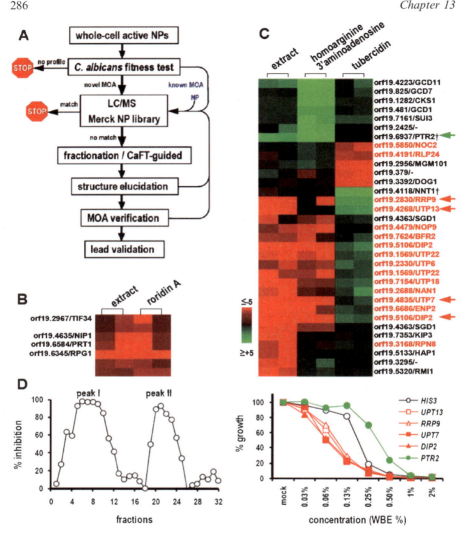

Figure 13.2 Application of the CaFT to biological de-replication of natural products. (A) Overall de-replication strategy. Note that the designation 'known MOA' refers to extracts whose profiles match those in the CaFT MOA profile compendium and those deduced from literature searches, and that of 'NP' (in blue) natural products which were not included in the library, but whose MOAs were subsequently identified by CaFT screening. (B) CaFT profiling of pure roridin A and a fungal extract containing roridin A. The active NP in the extract was deduced based on the CaFT profile and literature search, and confirmed by LC/MS subsequently. Key strain depletions correspond to heterozygotes of *TIF34*, *NIP1*, *PRT1*, and *RPG1*, whose protein products participate with the translation initiation factor eIF3, and is the reported MOA of roridin A-mediated inhibition of protein synthesis. The scale of heat map is shown in part C. (C) CaFT profiling of tubercidin (pure), a

13.4 Discovery and Characterization of Parnafungin

13.4.1 MOA Determination by CaFT Profiling

CaFT profiling of extract ECC577, a fermentation broth of *Fusarium larvarum*, showed that its active component caused hypersensitivity in a set of heterozygous strains, whose corresponding genes were linked to the eukaryotic mRNA cleavage and polyadenylation process[15] (Figure13.3A). To further characterize this extract, we compared its profile with that of cordycepin (3′-deoxyadenosine), a known 3′ mRNA processing inhibitor, which upon conversion to 3′-dATP *in vivo* acts as both an RNA molecule chain terminator and a substrate competitive inhibitor of the poly(A) polymerase (PAP). Due to a technical issue with the barcodes of *PAP1*, the *PAP1* heterozygote was absent from these profiles; however, its hypersensitivity to both ECC577 and cordycepin was confirmed alternatively.[15] While both profiles displayed a similar set of hypersensitive heterozygotes, thus suggesting some mechanistic overlap between the two agents, distinctively different hyposensitive or resistant strains were reflected in their respective CaFT profiles. First, two heterozygous mutants (*NNT1* and *ADO1*) were significantly resistant only to cordycepin. *NNT1* codes for a Na⁺-independent, H⁺-coupled nucleoside symporter, which is involved in the uptake of nucleosides in *C. albicans*.[8,14] *ADO1* codes for adenosine kinase which is required to convert cordycepin into 3′-dAMP *in vivo*. It is reasonable to infer that heterozygote levels of *C. albicans* Nnt1p and Ado1p reduce the uptake of cordycepin, an analog of adenosine, and its conversion to 3′-dATP

fungal extract, from which a new NP, homoarginine 3′-aminoadenosine was isolated. The original extract yielded a mechanistic profile indicating that ribosome biogenesis (with the corresponding heterozygotes highlighted in red) is affected and that the active compound is biologically distinct from tubercidin. The latter is further vindicated by the relative resistance of two heterozygotes, *NNT1* and *PTR2* (highlighted by †). While Nnt1p is involved in the uptake of tubercidin (hence the resistance of the corresponding heterozygote), the specific resistance of the *PTR2* (a putative oligopeptide transporter) heterozygote suggests that the corresponding transporter is involved in uptake of the active compound in the extract. Five heterozygotes (indicated by arrows) were selected for bioassay to guide the isolation (see part D). Note that the isolated NP, homoarginine 3′-aminoadenosine, largely reproduced the original profile with some quantitative differences. (D) CaFT-guided isolation of the extract shown in part C. The first step (CSP207p separation) fractionation yielded two active peaks (left). However, when both were tested against a panel of heterozygotes reflective of the CaFT profile (arrows in part C), only Peak I displayed growth phenotypes that were consistent with the CaFT profile; i.e. hypersensitivity of the *UPT13*, *UPT7*, *RRP9* and *DIP2* heterozygotes, and resistance of the *PTR2* heterozygote. This peak was followed up, which eventually yielded homoarginine 3′-aminoadenosine.

Table 13.1   Summary of biological de-replication of NPs in the CaFT

Number of extracts profiled in the CaFT	*1801 (1231 fungal, and 570 actinomycetes)*
Number of extracts with informative mechanistic profiles	~380
Number of extracts undergone isolation	~60
Glucan/cell wall inhibitors	106
NPs de-replicated [a]	Antibiotics A 30912, arundifungin, ascosteroside, chaetiacandins, corynecandin, ergokonins A & B, echinocandin B, furanocandin, fusacandin, mulundocandin, pneumocandins, papulacandins
Novel NPs (total of 2)	Novel desulfated ergokonin A[b], new papulacandin[b]
Other antifungal NPs:	~270
NPs de-replicated [a]	7-hydroxylguanine, aspirochlorine, BE-04385A, borolactin A, brefeldin A, cerulenin, compactin/mevastatin, conocandin, cordycepin, cytochalasins, fumonsins, hyalodendrin, hymeglusin, illudin, lipoxomycin, lovastatin, mycophenolic acid, oxanthromycin, PF1163A, PF1140, preussomerins, preussomerin MB7056B, radicicol, rachelmycin, rapamycin, ravidomycin, restricticin, roridin A, sordarin, TAN1254, tanzawaic acid G, toyocamycin, trichothecene, tubercidin, tunicamycin, tyroscherin, verrucarin A, wortmannin, xylarin, yatakemycin
Novel NPs (MOA/target) (total of 14) [a]	campafungin (cAMP-PKA, hyphal growth)[b], dretamycin (DRE2, DIP5)[b], fellutamides C & D (proteasome)[b,c], lipogungimide (similar to phomafungin)[c], parnafungin (RNA poly[A] polymerase)[d], phomafungin (sphingolipid biosynthesis)[e], virgineone (heat shock response)[f], yefafungin (translation initiation factor Yef3p)[b], 11-desacetoxyl-wortmannin (phosphoribosyl complex)[b], 12-deoxo-hamigerone (microtubule)[b], homoarginine derivative of 3'-aminoadenosine (ribosome biogenesis)[b]; new compound (nuclear transport, MEX67)[b], new compound (structurally distinct from cotransin, signal peptidase complex)[b], novel polyacetylene (fatty acid biosynthesis)[b]

[a]Most of these NPs were present in more than one extract; [b]reference 22; [c]manuscripts in preparation; [d]reference 15; [e]reference 23; [f]reference 25.

respectively, thereby decreasing its potency. Also, heterozygotes corresponding to RNA PolII that were resistant to ECC577 were not detected in cordycepin CaFT profiles. Collectively, the results predict that the MOA of the predominant bioactivity within ECC577 and cordycepin are closely related – but not identical – and importantly, the bioactive component within ECC577 is structurally distinct from cordycepin since it does not require a nucleoside permease for uptake whereas cordycepin does. Indeed, LC/MS analyses of the

ECC577 extract did not find the match to cordycepin or its potential nucleoside analogs. Thus, bioassay-guided purification and structure elucidation for the active(s) in ECC577 were warranted.

13.4.2 Isolation of Parnafungin

With the striking CaFT profile of the *Fusarium larvarum* extract in hand, the isolation of the active component(s) from this sample was prioritized based on the likelihood that the extract contained a novel secondary metabolite that inhibits normal 3′ mRNA processing. After chromatography of the crude sample with a hydrophobic polymer resin, activity against wild-type *C. albicans* was used to identify fractions that contained antifungal components. After the combination of these fractions, further CaFT profiling confirmed that this enriched sample maintained the mRNA processing profile that was observed in the initial extract. Each subsequent isolation stage was similarly monitored to confirm that the appropriate CaFT profile was retained. Further purification of this sample by countercurrent chromatography provided a 'clean' mixture of four components that were directly correlated with anti-*C. albicans* activity. It was clear from a comparison of the absorbance spectra as well as MS data that these four components were related and belonged to the same structural family. Elucidation of the chemical structure of each individual component required separation and further analysis of purified materials.

Reversed-phase HPLC (C18) indicated that this mixture contained two major and two minor components. The initial targets for purification and structure elucidation were the two major components present in the sample. Preparative reversed-phase HPLC was used to isolate the major components and immediate re-injection of the fractions obtained from this process indicated that successful separation of these components had been achieved. However, the purified components readily re-equilibrated back within 20 hours at pH 3 to a mixture that was essentially identical to the starting mixture of components. This rapid equilibration process made NMR analysis of a single component difficult and it was impossible to acquire data with extended 2D NMR experiments without significant change in the state of the sample. Thus, NMR analysis of the mixture of components could not further resolve the structure of these natural products.

It was essential that the equilibration of these components be controlled and a stable form of the natural products be identified. Manipulation of the pH of the sample indicated that the equilibration process was slowed at pH 3 relative to higher pH media, where rapid equilibration and sample degradation was observed. Without knowledge of the specific structural features of these molecules, a derivatization strategy could only be based on this general stability profile. Since the components were more stable in acid, it was reasonable to hypothesize that an acidic group was present in the molecule and that chemically locking that acidic position might lead to the control of the observed equilibration. Methylation of the natural product mixture was accomplished with diazomethane and a series of product peaks could be directly related to the starting sample by HPLC analysis. LC/MS of this mixture confirmed that the predominant products observed contained one additional methyl group (along with smaller amounts of dimethylated material). Two major monomethylated components were then purified by reversed-phase HPLC and these purified materials were sufficiently stable for structure elucidation studies.

13.4.3   Structure Elucidation of Parnafungins

The structures of the parnafungin family of natural products are shown in Figure 13.4A. After the isolation of the stabilized monomethyl derivatives of the parnafungin mixture of components, these compounds were extensively studied by NMR. Uncovering the structure of these derivatives would then indicate the structures of the natural product mixture from which they were derived. However, while various substructures could be deduced from the NMR data of the methylated derivatives, the complete structures were not revealed due to the high density of contiguous quaternary centers in the central portion of the molecules. X-Ray crystallography of the less polar of the two methylated derivatives was able to provide the extended hexacyclic structure **5** (Figure 13.4B). From this structure, the identity of the members of the original mixture of natural products could easily be resolved. Methylation had occurred at the hydroxyl group of the A ring enol, thereby blocking the equilibration of parnafungin A (**1**) and parnafungin B (**2**). X-Ray crystallography of the 4-chloro-benzyl enol ether (**6**, Figure 13.4B), prepared by Mitsunobu reaction of the natural product mixture and purification of the derivative corresponding to parnafungin A1, provided the stereochemistry of the C15 hydroxyl as *S* and the major epimer of the C15a methyl carboxylate also as *S*.[29]

The most unique structural feature of this family of natural products is the isoxazolidinone ring, containing a highly labile nitrogen–oxygen bond. This ring system was previously unprecedented in an isolated natural product and it plays an important role in the observed antifungal activity. After opening of the isoxazolidinone ring and elimination to form the corresponding benzoquinolines (Figure 13.4C), no measurable antifungal activity was observed. Due to the instability of parnafungin, further development of these natural products as an antifungal agent would necessitate the identification of stabilized isoxazolidinone analogs or complete replacement of that ring system while maintaining the potency of the natural products. Elegant, convergent approaches to the synthesis of parnafungin model compounds have recently been described.[30,31] The synthesis of additional parnafungin analogs by these or other methods may identify potent antifungal compounds with improved physicochemical properties.

After the isolation of parnafungins A and B extract had been described from the initial *Fusarium larvarum*, extracts of several closely related *Fusarium* spp. also demonstrated a mRNA polyadenylation CaFT profile. These extracts were analyzed for the presence of the known parnafungins as well as additional members of this family of natural products. From MS analysis of one such extract, the presence of two additional parnafungin components were identified. Parnafungin C (**3**) included an additional methyl group on the C7 phenolic hydroxyl and parnafungin D (**4**) contained an A-ring epoxide along with the same C7 phenolic methyl ether present in parnafungin C (Figure 13.4A).[24] Since both of these components have the phenolic C7 blocked, neither of these structures can equilibrate to the 'bent' form of these structures that would be analogous to that of parnafungin B. However, parnafungins C and D are still able to epimerize at the C15a quaternary center by a mechanism identical to that observed for parnafungins A and B.

13.4.4 Biochemical and Genetic Verification of MOA

As expected, purified parnafungins reproduced the CaFT profiles (Figure 13.3A), which predicted that they inhibit the RNA cleavage and polyadenylation process (Figure 13.3B). In an *in vitro* RNA cleavage and polyadenylation assay, parnafungins inhibited predominantly the polyadenylation reaction while the cleavage step was only moderately affected (Figures 13.3B and 13.3C).[15] In an *in vitro* polyadenylation polymerization (PAP) assay, parnafungin was a potent inhibitor of native *C. albicans* PAP with an IC_{50} of 31nM; compared with IC_{50} of 145 nM and 114 nM to the *S. cerevisiae* and human PAP orthologs respectively. Unlike cordycepin, parnafungin was not an ATP-competitive inhibitor of PAP activity (Figure 13.3C), since its IC_{50} was not shifted by changes in ATP levels.[15] Consistent with these biochemical results, a resistant mutant containing a single nucleotide alteration of T to A, resulting in an F115I change in the nucleotidyl-transferase domain of the *C. albicans PAP* gene was isolated. This mutation was sufficient to confer resistance to parnafungin once introduced in either allele of *PAP* in a wild-type strain background.[15] We further demonstrated that reduced expression of *PAP1* in *A. fumigatus* caused selective hypersensitivity to parnafungins.[15] Collectively, these data establish PAP as the drug target of parnafungins.

13.4.5 Lead Validation

Establishing that parnafungins inhibit *C. albicans* via a novel MOA and that they represent a new NP structural class, their potential as a bona fide antifungal drug lead was further evaluated. Parnafungins were selectively active against a panel of clinically relevant *Candida* and *Aspergillus* pathogens, with no activity against bacteria,[15] as predicted by their specific MOA. Further, parnafungins were equipotent across existing antifungal drug resistant *Candida* spp., as expected for a new agent with novel antifungal MOA. Parnafungins also demonstrated strong anti-*Candida* activity (MIC = 1.1 µg/mL) in the presence of 50% mouse serum, in part due to their potent intrinsic activity (MIC=0.014 µg/mL) against *C. albicans*.[15] Importantly, in an animal model of candidiasis, parnafungins administered intraperitoneally

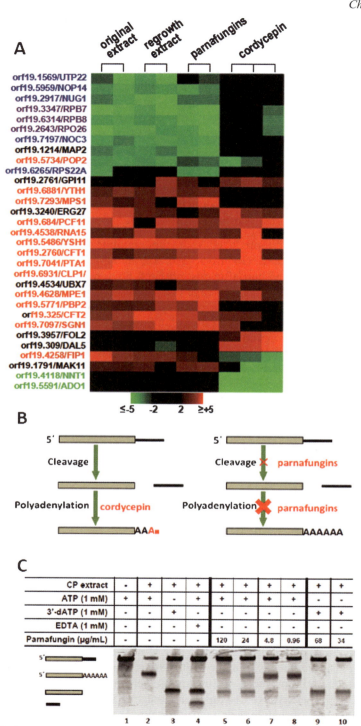

Figure 13.3   CaFT profiling of cordycepin, parnafungins, and parnafungin-containing extracts.

twice daily for two days produced a significant and dose-dependent reduction in fungal burden in the infected organ, with a ED_{99} between 20–30 mpk in two independent experiments and no obvious acute adverse effects observed.[15] Remarkably, the efficacy and spectrum ascribed to parnafungins are the result of purified NP without any semisynthetic modifications for improvement. Recently, the synthesis of parnafungin model compounds has been achieved.[30,31] This success potentially provides an exciting solution to addressing the chemical complexity and instability of the parnafungin class, as discussed above. Moreover, it also provides a medicinal chemistry opportunity to fully assess the structure–activity relationship (SAR) of the parnafungin scaffold, a critical first step in their development as an antifungal agent.

13.5 Additional Representative Novel Antifungals Identified by CaFT-Based Screening

13.5.1 Yefafungin

In addition to the parnafungins, other new natural products with novel MOAs were discovered during the course of this screening effort. CaFT analysis of

Figure 13.3 CaFT profiling of cordycepin, parnafungins, and parnafungin-containing extracts. (A) CaFT profiling of the original extract, a re-growth of the original extract, purified parnafungins, and purified cordycepin. Four groups of heterozygotes are highlighted, in red those corresponding to genes involved in RNA cleavage and polyadenylation, in blue ribosome biogenesis, in purple transcription by RNA polymerase II, and in green uptake (*NNT1*) and activation (*ADO1*) of cordycepin. (B) Schematic representation of RNA cleavage and polyadenylation and inhibition by cordycepin and parnafungins. Note the 5′ and 3′ cleavage fragments are depicted as the grey box and black bar, respectively. The incorporated 3′-dAMP is indicated by the symbol 'A ■' (in red). Note the CaFT profiles of parnafungin and the reference compound, cordycepin, used in the de-replication process, are related but distinct, prompting the isolation of parnafungins from their original bioactive crude extract. (C) *In vitro* assay of cleavage and polyadenylation (C/P) using the whole-cell extract (WCE). Adapted from reference 15. The ^{33}P-labeled precursor RNA (lane 1) was incubated with the WCE under different conditions for the production of poly(A)+ product (lane 2), the 5′ fragment (lane 3), with the polyadenylation reaction blocked by 3′-deoxyl-ATP (cordycepin), and both 5′ and 3′ fragments (lane 4). Different concentrations of parnafungin were added to the reactions under the standard conditions (lanes 5–8), or in the presence of 3′-deoxyl-ATP (lanes 9 and 10). Note that under the standard conditions poly(A)+ product was diminished, and that there was simultaneous accumulation of the 5′ fragment with increasing concentrations of parnafungin (lanes 5–8), suggesting a preferential inhibition of polyadenylation. Moreover, in the presence of 3′-deoxyl-ATP, which blocks polyadenylation (lane 3), the accumulation of the 5′ intermediate was also diminished, suggesting a partial inhibition of the cleavage by parnafungin.

A.   Structures of Parnafungins A-D

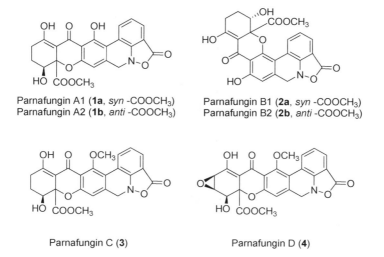

Parnafungin A1 (**1a**, *syn* -COOCH₃)
Parnafungin A2 (**1b**, *anti* -COOCH₃)

Parnafungin B1 (**2a**, *syn* -COOCH₃)
Parnafungin B2 (**2b**, *anti* -COOCH₃)

Parnafungin C (**3**)

Parnafungin D (**4**)

B.   Parnafungin analogs used for structure elucidation

5

6

C.   Ring-opened benzoquinoline forms of parnafungins A and B

7

8

D.   Synthetic parnafungin model compounds

Figure 13.4   Structures of parnafungins and their analogs.

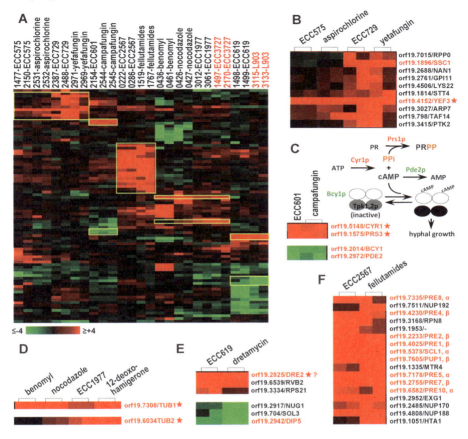

Figure 13.5 CaFT profiling of novel NPs isolated in the screening campaign. (A) Summary of CaFT profiles of novel NPs. For each extract (indicated with an ECC prefix) and compound, two independent experiments were selected. The experiment numbers (first four digits), z-scores and the hierarchical clustering are described in reference 22. Highlighted by yellow boxes are clusters of heterozygotes whose significant hypersensitivity or resistance is indicative of the MOAs of tested extracts and compounds. (B)–(F) Clusters of heterozygotes that display significant hypersensitivity or resistance to specific compounds and the source extracts of aspirochlorine and yefafungin (B), campafungin (C), 12-deoxo-hamigerone (D), dretamycin (E) and fellutamides C and D (F). Highlighted in red are heterozygotes predictive of their respective MOAs, with stars indicating known or potential targets. The insert in part C illustrates the part of the cAMP-PKA pathway that was detected in the CaFT profile, with green indicating resistance of the corresponding heterozygotes and red indicating hypersensitivity.

ECC729, a *streptomycete* extract, yielded a FT profile (Figures 13.5A and 13.5B) with strain depletions corresponding to YEF3, a fungal-specific elongation factor of protein synthesis,[32] and SSC1, a mitochondrial heat

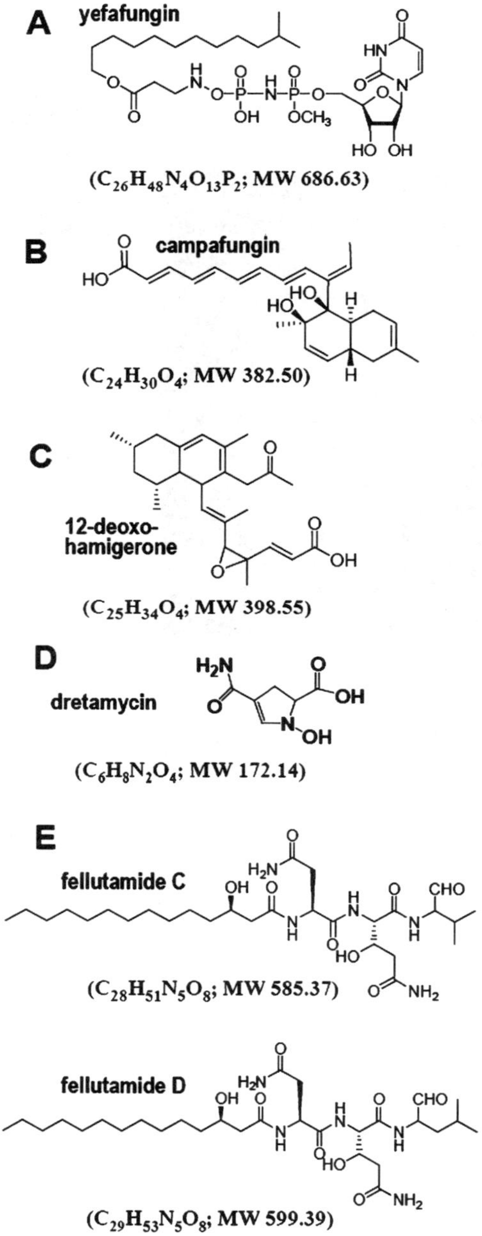

Figure 13.6 Chemical structures of novel NPs isolated.

shock chaperone protein (Figure 13.5B). The profile of aspirochlorine, a known YEF3 inhibitor repeatedly identified during the course of the CaFT screening, was compared with that of ECC729 and distinct differences were observed (Figure 13.5B). After LC/MS analysis failed to identify the presence of aspirochlorine in ECC729, bioassay-guided isolation and structural elucidation of the bioactivity revealed a novel uridine analog ($C_{26}H_{48}N_4O_{13}P_2$; MW 686.63) containing a unique amidodiphosphate linkage and aliphatic side chain (Figure 13.6A). The purified compound demonstrated potent activity against *C. albicans* (MIC = 0.2 µg/mL) and, consistent with its CaFT profile, specifically inhibited protein synthesis in whole cells. Based on these results we have named the compound yefafungin, for yeast elongation factor affecting fungin.[22]

13.5.2 Campafungin

In addition to detecting hypersensitive strains, CaFT profiles also reveal hyposensitive heterozygotes (i.e. partially resistant to a bioactive compound) that also provide clues as to the MOA of a bioactive agent, as illustrated in the profiling of the fungal-derived NP extract ECC601 and its purified bioactive compound. Their CaFT profiles highlighted not only two markedly hypersensitive strains heterozygous for *CYR1*, encoding adenylate cyclase, and *PRS3*, encoding phosphoribosylpyrophosphate (PRPP) synthetase, but also two hyposensitive heterozygotes corresponding to *PDE2*, a phosphodiesterase which converts cAMP to AMP, and *BCY1*, the negative regulatory subunit of protein kinase A (PKA) (Figures 13.5A and 13.5C). Remarkably, the chemical sensitivities of the *CYR1*, *PDE2*, and *BCY1* heterozygotes were congruent with the functional relationship between the steps in the cAMP-dependent PKA pathway (involved in yeast-to-hyphal transition, important for *C. albicans* pathogenicity), and suggested that the bioactive compound decreases the cellular cAMP levels. Furthermore, the *PRS3* mutant's hypersensitivity raised the possibility that the synthesis of cAMP is coupled with transfer of the pyrophosphate to form PRPP (Figure 13.5C). Bioassay-guided isolation led to the identification of a novel pentaene-containing compound ($C_{24}H_{30}O_4$; MW 382.50; Figure 13.6B), which we named campafungin, for cAMP-affecting fungin. According to the hypothetical MOA, campafungin inhibits (either) Cyr1p and/or Prs3p, hence the hypersensitivities of both corresponding heterozygous deletions. Furthermore, this inhibitory effect could be partially alleviated by a reduction in either the degradation of cAMP by phosphodiesterase or the inhibitory effect of the PKA regulatory subunit (which de-inhibits PKA when bound to cAMP). Thus, heterozygous mutants of either *PDE2* or *BCY1* are resistant to the cAMP-limiting effects of campafungin. Consistent with this hypothesis, campafungin effectively blocked the transition of *C. albicans* from yeast to hyphae, as did the known PKA inhibitor PKI(11-22).[33] Also, unlike PKI(11-22), campafungin's inhibitory effect, at a concentration near its MIC, was partially suppressed by exogenous cAMP. Suppression was not observed, however, at low concentrations of campafungin (Figure 13.3C). Given that *CYR1* is not essential in *C. albicans*,[22] our results suggest that Cyr1p is not the direct or only target of campafungin.

The possibility remains that it inhibits Prs3p, which is essential for viability and hyphal growth (data not shown). Nevertheless, campafungin is a NP with both anti-proliferative and anti-hyphal activities.[22]

13.5.3   12-Deoxo-hamigerone

Fungal extract ECC1977 produced a complex profile, yet featured two predominant hypersensitive heterozygotes (*TUB1* and *TUB2*) corresponding to α-tubulin and β-tubulin (Figures 13.5A and 13.5D). Structural elucidation determined the purified component ($C_{25}H_{34}O_4$, MW 398.55) to be a new analog of hamigerone.[34] Accordingly, we named it 12-deoxo-hamigerone (Figure 13.6C). Although the *TUB1* heterozygote strain is similarly hypersensitive to other known microtubule assembly inhibitors, including benomyl and nocodazole, only the *TUB2* heterozygote also displayed hypersensitivity to 12-deoxo-hamigerone (Figure 13.5D). This difference again reflects the de-replication potential of the CaFT in differentiating distinct chemical classes predicted to interdict a common cellular process (namely microtubular dynamics) as subsequently confirmed in an *in vitro* microtubule assembly assay.[22] 12-Deoxo-hamigerone is active against tubulin-containing eukaryotes such as *C. albicans* and *Trichophyton mentagrophytes* (with MICs of 4 and 1 μg/mL, respectively), as well as multiple mammalian cell lines (IC_{50} of 2–3 μg/mL), with no appreciable effects on bacteria.[22]

13.5.4   Dretamycin

Actinomycete extract ECC619 generated a CaFT profile with hypersensitivity in the *DRE2* heterozygote and resistance in that of *DIP5* (Figures 13.5A and 13.5E), corresponding to an essential Fe/S cluster protein implicated in oxidative stress-induced cell death[35] and the dicarboxylic acid permease,[36] respectively. The resistance of the latter indicated that the bioactive NP in ECC619 is likely an analog of a dicarboxylic acid. Indeed, the active compound, dretamycin, was determined to be a dicarboxylic amino acid analog of proline ($C_6H_8N_2O_4$, MW 172.14, Figure 13.6D). Consistent with its CaFT profile, tetracycline-based repression of *DRE2* using the corresponding GRACE strain amplified the hypersensitivity to dretamycin versus control strains, whereas marked resistance was observed in the absence of tetracycline. This resistance is presumably due to overexpression of *DRE2* from the heterologous tetracycline promoter as compared to its endogenous promoter. Accordingly, significant resistance rendered by transcriptional repression of *DIP5* demonstrated that the import of dretamycin into the cell is likely mediated directly by its protein product. This was further confirmed by the observation that exogenous L-glutamate and L-aspartate, native substrates of *DIP5*, noticeably reduced whole-cell sensitivity to dretamycin.[22]

13.5.5 Fellutamides C and D

The CaFT profiles of fungal extract ECC2567 reproducibly identified multiple hypersensitive strains corresponding to four α-(structural) and five ß-(catalytic) subunits of the 20S core particle, thus suggesting the proteasome to be the target of the bioactivity (Figures 13.5A and 13.5F). The proteasome is a large, multi-subunit, proteolytic complex that degrades ubiquinylated proteins to ensure protein homeostasis and to regulate diverse cellular processes.[37] Bioactivity isolation and structural elucidation revealed two highly related lipopeptide aldehydes with chemical formulas of $C_{28}H_{51}N_5O_8$ (MW 585.37) and $C_{29}H_{53}N_5O_8$ (MW 599.39) and differing only in having valine or leucine as the C-terminal amino acid (Figure 13.6E). Based on their structural similarity to the recently characterized NP proteasome inhibitor fellutamide B,[38] we named these new natural products fellutamide C and fellutamide D, respectively. Fellutamide C and fellutamide D, as well as the original ECC2567 extract, inhibited proteasome activity *in vitro* using a *C. albicans* whole-cell extract[22] and this inhibitory activity was reproduced using a human 20S proteasome assay (data not shown). Given their MOA and significantly higher anti-proliferative activity across multiple human cell lines (IC_{50} values range between 0.05 and 1.5 μM) than in fungal pathogens (MIC values of 4–32 μg/mL across *Candida* spp.), fellutamide C and D may provide additional chemotherapeutic starting points for oncology or other recently identified therapeutic applications for proteasome inhibitors.[39]

13.6 Conclusion and Prospects

Many interesting known and novel antifungal NPs, with structural and mechanistic diversity, were identified through our CaFT-based NP screening campaign (Table 13.1). These compounds target a wide range of cellular processes including: cell wall biogenesis, ergosterol biosynthesis, fatty acid biosynthesis, protein synthesis and secretion, actin cytoskeleton, proteasome function, rRNA biogenesis, and mRNA processing. While CaFT profiles reliably reflected the well-characterized mode of action for known NPs, the various MOA hypotheses for novel NPs annotated from CaFT profiles were verified only after independent biochemical and genetic evidence was obtained. Coupled with analytical methodologies, CaFT profiling has served as a robust de-replication strategy to avoid excessive rediscovery.

Noteworthy is the frequent identification of NPs targeting cell wall biogenesis, a historically favorable antifungal target. These include a new papulacandin and desulfated ergokonin A, as well as various known echinocandins (from which the therapeutic antifungal agent Cancidas, commonly known as caspofungin, is derived), pneumocandins, papulacandins, and acidic terpenoids (Table 13.1). This success can be attributed to the introduction of a set of 125 double heterozygote mutant strains with elevated hypersensitivity to glucan synthesis inhibitors,[22] as single heterozygote depletions of the 1,3-ß-glucan synthase enzyme complex did not exhibit pronounced hypersensitivity to these inhibitors, including caspofungin.[8] Each resulting *C. albicans* double heterozygote mutant strain was heterozygously deleted of one of a first set of four genes (*FKS1*, *FKS2*, *RHO1*, or *PKC1*) combined with the heterozygous deletion of one of a second set of genes (e.g. *FKS3*, *CHS5*, *GAS1*, etc.) known to participate in cell wall related functions. This second set of genes was further selected based on the demonstrated hypersensitivity of their mutant phenotypes to caspofungin or synthetic lethality in combination with *FKS1* or *FKS2* mutations in *S. cerevisiae*.[40]

More significantly, novel NP entities with antifungal lead potential were identified. Parnafungins, comprising a new NP structural class, exhibited broad spectrum activity across medically relevant *Candida* and *Aspergillus* pathogens, including existing antifungal drug resistant *Candida* spp. Furthermore, parnafungins were efficacious in a murine model of disseminated candidiasis, without displaying obvious cytotoxicity.[15] Thus the discovery of parnafungins and other novel compounds with lead potential for antifungal development provides a clear validation of our NP screening approach to identify new antifungal agents with unique MOAs from crude fermentation extracts. In fact, the discovery and characterization of this new NP class likely represents the most significant achievement over the last decade in identifying new antifungal leads since the discovery of sordarin[10] and enfumafungin;[11] the latter of which is presently in Phase I clinical trials.

Despite the advantages and successes of CaFT, the frequency of discovery of novel and druggable antifungal leads with unique MOAs was considerably lower than expected, and significantly decreased over time. It remains to be ascertained whether this issue is due to limitation of CaFT profiling, a limited chemical diversity within the screening source, or a combination thereof. One important limitation to CaFT profiling applies to extracts containing multiple mechanistically distinct bioactive compounds. In such cases, only the most dominant components contributing to the overall growth inhibitory activity are reliably profiled; minor but potentially interesting activities would be remain undetected.[15] In principle, this could be remedied by upfront fractionation of NP extracts prior to CaFT profiling so that the dominant 'offending' bioactivities of known NPs could be separated from minor activities with potential chemical and mechanistic novelty. Thus, a wide range of extracts may be screened, but the depth in which they can be evaluated is currently limited and likely addressable only after implementing significant NP fractionation efforts.

Alternatively, a more limited microbial diversity may have been screened despite our concerted effort to maximize chemical diversity by using fresh extracts and microbial isolates with broad taxonomic and geographic diversity[27] as well as employing nutritional array methodologies to stimulate normally unexpressed secondary metabolic pathways.[28] Perhaps a more limited biosynthetic potential exists to produce novel natural products than we assume from microbiota. Although, this seems highly unlikely, it could certainly be addressed (in part) by considering plant or marine extracts, both of which were omitted in our study. Most imperative is the need for innovative technological advances to augment the assayable NP chemical diversity. This may be made possible through advances in metagenomics[41,42] and/or novel approaches to culturing previously unculturable organisms.[43–45] Diversity-oriented synthesis (DOS) using existing NP scaffolds, for example applied to the parnafungins, may also provide important breakthroughs to expanding the assayable NP chemical diversity required to discover the next generation of NP-based antimicrobial agents.[46,47]

Acknowledgements

We thank all our present and past colleagues within Merck who have contributed to the development of our chemical genomics strategy as well as natural product-based screening. We also thank the Stanford Genome Center for the complete sequencing of the *C. albicans* genome. We are grateful to Genome Quebec and Genome Canada for research funding.

References

1. D. J. Newman, G. M. Cragg and K. M. Snader, *J. Nat. Prod.*, 2003, **66**, 1022.
2. C. Walsh, *Antibiotics: Actions, Origins, Resistance*, ASM Press, Washington, DC, 2003, p. 345.
3. L. Ostrosky-Zeichner, A. Casadevall, J. N. Galgiani, F. C. Odds and J. H. Rex, *Nat Rev. Drug. Discovery*, 2010, **9**, 719.
4. K. Baetz, L. McHardy, K. Gable, T. Tarling, D. Reberioux, J. Bryan, R. J. Andersen, T. Dunn, P. Hieter and M. Roberge, *Proc. Natl. Acad. Sci. USA*, 2004, **101**, 4525.
5. G. Giaever, D. D. Shoemaker, T. W. Jones, H. Liang, E. A. Winzeler, A. Astromoff and R. W. Davis, *Nat. Genet.*, 1999, **21**, 278.
6. G. Giaever, P. Flaherty, J. Kumm, M. Proctor, C. Nislow, D. F. Jaramillo, A. M. Chu, M. I. Jordan, A. P. Arkin and R.W. Davis, *Proc. Natl. Acad. Sci. USA*, 2004, **101**, 793.
7. A. M. Smith, R. Ammar, C. Nislow and G. Giaever, *Pharmacol. Ther.*, 2010, **127**, 156.
8. D. Xu, B. Jiang, T. Ketela, S. Lemieux, K. Veillette, N. Martel, J. Davison, S. Sillaots, S. Trosok, C. Bachewich, H. Bussey, P. Youngman and T. Roemer, *PLoS Pathogens*, 2007, **3**, e92.
9. P. Y. Lum, C. D. Armour, S. B. Stepaniants, G. Cavet, M. K. Wolf, J. S. Butler, J. C. Hinshaw, P. Garnier, G. D. Prestwich, A. Leonardson, P. Garrett-Engele, C. M. Rush, M. Bard, G. Schimmack, J. W. Phillips, C. J. Roberts and D. D. Shoemaker, *Cell*, 2004, **116**, 121.
10. J. M. Dominguez, V. A. Kelly, O. S. Kinsman, M. S. Marriott, F. Gomez de las Heras and J. J. Martin, *Antimicrob Agents Chemother.*, 1998, **42**, 2274.

11. F. Pelaez, A. Cabello, G. Platas, M. T. Diez, A. Gonzalez del Val, A. Basilio, I. Martan, F. Vicente, G. E. Bills, R. A. Giacobbe, R. E. Schwartz, J. C. Onish, M. S. Meinz, G. K. Abruzzo, A. M. Flattery, L. Kong and M. B. Kurtz, *Syst. Appl. Microbiol.*, 2000, **23**, 333.

12. M. A. Pfaller and D. J. Diekema, *Clin. Microbiol. Rev.*, 2007, **20**, 133.

13. T. Roemer, B. Jiang, J. Davison, T. Ketela, K. Veillette, A. Breton, F. Tandia, A. Linteau, S. Sillaots, C. Marta, N. Martel, S. Veronneau, S. Lemieux, S. Kauffman, J. Becker, R. Storms, C. Boone and H. Bussey, *Mol. Microbiol.*, 2003, **50**, 167.

14. R. Rodriguez-Suarez, D. Xu, K. Veillette, J. Davison, S. Sillaots, S. Kauffman, W. Hu, J. Bowman, N. Martel, S. Trosok, H. Wang, L. Zhang, L. Y. Huang, Y. Li, F. Rahkhoodaee, T. Ransom, D. Gauvin, C. Douglas, P. Youngman, J. Becker, B. Jiang and T. Roemer, *Chem. Biol.*, 2007, **14**, 1163.

15. B. Jiang, D. Xu, J. Allocco, C. Parish, J. Davison, K. Veillette, S. Sillaots, W. Hu, R. Rodriguez-Suarez, S. Trosok, L. Zhang, Y. Li, F. Rahkhoodaee, T. Ransom, N. Martel, H. Wang, D. Gauvin, J. Wiltsie, D. Wisniewski, S. Salowe, J. N. Kahn, M. J. Hsu, R. Giacobbe, G. Abruzzo, A. Flattery, C. Gill, P. Youngman, K. Wilson, G. Bills, G. Platas, F. Pelaez, M. T. Diez, S. Kauffman, J. Becker, G. Harris, P. Liberator and T. Roemer, *Chem Biol.*, 2008, **15**, 363.

16. D. Xu, S. Sillaots, J. Davison, W. Hu, B. Jiang, S. Kauffman, N. Martel, P. Ocampo, C. Oh, S. Trosok, K. Veillette, H. Wang, M. Yang, L. Zhang, J. Becker, C. E. Martin and T. Roemer, *J. Biol. Chem.*, 2009, **284**, 19754.

17. J. M. Becker, S. J. Kauffman, M. Hauser, L. Huang, M. Lin, S. Sillaots, B. Jiang, D. Xu and T. Roemer, *Proc. Natl. Acad. Sci. USA*, 2010, **107**, 22044.

18. S. M. Noble, S. French, L. A. Kohn, V. Chen and A. D. Johnson, *Nat. Genet.*, 2010, **42**, 590.

19. W. Hu, S. Sillaots, S. Lemieux, J. Davison, S. Kauffman, A. Breton, A. Linteau, C. Xin, J. Bowman, J. Becker, B. Jiang and T. Roemer, *PLoS Pathol.*, 2007, **3**, e24.

20. O. W. Liu, C. D. Chun, E. D. Chow, C. Chen, H. D. Madhani and S. M. Noble, *Cell*, 2008, **135**, 174.

21. P. D. Collopy, H. V. Colot, G. Park, C. Ringelberg, C. M. Crew, K. A. Borkovich and J. C. Dunlap, *Methods Mol. Biol.*, 2010, **638**, 33.

22. T. Roemer, D. Xu, S. B. Singh, C. A. Parish, G. Harris, H. Wang, J. E. Davies and G. F. Bills, *Chem. Biol.*, 2011, **18**, 148.

23. K. Herath, G. Harris, H. Jayasuriya, D. Zink, S. Smith, F. Vicente, G. Bills, J. Collado, A. Gonzalez, B. Jiang, J. N. Kahn, S. Galuska, R. Giacobbe, G. Abruzzo, E. Hickey, P. Liberator, D. Xu, T. Roemer and S. B. Singh, *Bioorg. Med. Chem.*, 2009, **17**, 1361.

24. D. Overy, K. Calati, J. N. Kahn, M. J. Hsu, J. Martin, J. Collado, T. Roemer, G. Harris and C. A. Parish, *Bioorg. Med. Chem. Lett.*, 2009, **19**, 1224.

25. J. Ondeyka, G. Harris, D. Zink, A. Basilio, F. Vicente, G. Bills, G. Platas, J. Collado, A. Gonzaez, M. de la Cruz, J. Martin, J. N. Kahn, S. Galuska, R. Giacobbe, G. Abruzzo, E. Hickey, P. Liberator, B. Jiang, D. Xu, T. Roemer and S. B. Singh, *J. Nat. Prod.*, 2009, **72**, 136.

26. O. Genilloud, I. Gonzalez, O. Salazar, J. Martin, J. R. Tormo and F. Vicente, *J. Ind. Microbiol. Biotechnol.*, 2011, **38**, 375.

27. G. F. Bills, J. Martín, J. Collado, G. Platas, D. Overy, J. R. Tormo, F. Vicente, G. Verkley and P. W. Crous, *Soc. Ind. Microbiol. News*, 2009, **59**, 133.

28. G. F. Bills, G. Platas, A. Fillola, M. R. Jimenez, J. Collado, F. Vicente, J. Martin, A. Gonzalez, J. Bur-Zimmermann, J. R. Tormo and F. Pelaez, *J. Appl. Microbiol.*, 2008, **104**, 1644.

29. C. A. Parish, S. K. Smith, K. Calati, D. Zink, K. Wilson, T. Roemer, B. Jiang, D. Xu, G. Bills, G. Platas, F. Pelaez, M. T. Diez, N. Tsou, A. E. McKeown, R. G. Ball, M. A. Powles, L. Yeung, P. Liberator and G. Harris, *J. Am. Chem. Soc.*, 2008, **130**, 7060.

30. Q. Zhou and B. B. Snider, *Org. Lett.*, 2009, **11**, 2936.

31. Q. Zhou and B. B. Snider, *J Org Chem.*, 2010, **75**, 8224.

32. J. Sturtevant, *Expert Opin. Ther. Targets*, 2002, **6**, 545.

33. R. Castilla, S. Passeron and M. L. Cantore, *Cell Signal.*, 1998, **10**, 713.

34. J. Breinholt, A. Kjoer, C. E. Olsen, B. R. Rassing and C. N. Rosendahl, *Acta Chem. Scand.*, 1997, **51**, 1241.

35. L. Vernis, C. Facca, E. Delagoutte, N. Soler, R. Chanet, B. Guiard, G. Faye and G. Baldacci, *PLoS One*, 2009, **4**, e4376.

36. B. Regenberg, S. Holmberg, L. D. Olsen and M. C. Kielland-Brandt, *Curr. Genet.*, 1998, **33**, 171.

37. S. Park, G. Tian, J. Roelofs and D. Finley, *Biochem. Soc. Trans.*, 2010, **38**, 6.

38. J. Hines, M. Groll, M. Fahnestock and C. M. Crews, *Chem. Biol.*, 2008, **15**, 501.

39. A. F. Kisselev, *Chem. Biol.*, 2008, **15**, 419.

40. G. Lesage, A. M. Sdicu, P. Menard, J. Shapiro, S. Hussein and H. Bussey, *Genetics*, 2004, **167**, 35.

41. F. Lefevre, P. Robe, C. Jarrin, A. Ginolhac, C. Zago, D. Auriol, T. M. Vogel, P. Simonet and R. Nalin, *Res. Microbiol.*, 2008, **159**, 153.

42. X. Li, J. Guo, S. Dai, Y. Ouyang, H. Wu, W. Sun and G. Wang, *Curr. Top. Med. Chem.*, 2009, **9**, 1525.

43. T. Kaeberlein, K. Lewis and S. S. Epstein, *Science*, 2002, **296**, 1127.

44. A. D'Onofrio, J. M. Crawford, E. J. Stewart, K. Witt, E. Gavrish, S. Epstein, J. Clardy and K. Lewis, *Chem. Biol.*, 2010, **17**, 254.

45. K. Lewis, S. Epstein, A. D'Onofrio and L. L. Ling, *J. Antibiot. (Tokyo)*, 2010, **63**, 468.

46. L. A. Marcaurelle and C. W. Johannes, *Prog. Drug Res.*, 2008, **66**, 189.

47. W. R. Galloway, A. Bender, M. Welch and D. R. Spring, *Chem. Commun (Camb)*, 2009, **18**, 2446.

Section 3
Novel Microbial Natural Products and Derivatives

CHAPTER 14

Natural Products: New Agents Against MDR Tuberculosis

UJJINI MANJUNATHA*, FUMIAKI YOKOKAWA,
MEERA GURUMURTHY AND THOMAS DICK

Novartis Institute for Tropical Diseases, 05-01 Chromos, 10 Biopolis Road,
138670, Singapore
*E-mail: manjunatha.ujjini@novartis.com

14.1 Introduction

Tuberculosis (TB), caused by *Mycobacterium tuberculosis* (Mtb), is an airborne infectious disease that is estimated to infect one-third of the world's population.[1] TB commonly manifests as an infection of the lungs with symptoms of fever, severe coughing and weight loss. Mtb's ability to remain dormant within the host for several years without manifesting as active disease (described as latency) makes diagnosis and treatment of TB particularly challenging.[2] The Bacillus Calmette-Guerin (BCG) vaccine provides immunity against TB in children; however, there is currently no vaccine that provides reliable protection against adult pulmonary TB.[3] This situation has further been exacerbated by the advent of the acquired immunodeficiency syndrome (AIDS), caused by the human immunodeficiency virus (HIV). Amongst HIV-positive immune-compromised patients, TB is a leading cause of death.[3] Moreover, interaction between anti-retroviral drugs and those that constitute standard TB chemotherapy is of serious concern.[4] All these factors in combination make TB a severe threat due to which the World Health Organization (WHO) declared it as a global health emergency. In 2009, TB

RSC Drug Discovery Series No. 25
Drug Discovery from Natural Products
Edited by Olga Genilloud and Francisca Vicente
© The Royal Society of Chemistry 2012
Published by the Royal Society of Chemistry, www.rsc.org

prevalence worldwide was 14 million. There were 9.4 million new cases of active TB including 1.1 million cases amongst HIV-positive patients.[1] TB takes a huge toll on mankind with nearly 2 million deaths every year.

The current WHO recommended standard treatment for active TB is a lengthy multi-drug regimen. It comprises an initial 2-month intensive phase with isoniazid, rifampicin, pyrazinamide and ethambutol, followed by a 4-month continuation phase with isoniazid and rifampicin (Table 14.1). To improve adherence to this lengthy regimen, WHO promoted the directly observed treatment and short-course drug therapy (DOTS) program in 1993. Under this program, healthcare workers directly observe patients for treatment compliance. So far, DOTS has proven extremely effective, with nearly 95% cure rates. In the early 1990s, significant clinical resistance to isoniazid and rifampicin, the two most potent first-line TB drugs, began to emerge.[5] TB caused by Mtb that is resistant at least to isoniazid and rifampicin, is referred to as multi-drug resistant tuberculosis (MDR-TB). Drug resistance develops through spontaneous mutations in the Mtb genome when treatment is inconsistent or incomplete owing to incorrect regimens, lack of patient compliance and/or unreliable drug supply.

Patients with MDR-TB are treated with the remaining first-line drugs to which Mtb is sensitive, in combination with second-line and other drugs used as TB therapeutics (Table 14.1). In the last three decades there has been no new drug specifically developed for TB introduced into this regimen. Extensively drug resistant (XDR) TB results from Mtb strains causing MDR-TB that have acquired additional resistance to a fluoroquinolone and to at least one of the three most potent injectables (kanamycin, amikacin and capreomycin).[6] Nearly 10% of MDR-TB cases acquire such resistance to become XDR-TB.[7] Recently, the emergence of totally drug resistant (TDR) TB strains have been reported.[8] These are XDR-TB strains that are resistant to all the tested additional second-line TB drugs (thioamides, cycloserine and salicylic acid derivatives). Like XDR-TB, strains causing TDR-TB emerge from MDR-TB strains. In 2009, WHO reported that 3.3% of the TB incident cases diagnosed worldwide were MDR-TB.[1] MDR-TB has been reported in

Table 14.1   List of Drugs Available to Treat TB

First-line TB Drugs	Second-line TB Drugs	Other TB Drugs
Isoniazid*	**Streptomycin/Kanamycin /Amikacin**	**Clofazimine**
Rifampicin	**Capreomycin**	**Amoxicilin with Clavulanate**
Ethambutol,	Ofloxacin/Levofloxacin/Moxifloxacin	**Imipenem**
Pyrazinamide	**Cycloserine**	Linezolid
	Para-aminosalicylic acid*	**Clarithromycin**
	Ethionamide / Prothionamide	Thioridazine
	Thioacetazone	

*Inspired by a natural product; natural products and derivatives in bold.

nearly 90 countries and regions worldwide, with especially high rates in India, China, Russia and other countries of the former Soviet Union. In 2008, approximately 440 000 cases of MDR-TB emerged worldwide, causing the death of nearly 150 000 people.[9] MDR-TB is therefore a severe public health concern in both the developing and the developed world. Existing treatment for MDR-TB is not only longer and more extensive (up to 2 years) but is also more expensive. Despite the long treatment, MDR-TB is associated with higher mortality rates and the cure rates for range from 6 to 59% as opposed to 95% or above in the case of drug-sensitive TB.[10] Moreover, second-line drugs used in the treatment of MDR-TB are not only less effective but are also often less tolerated with severe toxicity issues, therefore frequently leading to discontinuation of treatment.

The *Mycobacterium avium* complex (MAC) is a group of atypical *Mycobacterial* spp. consisting of *M. intracellulare*, *M. avium* and other related species. MAC infections are usually associated with disseminated disease in immune-compromised individuals and with pulmonary disease in some immune-competent adults.[11] Dissemination of MAC is one of the most common complications of AIDS. MAC is far less susceptible than Mtb to most anti-tubercular agents and hence treatment options are very limited.[12] Unlike MDR-TB, where resistance is usually associated with specific genetic mutations resulting from inadequate treatment, drug resistance in MAC is an intrinsic property likely due to the differential permeability barrier of the bacteria.[13] The most efficacious drugs against MAC are clarithromycin, azithromycin and amikacin; they are part of a multi-drug regimen that includes rifampicin and ethambutol.[14] In general, MAC infections are treated with two or three antimicrobials for at least 12 months.

Therefore, in the treatment of MDR-TB and MAC infections, and in controlling mycobacterial resistance, it is critical to have new chemical entities with a mechanism of action different from existing anti-tuberculars. In addition, it would be important for these novel candidates to possess attributes that are desirable for treatment of MDR-TB such as oral bioavailability and no interaction with P450 enzymes. Lastly, there certainly exist critical economic reasons that make progress in TB drug development challenging. Therefore, the development of any new drug or regimen for TB needs to be affordable and cost effective. Currently, the pipeline of anti-tuberculars in clinical development consists of ten compounds belonging to six different classes – diarylquinolines (TMC207), fluoroquinolones (gatifloxacin and moxifloxacin), nitroimidazoles (PA-824 and OPC67683), rifamycins (rifapentine), oxazolidinones (Linezolid, PNU-100480 and AZD5847) and ethylenediamines (SQ-109).[15] Of these, compounds belonging to only two classes – nitroimidazoles and diarylquinolines – offer new mechanisms of action against Mtb. Therefore, keeping in mind the high attrition rates in drug development, the drug pipeline is still insufficient to tackle the unmet needs for TB treatment and, importantly, the emergence of resistance.

14.2   The Importance of Natural Products in TB Treatment

Natural products have always been a rich source of novel chemical structures that have played important roles in drug discovery. Despite enormous efforts that have gone into screening large synthetic chemical compound libraries, roughly half of the new chemical entities introduced as drugs worldwide between 1981 and 2006 were derived or inspired from a natural product. The success of natural products and natural product derivatives as therapeutics is particularly high (78%) in the area of infectious diseases.[16]

Currently, there are a number of natural products or their synthetic derivatives that are used as first- and second-line TB drugs (Table 14.1). Streptomycin, isolated from the actinobacterium *Streptomyces griseus*, was the first drug discovered against TB.[17] Isoniazid, although first synthesized in 1912, was introduced into the TB regimen only in 1952. Interestingly, this was because it was reinvestigated as a TB therapeutic following an observation that nicotinamide, a natural product, exhibited anti-TB activity.[18] Similarly, reports allude to the discovery of *p*-aminosalicylic acid (PAS) as an agent against TB as a follow up of previous findings that identified the rapid metabolism of the natural product salicylic acid by Mtb.[19] Rifamycin, first isolated as a polyketide from the soil bacterium *Amycolatopsis mediterranei*, is to date the most important natural product as a TB therapeutic.[20] Rifampicin, a potent RNA polymerase (RNAP) inhibitor, is a semi-synthetic derivative of rifamycin. Table 14.1 highlights several other drugs that are either derived from or inspired by a natural product and are currently used in the treatment of TB.

There are multiple reviews and scientific documents reporting anti-mycobacterial activity of a large number of natural products.[21–24] Likewise, several articles report anti-mycobacterial activity of natural product extracts. However, without knowing the structures of the active ingredients of these extracts, it is premature to speculate their potential role in treating MDR-TB infection.[25–29] This chapter briefly reviews the role of natural products in the current treatment of MDR-TB and provides an overview of selected promising natural products against MDR-TB and MAC infections, highlighting key liabilities that need to be addressed during their development.

14.3   Role of Natural Products in MDR-TB Treatment

Emergence of MDR-TB and XDR-TB has sparked renewed interest in exploring the anti-tubercular activity of both historical drugs for TB such as PAS and thioamides, as well as other commonly used antibacterials such as β-lactams, fluoroquinolones and macrolides. Drugs generally used to treat MDR-TB are ethambutol, pyrazinamide, aminoglycosides (kanamycin, amikacin), capreomycin, fluoroquinolones (ofloxacin, ciprofloxacin), cyclo-serine, PAS and thioamides (ethionamide, prothionamide). Table 14.1 summarizes first- and second-line TB therapeutics as well as other drugs that are occasionally used to treat MDR-TB – of these a number are highlighted as being derived from or inspired by a natural product (Figure 14.1). The following section provides a brief summary of these natural product derivatives that are well studied and are clinically used anti-tuberculars.

14.3.1 Natural Products in the Existing Treatment for MDR-TB

Capreomycin and aminoglycosides, in combination with fluoroquinolones, are among the most effective drugs against MDR-TB. Streptomycin (**1**), kanamycin (**2**) and amikacin (**3**) are all aminoglycosides and they inhibit protein biosynthesis by binding to the bacterial 30S ribosomal subunit. Owing to their poor absorption in the gastrointestinal tract, they are administered via the intravenous route. In general, cross-resistance across aminoglycosides is observed. Even though aminoglycosides are quite effective against MDR-TB, they have severe side effects. Streptomycin has toxic manifestations in the peripheral and central nervous system (CNS) whereas amikacin and kanamycin can cause kidney damage and hearing loss.[30] Capreomycin (**4**) belongs to the tuberactinomycin family which contains a pentapeptide core composed of L-serine and the non-proteinogenic residues 2,3-diaminopropionate, L-capreomycidine and β-ureidodehydroalanine. Capreomycin binds to the ribosome and inhibits prokaryotic protein synthesis; a co-crystal structure

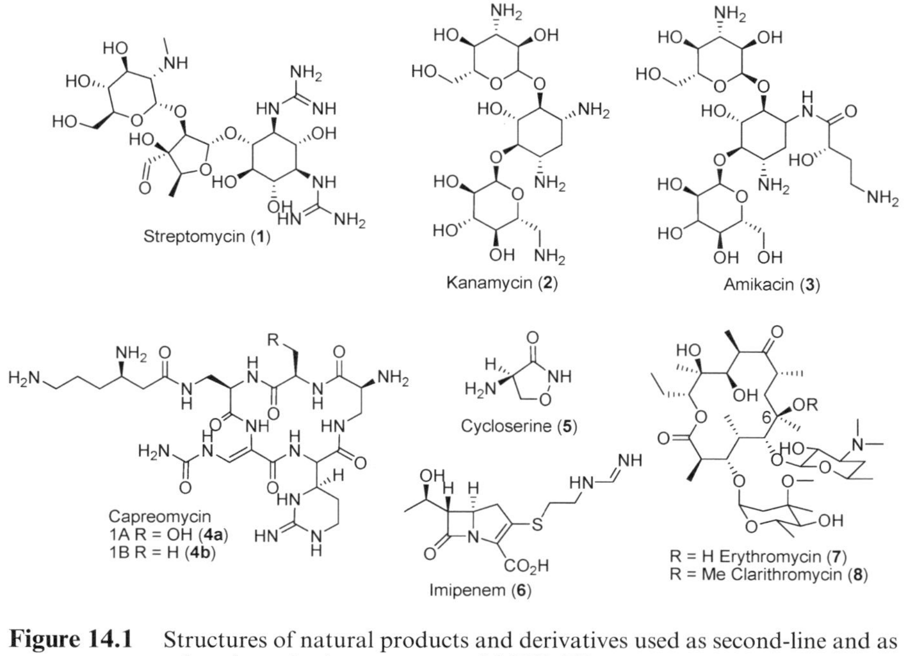

Figure 14.1 Structures of natural products and derivatives used as second-line and as other drugs in the treatment of MDR-TB.

of the drug with the 70S ribosomal complex has been solved.[31] Similar to the aminoglycosides, capreomycin is an injectable and it exhibits adverse effects including nephrotoxicity.[30]

Cycloserine (5) is a structural analogue of D-alanine which is an essential component in the peptidoglycan layer of the mycobacterial cell wall.[32] The side effects of this drug are mainly CNS manifestations such as depression and convulsions.[30]

Mtb is resistant to β-lactams because of a chromosomally encoded β-lactamase *blaC* gene.[33] More than two decades ago, Cynamon *et al.* observed the effect of amoxicillin, a β-lactam, on Mtb, in combination with clavulinic acid (inhibitor of β-lactamase).[34] Subsequently, in 2005, Chambers and colleagues described the therapeutic efficacy of imipenem (6) in patients with MDR-TB infection.[35] Recently, meropenem (a carbapenem with a methyl-substituted pyrroline ring) showed *in vitro* activity, in combination with clavulanate, against XDR-TB strains.[36,37]

Clarithromycin (8) is a macrolide antibiotic used to treat MAC infections in MDR-TB and HIV-infected TB patients. It binds to the 50S ribosomal subunit and inhibits translation. Clarithromycin differs from erythromycin (7) by the methylation of a hydroxyl group at the position 6 of the lactone ring. The methylation confers acid stability in the stomach, thereby contributing to improved oral bioavailability.

14.3.2 Promising Natural Products for MDR-TB Treatment and their Key Liabilities

Several efforts are in progress to identify and develop new molecules from nature with novel mechanisms of action against drug resistant TB (Figure 14.2).

- Riminophenazines. Clofazimine (9) is a lichen-derived, orange-red fat-soluble riminophenazine, first synthesized as B663.[38] Ever since its effectiveness against *Mycobacterium leprae* was demonstrated, clofazimine has been marketed as Lamprene against leprosy.[39] Clofazimine is a bactericidal agent shown to preferentially accumulate in macrophages and is active against MDR-TB strains *in vitro*.[38,40] The spread of MDR-TB and XDR-TB has renewed interest in revisiting clofazimine and the riminophenazine class for TB drug discovery.

The *in vivo* efficacy of clofazimine was investigated in various animal models including mice, guinea pigs, rabbits, hamsters and monkeys.[38,41] Clofazimine is highly active in the TB mouse model with moderate activity in hamsters and rabbits and little anti-TB activity in guinea pigs and monkeys. The varied spectrum of activity in different animal models has been attributed mainly to the retention of clofazimine in macrophages and to differences in absorption. The chemistry and pharmacology of clofazimine have been extensively reviewed.[42] Clofazimine is highly hydrophobic with a log P (octanol/water partition coefficient) of ~ 7.48, has very low aqueous solubility and a pK_a value of 8.51.

Clofazimine is also active against MAC; however, it is no longer used in the clinic after observation of increased mortality potentially associated with its use in one clinical study.[43] Currently, new analogues of clofazimine are being tested for treatment of MAC infections.[38] A recent study showed that the reduction of clofazimine by mycobacterial NDH-2 enzyme (Type 2 NADH-quinone oxidoreductase) results in generation of bactericidal levels of reactive oxygen species.[44] The TB Alliance has an active lead optimization program for clofazimine to address liabilities and further improve drug-like properties of the compound (www.gatb.org).

- Pleuromutilins. Pleuromutilin (**10**) is a tricyclic diterpenoid, a natural product that was first isolated in 1951 from the basidiomycetes fungi, *Pleurotus mutilus*.[45] Several semi-synthetic derivatives of pleuromutilin are marketed as drugs in veterinary and human medicine. Tiamulin is used as a prophylactic and therapeutic agent against swine dysentery, valnemulin (**10**) is used against enzootic pneumonia and swine dysentery and retapamulin is the first pleuromutilin derivative to be used (topically) in humans against infections caused by *Staphylococcus aureus* and *Streptococcus pyogenes*.[46] Recently two new pleuromutilin analogues have entered Phase I clinical trials against methicillin-resistant *Staphylococcus aureus*, MRSA (discussed elsewhere in this book).[47] Pleuromutilin derivatives interact with the L3 protein of the 50S subunit of bacterial ribosome near the peptidyl transferase centre of the 23S rRNA peptidyl transferase region.[48] The crystal structure of the 50S ribosomal subunit from *Deinococcus radiodurans* in complex with tiamulin demonstrated binding of the tricyclic mutilin to the tRNA binding site of the peptidyl transferase center.[49,50] In *M. smegmatis*, valnemulin-resistance mutations have been mapped to the 23S rRNA. The mechanism of action of valnemulin is different from that of other known mycobacterial ribosomal inhibitors such as streptomycin, kanamycin, capreomycin and linezolid.[48] Valnemulin is active against clinical MDR-TB strains obtained from different geographic locations.[51] The pleuromutilin class may therefore present attractive leads against MDR-TB.

Pleuromutilins with their unusual tricyclic structure, featuring a highly functionalized eight-membered ring and nine contiguous stereogenic centres, have attracted the attention of several synthetic groups. The total synthesis of pleuromutilin is reported in the literature.[52] Several studies have demonstrated the importance of the C14 substitution for activity against *S. aureus*, Mtb and other organisms.[47,53] The two predominant types of C14 side chains are thioethers and carbamates. All marketed pleuromutilin derivatives, either for human or veterinary use, contain thioether side chains.[47] Very limited information is available about analogues bearing structural variations in the tricyclic core structure. Two important liabilities that need to be addressed while developing oral pleuromutilin analogues for MDR-TB treatment are the rapid and extensive metabolism of these compounds *in vivo* and their selective inhibition of CYP3A4.

- Rifamycin analogues. The rifamycins are characterized by a cyclic structure in which a long aliphatic chain forms a bridge between two non-adjacent positions of the tricyclic ring. Rifamycin A was originally isolated from *Streptomyces mediterranei*.[54] Rifampicin (**11a**), also named rifampin, is an indispensable component of first-line TB treatment. It is a semi-synthetic derivative of rifamycin A with a modification on the C3 centre of the naphthaquinone core which enhances its membrane permeability and ADME properties (absorption, distribution, metabolism, excretion).[54] MDR-TB strains, as discussed previously, are resistant to rifampicin and isoniazid. Rifabutin (**11b**), another rifamycin analogue that is currently an approved first-line TB drug, has modifications on the C3 and C4 positions and is active against certain rifampicin-resistant clinical pathogens.[55] An exciting new improvement in rifabutin analogues comes from Berluenga *et al.* who showed that rifabutin congeners substituted on the piperidine ring showed improved activity against rifampicin-resistant and rifabutin-resistant Mtb strains.[56,57] Similarly, rifalazil (**11c**) analogues synthesized by the ActivBiotics group displayed improved activity against rifampicin-resistant strains. This gives renewed hope for the discovery of rifamycin analogues that can be used against MDR-TB.[58–60] Recently, however, development of rifalazil was terminated due to major side effects observed in the clinic.[61] Nevertheless, safety data on the rifalazil derivatives from the ActivBiotics group is yet to be seen, before which it is premature to comment on the future development of rifalazil derivatives.

The core RNAP complex of *Thermus aquaticus* has been co-crystallized with rifampicin and the structure of the complex has been solved with 3.2 Å resolution.[62] The RNAP holoenzyme from *Thermus thermophilus* has been co-crystallized with rifapentine and rifabutin leading to a structure with 2.5 Å resolution.[63] The data on rifabutin and rifalazil analogues along with information from co-crystal structures with RNAP provide an outstanding opportunity for the discovery of new rifamycin analogues to overcome MDR-TB infections.[64] However, future lead optimization activities should be undertaken keeping in view other shortfalls such as hepatotoxicity and induction of cytochromes P450 that are attributed to the rifamycin class.

- Lipiarmycin A3 (**12**), also known as tiacumicin B, OPT-80, fidaxomicin or difimicin, is an 18-membered macrocyclic compound produced by *Actinoplanes deccanensis*.[65] Lipiarmycin has a narrow spectrum of antibacterial activity. It is active against Gram-positive aerobes and anaerobes with little activity against Gram-negative pathogens.[66] It is active against Mtb and lacks cross-resistance to standard anti-TB drugs.[67] Lipiarmycins are transcription inhibitors which target RNAP, similar to rifampicin. However, mutations in lipiarmycin-resistant Mtb strains have been mapped to the C-terminus of the rpoB protein, therefore making rifampicin-resistant mutants sensitive to lipiarmycin. Recent mechanistic studies showed a novel mode of action for lipiarmycin; the compound blocks promoter melting and is likely to trap RNAP in a closed intermediate.[68] Accordingly, rifampicin-resistant and lipiarmycin-resistant mutants were not cross-resistant to each other, emphasizing lipiarmycin's potential for MDR-TB treatment.[67] Currently, lipiarmycin has shown promising results in Phase III clinical studies for its use in diarrhoea caused by *Clostridium difficile* infection.[69] Other than rifampicin, lipiarmycin is the only other RNAP inhibitor that has advanced to clinical development. The key issues concerning the development of lipiarmycin are the compound's large macrocyclic ring structure (MW > 1000 Da), lack of solubility in water and most notably, its very low systemic absorption after oral

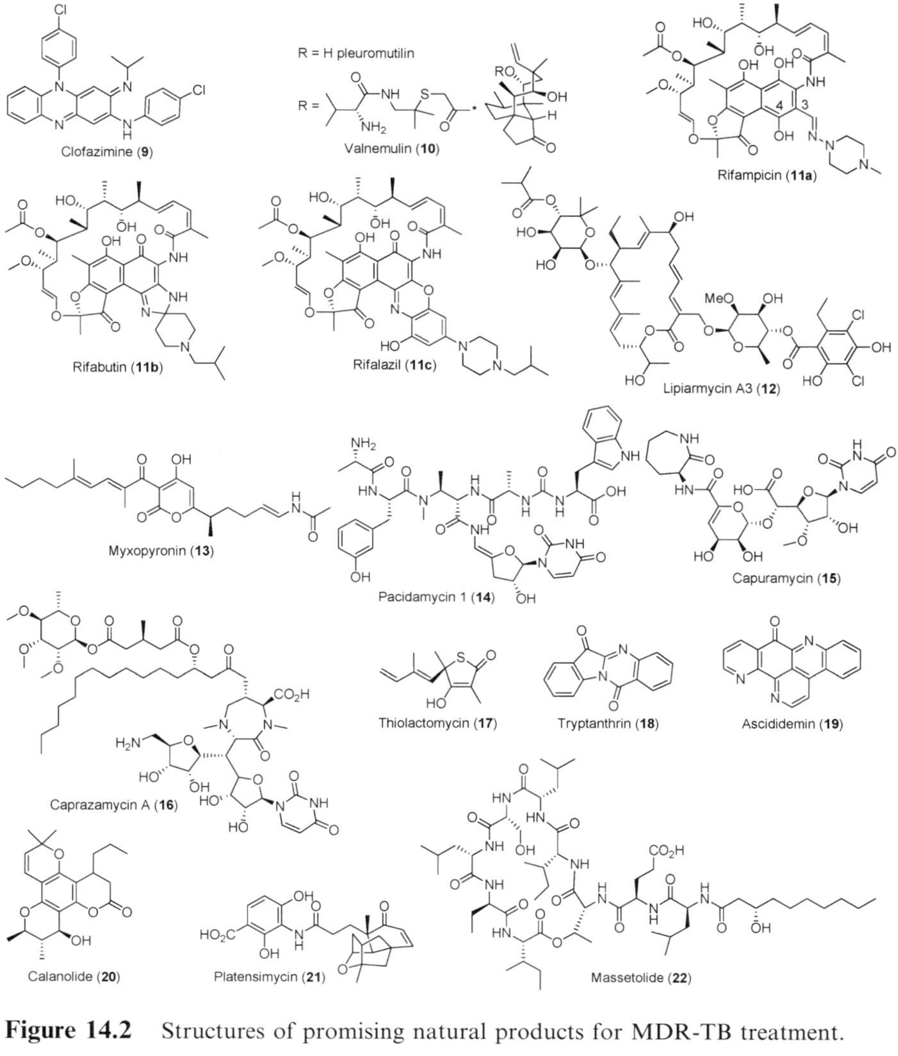

Figure 14.2 Structures of promising natural products for MDR-TB treatment.

administration. Among six healthy subjects who underwent multiple oral dosing (at 450 mg/day) for 10 days, only three individuals had detectable levels of the drug in plasma, the highest recording being 6.7 ng/mL.[70]

In addition to rifamycin analogues and lipiarmycin, there are other potential RNAP inhibitors like myxopyronin (**13**), corollopyronin and ripostatin.[71] Myxopyronin is a α-pyrone antibiotic, the first in a new class of RNAP inhibitors which target the switch region in the β-subunit.[72] A homology model of the Mtb RNAP-myxopyronin complex is a useful tool for designing and screening novel Mtb RNAP inhibitors.[73] With a mechanism of action being different from rifamycins, myxopyronin could be developed for MDR-TB. However, they have various structural and physicochemical deficiencies that require significant medicinal chemistry efforts to rectify.

14.3.2.1 *Other Potential Anti-Tuberculars from Natural Products*

- Pacidamycin (**14**), isolated from *Streptomyces coeruleorubidus*, belongs to a uridyl peptide antibiotic class. It inhibits MraY translocase I (phospho-muramyl *N*-acetyl-pentapeptide translocase), an essential enzyme in peptidoglycan biosynthesis.[74] Some of the dihydropacidamycin analogues have been reported as active against MDR-TB.[75] Recently, a library of 1000 uridyl branched peptidomimetics have been synthesized and several analogues from this collection have shown potent anti-tubercular activity.[76]
- Capuramycin (**15**), a uracil nucleoside antibiotic with a caprolactam substituent, was isolated from *Streptomyces griseus* 446-S3.[77] Like pacidamycin, capuramycin's mechanism of action is also based on inhibition of translocase I.[78] Capuramycin and its derivatives are bactericidal and show activity not only against MDR-TB, but also against MAC.[79,80] None of the capuramycin analogues are absorbed well by oral administration, likely due to their high PSA (polar surface area) and low c log *P* (calculated log *P*) (Table 14.2). However, *in vivo* efficacy in mouse has been demonstrated with intranasal drug treatment.[81] The delineation of the capuramycin biosynthetic pathway offers an opportunity for a genetic engineering approach to generate new capuramycin analogues through biosynthetic variations.[82] In 1994, a 21-step synthesis of capuramycin was reported.[83] Recently, simpler synthetic schemes have been described for the generation of capuramycin analogues with desirable properties.[84]
- Caprazamycin (**16**) represents a novel class of liponucleoside antibiotics isolated from *Streptomyces* spp. MK730-62F2 and is bactericidal against Mtb.[85] Similar to pacidamycin and capuramycin, caprazamycin targets MraY translocase I, thereby affecting cell wall biosynthesis at different step when compared to known cell well synthesis inhibiting anti-tuberculars.[86] CPZEN-45 is a semi-synthetic, water-soluble derivative of caprazamycin that is active against MDR and XDR-TB strains and has shown *in vivo* efficacy in mouse (http://www.newtbdrugs.org/pipeline.php).

- Thiolactomycin (**17**) is a thiotetronic acid-containing natural product with a broad spectrum antibiotic activity. It inhibits mycobacterial type II fatty acid synthase (FAS-II) via inhibition of the β-ketoacyl-ACP synthases, key condensing enzymes involved in chain elongation.[87,88] Thiolactomycin has several characteristics such as low molecular weight, high water solubility and appropriate lipophilicity (log *P*) to make it an attractive candidate for further optimization for TB. It is a promising lead compound for the development of potent FAS-II inhibitors. So far there are no drugs in the clinic targeting FAS-II. Derivatization of thiolactomycin to improve potency against Mtb has been the focus of many recent investigations.[89,90]

- Tryptanthrin (**18**) is a potent indolo-quinazolinone alkaloid obtained from the Chinese herb *Strobilanthes cusia*.[24] This compound has been pursued intensively because of its simple structure, good *in vitro* properties and ease of synthesis.[91] It has potent bacteriostatic activity against drug-sensitive as well as drug-resistant strains of Mtb. A tryptanthrin analogue, PA-505 (oral bioavailability, 45%), failed to show significant efficacy *in vivo* against Mtb infection in mouse.[24] Understanding the mechanism of action of tryptanthrin and synthesizing analogues with better *in vitro* and *in vivo* pharmacological profiles will be necessary to explore the potential of this class as anti-tuberculars.

- *Ascididemin* (**19**) is a marine pyridoacridine alkaloid that has potent anti-TB activity, anti-parasitic activity and is cytotoxic to tumour cells.[92,93] Based on genome-wide microarray gene expression analysis, its anti-mycobacterial activity has been attributed to iron sequestering, a mechanism different from that of known TB drugs.[92]

- *Calanolide* (**20**), a plant-derived coumarin analogue from *Calophyllum lanigerum*, is active against Mtb, including MDR-TB strains and is also known to possess anti-retroviral activity.[94]

In addition to those discussed, natural products such as platensimycin (**21**) and massetolide (**22**) have shown promising activity against drug-resistant Mtb.[22–24] These natural products, covering such a diverse chemical space, are all promising hits for MDR-TB. However, they have various biological, pharmacological and physicochemical limitations which need to be addressed through significant efforts in medicinal chemistry.

14.3.3 Natural Products as Chemical Probes

Natural products, as a result of long evolutionary selection processes, have been optimized by nature for interaction with biological target macromolecules.[95,96] Therefore, in addition to their value as candidate therapeutic agents, natural products have found immense utility as chemical probes in identifying molecular targets and in the functional dissection of biological processes.[97] Recent structural studies of various natural products such as rifampicin, myxopyronin, capreomycin and viomycin have not only helped

Table 14.2   Properties of natural products and natural product derivatives in MDR-TB treatment

#	Compound	Natural source	MW	c log P	PSA	# of H-bond Donors/ Acceptors
1	Streptomycin	*Streptomyces griseus*	581.6	−4.3	331.4	14/19
2	Kanamycin	*Streptomyces kanamyceticus*	484.5	−5.2	282.6	11/15
3	Amikacin	Semi-synthetic	585.6	−6.4	326.1	13/18
4a	Capreomycin	*Streptomyces*	668.7	−7.1	375.9	15/22
4b		*capreolus*	652.7	−6.0	355.7	14/21
5	Cycloserine	*Streptococcus orchidaceus*	102.1	−1.2	64.3	2/4
6	Imipenem	*Penicillin* spp.	299.4	−1.4	113.7	4/7
7	Erythromycin	*Streptomyces erythreus*	734.0	1.6	193.9	5/14
8	Clarithromycin	Semi-synthetic	748.0	2.4	182.9	4/14
9	Clofazimine	Lichens	473.4	7.7	44.7	1/4
10	Valnemulin	*Pleurotus mutilis*	564.8	5.7	118.7	3/7
11a	Rifampicin	*Streptomyces mediterranei*	823.0	3.7	220.2	6/16
11b	Rifabutin	Semi-synthetic	847.1	4.7	205.6	5/15
11c	Rifalazil	Semi-synthetic	941.1	8.2	230.7	5/17
12	Lipiarmycin A3	*Actinoplanes deccanensis*	1058.1	7.2	266.7	7/18
13	Myxopyronin	*Myxococcus fulvus*	415.5	5.7	96.6	2/6
14	Pacidamycin 1	*Streptomyces coeruleorubidus*	874.9	−2.5	334.9	11/22
15	Capuramycin	*Streptomyces griseus* 446-S3	570.5	−4.3	250.4	7/17
16	Caprazamycin A	*Streptomyces* spp. MK730-62F2	1114.3	0.917	350.2	7/25
17	Thiolactomycin	*Nocardia* spp.	210.3	2.8	37.3	1/2
18	Tryptanthrin	*Strobilanthes cusia* (plant)	248.2	2.3	54.4	0/4
19	Ascididemin	Natural marine	283.3	2.9	55.7	0/4
20	Calanolide	*Calophyllum lanigerum* (plant)	372.5	4.7	65.0	1/5
21	Platensimycin	*Streptomyces platensis*	441.5	2.3	133.2	4/8
22	Massetolide	*Pseudomonas* spp.	1138.5	8.3	366.0	12/24

to elucidate mechanistic details of complex macromolecular processes but have also facilitated rational structure based design of these complex natural products.[31,62,72,98] Screening natural products or their derivatives for anti-tubercular activity will not only be valuable for identifying lead molecules but will also prove useful in the identification of new molecular targets for target-based drug discovery programs and in the understanding of mycobacterial biology.

14.4 Natural Products: Challenges, Opportunities and Future Outlook

Although compound collections of pharmaceutical companies are generally biased towards structures satisfying the Lipinski 'rule of five', existing antibacterials are often recognized for deviating from this rule.[99,100] Most of the large pharmaceutical companies have had limited success when screening their chemical libraries for attractive hits with antibacterial activity. When compared to synthetic compounds, natural products have broader distribution of molecular properties such as molecular weight (MW), octanol/water partition coefficient (log P) and polar surface area (PSA) (Table 14.2 and Figure 14.3).[101] Natural products therefore occupy a broader chemical space that is frequently outside the drug space for synthetic oral compounds, often limiting their use to acute situations where non-oral treatment routes become

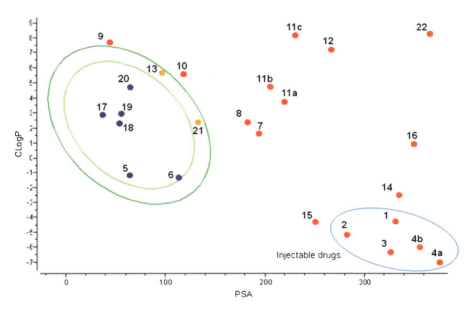

Figure 14.3 The chemical diversity of natural products and derivatives (compound numbers as in Table 14.2) with MDR-TB Treatment Potential. A plot of c log P and PSA shows that the majority of well absorbed compounds (represented by blue/orange dots) fall within the Egan egg as represented by the green ellipsoids. Compounds that fall outside the outer ellipsoid are expected to have poor passive gut absorption (represented by red dots); however they can be absorbed by active transport processes.

acceptable.[102,103] While nature optimizes compounds for potency related chemical properties, it does not do so for those that make it orally bioavailable. A key challenge in the development of natural products as drugs is to therefore combine their inherent antibacterial properties with physicochemical properties that confer oral bioavailability, an attribute that is highly desirable for treatment of MDR-TB. Moreover, in TB drug development, affordability (cost of goods) is a major limiting factor. Not all natural products can be totally synthesized and many of them have extremely complex structures that are too difficult and expensive to synthesize on a large scale.

Screening of natural products is anticipated to offer higher hit rates over synthetic compounds in discovery of novel antibacterial drugs.[104] Nature has not been tapped fully so far; it is estimated that less than 1% of all microorganisms have yet been cultured and examined for production of bioactive compounds.[105] These untapped biological resources, in combination with novel screening methods, rapid and inexpensive whole genome sequencing for target deconvolution, sophisticated structure elucidation techniques, metabolic engineering, synthetic biology and techniques for total synthesis of complex secondary metabolites, offer novel ideas and technologies for drug discovery from natural products.[106]

The various grave issues surrounding control of TB and of mycobacterial resistance demand new chemical entities with unexploited modes of action; the success stories of natural products as antibiotics in general and more specifically as anti-tuberculars encourage devoted attention to the discovery of new molecules from nature to fight MDR-TB as well as MAC infections. During the past 15 years, research of natural products in the pharmaceutical industry has declined due to reasons such as difficulties in access and supply of chemical matter, structural complexity for lead optimization and concerns regarding intellectual property rights. In addition to these disadvantages of natural products, most of the natural products that are potential anti-tuberculars violate two or more parameters of the 'rule-of five' (Table 14.2), which are primary considerations for synthetic molecules that are intended for oral administration. Ideally, new anti-TB drugs must not only exhibit improved pharmacokinetic properties but must also result in the reduction of the total length of treatment with fewer toxic side-effects. In addition to pursuing promising natural products, new antibacterial assays which combine cell-based screening with clinically validated targets using natural product libraries have the potential to offer better prospects for lead discovery. In conclusion, work with natural products no doubt remains challenging. However, their excellent track record, together with the huge medical need for new anti-mycobacterials, strongly advocates efforts to focus on this segment of chemical diversity.

Acknowledgements

We would like to thank the following for their valuable comments and discussion: Paul Smith, Kevin Pethe, Paul Herrling and other colleagues from

NITD, Singapore; Silvio Roggo from the Novartis Institutes for Biomedical Research, Basel, Switzerland.

References

1. WHO, *2010 Global Tuberculosis Control*, 2010.
2. C. E. Barry, III, H. I. Boshoff, V. Dartois, T. Dick, S. Ehrt, J. Flynn, D. Schnappinger, R. J. Wilkinson and D. Young, *Nat. Rev. Microbiol.*, 2009, **7**, 845.
3. S. H. Kaufmann, *Immunity*, 2010, **33**, 567.
4. H. Khachi, R. O'Connell, D. Ladenheim and C. Orkin, *J. Antimicrob. Chemother.*, 2009, **64**, 871.
5. B. R. Edlin, J. I. Tokars, M. H. Grieco, J. T. Crawford, J. Williams, E. M. Sordillo, K. R. Ong, J. O. Kilburn, S. W. Dooley, K. G. Castro, W. R. Jarvis and S. D. Holmberg, *N. Engl. J. Med.*, 1992, **326**, 1514.
6. N. R. Gandhi, A. Moll, A. W. Sturm, R. Pawinski, T. Govender, U. Lalloo, K. Zeller, J. Andrews and G. Friedland, *Lancet*, 2006, **368**, 1575.
7. M. Zignol, M. S. Hosseini, A. Wright, C. L. Weezenbeek, P. Nunn, C. J. Watt, B. G. Williams and C. Dye, *J. Infect. Dis.*, 2006, **194**, 479.
8. A. A. Velayati, M. R. Masjedi, P. Farnia, P. Tabarsi, J. Ghanavi, A. H. Ziazarifi and S. E. Hoffner, *Chest*, 2009, **136**, 420.
9. WHO, *Multidrug and Extensively Drug-Resistant TB*, 2010 Global Report on Surveillance and Response, 2010.
10. M. A. Aziz, A. Wright, A. Laszlo, M. A. De, F. Portaels, D. A. Van, C. Wells, P. Nunn, L. Blanc and M. Raviglione, *Lancet*, 2006, **368**, 2142.
11. D. Y. Rosenzweig, *Chest*, 1979, **75**, 115.
12. M. D. Iseman, *Clin. Chest Med.*, 2002, **23**, 633.
13. N. Rastogi, *Res. Microbiol.*, 1991, **142**, 464.
14. J. Esteban and A. Ortiz-Perez, *Expert Opin. Pharm.*, 2009, **10**, 2787.
15. Z. Ma and C. Lienhardt, *Clin. Chest Med.*, 2009, **30**, 755.
16. D. J. Newman and G. M. Cragg, *J. Nat. Prod.*, 2007, **70**, 461.
17. K. H. Pfuetze, M. M. Pyle, H. C. Hinshaw and W. H. Feldman, *Am. Rev. Tuberc.*, 1955, **71**, 752.
18. G. Domagk, *Ann. R. Acad. Nac. Med. (Madr.)*, 1952, **69**, 504.
19. J. Lehman, *Experientia*, 1949, **5**, 365.
20. H. G. Floss and T. W. Yu, *Chem. Rev.*, 2005, **105**, 621.
21. M. V. de Souza, *Curr. Opin. Pulm. Med.*, 2006, **12**, 167.
22. B. R. Copp and A. N. Pearce, *Nat. Prod. Rep.*, 2007, **24**, 278.
23. M. V. de Souza, *Fitoterapia*, 2009, **80**, 453.
24. L. A. Mitscher and W. Baker, *Med. Res. Rev.*, 1998, **18**, 363.
25. L. J. McGaw, N. Lall, J. J. Meyer and J. N. Eloff, *J. Ethnopharmacol.*, 2008, **119**, 482.
26. K. Bamuamba, D. W. Gammon, P. Meyers, M. G. Dijoux-Franca and G. Scott, *J. Ethnopharmacol.*, 2008, **117**, 385.

27. T. S. A. Thring, E. P. Springfield and F. M. Weitz, *African J. Biotechnol.*, 2007, **6**, 1779.
28. R. Gautam, A. Saklani and S. M. Jachak, *J. Ethnopharmacol.*, 2007, **110**, 200.
29. S. P. N. Mativandlela, N. Lall and J. J. M. Meyer, *South African J. Bot.*, 2006, **72**, 232.
30. S. S. Shin, F. Alcantara and P. E. Farmer, in *Handbook of Tuberculosis: Clinics, Diagnostics, Therapy and Epidemiology*, ed. S. H. E. Kaufmann and P. van Helden, Wiley, 2008, Chapter 10, pp. 181–212.
31. R. E. Stanley, G. Blaha, R. L. Grodzicki, M. D. Strickler and T. A. Steitz, *Nat. Struct. Mol. Biol.*, 2010, **17**, 289.
32. Y. Zhang, *Annu. Rev. Pharmacol. Toxicol.*, 2005, **45**, 529.
33. S. T. Cole, R. Brosch, J. Parkhill, T. Garnier, C. Churcher, D. Harris, S. V. Gordon, K. Eiglmeier, S. Gas, C. E. Barry, III, F. Tekaia, K. Badcock, D. Basham, D. Brown, T. Chillingworth, R. Connor, R. Davies, K. Devlin, T. Feltwell, S. Gentles, N. Hamlin, S. Holroyd, T. Hornsby, K. Jagels, A. Krogh, J. McLean, S. Moule, L. Murphy, K. Oliver, J. Osborne, M. A. Quail, M. A. Rajandream, J. Rogers, S. Rutter, K. Seeger, J. Skelton, R. Squares, S. Squares, J. E. Sulston, K. Taylor, S. Whitehead and B. G. Barrell, *Nature*, 1998, **393**, 537.
34. M. H. Cynamon and G. S. Palmer, *Antimicrob. Agents Chemother.*, 1983, **24**, 429.
35. H. F. Chambers, J. Turner, G. F. Schecter, M. Kawamura and P. C. Hopewell, *Antimicrob. Agents Chemother.*, 2005, **49**, 2816.
36. J. E. Hugonnet, L. W. Tremblay, H. I. Boshoff, C. E. Barry and J. S. Blanchard, *Science*, 2009, **323**, 1215.
37. C. E. Barry, III and J. S. Blanchard, *Curr. Opin. Chem. Biol.*, 2010, **14**, 456.
38. V. M. Reddy, J. F. O'Sullivan and P. R. Gangadharam, *J. Antimicrob. Chemother.*, 1999, **43**, 615.
39. Y. T. Chang, *Int.J. Lepr. Other Mycobact. Dis.*, 1967, **35**, 78.
40. M. L. Conalty, in *Methods of Preclinical Evaluation of Antitubercular Drugs*, Butterworths, London, 1964, pp. 150.
41. V. C. Barry, *Ir. J. Med. Sci.*, 1960, **416**, 345.
42. R. O'Connor, J. F. O'Sullivan and R. O'Kennedy, *Drug Metab. Rev.*, 1995, **27**, 591.
43. R. E. Chaisson, P. Keiser, M. Pierce, W. J. Fessel, J. Ruskin, C. Lahart, C. A. Benson, K. Meek, N. Siepman and J. C. Craft, *AIDS*, 1997, **11**, 311.
44. T. Yano, S. Kassovska-Bratinova, J. S. Teh, J. Winkler, K. Sullivan, A. Isaacs, N. M. Schechter and H. Rubin, *J. Biol. Chem*, 2011, **286**, 10276.
45. F. Kavanagh, A. Hervey and W. J. Robbins, *Proc. Natl. Acad. Sci. USA*, 1951, **37**, 570.
46. C. Hu and Y. Zou, *Mini. Rev. Med. Chem.*, 2009, **9**, 1397.
47. R. Novak and D. M. Shlaes, *Curr. Opin. Investig. Drugs*, 2010, **11**, 182.

48. K. S. Long, J. Poehlsgaard, L. H. Hansen, S. N. Hobbie, E. C. Bottger and B. Vester, *Mol. Microbiol.*, 2009, **71**, 1218.
49. F. Schlunzen, E. Pyetan, P. Fucini, A. Yonath and J. M. Harms, *Mol. Microbiol.*, 2004, **54**, 1287.
50. G. Gurel, G. Blaha, P. B. Moore and T. A. Steitz, *J. Mol. Biol.*, 2009, **389**, 146.
51. U. H. Manjunatha. 2010, unpublished work.
52. E. G. Gibbons, *J. Am. Chem. Soc.*, 1982, **104**, 1767.
53. H. Egger and H. Reinshagen, *J. Antibiot.*, 1976, **29**, 923.
54. P. Sensi, *Rev. Infect. Dis.*, 1983, **5** (Suppl. 3), S402.
55. R. N. Brogden and A. Fitton, *Drugs*, 1994, **47**, 983.
56. J. Barluenga, F. Aznar, A. B. Garcia, M. P. Cabal, J. J. Palacios and M. A. Menendez, *Bioorg. Med. Chem. Lett.*, 2006, **16**, 5717.
57. A. B. Garcia, J. J. Palacios, M. J. Ruiz, J. Barluenga, F. Aznar, M. P. Cabal, J. M. Garcia and N. Diaz, *Antimicrob. Agents Chemother.*, 2010, **54**, 5363.
58. J. H. van Duzer, A. F. Michaelis, W. B. Geiss, D. G. Stafford and J. Raker, Rifamycin analogs and uses thereof. [2005/0043298]. 2-24-2005. U.S. Patent.
59. J. H. van Duzer, A. F. Michaelis, W. B. Geiss, D. G. Stafford, X. Y. Yu, J. M. Siedlecki and Y. Yang. Rifamycin analogs and uses thereof. [2005/0137189]. 6-23-2005. U.S. Patent.
60. J. H. van Duzer, A. F. Michaelis, W. B. Geiss, D. G. Stafford, J. Raker, X. Y. Yu, J. M. Siedlecki and Y. Yang. Rifamycin analogs and uses thereof. [2005/0197333]. 9-8-2005. U.S. Patent.
61. R. J. O'Brien and M. Spigelman, *Clin. Chest Med.*, 2005, **26**, 327.
62. E. A. Campbell, N. Korzheva, A. Mustaev, K. Murakami, S. Nair, A. Goldfarb and S. A. Darst, *Cell*, 2001, **104**, 901.
63. I. Artsimovitch, M. N. Vassylyeva, D. Svetlov, V. Svetlov, A. Perederina, N. Igarashi, N. Matsugaki, S. Wakatsuki, T. H. Tahirov and D. G. Vassylyev, *Cell*, 2005, **122**, 351.
64. P. A. Aristoff, G. A. Garcia, P. D. Kirchhoff and H. D. Hollis Showalter, *Tuberculosis (Edinb.)*, 2010, **90**, 94.
65. C. Coronelli, R. J. White, G. C. Lancini and F. Parenti, *J. Antibiot. (Tokyo)*, 1975, **28**, 253.
66. K. M. Sullivan and L. M. Spooner, *Ann. Pharmacother.*, 2010, **44**, 352.
67. M. Kurabachew, S. H. Lu, P. Krastel, E. K. Schmitt, B. L. Suresh, A. Goh, J. E. Knox, N. L. Ma, J. Jiricek, D. Beer, M. Cynamon, F. Petersen, V. Dartois, T. Keller, T. Dick and V. K. Sambandamurthy, *J. Antimicrob. Chemother.*, 2008, **62**, 713.
68. A. Tupin, M. Gualtieri, J. P. Leonetti and K. Brodolin, *EMBO J.*, 2010, **29**, 2527.
69. M. Miller, *Expert Opin. Pharmacother.*, 2010, **11**, 1569.
70. Y. K. Shue, P. S. Sears, S. Shangle, R. B. Walsh, C. Lee, S. L. Gorbach, F. Okumu and R. A. Preston, *Antimicrob. Agents Chemother.*, 2008, **52**, 1391.

71. D. Haebich and N. F. von, *Angew. Chem., Int. Ed.*, 2009, **48**, 3397.

72. J. Mukhopadhyay, K. Das, S. Ismail, D. Koppstein, M. Jang, B. Hudson, S. Sarafianos, S. Tuske, J. Patel, R. Jansen, H. Irschik, E. Arnold and R. H. Ebright, *Cell*, 2008, **135**, 295.

73. M. X. Ho, B. P. Hudson, K. Das, E. Arnold and R. H. Ebright, *Curr. Opin. Struct. Biol.*, 2009, **19**, 715.

74. M. Winn, R. J. M. Goss, K. Kimura and T. D. H. Bugg, *Nat. Prod. Rep.*, 2010, **27**, 279.

75. C. G. Boojamra, R. C. Lemoine, J. Blais, N. G. Vernier, K. A. Stein, A. Magon, S. Chamberland, S. J. Hecker and V. J. Lee, *Bioorg. Med. Chem. Lett.*, 2003, **13**, 3305.

76. D. Sun, V. Jones, E. I. Carson, R. E. Lee, M. S. Scherman, M. R. Mcneil and R. E. Lee, *Bioorg. Med. Chem. Lett.*, 2007, **17**, 6899.

77. H. Yamaguchi, S. Sato, S. Yoshida, K. Takada, M. Itoh, H. Seto and N. Otake, *J Antibiot. (Tokyo)*, 1986, **39**, 1047.

78. T. D. Bugg, A. J. Lloyd and D. I. Roper, *Infect. Disord. Drug Targets*, 2006, **6**, 85.

79. T. Koga, T. Fukuoka, N. Doi, T. Harasaki, H. Inoue, H. Hotoda, M. Kakuta, Y. Muramatsu, N. Yamamura, M. Hoshi and T. Hirota, *J. Antimicrob. Chemother.*, 2004, **54**, 755.

80. E. Bogatcheva, T. Dubuisson, M. Protopopova, L. Einck, C. A. Nacy and V. M. Reddy, *J. Antimicrob. Chemother*, 2011, **66**, 578.

81. V. M. Reddy, L. Einck and C. A. Nacy, *Antimicrob. Agents Chemother.*, 2008, **52**, 719.

82. M. Funabashi, Z. Yang, K. Nonaka, M. Hosobuchi, Y. Fujita, T. Shibata, X. Chi and S. G. Van Lanen, *Nat. Chem. Biol.*, 2010, **6**, 581.

83. S. Knapp and S. R. Nandan, *J. Org. Chem.*, 1994, **59**, 281.

84. M. Kurosu, K. Li and D. C. Crick, *Org. Lett.*, 2009, **11**, 2393.

85. M. Igarashi, Y. Takahashi, T. Shitara, H. Nakamura, H. Naganawa, T. Miyake and Y. Akamatsu, *J. Antibiot. (Tokyo)*, 2005, **58**, 327.

86. S. Hirano, S. Ichikawa and A. Matsuda, *Bioorg. Med. Chem.*, 2008, **16**, 5123.

87. R. A. Slayden, R. E. Lee, J. W. Armour, A. M. Cooper, I. M. Orme, P. J. Brennan and G. S. Besra, *Antimicrob. Agents Chemother.*, 1996, **40**, 2813.

88. L. Kremer, J. D. Douglas, A. R. Baulard, C. Morehouse, M. R. Guy, D. Alland, L. G. Dover, J. H. Lakey, W. R. Jacobs, P. J. Brennan, D. E. Minnikin and G. S. Besra, *J. Biol. Chem.*, 2000, **275**, 16857.

89. P. Kim, Y. M. Zhang, G. Shenoy, Q. A. Nguyen, H. I. Boshoff, U. H. Manjunatha, M. B. Goodwin, J. Lonsdale, A. C. Price, D. J. Miller, K. Duncan, S. W. White, C. O. Rock, C. E. Barry and C. S. Dowd, *J. Med. Chem.*, 2006, **49**, 159.

90. J. D. Douglas, S. J. Senior, C. Morehouse, B. Phetsukiri, I. B. Campbell, G. S. Besra and D. E. Minnikin, *Microbiology*, 2002, **148**, 3101.

91. T. M. Potewar, S. A. Ingale and K. V. Srinivasan, *Arkivoc*, 2008, **xiv**, 100.

92. H. I. M. Boshoff, T. G. Myers, B. R. Copp, M. R. Mcneil, M. A. Wilson and C. E. Barry, *J. Biol. Chem.*, 2004, **279**, 40174.

93. B. R. Copp, O. Kayser, R. Brun and A. F. Kiderlen, *Planta Med.*, 2003, **69**, 527.

94. Z. Q. Xu, W. W. Barrow, W. J. Suling, L. Westbrook, E. Barrow, Y. M. Lin and M. T. Flavin, *Bioorg. Med. Chem.*, 2004, **12**, 1199.

95. D. G. Kingston and D. J. Newman, *Curr. Opin. Drug Discov. Devel.*, 2002, **5**, 304.

96. P. Ertl and A. Schuffenhauer, *Prog. Drug Res.*, 2008, **66**, 217, 219.

97. V. P. Gullo, J. McAlpine, K. S. Lam, D. Baker and F. Petersen, *J. Ind. Microbiol. Biotechnol.*, 2006, **33**, 523.

98. G. A. Belogurov, M. N. Vassylyeva, A. Sevostyanova, J. R. Appleman, A. X. Xiang, R. Lira, S. E. Webber, S. Klyuyev, E. Nudler, I. Artsimovitch and D. G. Vassylyev, *Nature*, 2009, **457**, 332.

99. D. J. Payne, M. N. Gwynn, D. J. Holmes and D. L. Pompliano, *Nat. Rev. Drug Discov.*, 2007, **6**, 29.

100. R. O'Shea and H. E. Moser, *J. Med. Chem.*, 2008, **51**, 2871.

101. M. Feher and J. M. Schmidt, *J. Chem. Inf. Comput. Sci.*, 2003, **43**, 218.

102. R. D. Firn and C. G. Jones, *Nat. Prod. Rep.*, 2003, **20**, 382.

103. S. Wetzel, A. Schuffenhauer, S. Roggo, P. Ertl and H. Waldmann, *Chimia*, 2007, **61**, 355.

104. K. J. Weissman and P. F. Leadlay, *Nat. Rev. Microbiol.*, 2005, **3**, 925.

105. A. Saklani and S. K. Kutty, *Drug Discov. Today*, 2008, **13**, 161.

106. J. W. Li and J. C. Vederas, *Science*, 2009, **325**, 161.

CHAPTER 15

Retapamulin: a First-in-Class Pleuromutilin Antibiotic

RODGER NOVAK

Nabriva Therapeutics AG, Leberstrasse 20, 1110 Vienna, Austria
E-mail: rodger.novak@nabriva.com

15.1 Introduction

The first mention of antibacterial activity of a pleuromutilin dates back to the early 1950s.[1–3] In 1951 Kavanagh and colleagues from Columbia University and The New York Botanical Garden described for the first time a substance with antibacterial activity isolated from two basiomycetes species, *Pleurotus mutilis* (now known as *Clitopilus scyphoides*) and *Pleurotus passeckeranius*, named pleuromutilin. While the new substance showed decent activity against some Gram-positive cocci, it did not receive too much attention also because *in vivo* efficacy experiments revealed only limited activity.[1,2] In addition, the discovery was made during a time where a significant number of antibiotics had already been made available to the market; namely, sulfonamides (1930s), penicillin (1942), streptomycin (1943), chloramphenicol (1947), tetracycline (1948), and erythromycin (1952).[4–11] In other words, the first pleuromutilin saw the light of day at a time where many physicians contemplated that most bacterial infections were certainly manageable, if not the eradication of key pathogens could be possible. One has to consider that this mindset primarily originated from the perspective of the just recently concluded pre-antibiotic era, where bacterial infections were the number one killer – taking in particular a major toll among the youth.[4] Although it became rapidly clear that the introduction of any new class of antibiotic would lead to widespread use and in

RSC Drug Discovery Series No. 25
Drug Discovery from Natural Products
Edited by Olga Genilloud and Francisca Vicente
© The Royal Society of Chemistry 2012
Published by the Royal Society of Chemistry, www.rsc.org

turn to the development of resistance, many researchers, in particular medicinal chemists, were of the opinion that synthesis efforts pursuing the classical 'magic bullet' approach would finally lead to pure, fully synthetic compounds which would eventually overcome the issue of resistance development.

Since the introduction of quinolones, one out of three synthetic antibiotic classes in clinical use, it has became clear that in particular for patients infected with multi-drug resistant bacteria, there simply is no magic bullet. Today we are witnessing a highly worrisome development where infections caused by antibiotic-resistant bacteria are a major challenge for physicians and constitute a significant threat to public health and the society at large.[12,13] The reasons for this development are diverse and include the over- and misuse of antibiotics, a scarcity of new antibiotic classes available for physicians and patients, high and sometimes politically motivated regulatory hurdles, which, in combination with limited return on investment, have led to a significant drain of pharmaceutical development activities into other, more lucrative, disease areas.

15.2 The Urgent Need of Antibiotics With New Modes of Action

Data from the Centers for Disease Control and Prevention (CDC) show rapidly increasing rates of infection due to methicillin-resistant *Staphylococcus aureus* (MRSA), vancomycin-resistant *Enterococcus faecium* (VRE), and fluoroquinolone-resistant *Pseudomonas aeruginosa*.[14] While in the past development of resistance was a major issue among nosocomial, hospital-acquired infections, there are more and more community-acquired resistances that have been observed during the last decade, with, for example, approximately 70% of *S. aureus* strains identified in emergency rooms being MRSA.[15] As of today, more people die of MRSA infection in US hospitals than of HIV/AIDS and tuberculosis combined.[16,17] Another area of major concern is the emergence of multi-drug resistant Gram-negative pathogens, including carbapenem-resistant *Escherichia coli* and *Klebsiella* species and multi-drug resistant non-fermenters like *Pseudomonas aeruginosa* and *Acinetobacter* species.[18–22]

What adds to this 'mishap' is the fact that during the past 25 years the number of new antibiotics that have been approved by the US Food and Drug Administration (FDA) has steadily decreased, with only two novel classes being approved; linezolid, an oxazolidinone approved in 2000, and daptomycin, a lipopeptide approved in 2003.[23] All other approvals were made for derivatives of already established antibiotic classes. One has to mention that derivatives of a given class often differ substantially regarding their basic physicochemical properties leading to sometimes significant differences among pharmacology and side-effect profiles, a feature required for appropriate differentiation. Nonetheless, antibiotics of the same class usually share one important feature – cross-resistance and, potentially even more relevant, selection of resistance. Textbook examples for these features are cephalosporins, more specifically third-generation cephalosporins, and quinolones.[24-27] The widespread use of these antibiotics for very different indications targeting different causative pathogens has led to significant non-clonal quinolone resistance among *Enterobacteriaceae* species and a dramatic increase in quinolone-resistant *Clostridium difficile* outbreaks of highly virulent NAP-1 strains.[28,29] Similarly, the widespread use of ceftriaxone and other third-generation cephalosporins has not only led to an increase in resistance of *Streptococcus pneumoniae* but also contributed significantly to the selection of extended-spectrum β-lactamases (ESBL) and *Klebsiella pneumoniae* carbapenemases (KPC) among *Enterobacteriaceae* species.[30-33]

Until recently, the only topical antibiotics recommended for the treatment of uncomplicated skin infections with superficial lesions, limited in number, were fusidic acid (not US) and mupirocin.[34,35] With the emergence of resistance against fusidic acid and mupirocin, novel mode of action antibiotics are urgently needed.[36-38] In this context, retapamulin is not to be considered as just another antibiotic, a topical one you may add, but as the first member of a class which might offer an important opportunity to establish an urgently needed novel class of antibiotics for human use, including systemic use. While the antibacterial spectrum of pleuromutilins is well suited for skin and skin structure infections (SSSI) and respiratory tract infections (RTI), pleuromutilins are in particular to be considered as a new mode of action, new class antibiotic, which not only lack considerable cross-resistance to alternative classes but certainly do not contribute to the dramatic selection of resistance among *Enterobacteriaceae* species and non-fermenters like *P. aeruginosa*.[39]

15.3 Changing Epidemiology of MRSA: Are Pleuromutilins Finally Fit for Changing Needs?

Although, like almost all classes of antibiotics, pleuromutilins made their way into veterinary medicine, the question arises why it took more than 50 years for the first pleuromutilin to be made available for human use. Apparently, after their first discovery in the early 1950s, pleuromutilins appeared to have received little attention after initial biological and chemical characterization revealed that *in vivo* efficacy experiments in mice, using *Streptococcus hemolyticus* (*pyogenes*), only resulted in low rates of survival.[1]

Structure elucidation in the early 1960s and some insights into the basic biosynthesis by the groups of Arigoni and Birch provided the basis for first research and development considerations.[40-42] However, it required the finding by Brandl and co-workers that pleuromutilins were highly active against *Mycoplasma* species to make the class of pleuromutilins attractive for the Sandoz group, with a primary focus on veterinary medicine at the time.[43] It became rapidly clear that further improvements of the antimicrobial activity was needed, which led between 1963 and 1966 to a number of new derivatives with a strong emphasis on the C(14) side chain.[44] Of the derivatives generated, the mutilin esters of substituted thioglycolic acids demonstrated superior anti-microbial activity.[45] Further alterations within this group led to the development of the first veterinary pleuromutilin, tiamulin (81.723 hfu), which was approved in 1979.[46] Valnemulin, another pleuromutilin derivative, was approved in 1999 for the prevention and treatment of swine dysentery caused by *Brachyspira hyodysenteriae* and enzootic pneumonia caused by *Mycoplasma hyopneumoniae*.

While the first development successes were achieved with veterinary products, a specific property of the antimicrobial profile of pleuromutilins offered an opportunity for human medicine – their excellent activity against MRSA.[39]

The first strains of MRSA were detected in 1961, shortly after methicillin came into clinical use.[47] These strains were resistant only to β-lactam antibiotics, very much in contrast to strains that appeared in Australia in the late 1970s which were resistant to almost all classes of antibiotics, except glycopeptides.[48–51] Over the past 50 years, MRSA infections have become endemic in most hospitals worldwide, at the time classifying MRSA infections almost exclusively as hospital-acquired infections.[48,52] More recently, however, MRSA are frequently encountered among patients in the community. These patients are often young, healthy people and frequently have no established risk factors as usually seen in patients suffering from hospital-acquired MRSA (HA-MRSA).[15] Lately, reports have emerged suggesting that community-acquired *S. aureus* (CA-MRSA) clones are beginning to supplant HA-MRSA clones as cause of endemic hospital-acquired infections.[53,54] While in the past the majority of CA-MRSA isolates remained sensitive to non-β-lactam antibiotics, the epidemiology of MRSA infections becomes less clear-cut with more healthcare exposures occurring among the general population leading to hospital-associated strains entering the community and community-associated strains encroaching on hospitals.[16,54–57]

The question arises if the convincing *in vitro* antibacterial properties of pleuromutilin antibiotics against MRSA can be translated successfully into a valuable clinical asset addressing the needs of physicians and patients.[58,59]

15.4 Setting the Stage for Retapamulin

A significant number of community-acquired SSSI can be classified as uncomplicated skin infections with superficial lesions. These infections account for almost 200 million physician-office visits in the USA annually and comprise secondarily infected traumatic lesions (SITL), cellulitis or abscesses, secondarily infected dermatoses (SID), impetigo, and carbuncles or furunculosis. Although often oral antibiotics are used for the treatment of uncomplicated skin infections, the proportion of topical antibiotics is significant; about 80% of SID and 50% of impetigo cases are treated topically.[60]

A study by Tillotson and colleagues evaluated the incidence of MRSA from community-acquired skin and wound infections cultured in the USA in 2007 and found that the rate of MRSA did not vary markedly between intensive care, in patient-acquired or community-acquired *S. aureus*, with almost 60% of isolates being methicillin-resistant.[60] This finding is in strong support of a widely observed trend; the continued increase of MRSA, in particular CA-MRSA among patients suffering from SSSI, has significant implications for the empiric treatment of complicated SSSI in particular.[15,61]

Although, as of today, the majority of staphylococcal isolates causing uncomplicated SSSI have remained sensitive to methicillin, it is probably only a question of time that in uncomplicated SSSI the epidemiology will also shift eventually towards a significant increase in CA-MRSA cases.

15.4.1 Indications and Usage of Retapamulin

Retapamulin (Altabax®) is indicated in the USA for use in adults and pediatric patients aged 9 months and older for the topical treatment of impetigo due to *S. aureus* (methicillin-susceptible isolates only) or *Streptococcus pyogenes*. Although the FDA also issued an approvable letter for the treatment of SITL in December 2006, it did not approve this indication later on.[62]

In June 2006, a market authorization was filed in Europe. The Committee for Medicinal Products for Human Use (CHMP) adopted a positive opinion for the application in March 2007 and in contrast to FDA granted in addition to impetigo an approval for infected small lacerations, abrasions or sutured wounds (no abscesses). Again, MRSA isolates were excluded from the indication for reasons provided later.[63]

So far GlaxoSmithKline (GSK) has published five Phase III trials in patients with impetigo, SITL, and SID.[64–67] One trial in patients with impetigo was a randomized, double-blind, placebo-controlled superiority trial.[64] The other trials were active comparator non-inferiority trials. Comparators included topical fusidic acid in an observer-blind trial in patients with impetigo or oral cephalexin in two double-blind trials in patients with secondarily infected traumatic lesions. In all trials, topical retapamulin 1% ointment was administered twice daily for 5 days. Oral cephalexin 500 mg capsules or oral placebo was administered twice daily for 10 days. Topical fusidic acid 2% ointment was administered three times daily for 7 days.

Details of study outcomes have been reviewed elsewhere in detail, thus only a brief summary is provided.[68]

15.4.1.1 Impetigo

In both clinical trials in patients with impetigo, topical retapamulin either was superior to placebo ointment or was non-inferior to fusidic acid (Table 15.1).[64,65] The superiority trial comparing the efficacy of retapamulin with placebo (both applied twice daily for 5 days) enrolled 213 patients with 139 evaluable patients in the retapamulin group and 71 in the placebo group (ITT population).[64] The primary outcome measure was the clinical response at the end-of-therapy (EOT) on day 7 in the intent-to-treat (ITT) population,

Table 15.1 Clinical response of retapamulin in patients with impetigo or SITL

Trial	Treatment regimen	ITT[a]			PP[b]		
		Pat. No.	Response rate (%)	Diff. in success (%) [$CI_{95\%}$]	Pat. No.	Response rate (%)	Diff. in success (%) [$CI_{95\%}$]
Patients with impetigo							
Placebo contr.[64]	Retapamulin	139	85.6[3]	33.5 [20.5, 46.5]	124	89.5	36.3 [22.8, 49.8]
	Placebo	71	52.1[3]		62	53.2	
Fusidic acid contr.[65]	Retapamulin	345	94.8	4.7 [−0.4, 9.7]	317	99.1[c]	5.1 [1.1, 9.0]
	Fusidic acid	172	90.1		150	94.0[c]	
Patients with SITL							
Study A[66]	Retapamulin	662	85.2	1.1 [−3.7, 6.0]	592	88.7[c]	−3.2 [−7.4, 0.9]
	Cephalexin	326	84.0		260	91.9[c]	
Study B[67]	Retapamulin	606	87.5	0.0 [−4.5, 4.6]	540	90.4[c]	−1.6 [−5.8, 2.6]
	Cephalexin	310	87.4		249	92.0[c]	

[a]ITT population consisted of all randomized patients who took at least 1 dose of study medication; [b] PP population included all ITT patients who adhered to the protocol; [c]primary efficacy endpoint.

which was defined as all randomized patients who took at least one dose of study medication. Clinical response was assessed according to an analysis if patients had a total absence of treated lesions, if lesions became dry without crusting, or if infected lesions improved so that further antibacterial treatment was not required. Based on the primary efficacy endpoint, retapamulin was superior to placebo (success rate 85.6% versus 52.1%; $P < 0.0001$).[64]

Similar results were found in the per-protocol (PP) analysis and in patients who had a pathogen isolated at baseline (Table 15.1). Since none of the *S. aureus* isolates were resistant to methicillin, retapamulin is currently only indicated for MSSA. Most common adverse effects were pruritus at the application site, which was reported by 6% of patients receiving retapamulin and 1% of patients receiving placebo.[64]

The non-inferiority trial comparing the efficacy of retapamulin with fusidic acid had 517 patients in the ITT population and 467 patients in the PP population (all ITT patients who adhered to the treatment protocol).[65] While retapamulin was administered twice daily for 5 days, fusidic acid 2% ointment was administered three times daily for 7 days. The primary response rate, defined as clinical success as an improvement or absence of the treated lesion with no further treatment being required, was 99.1% for retapamulin and 94.0% for fusidic acid in the PP population (Table 15.1).

In both trials a significant number of baseline pathogens could be isolated (>75%) with *S. aureus* being the most frequently isolated bacterium (63 to

68%) followed by *S. pyogenes* (13 to 28%). As to be expected, in contrast to complicated SSSI where as of today CA-MRSA has in the USA a prevalence of >50%, only a small number of MRSA could be identified in the described impetigo trials; no single isolate in the placebo trial and only 8 isolates in the fusidic acid trial. Since numbers of MRSA were not sufficient, MRSA was excluded from the impetigo label for indicated pathogens.[63,69]

15.4.1.2 Secondarily Infected Traumatic Lesions

Topical retapamulin was compared with oral cephalexin in two identical non-inferiority trials in patients with SITL.[66,67] The primary endpoint was clinical response which was defined as improvement of infected lesions so that further antibacterial treatment is not required. The secondary endpoint was bacteriological success at follow-up. Both endpoints were evaluated in the PP population. Study A enrolled 996 patients, of which 988 patients were treated; the PP population comprised 826 patients (83.6%). Most of the patients (87% retapamulin and 86% cephalexin of the ITT population) had secondarily infected open wounds while the remainder had simple abscesses. Study B enrolled 922 patients, of which 916 were treated; the PP population consisted of 766 patients (83.6%). Again about 85% of patients had secondarily infected open wounds rather than simple abscesses.[69] Table 15.1 lists the efficacy outcome for retapamulin for studies A and B.

Pooled data from both studies showed that the overall clinical success rates at follow-up in the PP population were 89.5% for retapamulin and 91.9% for cephalexin, corresponding to a treatment difference of –2.4 with a $CI_{95\%}$ of –5.4 to 0.5. Interestingly, a treatment difference of –4.2 for simple abscesses was observed with a $CI_{95\%}$ of –12.8 to 4.4. Although the CHMP acknowledged that the studies were neither stratified by diagnosis nor powered to provide reliable comparisons between treatments within each diagnostic entity, abscesses were considered as problematic to treat with retapamulin and excluded from the label.[69]

Clinical success rates at follow-up by pathogen isolated at baseline in the PP population were almost identical considering all pathogens; 89.5% for retapamulin and 89.1% for cephalexin. However, a significant difference was observed for MRSA with only 68.6% of isolates defined as success in the retapamulin group compared to 88.5% in the cephalexin group. In the scientific discussion document of the retapamulin European Public Assessment Report (EPAR) it is noted that success rates for retapamulin were in particular lower among patients enrolled in North America (56%) than elsewhere (80 to 92%).[69] Although overall MRSA numbers were relatively low, around 10%, and consequently, a solid statistical analysis is difficult to achieve, it is not obvious why retapamulin performed so poorly in this subgroup; in particular since MRSA are deemed to be resistant to cephalexin. Considering the fact that the pooled SITL studies demonstrated an association between a clinical presentation of abscess and infection with MRSA, retapamulin was not approved for treating abscesses (as mentioned before) and infection due to MRSA.

Table 15.2 Clinical response of retapamulin and placebo at follow-up (day 12 to 14) by analysis population

Analysis population	Retapamulin		Placebo		Diff. in success (%)	[CI$_{95\%}$]
	Pat. No.	Response rate (%)	Pat. No.	Response rate (%)		
ITT[a,b]	246	74.8	113	66.4	8.4	[−1.6, 18.4]
PP[c]	215	79.1	97	74.2	4.8	[−5.2, 14.8]
ITTB[d]	182	76.4	84	64.3	12.1	[0.6, 23.6]
PPB[e]	158	81.0	69	73.9	7.1	[−4.4, 18.6]

[a]Primary efficacy endpoint; [b]ITT population consisted of all randomized patients who took at least 1 dose of study medication; [c]PP population included all ITT patients who adhered to the protocol; [d]ITTB intend-to-treat bacteriological evaluable population; [e]PPB per-protocol bacteriologically evaluable population.

As a follow-up measure of the above described results, an additional study was performed by GSK which intended to demonstrate superiority of retapamulin over placebo in the treatment of SITL (NCT00684177).[70] However, this randomized, double-blind study failed to meet its pre-defined primary endpoint, which was clinical success at follow-up (day 12 to 14) for subjects in the ITT population (Table 15.2). When adjusted for baseline wound characteristics, including pathogen, wound size and severity, the clinical success rate of retapamulin was superior to placebo for the primary efficacy endpoint ($p = 0.0336$). Also, when considering an analysis at the EOT visit (day 7 to 9) the healing process was faster and more pronounced in patients treated with retapamulin compared to patients treated with placebo (77.3% versus 43.5%). Not too surprisingly this effect was less pronounced at the follow-up visit, with 88.6% versus 81% for retapamulin and placebo-treated patients, respectively.[70]

With regard to the ITT bacteriological evaluable population, the clinical success rate of retapamulin (76.4%) remained statistically superior even at the follow-up visit to that of placebo (64.3%). The difference was driven by higher success rates in cases of *S. aureus* infection. Although MRSA sample numbers were again low, 24 out of 147 *S. aureus* strains in the retapamulin group, and 8 out 65 in the placebo group, retapamulin performed clearly better in this subgroup (62.5% versus 25%). Finally, retapamulin did not show any advantage over placebo in subjects with *S. pyogenes* isolated at baseline.[70]

15.5 MRSA Coverage in SITL as a Difficult to Achieve Task: What Has Happened?

Topical antibiotic treatment of uncomplicated skin and skin structure infections, such as SITL, can be justified in smaller and less deep lesions with no signs or symptoms of systemic toxicity.[71] In less severe, uncomplicated disease settings topical treatment might offer a number of advantages which contributes to the partial replacement of systemic, oral antibiotic therapy.[11] For instance in case of limited systemic exposure, a number of possible side-effects caused by systemic treatment are avoided.[72] Since a significant number of patients suffering from uncomplicated skin and wound infections have co-medications, considerations of possible drug–drug interactions are less relevant. As shown in the SITL studies conducted with retapamulin, the compliance with topical therapy can be potentially greater than that for oral therapy.[11]

It is important to point out that although the antibiotic resistance pattern of *S. aureus* in uncomplicated skin infections is currently less broad than in complicated disease, a clearly increased prevalence of CA-MRSA has been observed over the last couple of years.[11,73] In addition, the emergence of CA-MRSA with a reduced susceptibility to glycopeptides and possibly daptomycin, and increasing resistance to mupirocin has to be considered.[11,74–76] Guidance issued by the Infectious Diseases Society of America (IDSA) recommend that minor and uncomplicated SSSI may be empirically treated.[11,77] While current treatment recommendations still include anti-staphylococcal penicillins, first- or second-generation oral cephalosporins, macrolides, or clindamycin, some of these treatment options might have to be reconsidered in the future assuming a further increase in resistant pathogens, in particular CA-MRSA.

Unfortunately, clinical data provided so far for retapamulin shows an ambiguous picture. In cases of superficial lesions, like those usually encountered in patients suffering from impetigo, clinical studies have shown that retapamulin is highly efficient – providing a very convincing treatment effect over placebo.[64,65] Assuming that lesion characteristics induced by MRSA are similar to MSSA, a comparable efficacy of retapamulin seems to be realistic in this disease setting (eight out of eight MRSA isolates were treated successfully by retapamulin in the fusidic acid impetigo trial).[65]

However, both cephalexin SITL trials led to difficult to interpret data; it is no surprise that retapamulin did not do too well in patients suffering from abscesses, in particular considering the fact that the antibiotic treatment effect in small abscesses is considered to be marginal anyhow. This assumption is supported by the initially contradictory finding that cephalexin did much better in cases where MRSA was isolated as a baseline pathogen.[66,67] Undoubtedly, cephalexin is not active *in vitro* against MRSA and therefore it is highly unlikely that the observed treatment effect is linked directly to bacterial killing of cephalexin. Rather one might assume that the observed difference between retapamulin and cephalexin is a function of small treatment effects in abscesses, linked to variable outcome rates, different clonal origin and/or differing non-antibacterial treatment regimens.

Another lesson might be learned from the very recently conducted trial in SITL patients comparing the efficacy of retapamulin with placebo. Similar to endpoint considerations currently given by the FDA to early primary endpoints, 48 to 72 hours after onset of treatment in acute bacterial skin and skin structure infections (ABSSSI) requiring systemic treatment, one also might want to give primary endpoint selection in less severe disease a special consideration. It probably does not come as a surprise that in diseases with inherently lower treatment effects, like presumably uncomplicated SSSI, later efficacy endpoints are suited less to determine minor differences in efficacy or related parameters. Later efficacy endpoints might not only be more error prone due to additional and difficult to control confounding factors, but ultimately might also require higher enrollment numbers. A similar view could be taken as to the specifics of inclusion criteria; in light of the relatively low prevalence of MRSA in SITL, at least currently, an entire exclusion of abscesses might be warranted, simply to avoid difficult to interpret data.

15.6 Perspectives

Retapamulin is a first-in-class topical pleuromutilin which, although probably perfectly suited to treat uncomplicated SSSI caused by MRSA, has failed so far to provide sufficient evidence justifying the inclusion of MRSA on its label. While one could consider this as a set-back to the class as such, it seems more likely that the low prevalence of MRSA in SITL and impetigo as well as possibly some aspects of clinical trial design have led to this mishap. Nevertheless, given the novel mode of action, low potential for cross-resistance and low propensity of developing resistance, in combination with the overall high efficacy demonstrated in clinical trials, retapamulin should be considered as a valuable new option for the management of SITL and impetigo.

More recent data published on a successful clinical Phase II trial of a systemically available pleuromutilin in patients suffering from ABSSSI might be a further stepping stone toward the successful establishment of this new class of antibiotics in humans.[78] MRSA accounted for 70% of the treated staphylococcal skin infections in this trial. Upcoming Phase III trials will have to demonstrate if this potentially exciting and clinically valuable class can live up to its promise to treat severe infections caused by resistant pathogens, including MRSA.

References

1. F. Kavanagh, A. Hervey and W. J. Robbins, *Proc. Natl. Acad. Sci. USA*, 1951, **37**, 570.
2. F. Kavanagh, A. Hervey and W. J. Robbins, *Proc. Natl. Acad. Sci. USA*, 1952, **38**, 555.
3. M. Anchel, *J. Biol. Chem.*, 1952, **199**, 133.
4. S. G. Amyes, *Magic Bullets, Cost Horizons: The Rise and Fall of Antibiotics*, Taylor and Francis, New York, NY, 2001.
5. H. F. Smyth and C. P. Carpenter, *Science*, 1938, **15**, 350.

6. K. Meyer, E. Chaffee, G. L. Hobby, M. H. Dawson, E. Schwenk and G. Fleischer, *Science*, 1942, **96**, 20.
7. D. Jones, H. J. Metzger, A. Schatz and S. A. Waksman, *Science*, 1944, **100**, 103.
8. F. R. Heilman, W. E. Herrell, W. E. Wellmann and J. E. Geraci, *Proc. Staff. Meet. Mayo Clin.*, 1952, **27**, 285.
9. M. S. Bryer and E. B. Schoenbach, *Ann. NY Acad. Sci.*, 1948, **51**, 254.
10. Q. R. Bartz, *J. Biol. Chem.*, 1948, **172**, 445.
11. R. Shawar, N. Scangarella-Oman, M. Dalessandro, J. Breton, M. Twynholm, G. Li and H. Garges, *Ther. Clin. Risk Manag.*, 2009, **5**, 41.
12. H. W. Boucher, G. H. Talbot, J. S. Bradley, J. E. Edwards, D. Gilbert, L. B. Rice, M. Scheld, B. Spellberg and J. Bartlett, *Clin. Infect. Dis.*, 2009, **48**, 1.
13. B. Spellberg, R. Guidos, D. Gilbert, J. Bradley, H. W. Boucher, W. M. Scheld, J. G. Bartlett and J. Edwards, Jr., *Clin. Infect. Dis.*, 2008, **46**, 155.
14. National Nosocomial Infections Surveillance System Report, *Am. J. Infect. Control*, 2004, **32**, 470.
15. G. J. Moran, A. Krishnadasan, R. J. Gorwitz, G. E. Fosheim, L. K. McDougal, R. B. Carey and D. A. Talan, *N. Engl. J. Med.*, 2006, **355**, 666.
16. R. M. Klevens, J. R. Edwards, F. C. Tenover, L. C. McDonald, T. Horan and R. Gaynes, *Clin. Infect. Dis.*, 2006, **42**, 389.
17. H. W. Boucher and G. R. Corey, *Clin. Infect. Dis.*, 2008, **46**, S344.
18. M. M. Neuhauser, R. A. Weinstein, R. Rydman, L. H. Danziger, G. Karam and J. P. Quinn, *JAMA*, 2003, **289**, 885.
19. L. S. Munoz-Price and R. A. Weinstein, *N. Engl. J. Med.*, 2008, **358**, 1271.
20. L. Poirel, P. Nordmann, E. Lagrutta, T. Cleary and L. S. Munoz-Price, *Antimicrob. Agents Chemother.*, 2010, **54**, 3072.
21. D. Landman, S. Bratu, M. Alam and J. Quale, *J. Antimicrob. Chemother.*, 2005, **55**, 954.
22. S. Bratu, D. Landman, R. Haag, R. Recco, A. Eramo, M. Alam and J. Quale, *Arch. Intern. Med.*, 2005, **165**, 1430.
23. G. H. Talbot, J. Bradley, J. E. Edwards, Jr., D. Gilbert, M. Scheld and J. G. Bartlett, *Clin. Infect. Dis.*, 2006, **42**, 657.
24. A. Philippon, R. Labia and G. Jacoby, *Antimicrob. Agents Chemother.*, 1989, **33**, 1131.
25. C. Kliebe, B. A. Nies, J. F. Meyer, R. M. Tolxdorff-Neutzling and B. Wiedemann, *Antimicrob. Agents Chemother.*, 1985, **28**, 302.
26. P. Komp Lindgren, Å. Karlsson and D. Hughes, *Antimicrob. Agents Chemother.*, 2003, **47**, 3222.
27. A. M. Emmerson and A. M. Jones, *J. Antimicrob. Chemother.*, 2003, **51**, 13.
28. J. G. Bartlett, *Clin. Infect. Dis.*, 2006, **43**, 428.
29. D. R. Pillai, J. Longtin and D. E. Low, *Clin. Infect. Dis.*, 2010, **50**, 1685.
30. S. Bratu, M. Mooty, S. Nichani, D. Landman, C. Gullans, B. Pettinato, U. Karumudi, P. Tolaney and J. Quale, *Antimicrob. Agents Chemother.*, 2005, **49**, 3018.

31. S. Bratu, D. Landman, M. Alam, E. Tolentino and J. Quale, *Antimicrob. Agents Chemother.*, 2005, **49**, 776.
32. S. Bratu, S. Brooks, S. Burney, S. Kochar, J. Gupta, D. Landman and J. Quale, *Clin. Infect. Dis.*, 2007, **44**, 972.
33. K. Bush, *Clin. Infect. Dis.*, 2001, **32**, 1085.
34. Available from URL:http://emc.medicines.org.uk, 2010.
35. Available from URL: http://emc.medicines.org.uk, 2010.
36. F. McLaws, I. Chopra and A. J. O'Neill, *J. Antimicrob. Chemother.*, 2008, **61**, 1040.
37. A. J. O'Neill, A. R. Larsen, R. Skov, A. S. Henriksen and I. Chopra, *J. Clin. Microbiol.*, 2007, **45**, 1505.
38. J. B. Patel, R. J. Gorwitz and J. A. Jernigan, *Clin. Infect. Dis.*, 2009, **49**, 935.
39. R. Novak and D. M. Shlaes, *Curr. Opin. Investig. Drugs*, 2010, **11**, 182.
40. A. J. Birch, C. W. Holzapfel and R. W. Richards, *Tetrahedron (Suppl.)*, 1966, 359.
41. A. J. Birch, D. W. Cameron, C. W. Holzapfel and R. W. Richards, *Chem. Ind.*, 1963, **8**, 374.
42. D. Arigono, *Gazz Chim. Ital.*, 1962, **92**, 884.
43. F. Knauseder and E. Brandl, *J. Antibiot. (Tokyo)*, 1976, **29**, 125.
44. K. Riedl, *J. Antibiot. (Tokyo)*, 1976, **29**, 132.
45. H. Egger and H. Reinshagen, *J. Antibiot. (Tokyo)*, 1976, **29**, 923.
46. J. Drews, A. Georgopoulos, G. Laber, E. Schutze and J. Unger, *Antimicrob. Agents Chemother.*, 1975, **7**, 507.
47. M. P. Jevons, A. W. Coe and M. T. Parker, *Lancet*, 1963, **1**, 904.
48. W. Brumfitt and J. Hamilton-Miller, *N. Engl. J. Med.*, 1989, **320**, 1188.
49. H. F. Chambers, *Emerg. Infect. Dis.*, 2001, **7**, 178.
50. G. L. Archer, *Clin. Infect. Dis.*, 1998, **26**, 1179.
51. F. F. Barrett, R. F. McGehee, Jr. and M. Finland, *N. Engl. J. Med.*, 1968, **279**, 441.
52. F. D. Lowy, *N. Engl. J. Med.*, 1998, **339**, 520.
53. K. J. Popovich and R. A. Weinstein, *Infect. Control Hosp. Epidemiol.*, 2009, **30**, 9.
54. J. M. Boyce, *Clin. Infect. Dis.*, 2008, **46**, 795.
55. J. A. Otter and G. L. French, *Lancet Infect. Dis.*, 2010, **10**, 227.
56. J. A. Otter and G. L. French, *J. Hosp. Infect.*, 2011, **79**, 189.
57. R. M. Klevens, M. A. Morrison, J. Nadle, S. Petit, K. Gershman, S. Ray, L. H. Harrison, R. Lynfield, G. Dumyati, J. M. Townes, A. S. Craig, E. R. Zell, G. E. Fosheim, L. K. McDougal, R. B. Carey and S. K. Fridkin, *JAMA*, 2007, **298**, 1763.
58. R. N. Jones, T. R. Fritsche, H. S. Sader and J. E. Ross, *Antimicrob. Agents Chemother.*, 2006, **50**, 2583.
59. E. Perez-Trallero, E. Tamayo, M. Montes, J. M. Garcia-Arenzana and V. Iriarte, *Antimicrob. Agents Chemother.*, 2011, **55**, 2406.

60. R. Pangilinan, A. Tice and G. Tillotson, *Expert. Rev. Anti. Infect. Ther.*, 2009, **7**, 957.
61. D. Styers, D. J. Sheehan, P. Hogan and D. F. Sahm, *Ann. Clin. Microbiol. Antimicrob.*, 2006, **5**, 2.
62. Available from URL: http://us.gsk.com/products/assets/us_altabax.pdf, 2011.
63. Available from URL: http://ema.europe.eu, 2007.
64. S. Koning, J. C. van der Wouden, O. Chosidow, M. Twynholm, K. P. Singh, N. Scangarella and A. P. Oranje, *Br. J. Dermatol.*, 2008, **158**, 1077.
65. A. P. Oranje, O. Chosidow, S. Sacchidanand, G. Todd, K. Singh, N. Scangarella, R. Shawar and M. Twynholm, *Dermatology*, 2007, **215**, 331.
66. A. Free, E. Roth, M. Dalessandro, J. Hirman, N. Scangarella, R. Shawar and S. White, *Skinmed*, 2006, **5**, 224.
67. L. C. Parish, J. L. Jorizzo, J. J. Breton, J. W. Hirman, N. E. Scangarella, R. M. Shawar and S. M. White, *J. Am. Acad. Dermatol.*, 2006, **55**, 1003.
68. L. P. Yang and S. J. Keam, *Drugs*, 2008, **68**, 855.
69. Available from URL: http://ema.europe.eu, 2007.
70. Available from URL: http://ema.europe.eu, 2011.
71. D. L. Stevens, A. L. Bisno, H. F. Chambers, E. D. Everett, P. Dellinger, E. J. Goldstein, S. L. Gorbach, J. V. Hirschmann, E. L. Kaplan, J. G. Montoya and J. C. Wade, *Clin. Infect. Dis.*, 2005, **41**, 1373.
72. S. Koning, A. P. Verhagen, L. W. van Suijlekom-Smit, A. Morris, C. C. Butler and J. C. van der Wouden, *Cochrane. Database. Syst. Rev.*, 2004, **2**, CD003261.
73. G. Perera and R. Hay, *J. Eur. Acad. Dermatol. Venereol.*, 2005, **19**, 531.
74. P. C. Appelbaum, *Clin. Microbiol. Infect.*, 2006, **12**, 16.
75. L. M. Deshpande, A. M. Fix, M. A. Pfaller and R. N. Jones, *Diagn. Microbiol. Infect. Dis.*, 2002, **42**, 283.
76. P. A. Moise, D. North, J. N. Steenbergen and G. Sakoulas, *Lancet Infect. Dis.*, 2009, **9**, 617.
77. D. L. Stevens, A. L. Bisno, H. F. Chambers, E. D. Everett, P. Dellinger, E. J. Goldstein, S. L. Gorbach, J. V. Hirschmann, E. L. Kaplan, J. G. Montoya and J. C. Wade, *Clin. Infect. Dis.*, 2005, **41**, 1373.
78. W. T. Prince, F. Obermayr, Z. Ivezic-Schoenfeld, C. Lell, W. W. Wicha, D. Stricklmann, K. J. Tack and R. Novak, *ICAAC*, 2011, L 966.

CHAPTER 16

Epothilones as Lead Structures for New Anticancer Drugs

BERNHARD PFEIFFER, FABIENNE ZDENKA GAUGAZ, RAPHAEL SCHIESS AND KARL-HEINZ ALTMANN*

Department of Chemistry and Applied Biosciences, Institute of Pharmaceutical Sciences, Swiss Federal Institute of Technology (ETH), Zürich, Switzerland
*E-mail: karl-heinz.altmann@pharma.ethz.ch

16.1 Introduction

Drugs targeting cellular microtubules are an important element of our clinical armamentarium for cancer therapy,[1] with applications in the treatment of a variety of cancer types, either as single agents or as part of different combination regimens.[2,3] Microtubule-interacting agents can be grouped into two distinct classes, depending on the exact molecular mechanisms that lead to interference with microtubule function. Thus, they can either inhibit the assembly of tubulin heterodimers into microtubule polymers ('tubulin polymerization inhibitors') or they can stabilize microtubules under normally destabilizing conditions ('microtubule stabilizers').[4] Microtubule stabilizers (or microtubule-stabilizing agents, MSA) also promote the assembly of tubulin heterodimers into microtubule polymers, with the induction of tubulin polymerization often being used as a biochemical readout for a quantitative assessment of the interaction of MSA with tubulin. Tubulin polymerization inhibitors such as vincristine and vinblastine have been employed in cancer therapy for more than 40 years (vincristine and vinblastine received FDA

RSC Drug Discovery Series No. 25
Drug Discovery from Natural Products
Edited by Olga Genilloud and Francisca Vicente
© The Royal Society of Chemistry 2012
Published by the Royal Society of Chemistry, www.rsc.org

approval in 1963 and 1965, respectively), but new agents continue to be evaluated clinically[1,5] and the latest tubulin polymerization inhibitor (eribulin) was approved by the FDA as recently as 2010.[6] In contrast, the clinical history of MSA dates back only to 1993, which marks the introduction into clinical practice of the natural product taxol (paclitaxel; Taxol®),[7] followed three years later by the FDA approval for the semi-synthetic taxol analog docetaxel (Taxotere®) (Figure 16.1). Again, a new MSA has been approved by the FDA in 2010, namely the docetaxel derivative cabazitaxel.[8]

Interestingly, while a number of small synthetic molecules are able to effect inhibition of tubulin polymerization (apart from complex natural products like vincristine or vinblastine),[1,5] all potent MSA identified to date are natural products or natural product-derived (for reviews see refs. 8–11). After the elucidation of taxol's mode of action in 1979[12] more than a decade passed before the discovery of MSA with structures that were not based on a taxane scaffold. Most prominent among these new MSA are the bacterial natural products epothilone A and B (Epo A and B) that were first isolated in 1987 by Reichenbach and Höfle from the myxobacterium *Sorangium cellulosum Sc 90* (Figure 16.2).[13,14] In addition, numerous related structures have subsequently been identified as minor components in fermentation broths from myxobacteria.[15] The microtubule-stabilizing properties of Epo A and B were not immediately recognized, however, and were only discovered eight years after their first discovery by a group at Merck Research Laboratories.[16] In the meantime a growing number of additional natural products have been recognized to be MSA (for reviews see refs. 8–11; for the most recent example see ref. 17), thus providing a whole new set of potential lead structures for anticancer drug discovery.

Epothilones exert their antiproliferative activity through interference with the same molecular target as taxol, but in contrast to taxol they also inhibit the growth of multidrug-resistant cancer cell lines with excellent potency, including cells whose taxol resistance is mediated by specific tubulin mutations.[18,19]

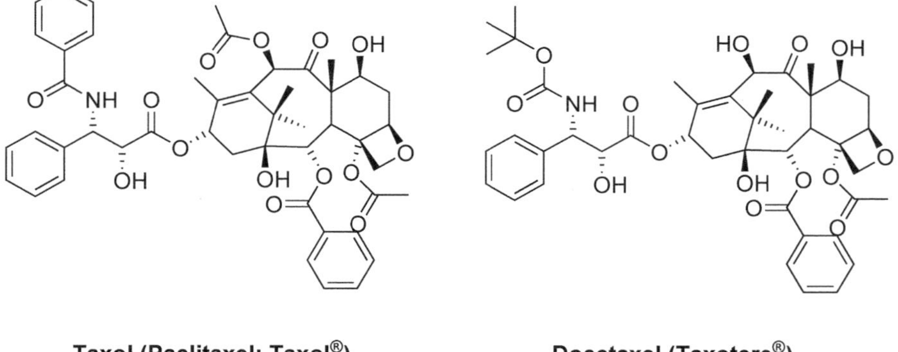

Taxol (Paclitaxel; Taxol®)   **Docetaxel (Taxotere®)**

Figure 16.1   Molecular structures of taxol and docetaxel.

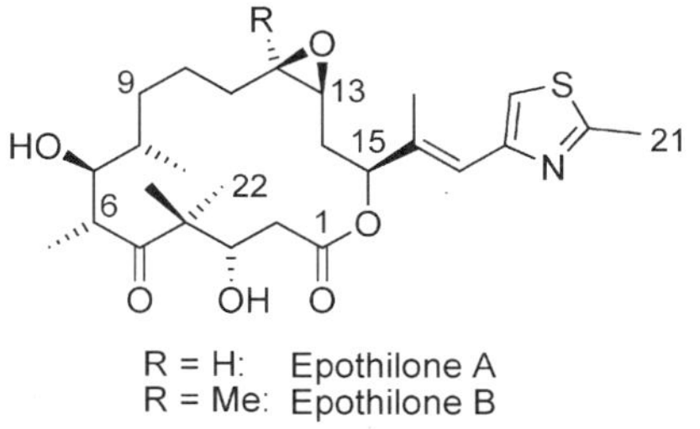

R = H: Epothilone A
R = Me: Epothilone B

Figure 16.2 Molecular structure of epothilones.

Thus, IC_{50} values are often identical with, or at least close to, those that are observed for drug-sensitive cancer cells.[16,18-20] At the same time, epothilones possess more favorable biopharmaceutical properties than taxol (e.g. improved water solubility[21]), which, at least in principle, enables the use of more benign clinical formulation vehicles than are required for taxol.[7] Epo B and a number of its analogs have been demonstrated to possess potent *in vivo* antitumor activity[20,22] and up to this point at least eight compounds based on the epothilone structural scaffold have entered clinical evaluation in humans. These include Epo B itself (EPO906, patupilone; developed by Novartis), Epo D (deoxyEpo B, KOS-862; Kosan/Roche/BMS), BMS-247550 (ixabepilone, the lactam analog of Epo B; BMS), BMS-310705 (C21-amino-Epo B; BMS), ABJ879 (C20-desmethyl-C20-methylsulfanyl-Epo B; Novartis), 9,10-didehydro-Epo B (KOS-1584 (Kosan/Roche/BMS)), the fully synthetic analog ZK-Epo (sagopilone, Bayer Schering),[23] and, most recently, the folic acid conjugate BMS-753493 (BMS).[24] Of these, ixabepilone has obtained regulatory approval in the US for the treatment of advanced and metastatic breast cancer,[25] while approval has been denied by the European Medicines Agency (EMEA); patupilone and sagopilone still appear to be under active development, although Novartis has recently announced they will not apply for regulatory approval of patupilone in ovarian cancer based on its non-superiority over standard therapy in Phase III trials.[26] Clinical development of ABJ-879,[27] BMS-310705, KOS-862, KOS-1584, and also BMS-753493[28] appears to have been terminated or at least put on hold.[29] With the exception of BMS-753493 these latter compounds will not be discussed in this article in any detail.

Epothilones have been attractive targets for total chemical synthesis and numerous syntheses of Epo A and B have been developed since the elucidation of their relative and absolute configuration by Höfle and co-workers in 1996[21] (for reviews see refs. 30–36). In addition, the methodology established in the course of the total synthesis work has provided the foundation for the preparation of a host of synthetic analogs (reviewed in refs. 13, 30, 31, 34, 37–41), which, in turn, has enabled the comprehensive experimental assessment of the structural parameters that are important for biological activity. In one case (sagopilone), chemical synthesis has even been the means for the production of drug substance for clinical trials.[42] This highlights the difference in structural complexity (which is reflected in synthetic accessibility) between epothilone-type structures and taxol, for which an industrial-scale synthesis is clearly out of reach.

The chemistry, biology, and SAR (structure–activity relationship) of epothilones have been extensively discussed in a number of review articles[20,30,31,34,37–41,43] and book chapters[13,44] and the information available from these previous accounts will not be reiterated here. Instead, this chapter, with the exception of a summary of the most pertinent aspects of the *in vitro* and *in vivo* pharmacology of Epo B, will mostly focus on SAR data, structural studies, and biological and pharmacological findings that have been reported in the literature for epothilone-type structures subsequent to 2007.

16.2   Epothilone B

16.2.1   *In vitro* Biochemistry and Pharmacology

The antitumor activity of Epo B is based on its ability to bind to microtubules and to alter their intrinsic stability and dynamic properties. Microtubules are hollow filaments of ca. 240 Å outer diameter, which are composed of the 55 kD proteins α- and β-tubulin as the constituent subunits (Figure 16.3). Together with actin polymers and intermediate filaments microtubules form the cytoskeleton and they are of critical importance for the development and maintenance of cell shape and motility, intracellular transport of vesicles, mitochondria and other components and, in the context of cell division, for the formation of the mitotic spindle. (For excellent recent reviews on microtubule structure and function see refs. 4, 45–47).

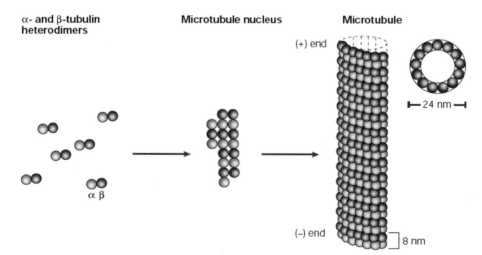

Figure 16.3   Schematic representation of the α/β-tubulin/microtubule equilibrium and the structure of microtubules. (Reproduced from ref. 10 with permission.)

Cellular microtubules shorten and lengthen in a stochastic fashion through loss or addition of α/β-tubulin heterodimers from microtubule ends, a phenomenon referred to as 'dynamic instability'.[4,45–48] Dynamic instability is essential for the proper assembly of the mitotic spindle and the subsequent movement of sister chromatids to the spindle poles, with spindle microtubules being significantly (ca. 4–100-fold) more dynamic than those of the interphase cytoskeleton.[49,50]

Epothilones prevent the Ca^{2+}- or cold-induced depolymerization of pre-existing microtubule polymers in cell-free systems[18] and they promote the polymerization of soluble tubulin into microtubule-like polymers under conditions that would normally destabilize microtubules.[16,18] Epo B is a more potent tubulin-polymerizing agent than taxol or Epo A; e.g. EC_{50}-values for the polymerization of microtubule protein by Epo A, Epo B, and taxol have been determined as 1.12, 0.67, and 1.88 μM, respectively.[51] As demonstrated by kinetic experiments, epothilones inhibit the binding of taxol to microtubules in a competitive manner;[16,18] subsequent structural studies on a complex between Zn^{2+}-stabilized tubulin polymer sheets and Epo A[52] have confirmed that taxol and epothilones bind to the same site on β-tubulin (see below). Most recently, the binding constants of Epo A and B to stabilized (cross-linked) microtubules *in vitro* have been determined as $2.93 \times 10^7 \ M^{-1}$ (Epo A) and $6.08 \times 10^8 \ M^{-1}$ (37 °C).[53] Interestingly, inhibition of cancer cell growth occurs at concentrations that are significantly lower than those required for the induction of tubulin polymerization *in vitro*.[38] This apparent discrepancy can be explained by the fact that Epo A and B accumulate inside cells several-hundred fold over external medium concentrations.[20,38,54]

The growth inhibitory effect of epothilones (and also other microtubule-interacting agents) is generally ascribed to the suppression of microtubule dynamics rather than an overall increase in microtubule polymer mass due to massive induction of tubulin polymerization.[4,48] The concentration-dependent inhibition of the dynamics of interphase microtubules by Epo B has been experimentally demonstrated by Kamath and Jordan[55] by means of time-lapse microscopy in MCF-7 cells stably transfected with GFP (green fluorescent protein)-α-tubulin. As pointed out above, microtubule dynamics is significantly increased in spindle microtubules (over interphase microtubules).

Treatment of human cancer cells with low nM concentrations of Epo B produces aberrant mitotic spindles, results in cell cycle arrest in mitosis, and eventually leads to apoptotic cell death.[16,18] Depending on the specific cell type, IC_{50} values for cancer cell growth inhibition by Epo B are in the nM or even sub-nM range.[16,18,20] Apoptosis is often assumed to be a direct consequence of G2/M arrest, but Horwitz and co-workers have shown that low concentrations of Epo B (and also taxol or discodermolide) produce a large aneuploid cell population in A549 lung carcinoma cells in the *absence* of a mitotic block.[56,57] These cells arise from aberrant mitosis after formation of multipolar spindles, are arrested in the G1 phase of the cell cycle, and will eventually undergo apoptosis. In contrast, higher drug concentrations lead to a protracted mitotic block from which the cells exit without division, thus forming tetraploid G1 cells.[57] These results indicate that *entry* of cells into mitosis is a fundamental prerequisite for cell killing by Epo B, but that cell death itself does not necessarily require prior mitotic arrest.

More recent studies have suggested that apoptosis induction by Epo B may also involve the formation of reactive oxygen species (ROS) in mitochondria and an associated enhancement of the mitochondrial localization of the pro-apoptotic protein Bim.[58] Furthermore, the cytotoxicity of Epo B has been linked to the activation of the nuclear translocation of the transcription factor NFκB in a process that involves activation of IKK (inhibitor of κB kinase) and that occurs independently of the microtubule effects of the compound.[59] The true significance of these findings for the antitumor activity of Epo B still needs to be established.

The cellular response to epothilones, as for other anticancer agents, can be modulated by acquired drug resistance; alternatively, cells may be inherently protected from the antiproliferative effects of cytotoxic agents by a variety of mechanisms. As indicated above, Epo A/B, in contrast to taxol and other standard cytotoxic anticancer agents, are largely insensitive to Permeability glycoprotein-170 (Pgp)-mediated drug efflux in cellular systems *in vitro*.[16,18,20,51] At the same time, it is now well established that resistance to epothilones can arise via Pgp-independent mechanisms, such as tubulin mutations.[38,59–61] For example, Horwitz *et al.* have generated a variant of the A549 human lung carcinoma cell line, A549.EpoB40, with 95-fold resistance to Epo B, resulting from a mutation of Gln292 to Glu in β-tubulin.[61] Further selection yielded an even more resistant cell line, A549.EpoB480,[62] which is ca. 900-fold resistant to Epo B. In addition to the βGln292Glu mutation, the EpoB480 line is characterized by mutations of β60 from Val to Phe and of α195 (α-tubulin) from Leu to Met. As βVal60 is located at the end of the H1-S2 loop, which is assumed to be involved in M-loop contacts between protofilaments, amino acid changes at this position could affect lateral contacts between protofilaments and thus lead to less stable microtubules. This hypothesis is supported by the fact that A549.EpoB40 as well as EpoB480 cells are *hyper*sensitive towards tubulin polymerization inhibitors, with EpoB480 cells even being dependent on (low concentrations of) Epo B for survival[62] (see also ref. 63). The βGln292Glu mutation was also identified by Verrills *et al.*[64] in a highly resistant subline of the human T cell acute leukemia cell line CCRF-CEM in combination with a second mutation at position 231 of β-tubulin (Thr → Ala).

Apart from tubulin mutations, another non-Pgp resistance mechanism against Epo B may be (over)expression of the MRP7 efflux pump.[65] However, the clinical significance of either of these alternative mechanisms has not been established.

16.2.2 *In vivo* Pharmacology

Early experiments with Epo B at the Sloan-Kettering Cancer Center had indicated promising antitumor activity in xenograft models of human leukemia (drug-sensitive as well as multidrug-resistant tumors);[66] however, only limited effects on tumor growth in human MX-1 breast or SKOV-3 ovarian tumors in mice were reported for subsequent experiments together with considerable toxicity.[67] Based on these latter data Epo B was concluded to be too toxic for a clinically useful anticancer agent. In contrast to these findings, studies by the Novartis group have clearly demonstrated potent antitumor activity of Epo B in a number of human tumor models in mice[20,22,68,69] and also in syngeneic rat models.[22] These experiments have addressed all four major types of solid human tumors (lung,[20] breast,[20] colon,[22,69] prostate[68]) and they have included drug-sensitive as well as multidrug-resistant models. Therapeutic effects were manifest either as profound growth inhibition (stable disease) or significant tumor regression. In particular, regressions were observed in a HCT-15 colon carcinoma[22,69] and a KB-8511 cervix carcinoma[20,22] model, which are either poorly responsive or completely non-responsive to treatment with Taxol®. Interestingly, in a HT-29 colon carcinoma model treatment outcome was shown to be independent of starting tumor size (i.e. for initial tumor volumes of 100 and 500 mm^3, respectively).[22] Although Epo B treatment was frequently accompanied by significant body weight loss, these effects were reversible and therapeutic effects could generally be achieved at tolerated dose levels. Tumor cell kill by Epo B *in vivo* may not only occur by a direct mechanism (i.e. apoptosis induction in cancer cells), but it may also be related to damage of the tumor vasculature. This idea is supported by the observation of protracted effects of low doses of Epo B on endothelial cell proliferation *in vitro*[70] and in tumor explants;[71] *in vivo*, Epo B has been found to induce vascular disruption in syngeneic tumors in rats.[72]

After administration of a single i.v. dose to tumor-bearing mice or rats, Epo B rapidly distributes to all tissues.[22] Subsequent elimination from the tissue compartment is uniformly slow, but the longest retention clearly occurs in tumor and brain, with virtually none of the compound being eliminated for the first 24 h after dosing. Epo B is rapidly degraded in rodent plasma *in vitro*,[73] but based on the available *in vivo* data, tissue distribution must be significantly more rapid than plasma metabolism. In humans, blood concentrations of Epo B after a short infusion were found to decline in a biphasic manner with a terminal half-life of four days,[74] which clearly indicates that plasma stability is not critical for human therapy.

Epo B (patupilone) is currently being evaluated in a number of Phase I and Phase II clinical trials, either in combination or as a single agent.[75] However, Phase III trials in ovarian cancer have not shown superiority over liposomal doxorubicin and Novartis has decided not to file for registration of the compound in this indication.[75]

16.3 Epothilone Analogs and SAR Studies

As pointed out above, the epothilone SAR has been reviewed extensively and this chapter is not intended to provide a detailed overview on the activity of the numerous analogs that have been investigated. In the following, some of the most pertinent features of the epothilone SAR are summarily reviewed; this will be followed by a more detailed discussion of a number of selected specific modifications.

(i)   The presence of a C12-C13 epoxide moiety is not an absolute requirement for efficient microtubule stabilization and potent antiproliferative activity, which decreases only about 10–30-fold upon reduction of the epoxide moiety to a *cis* olefin ('deoxyepothilones'; reviewed in ref. 31). Likewise, the replacement of the oxirane ring by a cyclopropane or variously *N*-substituted aziridine moieties is well tolerated.[76]

(ii)   Changing the geometry of the epoxide moiety in Epo A from *cis* to *trans* is well tolerated for a 12*S*,13*S* stereochemistry, while the corresponding 12*R*,13*R*-isomer is substantially less active than Epo A.[77] For Epo B a change in epoxide geometry leads to a significant loss in biological potency for both *trans*-configured diastereoisomers.[78]

(iii)   The incorporation of a *trans*-double bond between C10 and C11 in Epo D is well tolerated, with the activity of the parent compound essentially being retained.[79] The presence of a double bond between C9 and C10 in Epo B or D results in enhanced biological potency.[80]

(iv)   The ester group in epothilones can be replaced by a secondary lactam moiety with only a ca. 10-fold reduction in *in vitro* potency.[73,81] The lactam analog of Epo B will be discussed in more detail below.

(v)   The replacement of the thiazole ring either by an oxazole or various pyridine moieties is well tolerated. Epo (A or B)-like activity in these heterocycle-modified analogs requires the presence of an aromatic nitrogen atom *ortho* to the attachment point of the linker between the heterocycle and the macrocyclic skeleton.[82]

(vi)   Rigidification of the side chain through conversion of the allylic structure to a fused aromatic ring system results in a pronounced increase in antiproliferative activity, in particular at the level of the deoxy analogs.[83]

16.3.1   Ixabepilone and Related Structures

16.3.1.1   Chemical Synthesis

As lactone ring opening by esterases had been observed in rodent plasma for natural epothilones, thus raising concerns about the metabolic stability of these compounds in humans, 15-aza epothilone analogs were developed as metabolically more stable alternatives. Early studies on epothilone lactams by the BMS and Schinzer groups were directed at the total chemical synthesis of these analogs.[73,84] While BMS abandoned their efforts in favor of a more efficient semi-synthetic approach (see below), Schinzer and co-workers have reported a successful synthesis of the Epo C lactam (**1**; Figure 16.4) and its *N*-

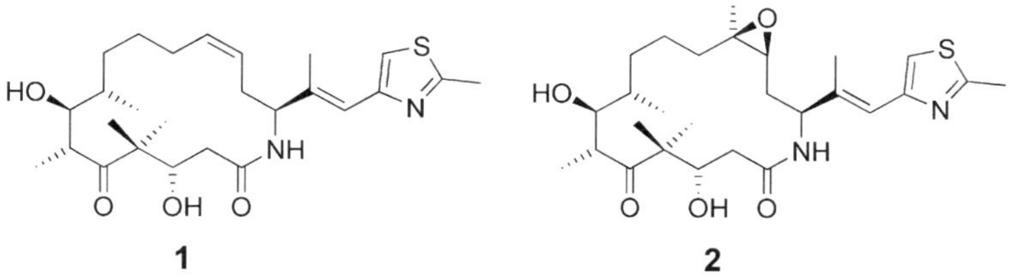

Figure 16.4 Molecular structure of epothilone lactams.

methyl derivative employing a ring-closing olefin metathesis (RCM)-based macrocylization approach.[84]

The work by the BMS group eventually led to an efficient synthesis of the Epo B lactam (**2**), and this compound (ixabepilone, BMS-247550) was then developed into a clinical anticancer drug which is marketed under the trade name Ixempra® (Figure 16.4).

The development of an efficient and scalable semi-synthetic route to ixabepilone at BMS had been enabled by the collaboration with the Gesellschaft für Biotechnologische Forschung in Braunschweig, Germany (GBF, now Helmholtz Institute for Infectious Research), the home institution of the epothilone discoverers Reichenbach and Höfle, which provided the BMS group with an optimized epothilone-producing strain of *S. cellulosum*, with production levels of 120 mg/L Epo A and 60 mg/L Epo B.[85]

The semi-synthesis of ixabepilone as depicted in Scheme 16.1[73] is based on the recognition of the allylic nature of the epothilone lactone moiety, which allows the conversion of unprotected Epo B into azide **3** with complete retention of configuration by treatment with Pd(PPh$_3$)$_4$ in the presence of sodium azide. Subsequent reduction of the azide moiety with trimethylphosphine followed by macrolactamization under peptide-coupling conditions gives ixabepilone. The semi-synthesis was optimized as a one-pot procedure yielding 23% ixabepilone in one day, purification included.[73]

Subsequent to the development of a semi-synthetic route by the BMS group Danishefsky and co-workers have also reported the chemical synthesis of ixabepilone (**2**).[86] As illustrated in Scheme 16.2, this has involved Suzuki cross-coupling of olefin **6** with azide **5** to afford **7**. The latter was then reduced under Staudinger conditions and the resulting amine was Boc-protected to obtain **8**. After liberation of the β-keto ester by transfer hydrolysis, a modified Noyori catalyst was used for the asymmetric hydrogenation to alcohol **9**. After simultaneous removal of the amino and carboxy protecting groups, HATU-mediated macrolactamization afforded **10**. Subsequent deprotection and epoxidation furnished ixabepilone (**2**) as a single diastereomer.[86]

16.3.1.2 *In vitro Activity*

Ixabepilone induces tubulin polymerization with similar efficiency as Epo B (Table 16.1), but its antiproliferative activity is lower than that of Epo B by a

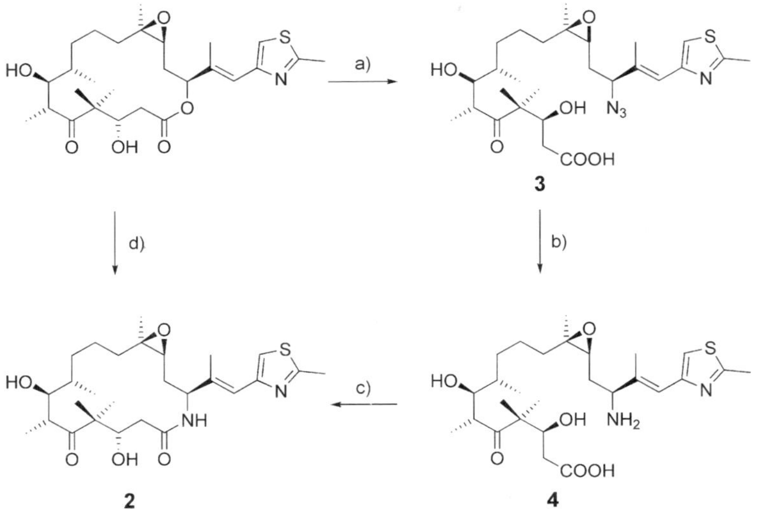

Scheme 16.1   (a) Pd(PPh$_3$)$_4$ 10 mol%, NaN$_3$, THF/H$_2$O, 45 °C, 70%; (b) PMe$_3$, THF/
H$_2$O, 71%; (c) DPPA, NaHCO$_3$, DMF, 4 °C, 24 h, 43% or EDCI,
HOBt, MeCN/DMF, 65%; (d) 'one pot' Pd(PPh$_3$)$_4$ 10 mol%, NaN$_3$,
degassed THF/H$_2$O, 45 °C, 20 min, then PMe$_3$, 1–2 h, then MeCN/
DMF (20:1), EDCI/HOBt, 25 °C, 4–12 h, 23%. DPPA: diphenyl
phosphoryl azide; EDCI: 3-((ethylimino)methylenamino)-*N*,*N*,*N*-tri-
methylpropan-1-aminium chloride; HOBt: 1-hydroxy benzotriazol.

factor of ca. 10.[87,88] Ixabepilone is clearly more susceptible to Pgp-mediated
drug efflux than Epo B,[38,42,86] but it retains activity against cell lines which are
resistant to taxol and other chemotherapeutic agents due to overexpression of
βIII-tubulin.[89] Reduced uptake of ixabepilone has been observed, e.g. for
taxol-resistant Pgp-overexpressing HCT116/VM46 cells, relative to the
HCT116 parental cells.[90] At the same time the compound was found to
accumulate in HCT116/VM46 cells significantly more effectively than taxol
and to be ca. 10-fold more potent against this cell line than taxol in a
clonogenic cell survival assay after 16 h of drug exposure.[90]

   As for taxol and other epothilones, ixabepilone blocks the cell cycle at the
G2/M transition, with the concentrations needed to arrest cells in mitosis
corresponding to those required for cell kill.[87] Cell cycle arrest by ixabepilone
has also been shown to lead to p53- and Bax-mediated cytochrome c release
and caspase-dependent apoptosis in MDA-MB-468 human breast carcinoma
cells. Apoptosis induction appears to be related to a conformational change of
the Bax protein and its subsequent translocation to the mitochondrial
membrane.[91] In line with a Bax-dependent mechanism of cell death,
ixabepilone-mediated apoptosis, but not G2/M arrest, is significantly delayed
in the presence of increased levels of the anti-apoptotic protein Bcl-2. This is
paralleled by a delay in Bax conformational change. Subsequent studies in
human ovarian cancer cells confirmed the caspase-dependent mechanism of

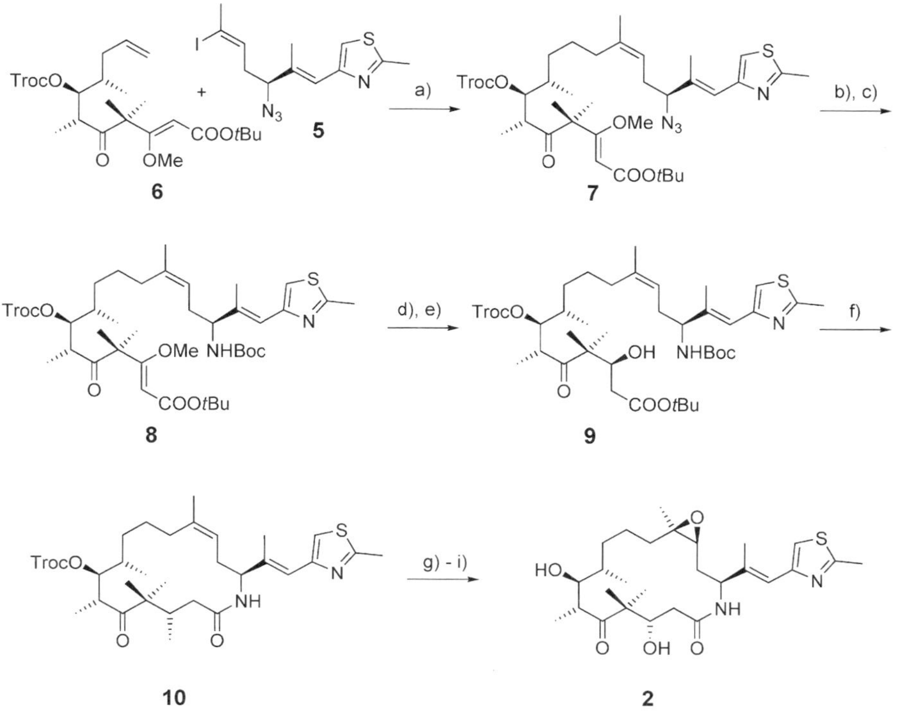

Scheme 16.2 (a) 1. **6**, 9-BBN, THF; 2. Pd(dppf)Cl$_2$, AsPh$_3$, Cs$_2$CO$_3$, DMF 78%; (b) PPh$_3$, H$_2$O/THF, 98%; (c) Boc$_2$O, MeCN, 70%; (d) *p*-TsOH, acetone, 82%; (e) 2% Et$_2$NH$_2$[{((*R*)-BINAP)RuCl}$_2$Cl$_3$], HCl/MeOH, H$_2$, 1250 psi, 78%; (f) TFA, DCM, used as such; (g) HATU, HOAt, DIPEA, DCM, 90%; (h) Zn dust, sonication, HOAc, 88%; (i) DMDO, DCM, –50 °C, 70%. Troc: 2,2,2-trichloroethyl carbonate; 9-BBN: 9-borabi-cyclononane; dppf: 1,1′-bis(diphenylphosphine)ferrocene; BINAP: 2,2′-bis(diphenylphosphino)-1,1′-binaphtyl; TFA: trifluoroacetic acid; HATU: 2-(7-aza-1H-benzotriazol-1-yl)-1,1,3,3-tetramethyluronium hexafluorophosphate; HOAt: 1-hydroxy-7-azabenzotriazole; DIPEA: diisopropylethylamine; DMDO: 3,3-dimethyldioxirane.

apoptosis and also showed that ixabepilone augments apoptosis induction by Apo-2L/TRAIL.[92]

16.3.1.3 In vivo Pharmacology

Ixabepilone has demonstrated *in vivo* efficacy in a wide range of human tumor xenograft models in mice.[87,90] This includes several taxol-resistant models (overexpressing Pgp or βIII-tubulin)[87] and also pediatric solid tumor models.[93] Ixabepilone was mostly administered i.v. at its MTD in ethanol/water (1:9, *v/v*) or in cremophor/ethanol/water (1:1:8, *v/v*). Oral administration of ixabepilone

Table 16.1   Ixabepilone: preclinical profile[87]

	Efficacy $(LCK)^a$	IC_{50} $[nM]^b$	$EC_{0.01}$ $[\mu M]^c$	Metabolic stability $[nmol/$ $min\,mg]^d$	Plasma protein binding $[\%]^e$	MDR $(IC_{50}$ $R/S)^f$
Epo B	0.4	0.41	1.80	1.02	92.0	1.48
Ixabepilone	2.1	2.60	2.00	0.01	79.4	7.77

^{a} LCK: log cell kill; ^{b} Inhibition of proliferation of HCT-116 cells; ^{c} EC$_{0.01}$: interpolated concentration of test compound capable of inducing an initial tubulin turbidity slope of 0.01 – measure for the ability of the compound to induce tubulin polymerization; ^{d} rate of hydrolysis after incubation with mouse S9 liver fractions; ^{e} fraction of compound unbound, mouse plasma; ^{f} multidrug resistance susceptibility: ratio of IC$_{50}$ values for inhibition of proliferation of the multidrug-resistant HCT116/VM46 colon carcinoma cell line and the drug-sensitive HCT116 line.

in phosphate buffer pH 8 (60–80 mg/kg) and i.v. administration (10 mg/kg) were compared in Pat-7 and HCT-116 xenograft models and equivalent antitumor efficacy was achieved in both models for both routes of administration.[87] A2780 xenografts still showed 40% sensitivity to ixabepilone after 3 years of exposure (cycles of implantation, treatment and re-implantation in another group of mice after regrowth), as compared to 25% sensitivity for taxol after 1 year.[88] Ixabepilone has shown stronger synergy with bevacizumab and sunitinib than taxol, especially in GEO and HCT116/VM46 xenografts (MDR1-expressing), probably because of its higher anti-angiogenic activity or lower susceptibility to drug efflux proteins.[94]

In nude mice ixabepilone is rapidly cleared (4.3–5.1 L/h/kg) and shows extensive tissue distribution with a dose-dependent mean half-life of 13 and 16 h at 6 and 10 mg/kg respectively.[90] The compound does not induce or inhibit any CYPs; it is, however, converted to several inactive metabolites by CYP3A4/5.[88]

16.3.1.4   Clinical Studies

Early clinical studies with ixabepilone have been reviewed by Lin *et al.*[95] and also Pivot *et al.*[96] A number of objective responses to single-agent treatment were observed in Phase I studies, with neutropenia as the major dose-limiting toxicity (DLT). Significantly, the compound was also shown to induce microtubule bundling in peripheral blood monocytes (PBMCs) of treated individuals and a good correlation was observed between the magnitude of this effect and plasma AUCs.[97] The most comprehensive clinical profiling of ixabepilone has occurred in metastatic breast cancer (MBC). In a multinational randomized Phase III trial comprising 752 patients, who were anthracycline pretreated and met predefined resistance criteria to taxanes, a combination of ixabepilone (40 mg/m^2 i.v. over 3 h every 3 weeks) with capecitabine (1000 mg/m^2 p.o. bid q14d) was found to be superior to capecitabine alone (1250 mg/m^2 p.o. bid q14d), with 'significant benefit [being] consistently maintained across predefined subgroups, including HER2-/ER-/PR- and HER2+'. These studies led to drug approval by the FDA in October 2007. Ixabepilone for injection is indicated in combination with capecitabine for patients with metastatic or locally advanced anthracycline- and taxane-resistant breast cancer, or with taxane resistance and anthracycline therapy contraindication, and as monotherapy for the treatment of metastatic or locally advanced anthracycline-, taxane-, and capecitabine-resistant or refractory breast cancer.[98] However, registration was not granted in the EU, because of concerns about neuropathy and the application was withdrawn in 2009.

Ixabepilone has also been evaluated in several other solid tumor types, including prostate cancer (33–48% response rate),[99,100] pancreatic cancer (21%),[101] NSCLC (11.6–14.3%),[102] urothelial carcinoma (11.9%),[103] gastric cancer (9%),[104] sarcoma (6%);[105] combination studies have been performed with gemcitabine, carboplatin, pegylated liposomal doxorubicin, mitoxantrone, and prednisone.[106,107] No significant antitumor activity was observed in melanoma and colorectal cancer patients, or against refractory childhood tumors.[107,108] Oral treatment studies with ixabepilone have been terminated.[109]

Numerous Phase I, II, and III studies with ixabepilone are active as of January 2011, including trials in breast cancer (alone or in combination with sorafenib, cetuximab, trastuzumab, cyclophosphamide, dasatinib, vorinostat (male breast cancer), carboplatin, or hydroxychloroquine), NSCLC (with carboplatin, cisplatin, cetuximab, or bevacizumab), solid tumors (with cisplatin, sunitinib, pazopanib, dasatinib, or lapatinib), kidney cancer (alone and with bevacizumab), endometrial cancer (alone, with bevacizumab and carboplatin), prostate cancer (alone or with radiation), uterine tumor (alone), gastric tumors (alone), cervical carcinoma (alone), and in several female tumors (with liposomal doxorubicin).[107]

16.3.2 Side Chain-Modified Analogs

Side chain modifications have occupied a prominent place in SAR studies on epothilones and numerous epothilone analogs with alternative side chain structures have been investigated over the course the last decade; of these, three compounds have been advanced to clinical trials in humans (sagopilone, BMS-310705, and ABJ879). The existing SAR on side chain-modified epothilone analogs has been reviewed extensively[20,30,31,34,37–41,43] and shall not be rediscussed in the chapter, except for a few specific examples.

16.3.2.1 Sagopilone

16.3.2.1.1 Chemical Synthesis

Sagopilone (ZK-EPO; Bayer Schering Pharma AG, Berlin, Germany, Figure 16.5) is a fully synthetic epothilone analog with high activity and efficacy, which is not susceptible to drug efflux.[42,110]

ZK-Epo was selected from about 350 active epothilone analogs produced by total synthesis in the laboratories of Bayer Schering Pharma AG, Berlin, Germany, as part of an extensive lead optimization program. Key steps in the

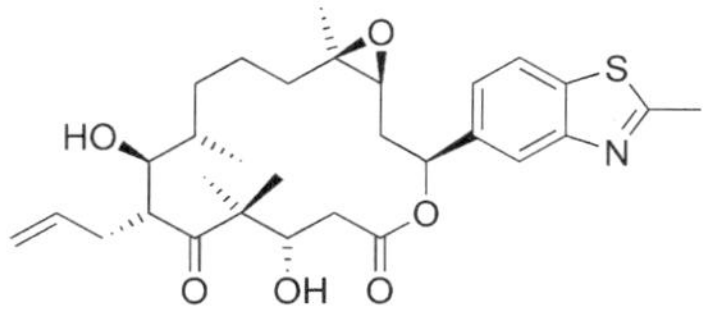

Figure 16.5   Molecular structure of sagopilone (ZK-Epo).

synthesis of the compound (Scheme 16.3) include the formation of the C12–C13 double bond via Wittig reaction between ketone **11** and phosphonium salt **12** (which is nonselective), the stereoselective aldol reaction of ketone **14** (for which an efficient synthesis has been developed from pantolactone)[111] with aldehyde **13**, Yamaguchi macrolactonization of the *seco* acid derived from **16** through TBS removal from O15, and epoxidation of Epo D analog **17** with DMDO at –78 °C. The mixture of 12,13-*cis*/*trans* isomers obtained in the Wittig step can be separated by chromatography after cleavage of the tetrahydropyranyl ether, and the undesired *trans* analog can be isomerized by irradiation at λ > 280 nm to a 3:2 mixture of *cis* and *trans* isomers. The synthesis comprises a total of 39 steps with a longest linear sequence of 22 steps. Further optimization based on this synthesis concept resulted in total yield of 5.5% over the 22-step linear sequence, requiring only four chromatographic steps.[42]

16.3.2.1.2 *In vitro* Activity

ZK-Epo consistently exhibits high antiproliferative activity with low nM or sub-nM IC_{50} values across a panel of more than 100 cancer cell lines.[112] Against drug-sensitive cell lines the compound is equally potent as, or even somewhat more potent than Epo B. While the activity of Epo B towards some multidrug-resistant cell lines is reduced about ten-fold, and the activity of ixabepilone and paclitaxel is almost lost, ZK-Epo retains full activity.[111]

This finding indicates that ZK-Epo is not susceptible to any (investigated) type of drug resistance, whereas Epo B may exhibit somewhat diminished activity against certain drug-resistant cell lines. In cellular uptake experiments, the intracellular levels of ZK-Epo in multidrug-resistant cells were found not to be affected by the presence or absence of verapamil, a Pgp inhibitor; in contrast, but unsurprisingly, the intracellular concentration of taxol was reduced below therapeutically relevant concentrations in the absence of the Pgp inhibitor.[112] Furthermore, ZK-Epo reaches its maximum antiproliferative activity after only 2 hours incubation; similar observations have also been made for Epo B.[20]

16.3.2.1.3 *In vivo* Activity

The *in vitro* activity of sagopilone is mirrored in its *in vivo* antitumor activity across different human xenograft tumor models, including models of breast, ovarian, lung, prostate, and pancreatic cancer. Antitumor activity extends to models that are resistant to taxol and is achievable in the absence of profound

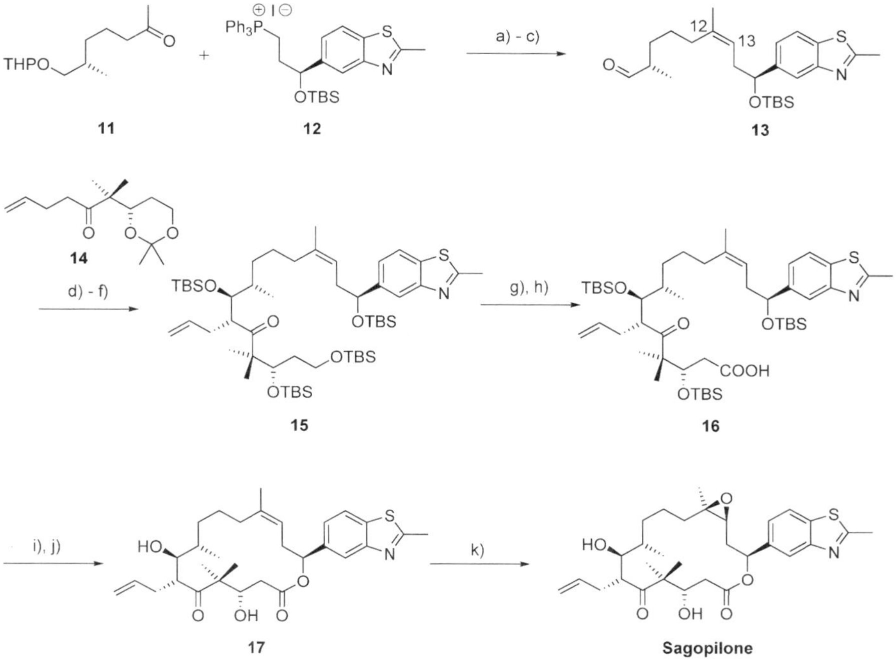

Scheme 16.3 (a) NaHMDS, THF, 0 °C → RT, 83% (1:1 mixture of *Z* and *E* isomers); (b) *p*-TsOH (cat.), EtOH, RT, 43% (86% for mixture of *Z* and *E* isomers); (c) (COCl)$_2$, DMSO, DCM, –78 °C; then Et$_3$N, –78 °C → 0 °C, crude; (d) **14**, LDA, ZnCl$_2$, THF, –70 °C, then **13**, THF, –70 °C, 64%; (e) *p*-TsOH (cat.), EtOH, RT, 97%; (f) TBSOTf, 2,6-lutidine, DCM, –70 °C → 0 °C, 96%; (g) CSA, DCM, MeOH, RT, 80%; (h) 1. (COCl)$_2$, DMSO, DCM, –78 °C; then Et$_3$N, –78 → 0 °C, crude; 2. NaOCl$_2$, NaH$_2$PO$_4$, 2-methyl-2-butene, THF/H$_2$O/*t*BuOH, 0 °C → 15 °C, 85%; (i) 1. TBAF, THF, RT, crude; 2. 2,4,6-Cl$_3$C$_6$H$_2$C(O)Cl, Et$_3$N, THF, 0 °C, 60%; (j) HF·pyr, hexafluorosilicic acid, THF, RT, 87%; (k) DMDO, acetone/CH$_2$Cl$_2$, –78 °C, 71% + 10% β-epoxide.

body weight loss.[110,113] Sagopilone accumulates to high concentrations in brain tissue of both tumor-bearing rats and mice with a half-life of several days; it significantly inhibited tumor growth in a human glioblastoma model in mice, whereas taxol only showed limited activity.[114] This indicates that sagopilone is able to cross the blood-brain barrier to effectively inhibit the growth of intracerebral tumors and could therefore be active against metastases in the brain, a site of metastatic spread common in patients with breast cancer.[114]

16.3.2.1.4 Clinical Studies

Based on its promising preclinical profile, sagopilone entered clinical studies in 2003. It was reported to be well tolerated in a Phase I trial, with peripheral neuropathy being the principal DLT.[115–117] The dose for efficacy evaluation in Phase II studies was chosen as 16 mg/m^2, administered as a 3 h infusion once every 3 weeks.[112] Sagopilone has been evaluated in a number of Phase II trials for several cancer indications, such as platinum-resistant[118] and platinum-sensitive ovarian cancer,[119] and in combination with prednisone in androgen-independent prostate cancer.[120] In these trials the success criteria are reported to have been met.[118–120] Furthermore, clinical activity in patients with non-small cell lung cancer (NSCLC)[121] and patients with melanoma[122] has been reported. Sagopilone appeared to have only limited activity in patients with metastatic breast cancer, who had been previously treated with taxanes and anthracyclines.[123] In spite of the promising reports of clinical activity, no recent development has been reported for the compound and it is not publicly known whether its development is still ongoing.

16.3.2.2 2-Substituted Thiazole- and Isoxazole-based Analogs

In a recent study, our own group has revisited the effects of the size of the 2-substituent on the thiazole ring in the epothilone side chain on tubulin-polymerizing and antiproliferative activity.[124] While previous studies on C21-modified analogs of epothilones had pointed to a clear size limitation for substituents at this position, these conclusions were derived from data scattered throughout the literature[38,41] and no systematic and directed study on this question had been performed. To provide a sound experimental basis for these studies, Epo A analogs **18–20** which incorporate a C20-propyl, -butyl, and -hydroxypropyl substituent, respectively, were investigated (Figure 16.6).

While the tubulin-polymerizing activity of analog **18** is similar to that of Epo A, the EC$_{50}$ values for the induction of tubulin polymerization of both **19** and **20** are increased by a factor of ~ 2. However, much larger differences between **18** and **19** or **20** are apparent at the cellular level, with the antiproliferative activity of **18** being similar to that of Epo A, while both **19** and **20** are 40 to 100-fold less active than **18**.[124] Thus, the cellular activity of differently C20-substituted epothilone analogs appears to be governed by strict steric requirements, with a very steep decline in activity occurring at the level of a (linear) four-(heavy) atom substituent.

In an attempt to improve the biopharmaceutical properties (such as solubility and metabolic stability) of their previously developed 9,10-dehydro

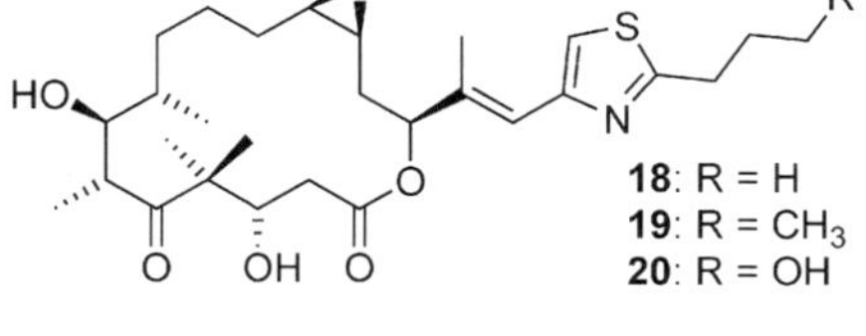

18: R = H
19: R = CH$_3$
20: R = OH

Figure 16.6   Epo A analogs with modified substituents at the 2-position of the thiazole ring.

derivatives of Epo D (dehydelone and fludelone, Figure 16.7) Danishefsky and co-workers have recently designed the isoxazole-based epothilone analogs iso-dehydelone (**21**) and iso-fludelone (**22**) (Figure 16.7).[125]

Both isoxazole derivatives **21** and **22** are more potent *in vitro* than their corresponding (thiazole-containing) parent compounds and both retain full activity against the multidrug resistant cell lines CCRF-CEM/VBL (IC_{50} = 0.94 nM (**21**); 2.08 nM (**22**)) and CCRF-CEM/taxol (IC_{50} = 0.39 nM (**21**); 0.89 nM (**22**)). Furthermore, they demonstrated greatly enhanced levels of metabolic stability in comparison to other epothilone analogs.[125] *In vivo*, **21** and **22** showed profound antitumor activity across a panel of different human xenografts models, including mammary, ovarian, and lung carcinoma models (some of which were taxol-resistant); both compounds were able to eradicate tumors without any sustained body weight loss.[125] Although iso-fludelone (**22**) displays a more favorable therapeutic index than **21**, it is not obvious which of the two compounds is really superior. Both would appear to be highly promising candidates for further development.

16.3.2.3 Hypermodified Epothilones

Based on earlier findings on the enhanced *in vitro* antiproliferative activity of Epo D and Epo B analogs with bicyclic heteroaromatic side chains (**23–28**)

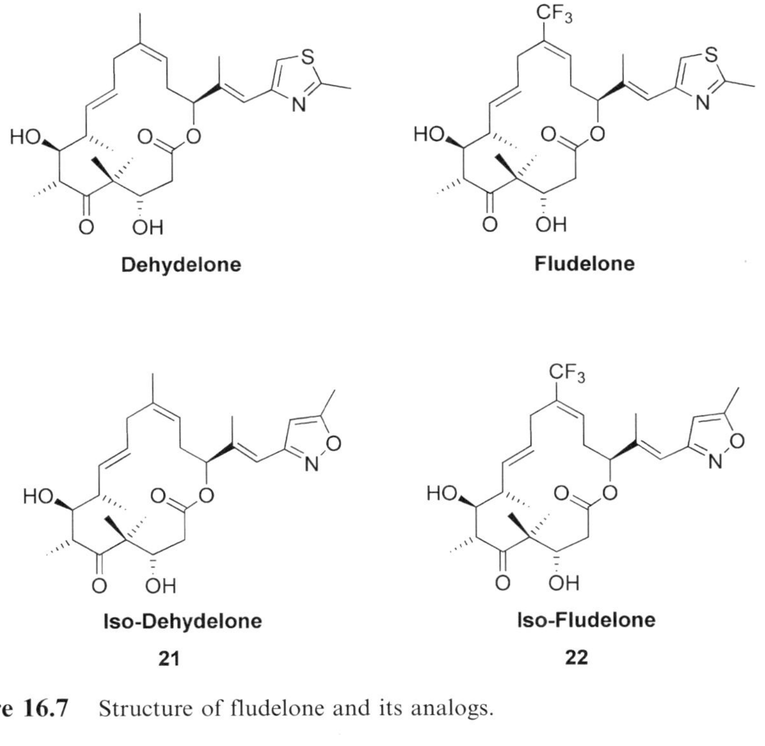

Figure 16.7 Structure of fludelone and its analogs.

(which were developed independent of the parallel work at Bayer Schering on sagopilone) (Figure 16.8),[83,126] our own group has investigated what we have termed 'hypermodified' epothilone analogs, i.e. epothilone-derived structures that are characterized by extensive modification of the side chain as well as the macrocyclic skeleton.[127]

In the context of this work we have prepared the first Epo A-derived epothilone analog (i.e. a compound without a methyl substituent at C13) with Epo B-like activity, i.e. 29 (Figure 16.9).[128] In attempt at further structural simplification, the removal of the 3-hydroxyl group from 29 led to 30, which exhibits comparable activity against drug-sensitive human cancer cell lines as Epo A or taxol, in spite of substantial structural differences from the original epothilone scaffold.[129] However, both analogs, 29 as well as 30, show reduced activity against Pgp-overexpressing cells; to overcome this drawback, the epoxide moiety in 29 was replaced by an isosteric cyclopropane ring, in order to reduce the polarity of the compound (31).[130]

The *in vitro* antiproliferative activity of 31 is comparable with that of 29 and the compound retains full activity against the multidrug-resistant cell line KB-8511 (IC_{50} = 0.17 nM versus 0.13 nM for the drug-sensitive parental KB-31 line).[130] This finding illustrates how polarity adjustments can efficiently eliminate the susceptibility of modified epothilones to Pgp-mediated drug efflux. Based on these results and also exploiting prior findings by the Danishefsky group on the activity-enhancing effects of an *E*-double bond between C9 and C10,[80c,131,132] we then went on to investigate hypermodified epothilone analogs 32 and 33 (Figure 16.10).[133]

Synthetically, the Eastern half of analogs 32 and 33 is derived from alcohol 39 (Scheme 16.4), which has been obtained from 2,3-dimethyl-benzimidazole 6-carbaldehyde via enantioselective allylation, homologation, and subsequent stereoselective cyclopropanation of an allylic alcohol.[133] Commercially available (*S*)-Roche ester served as the precursor for aldehyde 34; aldol reaction of the latter with γ-keto ester 35 installed the stereocenters at C6 and C7 (1.4:1 ratio of *syn* products only, in favor of 36). The aldol product 36 was further transformed into acid 38, which was then connected to alcohol 39 by esterification. The resulting ester was then elaborated into target structure 33 via RCM; 32 was accessible through hydrogenation, a reaction that required significant optimization.[133]

The tubulin-polymerizing activity of both 32 and 33 is comparable with that of Epo B. The antiproliferative activity against drug-sensitive KB-31 cells was

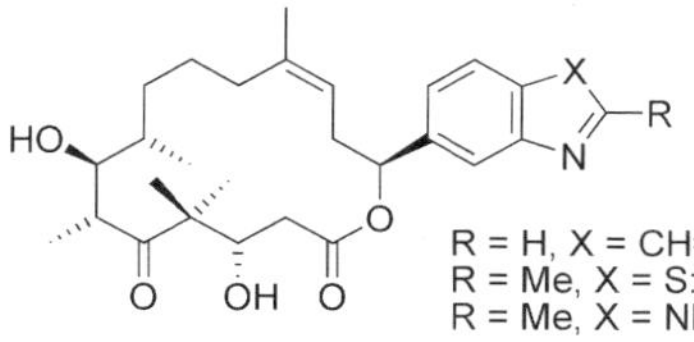

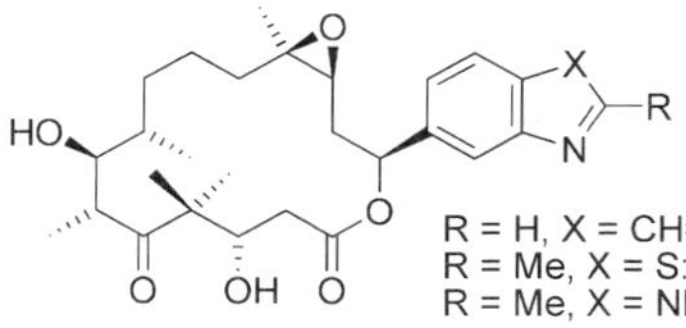

Figure 16.8 Epo D and B analogs with bicyclic heteroaromatic side chains.

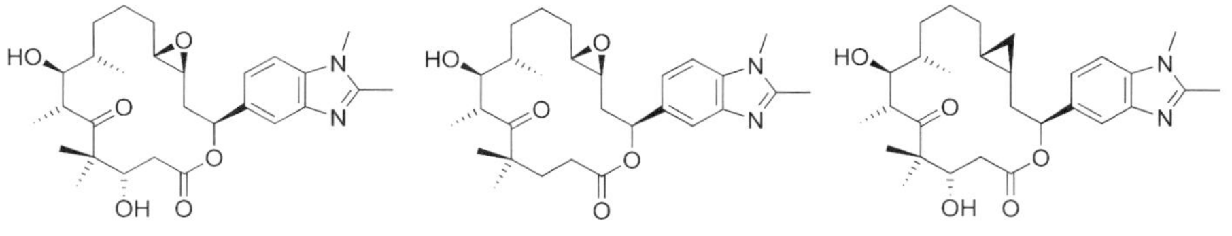

Figure 16.9 Highly modified epothilone analogs.

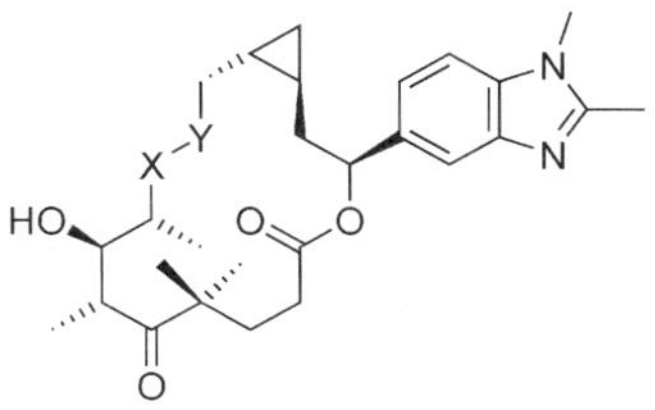

32: X-Y = CH$_2$-CH$_2$
33: X-Y = (E)-CH=CH

Figure 16.10 Hypermodified epothilone analogs.

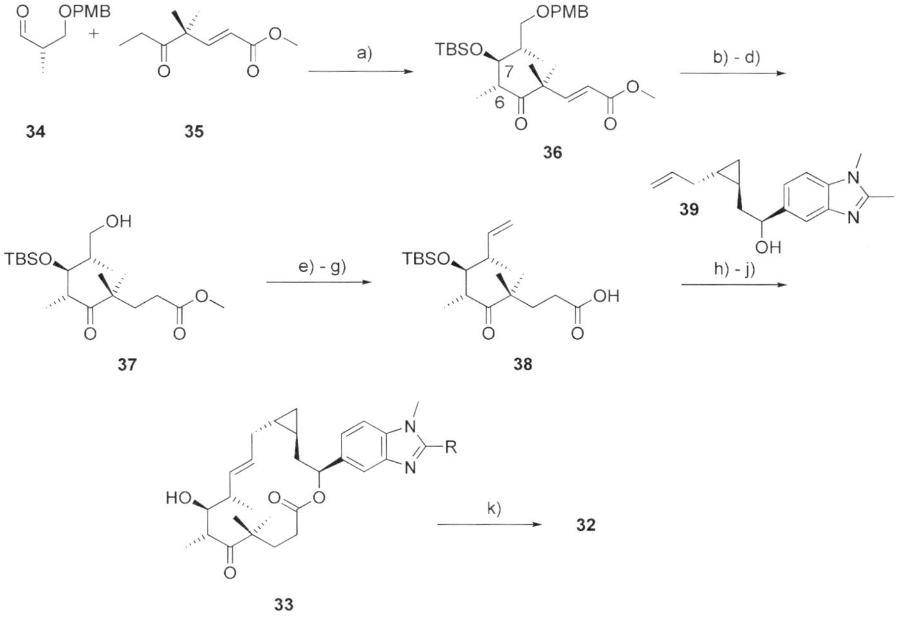

Scheme 16.4 (a) **35**, LDA, THF, −78 °C, then **34**, THF, −78 °C, 38%; (b) TBSOTf, DCM, −78 °C → RT, 94%; (c) H$_2$, Pd/C, 95%; (d) DDQ, DCM/H$_2$O 9/1, 74%; (e) NMO, TPAP (6 mol%), DCM; (f) MePPh$_3$Br, LiHMDS, THF, 0 °C → RT, 38% (2 steps); (g) LiOH, *i*PrOH/H$_2$O 4/1, 60 °C, quant; (h) **39**, EDC, DMAP, DCM, 87%; (i) Grubbs II (10 mol%), toluene, 110 °C, 94%; (j) CH$_2$Cl$_2$/TFA 4/1, 80%; (k) [Ir(cod)(PCy$_3$)(py)]PF$_6$, DCM, 0 °C → RT, 62% (after HPLC).

found to be in the sub-nM range for both compounds (**33**: IC_{50} (KB-31) = 0.43 ± 0.08 nM; **32**: IC_{50} (KB-31) = 0.25 ± 0.05 nM) and no loss in activity was detectable against drug-resistant KB-8511 cells (as it had been observed for **31**);[134] in fact, activity even distinctly increased against the multidrug-resistant cells (**33**: IC_{50} (KB-8511) = 0.082 ± 0.03 nM; **32**: IC_{50} (KB-31) = 0.024 ± 0.004 nM), although the reasons for this effect are currently unknown. Overall, the biological activity of compounds **32** and **33** make them attractive candidates for broader *in vitro* and eventual *in vivo* profiling.

16.3.3 Modifications of the Macrocycle

Besides the extensive work on side chain modifications and the replacement of the lactone moiety by a lactam group (see above), significant efforts have been spent on the elaboration of SAR features along the carbon chain of the macrocyclic lacton ring (reviewed in refs. 13, 30, 31, 34, 37–41). Since 2007 this work has focused mainly on the consolidation of the SAR of C12-C13 modified epothilones (including modification on C26 of Epo B) and the investigation of conformationally constrained, bridged analogs.

Other recent modifications of the macrocycle include the introduction of a *S*-methoxy functionality at the C14 position, a modification that was conceived as the result of a (hypothetical) alternative biogenic assembly process of polyketide extender units. The corresponding analog **40** (Figure 16.11) retained most of the activity of the parent compound Epo D (IC_{50} = 3.7 nM (MCF-7), 4.9 nM (H460)).[135] Hutt *et al.* have reported the replacement of the C8-methyl group in Epo C by a benzyl ether moiety, leading to analog **41** (Figure 16.11).[136] **41** was found to be virtually inactive, a finding that could be easily rationalized by molecular modeling. The C25-benzyloxy group is simply too big to be accommodated in the binding site.

A completely different idea was followed by the dimerization of Epo A with linkers of varying length as reported by Passarella and co-workers (Figure 16.12; **42**).[137] The design of these epothilone dimers was based on a finding that had been made by Nicolaou *et al.*[138] several years prior, namely that the dimeric epothilone macrolactone **43** (Figure 16.12) exhibited remarkable tubulin-polymerizing and antiproliferative activity.[138] However, none of

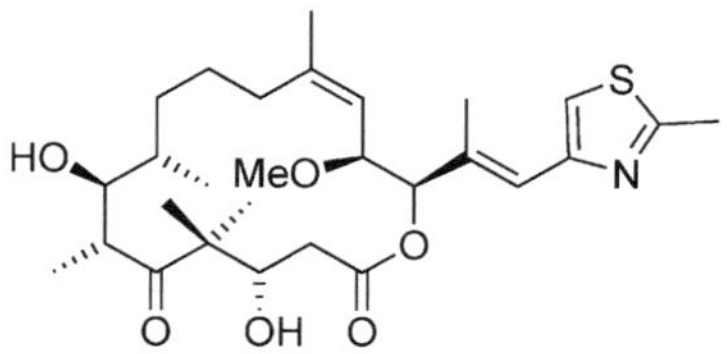

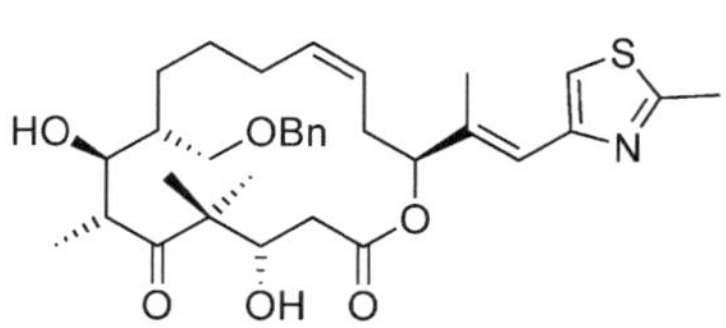

Figure 16.11 Structures of *S*-14-methoxy-epothilone D (**40**) and C25 benzyloxy-epothilone C (**41**).

the newly reported dimers showed any significant tubulin-polymerizing or antiproliferative activity. Some of the compounds investigated, however, seemed to have an inhibitory effect on endothelial cell differentiation and also migration; no such effects were observed for Epo A.[137]

One of the regions of the epothilone structure that has been investigated most intensively is the C12-C13 epoxide moiety. Based on the investigation of numerous derivatives, strong evidence exists that the oxirane ring system merely serves to stabilize the bioactive conformation of the macrocycle rather than to act as a hydrogen bond acceptor or a reactive electrophile. In light of this fact and following up on our own previous work on 12-aza-epothilones[134,139] we investigated 12-aza-epothilone ('azathilone') **44**, which combines an *E* double bond between C9 and C10 with a *t*-butyl-carbamoyl substituent on N12.[134,140]

Very surprisingly, and in strong contrast to the effects of C9-C10 desaturation in natural epothilone congeners,[80] **44** was found to be 30-fold less active than the C9-C10 saturated parent compound **45**.[134] The molecular origin for this discrepancy is not understood, but it might be assumed to be caused by differences in the bioactive conformation between the polyketide-based natural product and the azamacrolide-based azathilones.

Using **44** as a precursor we have recently synthesized the green-fluorescent NBD-azathilone **46**.[141] This compound proved to be a reasonably potent inhibitor of cell proliferation and thus represents a valuable tool for the investigation of the cellular effects associated with aza-epothilones and perhaps microtubule-stabilizing agents in general; the compound can also serve as a reference ligand in fluorescence-based binding studies.[141] Based on the cellular results, the fluorophor in **46** actually forms part of the pharmacophore, unlike the case of Flutax-1 and -2[142] which are standard fluorescent tools in biochemical and biophysical experiments with many microtubule-stabilizing agents (Figure 16.13).

In an extension of previous work on acetals derived from C12-C13-dihydroxy-C12-C13-dihydro-Epo C, we have recently also investigated the replacement of the C12-C13 oxirane ring in Epo A by *trans*-fused 2-aryl-1,3-oxazolines (Table 16.2, **47**). A clear SAR could be deduced in the 2-phenyl oxazoline series, such that both the tubulin-polymerizing as well as the

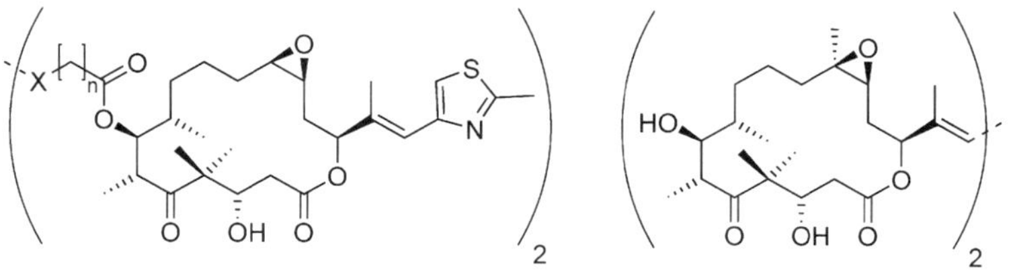

42   X = S, CH$_2$
         n = 1-3

43

Figure 16.12   Structure of dimeric epothilone analogs.

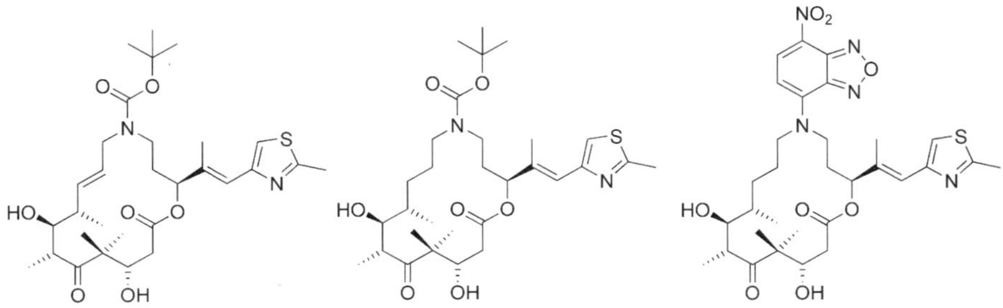

Figure 16.13   Structures of 12-aza-epothilones (azathilones).

antiproliferative activity of the compounds decreased with increasing steric demand of the *para*-substituent on the phenyl ring.[143] Thus, compound **47e** was found to be almost inactive; on the other hand the most potent derivative (**47b**) proved to be equipotent to its parent compound Epo A.

In earlier work, the BMS group had shown that the replacement of the C12-C13 oxirane ring system in Epo A by *N*-substituted aziridine moieties can produce analogs that are more potent than the parent natural product.[76] Based on these previous findings, the BMS group has recently reported the design and synthesis of a folate-based prodrug of *N*-hydroxyethyl-12,13-aziridyl-Epo

Table 16.2   Tubulin-polymerizing and antiproliferative activity of C12-C13-(2-phenyl)-oxazoline derivatives of Epo A (**47a–47e**)

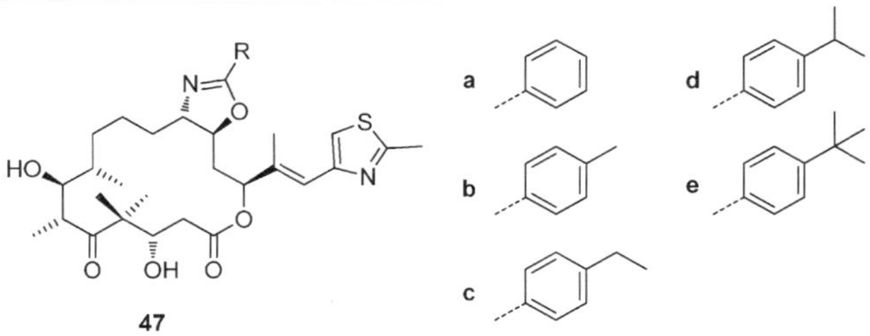

Compound	%-Tub-Pol. (5 µM / 2 µM)[a]	IC$_{50}$ [nM][b] A549	IC$_{50}$ [nM][b] MCF-7
Epo A	89 / 83	2.2 ± 0.3	2.6 ± 0.3
47a	80 / 65	16.4 ± 2.7	19.2 ± 1.4
47b	88 / 71	2.9 ± 0.3	1.9 ± 0.1
47c	36 / 32	378.5 ± 14.8	297 ± 5.2
47d	15 / 14	4873 ± 121	5144 ± 207
47e	7 / 0	> 10 000	> 10 000

[a] Induction of polymerization of porcine brain-derived microtubule protein by 5 µM and 2 µM of test compound relative to the effect of 25 µM Epo B, which gave maximal polymerization (85% of protein input); [b] concentration required to inhibit the growth of the human carcinoma cells lines A549 and MCF-7 by 50% (72 h exposure).

A (BMS-753493) (**48**) (Figure 16.4). The folate moiety in this construct serves to direct the epothilone derivative to the tumor, where its release is triggered by the intracellular cleavage of the disulfide bond. Folate conjugate **48** has entered Phase I clinical trials, but the development of the compound seems to have been discontinued.

Recent modifications of the epothilone structure have also included attempts to stabilize its bioactive conformation by designing bridged analogs (Figure 16.15).

Analog **49**, in which the mobility of the aromatic side chain was supposed to be restricted, lost all antiproliferative activity against cancer cell lines derived from solid tumors, such as breast (MCF-7) and ovarian (SK-OV-3) carcinomas.[144] Interestingly, however, the compound showed strong growth inhibitory activity against leukemia cell lines such as CCRF-CEM (IC_{50} = 2.7 nM) and SR (IC_{50} = 2.9 nM).

Attempts by Kingston, Snyder, and co-workers to constrain the electron crystallography-based, proposed bioactive conformation of Epo A[52] failed to provide any active analogs. Neither the C6-C8 bridged analog **50** nor various C4-C26-bridged variants (e.g. **51**) showed any significant antiproliferative activity.[145,146]

Most recently, however, the same group reported the highly active C3-C22-bridged internal γ-butyrolactone **52** (IC_{50} = 4.6 nM (A2780 cells)), which had initially been obtained as a side product in the attempted synthesis of C4-C26-bridged analog **51**.[147]

Based on molecular modeling studies, Hutt *et al.* designed the C22-(3-azidobenzoyl) derivative of Epo C (**53**) as a potential probe for photoaffinity labeling (Figure 16.16).[148]

Unfortunately, the compound showed only low cytotoxicity (ED_{50} = 1.2 μM (MCF-7)) and failed to deliver any conclusive labeling results. Based on findings by Nicolaou *et al.*[149] that substitution of C26 is well tolerated, Georg

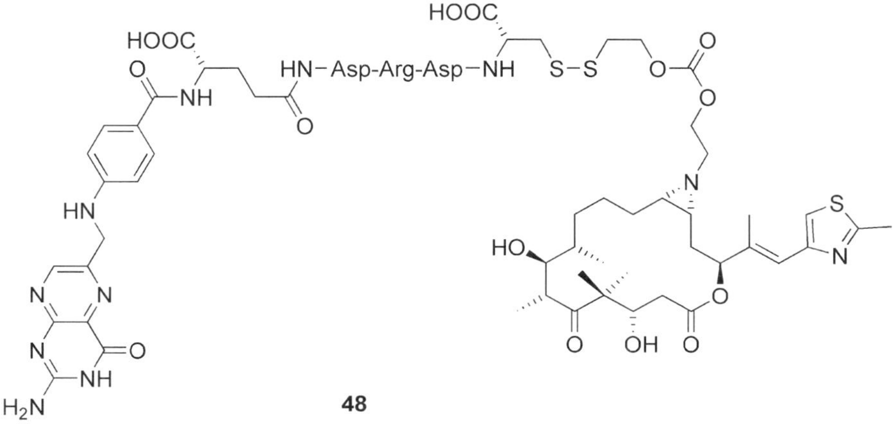

Figure 16.14 Structure of a folic acid epothilone conjugate.

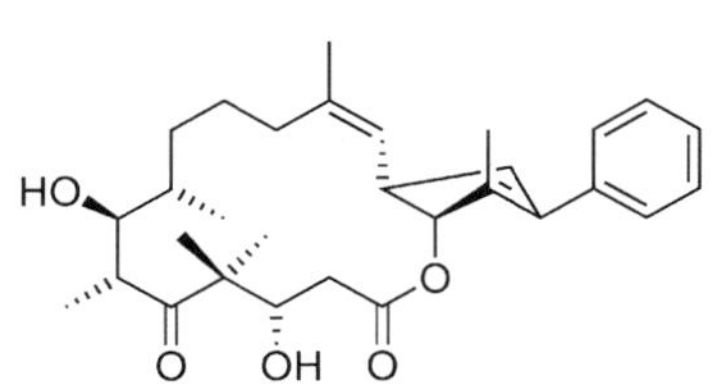

49 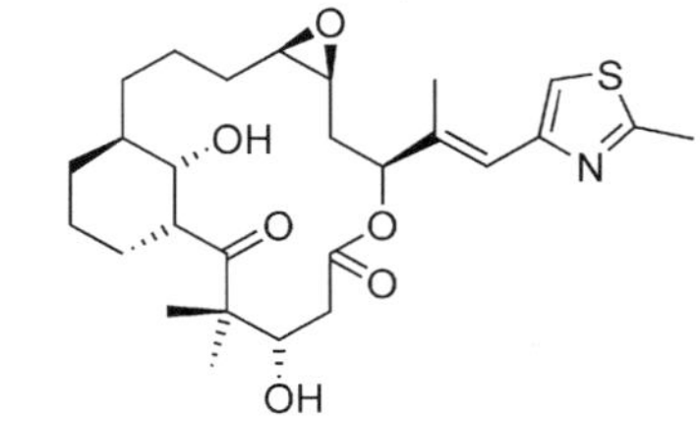   **50**

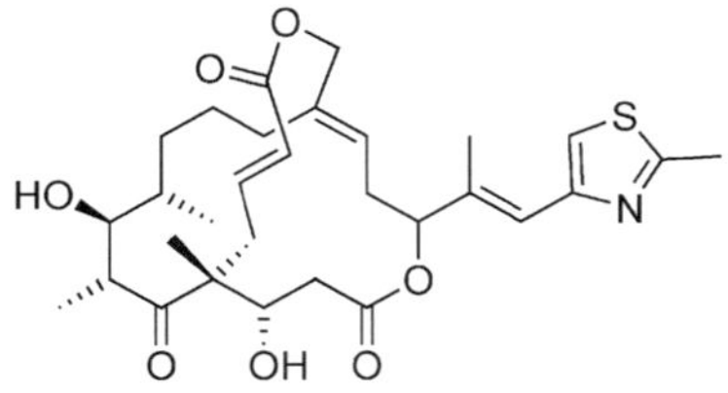

51 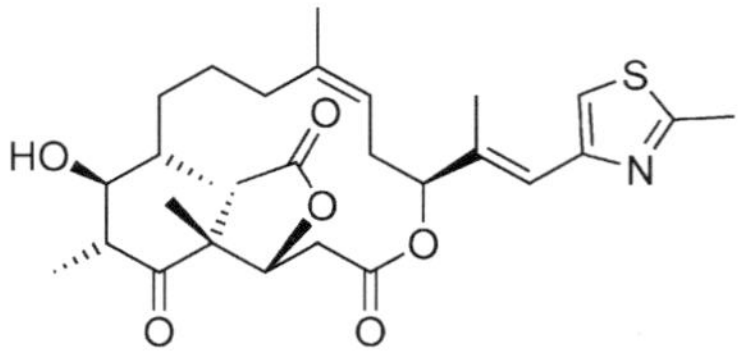   **52**

Figure 16.15   Structures of conformationally constrained epothilone analogs.

and co-workers also designed a series of C26-derivatives as potential photoaffinity probes (e.g. **54**).[150] However, as for **53** these compounds failed to provide any identifiable covalently labeled peptide fragments and were only moderately cytotoxic (**54**: $ED_{50} > 0.2$ μM (MCF-7)).[150]

16.4   Structural Studies and Pharmacophore Modeling

Since the discovery of the microtubule-stabilizing properties of epothilones, efforts have been made to develop a predictive pharmacophore model for this class of compounds. In the absence of any structural information on complexes

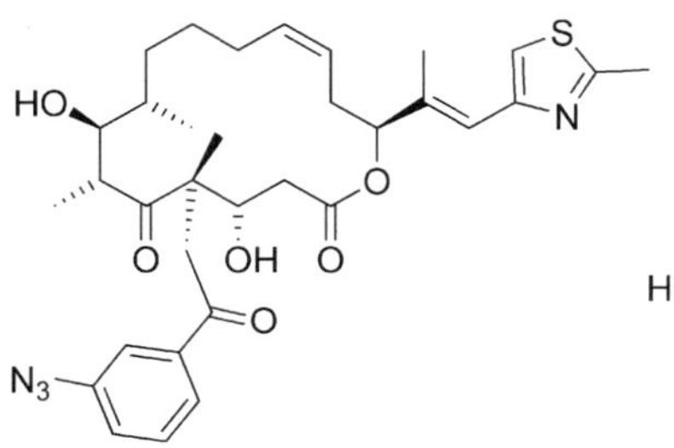

53 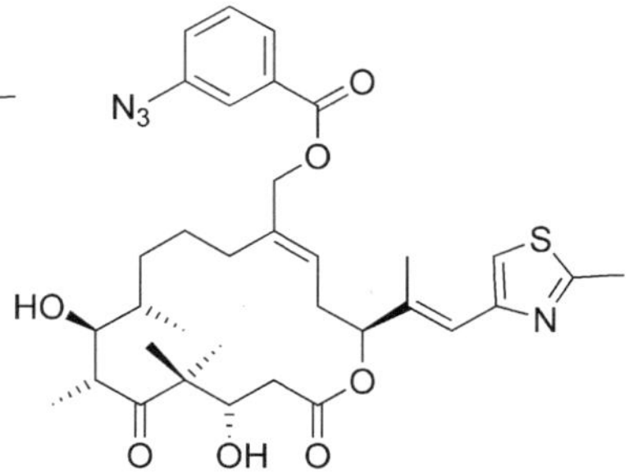   **54**

Figure 16.16   Epothilone analogs designed as photoaffinity labels for tubulin.

between tubulin/microtubules and epothilone-type ligands, early attempts on the development of an epothilone pharmacophore model[60,151] were generally based on the assumption of a common tubulin binding site for epothilones and taxol. However, all of these models had to be re-assessed in light of the structural studies on the tubulin-bound conformation of Epo A either by NMR spectroscopy on a soluble β-tubulin/Epo A complex[152,153] or by electron crystallography (EC) of a complex between Epo A and a Zn^{2+}-stabilized two-dimensional α,β-tubulin sheet (solved at 2.89 Å resolution).[52] Thus, the experimental structure derived by means of electron crystallography called into question the notion of a common pharmacophore between taxol and epothilones. While these data have confirmed that taxol and Epo A indeed occupy the same binding pocket on β-tubulin, the actual binding seems to be mediated through different sets of hydrogen bonding and hydrophobic interactions for the two compounds. Interestingly, the EC-derived tubulin-bound conformation of Epo A is significantly different from any of the above computational models,[60,151,154] and also from the NMR structure of (soluble) tubulin-bound Epo A as determined by Carlomagno *et al.*[152]

Attempts have been made to assess the validity of the various models by the investigation of specifically designed analogs. This has included bridged analogs **50** and **51**, which on the basis of the EC-derived model were anticipated to retain good activity (or even to show enhanced activity over the natural products). However, as indicated above, the compounds proved to be only poorly active.[146,147] With regard to the NMR-derived structure, two recent publications by Erdélyi *et al.*[124,154] not only showed that the NMR-derived tubulin-bound conformation of Epo A corresponds to a low energy conformation of the free ligand in aqueous solution, but also that neither the removal of the 3-OH group nor conformational restriction of the C2-C3 bond to a *trans* geometry significantly alters the overall conformation of the macrolide ring in the tubulin-bound state, relative to the NMR-derived bioactive conformation of Epo A.[154] This latter finding is in good agreement with the bioactivity data reported for these compounds, which, in turn, cannot easily be reconciled with the EC-derived conformation. Recent modeling studies by Botta and co-workers,[155] which were based on a previously developed pseudoreceptor model for β-tubulin,[151c,151d] also support the validity of the NMR-derived bioactive conformation of epothilones (and, in addition, suggests the existence of a common pharmacophore between epothilones and taxol).

In a recent solid-state NMR study, Baldus and co-workers have investigated the chemical shift perturbations caused by the binding of Epo B to microtubules.[156] From these data it was concluded that a unique tight interaction exists between the drug and the protein; however, the work fell short of explaining the differences between the electron crystallography- and NMR-derived models of the bioactive conformation of epothilones.

The question of the bioactive conformation of epothilones has also been addressed by density functional theory calculations.[157,158] One such study[157] has reached similar conclusions with regard to the conformational preferences of the macrolactone ring as had been derived previously by Erdélyi *et al.* in their NMR work.[154]

16.5 Conclusions

The discovery of the 'taxol-like' mechanism of action of epothilones together with their very low susceptibility to Pgp-mediated drug efflux has established these compounds as important lead structures for anticancer drug discovery. Among the various natural products that have been recognized as microtubule stabilizers to date, epothilones (together with taxanes) today represent the most widely explored leads for tubulin-directed anticancer drug discovery. Hundreds of epothilone analogs and derivatives have been prepared and their pharmacological properties evaluated *in vitro* and (in selected cases) also *in vivo*, thus establishing a comprehensive map of the structural elements associated with potent biological activity. The flow of structures has significantly decreased in the more recent past, but interesting new analogs still keep emerging. This includes hypermodified epothilones, such as azathilones, but also bridged analogs that were designed based on structural data for epothilone A/tubulin complexes. At this point in time, however, the question of the true bioactive conformation of epothilones (if there is in fact a unique single conformation), still awaits its final resolution.

Epothilone-based drug discovery research so far has delivered eight compounds that have entered clinical trials in humans. Of these, ixabepilone has received FDA approval in 2007, thus ultimately validating the lead potential of epothilones for anticancer drug discovery. On the other hand, for all the other candidate drugs their continued development is either uncertain (patupilone, sagopilone) or development seems to have been terminated. While there is no shortage of additional interesting preclinical candidates, it remains to be seen whether any of these will be taken to the clinic and, if so, whether it can be developed into a clinical drug.

References

1. C. Dumontet and M. A. Jordan, *Nature Rev. Drug Discov.*, 2010, **9**, 790.
2. T. M. Mekhail and M. Markman, *Exp. Opin. Pharmacother.*, 2002, **3**, 755.
3. C. Obasaju and G. R. Hudes, *Hematology/Oncology Clinics of North America*, 2001, **15**, 525.
4. M. A. Jordan and L. Wilson, *Nat. Rev. Cancer*, 2004, **4**, 253.
5. S.-M. Chen, L.-H. Meng and J. Ding, *Exp. Opin. Investig. Drugs*, 2010, **19**, 329.
6. http://www.cancer.gov/cancertopics/druginfo/fda-eribulinmesylate
7. E. K. Rowinsky, *Ann. Rev. Med.*, 1997, **48**, 35.
8. N. Agarwal, G. Sonpavde and O. Sartor, *Future Oncol.*, 2011, **7**, 15.

9. K.-H. Altmann, *Curr. Opin. Chem. Biol.*, 2001, **5**, 424.
10. K.-H. Altmann and J. Gertsch, *Nat. Prod. Rep.*, 2007, **24**, 327.
11. J. H. Miller, A. J. Singh and P. T. Northcote, *Marine Drugs*, 2010, **8**, 1059.
12. P. B. Schiff, J. Fant and S. B. Horwitz, *Nature*, 1979, **277**, 665.
13. G. Höfle and H. Reichenbach, *Epothilone, A Myxobacterial Metabolite with Promising Antitumor Activity*, in G. M. Cragg, D. G. I. Kingston and D. J. Newman (eds), Anticancer Agents from Natural Products, Taylor & Francis, Boca Raton, FL, 2005, p. 413.
14. K. Gerth, N. Bedorf, G. Höfle, H. Irschik and H. Reichenbach, *J. Antibiot.*, 1996, **49**, 560; see also G. Höfle, N. Bedorf, K. Gerth and H. Reichenbach, German Patent Disclosure, DE 4138042, 1993; *Chem. Abstr.*, 1993, **120**, 52841.
15. I. H. Hardt, H. Steinmetz, K. Gerth, F. Sasse, H. Reichenbach and G. Höfle, *J. Nat. Prod.*, 2001, **64**, 847.
16. D. M. Bollag, P. A. McQueney, J. Zhu, O. Hensens, L. Koupal, J. Liesch, M. Goetz, E. Lazarides and C. M. Woods, *Cancer Res.*, 1995, **55**, 2325.
17. J. J. Field, A. J. Singh, A. Kanakkanthara, T. Halafihi, P. T. Northcote and J. Miller, *J. Med. Chem.*, 2009, **52**, 7328.
18. R. J. Kowalski, P. Giannakakou and E. Hamel, *J. Biol. Chem.*, 1997, **272**, 2534.
19. P. Giannakakou, D. L. Sackett, Y. K. Kang, Z. R. Zhan, J. T. M. Buters, T. Fojo and M. S. Poruchynsky, *J. Biol. Chem.*, 1997, **272**, 17118.
20. K.-H. Altmann, M. Wartmann and T. O'Reilly, *Biochim. Biophys. Acta, Rev. Cancer*, 2000, **1470**, M79.
21. G. H. Höfle, N. Bedorf, H. Steinmetz, D. Schomburg, K. Gerth and H. Reichenbach, *Angew. Chem., Int. Ed.*, 1996, **35**, 1567.
22. T. O'Reilly, M. Wartmann, J. Brueggen, P. R. Allegrini, A. Flörsheimer, M. Maira and P. M. J. McSheehy, *Cancer Chemother. Pharmacol.*, 2008, **62**, 1045.
23. For a recent review on clinical trials with epothilones see H. M. Coley, *Cancer Treat. Rev.*, 2008, **34**, 378.
24. I. R. Vlahov, G. D. Vite, P. J. Kleindl, Y. Wang, H. K. R. Santhapuram, F. You, S. J. Howard, S.-H. Kim, F. F. Y. Lee and C. P. Leamon, *Bioorg. Med. Chem. Lett.*, 2010, **20**, 4578.
25. (a) E. Kaminskas, X. Jiang, R. Aziz, J. Bullock, R. Kasliwal, R. Harapanhalli, S. Pope, R. Sridhara, J. Leighton, B. Booth, R. Dagher, R. Justice and R. Pazdur, *Clin. Cancer Res.*, 2008, **14**, 4378; (b) for leading references on ixabepilone see E. S. Thomas, H. L. Gomez, R. K. Li, H.-C. Chung, L. E. Fein, V. F. Chan, J. Jassem, X. B. Pivot, J. Klimovsky, F. Hurtado de Mendoza, B. Xu, M. Campone, G. L. Lerzo, R. A. Peck, P. Mukhopadhyay, L. T. Vahdat and H. H. Roche, *J. Clin. Oncol.*, 2007, **25**, 5210.
26. http://www.novartis.com/newsroom/media-releases/en/2010/1419057.shtml

27. No publications for this compound have appeared subsequent to the announcement of Phase I trials in a 2004 AACR meeting abstract: M. Wartmann, J. Loretan, R. Reuter, M. Hattenberger, M. Muller, J. Vaxelaire, S.-M. Maira, A. Flörsheimer, T. O'Reilly, K. C. Nicolaou and K.-H. Altmann, *Proceedings of the American Association for Cancer Research*, 2004, **45**: Abstract 5440. The compound is no longer part of the (published) Novartis Oncology pipeline.

28. http://clinicaltrials.gov/ct2/results?intr=%22Epofolate%22

29. BMS-310705, KOS-862, or KOS-1584 are not part of BMS' (published) development pipeline: http://www.bms.com/research/pipeline/Pages/default.aspx. While KOS-862 and KOS-1584 were initially developed by Kosan Biosciences, the company has been acquired by BMS. No active clinical trials could be identified for any of these compounds in the NCI webspace.

30. K. C. Nicolaou, F. Roschangar and D. Vourloumis, *Angew. Chem., Int. Ed.*, 1998, **37**, 2015.

31. C. R. Harris and S. J. Danishefsky, *J. Org. Chem.*, 1999, **64**, 8434.

32. J. Mulzer, H. J. Martin and M. Berger, *J. Heterocycl. Chem.*, 1999, **36**, 1421.

33. K. C. Nicolaou, A. Ritzén and K. Namoto, *Chem. Commun.*, 2001, 1523.

34. K.-H. Altmann, *Org. Biomol. Chem.*, 2004, **2**, 2137.

35. E. B. Watkins, A. G. Chittiboyina, J. C. Jung and M. A. Avery, *Curr. Pharm. Des.*, 2005, **11**, 1615.

36. E. B. Watkins, A. G. Chittiboyina, J. C. Jung and M. A. Avery, *Eur. J. Org. Chem.*, 2006, **18**, 4071.

37. R. M. Borzilleri and G. D. Vite, *Drugs of the Future*, 2002, **27**, 1149.

38. M. Wartmann and K.-H. Altmann, *Curr. Med. Chem. Anticancer Agents*, 2002, **2**, 123.

39. K.-H. Altmann, *Mini-Rev. Med. Chem.*, 2003, **3**, 149.

40. K.-H. Altmann, *Curr. Pharm. Des.*, 2005, **11**, 1595.

41. K.-H. Altmann, B. Pfeiffer, S. Arseniyadis, B. A. Pratt and K. C. Nicolaou, *ChemMedChem*, 2007, **2**, 397.

42. U. Klar, B. Buchmann, W. Schwede, W. Skuballa, J. Hoffmann and R. B. Lichtner, *Angew. Chem., Int. Ed.*, 2006, **45**, 7942.

43. R. Altaha, T. Fojo, E. Reed and J. Abraham, *Curr. Pharm. Des.*, 2002, **8**, 1707.

44. For a recent book on epothilones see K.-H. Altmann, G. Höfle, R. Müller, J. Mulzer and K. Prantz, *The Epothilones: An Outstanding Family of Anti-Tumor Agents*ed. A. D. Kinghorn, H. Falk and J. Kobayashi, *Progress in Chemistry Organic Natural Products*, Springer, 2009, Vol. 90.

45. E. Nogales, *Annu. Rev. Biophys. Biomol. Struct.*, 2001, **30**, 397.

46. O. Valiron, N. Caudron and D. Job, *Cell. Mol. Life Sci.*, 2001, **58**, 2069.

47. L. Amos, *Org. Biomol. Chem.*, 2004, **2**, 2153.

48. M. A. Jordan, *Curr. Med. Chem: Anti-Cancer Agents*, 2002, **2**, 1.

49. Y. Zhai, P. J. Kronebusch, P. M. Simon and G. G. Borisy, *J. Cell. Biol.*, 1996, **135**, 201.

50. N. M. Rusan, C. J. Fagerstrom, A. M. C. Yvon and P. Wadsworth, *Mol. Biol. Cell*, 2001, **12**, 971.

51. K.-H. Altmann, G. Bold, G. Caravatti, N. End, A. Flörsheimer, V. Guagnano, T. O'Reilly and M. Wartmann, *Chimia*, 2000, **54**, 612.

52. J. H. Nettles, H. L. Li, B. Cornett, J. M. Krahn, J. P. Snyder and K. H. Downing, *Science*, 2004, **305**, 866.

53. R. M. Buey, J. F. Díaz, J. M. Andreu, A. O'Brate, P. Giannakakou, K. C. Nicolaou, P. K. Sasmal, A. Ritzen and K. Namoto, *Chem. Biol.*, 2004, **11**, 225.

54. R. B. Lichtner, A. Rotgeri, T. Bunte, B. Buchmann, J. Hoffmann, W. Schwede, W. Skuballa and U. Klar, *Proc. Natl. Acad. Sci. USA*, 2001, **98**, 11743.

55. K. Kamath and M. A. Jordan, *Cancer Res.*, 2003, **63**, 6026.

56. J.-G. Chen and S. B. Horwitz, *Cancer Res.*, 2002, **14**, 1935.

57. J.-G. Chen, C.-P. H. Yang, M. Cammer and S. B. Horwitz, *Cancer Res.*, 2003, **15**, 7891.

58. N. R. Khawaja, M. Carré, H. Kovacic, M. A. Estève and D. Braguer, *Mol. Pharmacol.*, 2008, **74**, 1072.

59. S. H. Lee, S. M. Son, D. J. Son, S. M. Kim, T. Kim, S. Song, D. C. Moon, H. W. Lee, J. C. Ryu, D.-Y. Yoon and J. T. Hong, *Mol. Cancer Ther.*, 2007, **6**, 2786.

60. P. Giannakakou, R. Gussio, E. Nogales, K. H. Downing, D. Zaharevitz, B. Bollbuck, G. Poy, D. Sackett, K. C. Nicolaou and T. Fojo, *Proc. Natl. Acad. Sci. USA*, 2000, **97**, 2904.

61. L. He, C.-P. Yang and S. B. Horwitz, *Mol. Cancer Ther.*, 2001, **1**, 3.

62. C. P. H. Yang, P. Verdier-Pinard, F. Wang, E. Lippaine-Horvath, L. F. He, D. S. Li, G. Höfle, I. Ojima, G. A. Orr and S. B. Horwitz, *Mol. Cancer Ther.*, 2005, **4**, 987.

63. F. Cabral and S. B. Barlow, *Pharmacol. Therapeut.*, 1991, **52**, 159.

64. N. M. Verrills, C. L. Flemming, M. Liu, M. T. Ivery, G. S. Cobon, M. D. Norris, M. Haber and M. Kavallaris, *Chem. Biol.*, 2003, **10**, 597.

65. E. Hopper-Borge, X. Xu, T. Shen, Z. Shi, Z.-S. Chen and G. D. Kruh, *Cancer Res.*, 2009, **69**, 178.

66. D. S. Su, A. Balog, D. F. Meng, P. Bertinato, S. J. Danishefsky, Y. H. Zheng, T. C. Chou, L. F. He and S. B. Horwitz, *Angew. Chem., Int. Ed.*, 1997, **36**, 2093.

67. T. C. Chou, X. G. Zhang, A. Balog, D. S. Su, D. F. Meng, K. Savin, J. R. Bertino and S. J. Danishefsky, *Proc. Natl. Acad. Sci. USA*, 1998, **95**, 9642.

68. T. O'Reilly, P. M. J. McSheehy, F. Wenger, M. Hattenberger, M. Muller, J. Vaxelaire, K.-H. Altmann and M. Wartmann, *Prostate*, 2005, **65**, 231.

69. J. Rothermel, M. Wartmann, T. L. Chen and J. Hohneker, *Semin. Oncol.*, 2003, **30**, 51.

70. G. Bocci, K. C. Nicolaou and R. S. Kerbel, *Cancer Res.*, 2002, **62**, 6938.

71. E. A. Woltering, J. M. Lewis, P. J. Maxwell 4th, D. J. Frey, Y.-Z. Wang, J. Rothermel, C. T. Anthony, D. A. Balster, J. P. O'Leary and L. H. Harrison, *Ann. Surgery*, 2003, **237**, 790.

72. S. Ferretti, P. R. Allegrini, T. O'Reilly, C. Schnell, M. Stumm, M. Wartmann, J. Wood, P. M. J. McSheehy and M. J. Paul, *Clin. Cancer Res.*, 2005, **11**, 7773.

73. R. M. Borzilleri, X. P. Zheng, R. J. Schmidt, J. A. Johnson, S. H. Kim, J. D. DiMarco, C. R. Fairchild, J. Z. Gougoutas, F. Y. F. Lee, B. H. Long and G. D. Vite, *J. Am. Chem. Soc.*, 2000, **122**, 8890.

74. E. H. Rubin, J. Rothermel, F. Tesfaye, T. Chen, M. Hubert, Y. Y. Ho, C.-H. Hsu and A. M. Oza, *J. Clin. Oncol.*, 2005, **23**, 9120.

75. For a recent review on clinical trials with Epo B (patupilone) see B. Bystricky and I. Chau, *Exp. Opin. Investig. Drugs*, 2011, **20**, 107.

76. A. Regueiro-Ren, R. M. Borzilleri, X. Zheng, S. Kim, J. A. Johnson, C. R. Fairchild, F. Y. F. Lee, B. H. Long and G. D. Vite, *Org. Lett.*, 2001, **3**, 2693.

77. K.-H. Altmann, G. Bold, G. Caravatti, D. Denni, A. Flörsheimer, A. Schmidt, G. Rihs and M. Wartmann, *Helv. Chim. Acta*, 2002, **85**, 4086.

78. K. C. Nicolaou, P. K. Sasmal, G. Rassias, M. V. Reddy, K.-H. Altmann, M. Wartmann, A. O'Brate and P. Giannakakou, *Angew. Chem., Int. Ed.*, 2003, **42**, 3515.

79. (a) R. L. Arslanian, L. Tang, S. Blough, W. Ma, R. G. Qiu, L. Katz and J. R. Carney, *J. Nat. Prod.*, 2002, **65**, 1061; (b) K. Biswas, H. Lin, J. T. Njardarson, M. D. Chappell, T. C. Chou, Y. B. Guan, W. P. Tong, L. F. He, S. B. Horwitz and S. J. Danishefsky, *J. Am. Chem. Soc.*, 2002, **124**, 9825.

80. (a) A. Rivkin, F. Yoshimura, A. E. Gabarda, T. C. Chou, H. J. Dong, W. P. Tong and S. J. Danishefsky, *J. Am. Chem. Soc.*, 2003, **125**, 2899; (b) A. Rivkin, F. Yoshimura, A. E. Gabarda, Y. S. Cho, T. C. Chou, H. J. Dong and S. J. Danishefsky, *J. Am. Chem. Soc.*, 2004, **126**, 10913; (c) F. Yoshimura, A. Rivkin, A. E. Gabarda, T. C. Chou, H. J. Dong, G. Sukenick, F. F. Morel, R. E. Taylor and S. J. Danishefsky, *Angew. Chem., Int. Ed.*, 2003, **42**, 2518.

81. F. Y. F. Lee, R. Borzilleri, C. R. Fairchild, S.-H. Kim, B. H. Long, C. Reventos-Suarez, G. D. Vite, W. C. Rose and R. A. Kramer, *Clin. Cancer Res.*, 2001, **7**, 1429.

82. K. C. Nicolaou, R. Scarpelli, B. Bollbuck, B. Werschkun, M. M. A. Pereira, M. Wartmann, K.-H. Altmann, D. Zaharevitz, R. Gussio and P. Giannakakou, *Chem. Biol.*, 2000, **7**, 593.

83. K.-H. Altmann, G. Bold, G. Caravatti, A. Flörsheimer, V. Guagnano and M. Wartmann, *Bioorg. Med. Chem. Lett.*, 2000, **10**, 2765.

84. D. Schinzer, K.-H. Altmann, F. Stuhlmann, A. Bauer and M. Wartmann, *ChemBioChem*, 2000, **1**, 67.

85. J. T. Hunt, *Mol. Cancer Ther.*, 2009, **8**, 275.

86. S. J. Stachel, C. B. Lee, M. Spassova, M. D. Chappell, W. G. Bornmann, S. J. Danishefsky, T.-C. Chou and Y. Guan, *J. Org. Chem.*, 2001, **66**, 4369.

87. F. Y. F. Lee, R. Borzilleri, C. R. Fairchild, S.-H. Kim, B. H. Long, C. Reventos-Suarez, G. D. Vite, W. C. Rose and R. A. Kramer, *Clin. Cancer Res.*, 2001, **7**, 1429.

88. F. Y. F. Lee, R. Borzilleri, C. R. Fairchild, A. V. Kamath, R. Smykla, R. A. Kramer and G. Vite, *Cancer Chemother. Pharmacol.*, 2008, **63**, 157.

89. C. Dumontet, M. A. Jordan and F. F. Y. Lee, *Mol. Cancer Ther.*, 2009, **8**, 17.

90. F. Y. F. Lee, R. Smykla, K. Johnston, K. Menard, K. McGlinchey, R. W. Peterson, A. Wiebesiek, G. Vite, C. R. Fairchild and R. A. Kramer, *Cancer Chemother. Pharmacol.*, 2009, **63**, 201.

91. H. Yamaguchi, S. R. Paranawithana, M. W. Lee, Z. Huang, K. N. Bhalla and H.-G. Wang, *Cancer Res.*, 2002, **62**, 466.

92. D. Griffin, S. Wittmann, F. Guo, R. Nimmanapalli, P. Bali, H. G. Wang and K. Bhalla, *Gynecol. Oncol.*, 2003, **89**, 37.

93. J. K. Peterson, C. Tucker, E. Favours, P. J. Cheshire, J. Creech, C. A. Billups, R. Smykla, F. Y. F. Lee and P. J. Houghton, *Clin. Cancer Res.*, 2005, **11**, 6950.

94. F. Y. F. Lee, K. L. Covello, S. Castaneda, D. R. Hawken, D. Kan, A. Lewin, M.-L. Wen, R.-P. Ryseck, C. R. Fairchild, J. Fargnoli and R. Kramer, *Clin. Cancer Res.*, 2008, **14**, 8123.

95. N. Lin, K. Brakora and M. Seiden, *Curr. Opin. Investig. Drugs*, 2003, **4**, 746.

96. X. Pivot, A. Dufresne and C. Villanueva, *Clin. Breast Cancer*, 2007, **7**, 543.

97. S. Mani, H. McDaid, A. Hamilton, H. Hochster, M. B. Cohen, D. Khabelle, T. Griffin, D. E. Lebwohl, L. Liebes, F. Muggia and S. B. Horwitz, *Clin. Cancer Res.*, 2004, **10**, 1289.

98. FDA, http://www.fda.gov/AboutFDA/CentersOffices/CDER/ucm129240.htm

99. M. Hussain, C. M. Tangen, P. N. Lara, U. N. Vaishampayan, D. P. Petrylak, A. D. Colevas, W. A. Sakr and E. D. Crawford, *J. Clin. Oncol.*, 2005, **23**, 8724.

100. M. D. Galsky, E. J. Small, W. K. Oh, I. Chen, D. C. Smith, A. D. Colevas, L. Martone, T. Curley, A. DeLaCruz, H. I. Scher and W. K. Kelly, *J. Clin. Oncol.*, 2005, **23**, 1439.

101. R. Whitehead, S. McCoy, S. Rivkin, H. Gross, M. Conrad, G. Doolittle, R. Wolff, J. Goodwin, S. Dakhil and J. Abbruzzese, *Invest. New Drug*, 2006, **24**, 515.

102. J. Vansteenkiste, P. N. Lara, T. Le Chevalier, J.-L. Breton, P. Bonomi, A. B. Sandler, M. A. Socinski, C. Delbaldo, B. McHenry, D. Lebwohl, R. Peck and M. Edelman, *J. Clin. Oncol.*, 2007, **25**, 3448.

103. R. Dreicer, S. Li, J. Manola, N. B. Haas, B. J. Roth and G. Wilding, *Cancer*, 2007, **110**, 759.

104. J. Ajani, H. Safran, C. Bokemeyer, M. Shah, H. J. Lenz, E. Van Cutsem, H. Burris, D. Lebwohl and B. Mullaney, *Invest. New Drug*, 2006, **24**, 441.

105. S. Okuno, W. J. Maples, M. R. Mahoney, T. Fitch, J. Stewart, P. M. Fracasso, M. Kraut, D. S. Ettinger, F. Dawkins and C. Erlichman, *J. Clin. Oncol.*, 2005, **23**, 3069.

106. J. J. Lee and S. M. Swain, *Clin. Cancer Res.*, 2008, **14**, 1618.

107. N. Denduluri and S. M. Swain, *Exp. Opin. Investig. Drugs*, 2008, **17**, 423.

108. S. Jacobs, E. Fox, M. Krailo, G. Hartley, F. Navid, L. Wexler, S. M. Blaney, A. Goodwin, W. Goodspeed, F. M. Balis, P. C. Adamson and B. C. Widemann, *Clin. Cancer Res.*, 2010, **16**, 750.

109. http://clinicaltrials.gov/ct2/home

110. J. Hoffmann, I. Vitale, B. Buchmann, L. Galluzzi, W. Schwede, L. Senovilla, W. Skuballa, S. Vivet, R. B. Lichtner, J. M. Vicencio, T. Panaretakis, G. Siemeister, H. Lage, L. Nanty, S. Hammer, K. Mittelstaedt, S. Winsel, J. Eschenbrenner, M. Castedo, C. Demarche, U. Klar and G. Kroemer, *Cancer Res.*, 2008, **68**, 5301.

111. U. Klar, B. Röhr, F. Kuczynski, W. Schwede, M. Berger, W. Skuballa and B. Buchmann, *Synthesis*, 2005, **2**, 301.

112. U. Klar, J. Hoffmann and M. Giurescu, *Exp. Opin. Investig. Drugs*, 2008, **17**, 11.

113. S. Hammer, A. Sommer and I. Fichtner, *Clin. Cancer Res.*, 2010, **16**, 1452.

114. J. Hoffmann, I. Fichtner, M. Lemm, Ph. Lienau, H. Hess, A. Rotgeri, B. Hofmann and U. Klar, *Neurol. Oncol.*, 2009, **11**, 158.

115. P. Schmid, P. Kiewe, K. Possinger, A. Korfel, S. Lindemann, M. Giurescu, S. Reif, H. Wiesinger, E. Thiel and D. Kühnhardt, *Ann. Oncol.*, 2010, **21**, 633.

116. P. Fumoleau, B. Coudert, N. Isambert and E. Ferrant, *Ann. Oncol.*, 2007, **18** (Suppl 5), v9.

117. D. Arnold, W. Voigt, P. Kiewe, C. Behrmann, S. Lindemann, S. Reif, M. Giurescu and E. Thiel, *Ann. Oncol.*, 2006, **17** (Suppl 9), abstract 416P.

118. G. J. Rustin, N. Reed, G. Jayson, J. Ledermann, M. Adams, C. Stredder, A. Wagner and M. Giurescu, *Abstract presented at the 15th International Meeting of the European Society for Gynaecological Oncology*, 2007, Berlin, Germany.

119. S. McMeekin, R. Patel, C. Verschraegen, P. Celano, J. Burke, S. Plaxe, P. Ghatage, M. Giurescu, C. Stredder and Y. Wang, *Ann. Oncol.*, 2008, **19** (Suppl 8), viii211, abstract 665.

120. T. Beer, D. C. Smith, A. Hussain, M. Alonso, L. Neerukonda, R. Hauke, Y. Wang and M. Giurescu, *Ann. Oncol.*, 2008, **19** (Suppl 8), viii198, abstract 616P.

121. U. Gatzemeier, J. V. von Pawel, E. Eschbach, A. Brune, A. Wagner, M. Giurescu and M. Reck, *Eur. J. Cancer*, 2007, **5**, 378, abstract 6538.

122. D. Wenk, R. C. DeConti, P. Urbas, S. Andrews, V. K. Sondak, N. Maker, J. S. Weber and A. I. Daud, *J. Clin. Oncol.*, 2008, **26**, abstract 9046.

123. P. K. Morrow, S. Divers, L. Provencher, S.-W. Luoh, T. M. Petrella, M. Giurescu, T. Schmelter, Y. Wang, G. N. Hortobagyi and L. T. Vahdat, *Breast Cancer Res. Treat.*, 2010, **123**, 837.

124. M. Erdélyi, A. Navarro-Vázquez, B. Pfeiffer, C. N. Kuzniewski, A. Felser, T. Widmer, J. Gertsch, B. Pera, J. F. Díaz, K.-H. Altmann and T. Carlomagno, *ChemMedChem*, 2010, **5**, 911.

125. T.-C. Chou, X.-G. Zhang, Z.-Y. Zhong, Y. Li, L. Feng, S. Eng, D. R. Myles, R. Johnson, Jr., N. Wu, Y. I. Yin, R. M. Wilson and S. J. Danishefsky, *Proc. Natl. Acac. Sci. USA*, 2008, **105**, 13157.

126. G. Bold, S. Wojeik, G. Caravatti, R. Lindauer, C. Stierlin, J. Gertsch, M. Wartmann and K.-H. Altmann, *ChemMedChem*, 2006, **1**, 37.

127. For a review see: F. Feyen, F. Cachoux, J. Gertsch, M. Wartmann and K.-H. Altmann, *Acc. Chem. Res.*, 2008, **41**, 21.

128. F. Cachoux, T. Isarno, M. Wartmann and K.-H. Altmann, *ChemBioChem*, 2006, **7**, 54.

129. F. Cachoux, T. Isarno, M. Wartmann and K.-H. Altmann, *Angew. Chem., Int. Ed.*, 2005, **44**, 7469.

130. F. Cachoux, T. Isarno, M. Wartmann and K.-H. Altmann, *Synlett*, 2006, **16**, 1384.

131. T.-C. Chou, H. J. Dong, X.-G. Zhang, W. P. Tong and S. J. Danishefsky, *Cancer Res.*, 2005, **65**, 9445.

132. A. Rivkin, F. Yoshimura, A. E. Gabarda, Y. S. Cho, T.-C. Chou, H. J. Dong and S. J. Danishefsky, *J. Am. Chem. Soc.*, 2004, **126**, 10913.

133. C. N. Kuzniewski, J. Gertsch, M. Wartmann and K.-H. Altmann, *Org. Lett.*, 2008, **10**, 1183.

134. F. Feyen, J. Gertsch, M. Wartmann and K.-H. Altmann, *Angew. Chem., Int. Ed.*, 2006, **45**, 5880.

135. J. D. Frein, R. E. Taylor and D. L. Sackett, *Org. Lett.*, 2009, **11**, 3186.

136. O. E. Hutt, B. S. Reddy, S. K. Nair, E. A. Reiff, J. T. Henri, J. F. Greiner, T. Chiu, D. G. VanderVelde, E. A. Amin, R. H. Himes and G. I. Georg, *Bioorg. Med. Chem. Lett.*, 2008, **18**, 4904.

137. D. Passarella, D. Comi, G. Cappelletti, D. Cartelli, J. Gertsch, A. R. Quesada, J. Borlak and K.-H. Altmann, *Bioorg. Med. Chem. Lett.*, 2009, **17**, 7435.

138. K. C. Nicolaou, D. Hepworth, N. P. King, M. R. V. Finlay, R. Scarpelli, M. M. A. Pereira, B. Bollbuck, A. Bogot, B. Werschkun and N. Winssigger, *Chem. Eur. J.*, 2000, **6**, 2783.

139. K.-H. Altmann, A. Flörsheimer, G. Bold, G. Caravatti and M. Wartmann, *Chimia*, 2004, **58**, 686.

140. F. Feyen, A. Jantsch, K. Hauenstein, B. Pfeiffer and K.-H. Altmann, *Tetrahedron*, 2008, **64**, 7920.

141. J. Gertsch, F. Feyen, A. Bützberger, B. Gerber, B. Pfeiffer and K.-H. Altmann, *ChemBioChem*, 2009, **10**, 2513.

142. See, for example, (a) J. F. Díaz, I. Barasoain, A. A. Souto, F. Amat-Guerri and J. M. Andreu, *J. Biol. Chem.*, 2005, **280**, 3928; (b) J. F. Díaz, I. Barasoain and J. M. Andreu, *J. Biol. Chem.*, 2003, **278**, 8407; (c) M. Abal, A. A. Souto, F. Amat-Guerri, A. U. Acuna, J. M. Andreu and I. Barasoain, *Cell Motil. Cytoskeleton*, 2001, **49**, 1; (d) J. F. Díaz, R. Strobe, Y. Engelborghs, A. A. Souto and J. M. Andreu, *J. Biol. Chem.*, 2000, **275**, 26265; (e) J. M. Andreu and I. Barasoain, *Biochemistry*, 2001, **40**, 11975; (f) J. A. Evangelio, M. Abal, I. Barasoain, A. A. Souto, M. P. Lillo, A. U. Acuna, F. Amat-Guerri and J. M. Andreu, *Cell Motil. Cytoskeleton*, 1998, **39**, 73.

143. B. Pfeiffer, K. Hauenstein, P. Merz, J. Gertsch and K.-H. Altmann, *Bioorg. Med. Chem. Lett.*, 2009, **19**, 3760.

144. M. M. Alhamadsheh, S. Gupta, R. A. Hudson, L. Perera and L. M. Viranga Tillekeratne, *Chem. Eur. J.*, 2008, **14**, 570.

145. W. Zhang, Y. Jiang, P. J. Brodie, D. G. I. Kingston, D. C. Liotta and J. P. Snyder, *Org. Lett.*, 2008, **10**, 1565.

146. Q.-H. Chen, T. Ganesh, P. Brodie, C. Slebodnick, Y. Jiang, A. Banerjee, S. Bane, J. P. Snyder and D. G. I. Kingston, *Org. Biomol. Chem.*, 2008, **6**, 4542.

147. Q.-H. Chen, T. Ganesh, Y. Jiang, A. Banerjee, S. Sharma, S. Bane, J. P. Snyder and D. G. I. Kingston, *Chem. Commun.*, 2010, **46**, 2019.

148. O. E. Hutt, J. Inagaki, B. S. Reddy, S. K. Nair, E. A. Reiff, J. T. Henri, J. F. Greiner, D. G. Vander Velde, T.-L. Chiu, E. A. Amin, R. H. Himes and G. I. Georg, *Bioorg. Med. Chem. Lett.*, 2009, **19**, 3293.

149. K. C. Nicolaou, S. Ninkovic, M. R. V. Finlay, F. Sarabia and T. Li, *Chem. Commun.*, 1997, 2343.

150. E. A. Reiff, S. K. Nair, J. T. Henri, J. F. Greiner, B. S. Reddy, R. Chakrasali, S. A. David, T.-L. Chiu, E. A. Amin, R. H. Himes, D. G. Vander Velde and G. I. Georg, *J. Org. Chem.*, 2010, **75**, 86.

151. (a) I. Ojima, S. Chakravarty, T. Inoue, S. Lin, L. He, S. B. Horwitz, S. D. Kuduk and S. J. Danishefsky, *Proc. Natl. Acad. Sci. USA*, 1999, **96**, 4256; (b) M. Wang, X. Xia, Y. Kim, D. Hwang, J. M. Jansen, M. Botta, D. C. Liotta and J. P. Snyder, *Org. Lett.*, 1999, **1**, 43; (c) F. Manetti, S. Forli, L. Maccari, F. Corelli and M. Botta, *Il Farmaco*, 2003, **58**, 357; (d) F. Manetti, L. Maccari, F. Corelli and M. Botta, *Curr. Topics. Med. Chem.*, 2004, **4**, 203.

152. T. Carlomagno, M. J. J. Blommers, J. Meiler, W. Jahnke, T. Schupp, F. Petersen, D. Schinzer, K.-H. Altmann and C. Griesinger, *Angew. Chem., Int. Ed.*, 2003, **42**, 2511.

153. M. Reese, M. V. Sánchez-Pedregal, K. Kubicek, J. Meiler, M. J. J. Blommers, C. Griesinger and T. Carlomagno, *Angew. Chem., Int. Ed.*, 2007, **46**, 1864.

154. M. Erdélyi, B. Pfeiffer, K. Hauenstein, J. Fohrer, J. Gertsch, K.-H. Altmann and T. Carlomagno, *J. Med. Chem.*, 2008, **51**, 1469.
155. S. Forli, F. Manetti, K.-H. Altmann and M. Botta, *ChemMedChem*, 2010, **5**, 35.
156. A. Kumar, H. Heise, M. J. J. Blommers, P. Krastel, E. Schmitt, F. Petersen, S. Jeganathan, E.-M. Mandelkow, T. Carlomagno, C. Griesinger and M. Baldus, *Angew. Chem., Int. Ed.*, 2010, **49**, 7504.
157. D. Rusinska-Roszak and M. Lozynski, *J. Mol. Model.*, 2009, **15**, 859.
158. V. A. Jiménez, *J. Chem. Inf. Model.*, 2010, **50**, 2176.

CHAPTER 17

The Contribution of Marine Chemistry in the Field of Antimalarial Research

ERNESTO FATTORUSSO* AND
ORAZIO TAGLIALATELA-SCAFATI

Neanat Group, Dipartmento di Chimica delle Sostanze Naturali Università di Napoli Federico II, Via Montesano 49, 80131, Napoli, Italy
*E-mail: fattoru@unina.it

17.1 Introduction

Malaria is an infectious disease caused by several protozoans belonging to the genus *Plasmodium*, with *P. falciparum* being the most widespread aetiological agent and the one responsible of the most severe cases. The protozoan comes in contact with humans through the vector contribution of female *Anopheles* mosquitoes. Within 30 minutes after injection into the bloodstream, the parasite (in the sporozoites form) infects hepatocytes, multiplying asexually and asymptomatically for a period of 6–15 days. When the protozoans reach the merozoite stage they are released from the liver and start to invade the erythrocytes, actively feeding on the haemoglobin. After proliferation, the rupture of the erythrocyte membrane and the consequent liberation of other merozoites, which invade other erythrocytes, cause the massive infection and the symptoms. A small portion of merozoites develop into the sexual stage of gametocytes, a form that is able to re-start the lifecycle of the malaria parasite when a mosquito takes a blood meal from an infected person.[1]

RSC Drug Discovery Series No. 25
Drug Discovery from Natural Products
Edited by Olga Genilloud and Francisca Vicente
© The Royal Society of Chemistry 2012
Published by the Royal Society of Chemistry, www.rsc.org

The most frequent clinical symptoms of malaria infections are splenomegaly and fever, and in the case of cerebral malaria, loss of consciousness and coma. If this form of cerebral malaria is not treated immediately, it is soon followed by death. These symptoms are almost exclusively attributable to parasites in the erythrocytic stage. The rupture of infected erythrocytes is associated with the release into the bloodstream of cell debris responsible for the characteristic fever spike patterns. In the lethal cases, a specific protein produced by the protozoan is embedded into the cell membrane of the infected erythrocyte and, as a consequence of this modification, the erythrocyte sticks to the walls of capillaries causing obstruction of vessels.

In the 21st century malaria still continues to be a dangerous threat to the health and economic prosperity of the human race, constituting a major cause of morbidity and mortality in tropical countries of Asia, Africa and South America. It is hard to believe, but about 40% of the world population lives in countries where the disease is endemic and causes 700 000 to 2.0 million deaths each year (over 75% of them are African children). These deaths are the unavoidable outcome of the 400 to 900 million acute febrile episodes per year among African children under the age of 5 living in malaria-endemic regions.[2] In the eastern and southern African regions, an estimated 30% of all recorded deaths during pregnancy are attributed to malaria infection.

These dramatic figures are the result of different problems and are undoubtedly related to the difficulties to guarantee a formal and organized healthcare to these populations. However, one of the most important explanations for the failure to control malaria in these areas comes from the emergence and spread of resistance to first-line antimalarial drugs, cross-resistance between the members of the limited number of drug families available, and in some areas, multi-drug resistance.[3] It has been estimated that in the last 30 years only 4 or 5 new drugs have been registered against malaria, thus rapidly highlighting the limited efforts of pharmaceutical companies in the field of tropical diseases, which can guarantee a too poor potential market.

The drug treatment of malaria infections holds an important place in the history of medicinal chemistry and of natural product chemistry. Indeed, malaria was the first disease to be treated with an active principle isolated from a natural source, namely quinine, isolated from the Cinchona bark in 1820, and, later again, the first human disease to be treated with a synthetic drug (methylene blue introduced in 1891). In the course of the 20th century, a series of effective synthetic antimalarial drugs have been developed, but, unfortunately, the available chemical weapons to treat malaria cases are largely based on these old molecules – chloroquine (**1**, the parent compound of 4-aminoquinolines), primaquine (**2**, the parent compound of 8-aminoquinolines), and atovaquone, (**3**, the parent compound of naphthoquinones) (Figure 17.1). Developed countries should thank the massive use of insecticides and the introduction of these molecules for the nearly complete eradication of malaria. Undoubtedly, a systematic use of these molecules would be still beneficial in tropical countries and, whenever possible, they are used in association with

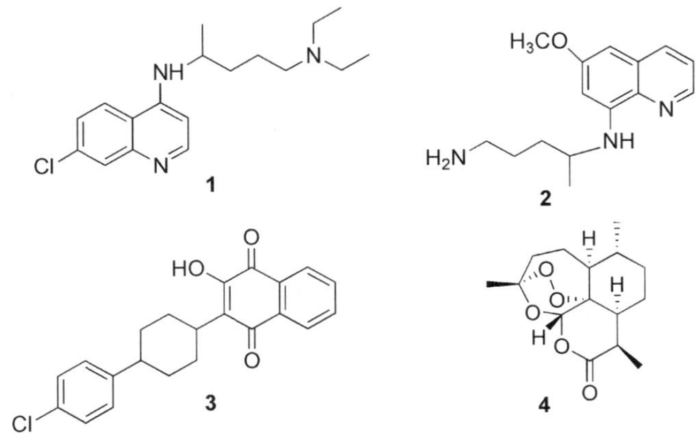

Figure 17.1   The most important existing weapons against malaria: chloroquine (**1**), primaquine (**2**), atovaquone (**3**), artemisinin (**4**).

artemisinin (**4**) (another natural product, see below for more detailed description) and its analogues, which represents the most advanced malaria treatment. However, the increasing presence of multi-drug resistant cases poses the problem of finding valid alternative therapeutic options, if possible through the use of drugs based on innovative mechanisms of action.

In this context, the recent sequencing of the entire genomes of both the malarial parasite and its insect vector disclosed a number of potential pharmaceutical targets to prevent and/or treat the disease.[4] In addition, it is also now widely accepted that the next significant advancement in the field of antimalarial drugs will not be obtained through the discovery of a single potent compound, but through the introduction of an innovative drug to be used in a combined therapy, preferably composed of molecules acting at different stages of the malaria parasite lifecycle. For example, studies conducted with the rodent model *P. chabaudi* revealed that the stress conditions imposed by drug treatment determine an increase in the gametocyte production, and gametocytes have been shown to have a much longer lifespan than the asexual stages. Moreover, an optimal strategy to combat against malaria infections cannot be exclusively based on drugs and should necessarily be multilateral, combining the treatment of infections (drugs) with all the different approaches based on prophylaxis and prevention (mosquito nets, insect repellents, vaccines). Unfortunately, both the number of different strains causing the pathology and their tendency to genetic mutation posed great obstacles to the development of the vaccine strategy. At the moment, RTS,S is heading Phase III trials and it appears to be the most promising malaria vaccine candidate.[5] Combined with a proprietary GSK adjuvant system, RTS,S induces the production of antibodies as well as of white blood cells that are believed to diminish the capacity of the malaria parasite to infect, survive and develop in the human liver. In addition to inducing partial protection against malaria, the RTS,S vaccine candidate stimulates a protective immune response to hepatitis B, a common infection in developing countries.

However, undoubtedly also this vaccine will not be the final solution against malaria and, therefore, the need for rapid development and introduction of safe and affordable drugs against malaria continues to be urgent. Hopefully, new breakthroughs in malaria treatment will come with the development of a marine lead compound.

17.2 Drugs from the Sea

The oceans are an unique environment that host a wealth of plants, animals and microorganisms, which largely base their adaptation to the habitat on the production of a wide variety of natural products, often characterized by peculiar structures. The chemistry of marine secondary metabolites was born around the end of the 1960s and now, after the first 50 years of work, it is a well-established science interesting a growing number of researchers. The number of novel marine natural products now discovered has been estimated to be more than 20 000, and, dividing these metabolites on the basis of the producer organisms, sponges play a dominating role as a source of new compounds, followed by coelenterates and microorganisms.[6] Although the biogenetical origin of this plethora of marine secondary metabolites can be conceived in the realm of the biosynthetic pathways commonly proposed for their terrestrial counterparts, they often embed in their structures functional groups uniquely or predominantly marine. As an illustrative example we can cite the abundance of halogenated compounds, most likely as a consequence of the relative abundance of halogen atoms in the marine environment, but also some functional groups as isonitrile, isothiocyanate (which we will see below in the structure of some antimalarial compounds), sulfamate and formamide that are much more abundant among marine metabolites than among the terrestrial ones.

The incredible biosynthetic potential of marine organisms can be rationalized by considering the common features of the secondary metabolism in all the living organisms as well as some peculiar features of the marine environment. There is a growing evidence that many marine invertebrates rely on their symbiotic population for a consistent part of the metabolic work. Marine invertebrates harbour in their tissues, in the extra- and intra-cellular spaces, microorganisms as bacteria, cyanobacteria and fungi which may constitute more than 50% of the biomass and whose presence appear to be essential for secondary metabolite production. In this view, sponges can be regarded as 'microbiological fermenters' containing species-specific marine microorganisms[7] which play a significant role in the biosynthesis of the natural products isolated from the invertebrate. The impressive progress in modern molecular biology has made it possible, in a reasonable time and at reasonable costs, to analyse a metagenomic DNA, to identify the biosynthetic genes in the producing organisms (a relatively easy task for bacterial genes which are commonly grouped in clusters), and finally to clone them in heterologous hosts suitable for large-scale fermentation.[8] The full exploitation of the metagenome could also allow the discovery of biosynthetic genes encoding for never isolated molecules and the triggering of biosynthetic pathways to obtain cryptic natural products. Thus, although marine chemists usually work with a limited supply of the biological material, often protected for environmental concerns, there is the reasonable hope that the full development and utilization of these techniques could overcome many obstacles to considering marine organisms as a potentially sustainable drug source.

Anyway, in spite of the traditional difficulties associated with the limited availability of the compounds under investigation, some interesting results have already been obtained. Through the combined efforts of marine natural product chemists and pharmacologists, an astounding array of pharmacologically promising compounds have been identified. In a recent review, Mayer *et al.* summarized the state of the art of the marine pharmaceuticals pipeline.[9] In brief, at the moment four marine drugs have been approved by FDA or EMA (the EU agency), two of them are nucleosides (cytarabine and vidarabine), one is a polypeptide (ziconotide), and the fourth is the complex alkaloid trabectedin (ET-743), an antitumour compound especially effective against soft tissue sarcoma and ovarian carcinoma. Another 13 compounds are in different phases of clinical trials and several hundreds of compounds are in the preclinical stages. The majority of these products fall within the area of cancer, pain and antimicrobial therapies. Just to cite an example, eribulin mesilate, a derivative of halichondrin B, a macrolide isolated from the sponge *Halichondria hokadai*, is undergoing Phase III clinical trials as an anticancer agent and constitutes one of the most advanced compounds among the marine natural products currently under clinical investigation.

The aim of this chapter is to highlight the contribution of marine chemistry in the field of antimalarial research. In recent years, a growing number of research groups has dedicated consistent efforts to the isolation and development of marine antimalarials and, as a result, more than one hundred active molecules have been published in the last decade. However, for some of them the toxicity toward *Plasmodium* strains is not specific and/or is clearly due to a general cytotoxicity, while for other compounds the activity is too weak. A series of recent reviews[10–12] can be consulted to provide a detailed account about all the marine compounds showing moderate to high antimalarial activity. We have decided to focus this chapter on four classes of compounds which have emerged as the most promising for future development and for which more detailed data (e.g. information on the mechanism of action) are available: (i) plakortin and related polyketide endoperoxides (Section 17.3); (ii) isonitriles and analogues (Section 17.4); (iii) manzamines (Section 17.5); and (iv) lepadins and salinosporamide (Section 17.6). In the following paragraphs we will highlight the potential of these molecules in the field of antimalarial research, with particular emphasis on recent discoveries.

17.3   Plakortin and Related Polyketide Endoperoxides

The discovery of the potent antimalarial activity exhibited by artemisinin (**4**, Figures 17.1 and 17.2), a low-nanomolar antimalarial active against chlor-

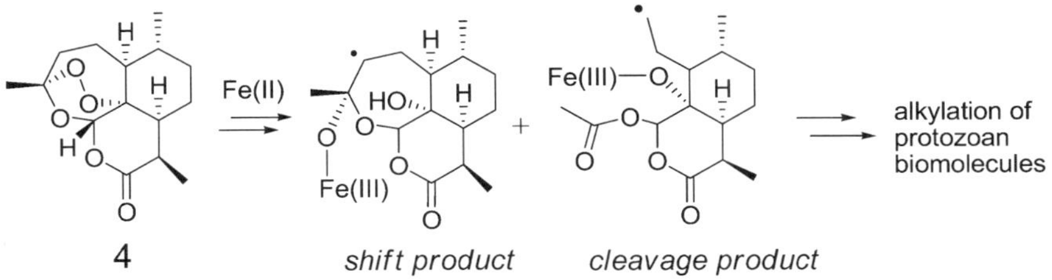

Figure 17.2 The C-centred radical species produced by interaction of artemisinin with Fe^II.

oquine-resistant strains of *Plasmodium*[13] and acting in the early form of the malarial blood stages, can be considered to be the most important breakthrough of the last three decades in the field of antimalarial research. Artemisinin was isolated from leaves of the sweet wormwood, *Artemisia annua* (Compositae), and it is a cadinane sesquiterpene lactone which includes a 1,2,4-trioxane moiety and shows an unique juxtaposition of peracetal, acetal and lactone functionalities.

The endoperoxide bond of artemisinin is essential for its antimalarial activity, as proved by the inactivity of deoxyartemisinin, an artemisinin derivative lacking the endoperoxide bridge. The most widely accepted hypothesis on the mechanism of action of artemisinins postulates that the endoperoxide bond of these molecules, upon interaction with the iron(II) centre of the heme unit, released in the food vacuole during the digestion of haemoglobin, gives rise to oxygen-centred radicals. As a result of an intramolecular rearrangement, these reactive species are converted into C-centred radicals, responsible of the alkylation of parasite macromolecular targets and, ultimately, for its death (Figure 17.2). Artemisinin activity seems not to be mediated by interaction with a specific target enzyme, in the light of the scarce tendency of this compound (and its analogues) to develop resistance. It is important to notice that artemisinin can produce C-radicals by means of two different rearrangement mechanisms of O-radicals, namely H-shift (O-radical links an H from a C–H bond) and C-cleavage (O-radical forms a carbonyl group with consequent cleavage of a carbon–carbon bond).

Secondary metabolites with the endoperoxide bond are not often elaborated by both terrestrial and marine organisms. As for compounds of marine origin, they constitute a rather large group of secondary metabolites, some of which exhibited a potent *in vitro* antimalarial activity. Many of the marine endoperoxides derive from a polyketide biogenesis, while very few marine endoperoxide-containing terpenoids have been tested for their antimalarial activity and none of them showed an antimalarial activity worthy of further investigation.

Generally, marine polyketide endoperoxides possess a six-membered 1,2-dioxygenated ring (1,2-dioxane) bearing two or three alkyl groups of different length and complexity (Figure 17.3). Based on the nature of the substituent R‴

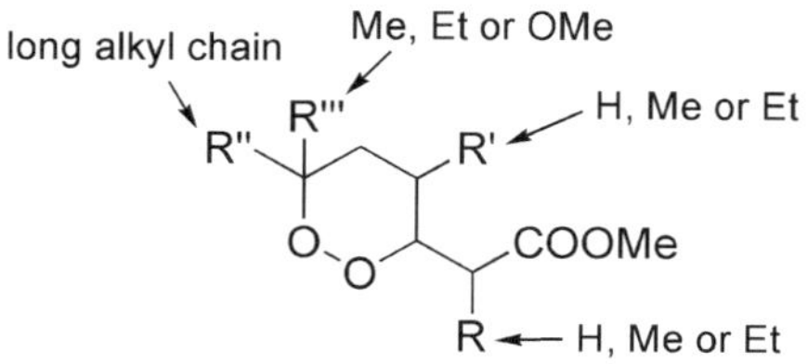

long alkyl chain   Me, Et or OMe

Figure 17.3   General structure of marine polyketide endoperoxides.

these compounds can be subdivided into simple 1,2-dioxanes (where R‴ is a methyl or an ethyl group) and peroxyketal dioxanes (where R‴ is a methoxy group). A small number of molecules show some variations on this general structure: (i) molecules showing a double bond between the two non-oxygenated carbons of the six-membered ring; (ii) molecules showing an additional ring fused to the dioxane ring; and (iii) five-membered endoperoxides.

Plakortin (**5**, Figure 17.4) is the lead compound in the class of marine endoperoxides, since it is the single compound for which more data about the antimalarial activity and the mechanism of action are available. It is a simple 1,2-dioxane metabolite whose polyketide biosynthesis is likely based on the assemblage of three butyrate, one propionate and one acetate units (Figure 17.4).

Plakortin was isolated about 30 years ago from *Plakortis halichondroides*, as a mild antibacterial agent,[14] but it was later re-isolated in remarkable amounts from the Caribbean sponge *Plakortis simplex* and the absolute configuration at the four stereogenic carbons of plakortin was determined.[15] A number of related endoperoxide derivatives were then isolated from the same source, e.g. dihydroplakortin (without the double bond in the alkyl chain)[15] and 3-epiplakortin.[16] These compounds exhibited a quite good antimalarial activity, and, similarly to artemisinin, they were more potent on the W2 strain (IC$_{50}$ ~ 0.3 µM) and devoid of cytotoxicity.[17]

The preparation of a number of plakortin semi-synthetic derivatives,[18] besides providing an unambiguous evidence of the crucial role of the endoperoxide bond for the antimalarial activity, allowed also a first investigation of the structure–activity relationships, which highlighted the role of the long alkyl chain. More recently, a multidisciplinary investigation, based on molecular dynamics/mechanics, *ab initio* calculations and analysis of

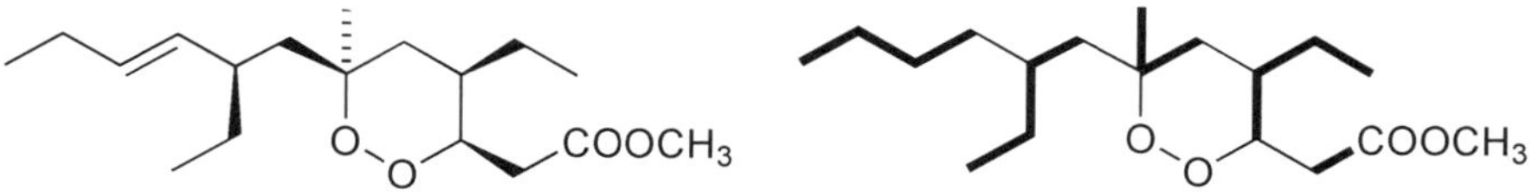

5

Figure 17.4   The structure of plakortin (left) and its likely polyketide biosynthetic origin. We have indicated in bold the ketide units (right).

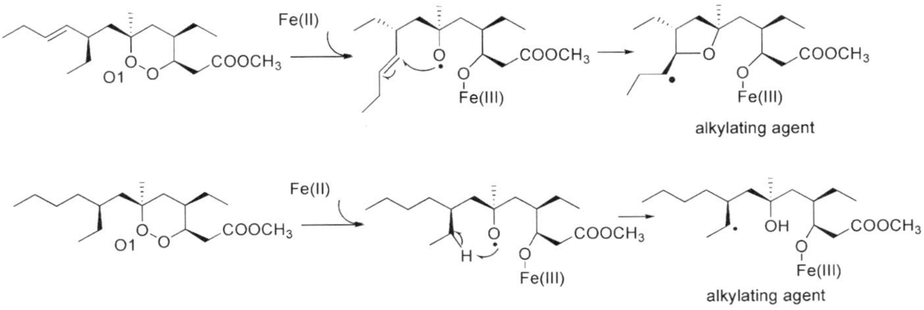

Figure 17.5 The mechanism proposed for the antimalarial actions of plakortin (top) and dihydroplakortin (bottom).

products obtained upon model reaction with $FeCl_2$, provided further interesting insights into the mechanism of the antimalarial action of plakortin compounds.[19] These molecules, as a result of the interaction of the endoperoxide bond with Fe^{II}, should give rise to the formation of the oxygen radical centred on the O1 oxygen atom (Figure 17.5). This event is immediately (or simultaneously) followed by rearrangement to give a C-radical centred on the 'western' alkyl side chain, which is hypothesized to be the toxic intermediate responsible for subsequent reactions leading to the plasmodium death (Figure 17.5). In order to exhibit antimalarial activity, a 1,2-dioxane of the plakortin family must experience a good accessibility of the endoperoxide bond to Fe^{II} and a significant conformational preference which could allow the correct orientation of the side chain bearing possible partners for the intramolecular shift from oxygen radical to carbon radical. In this regard, it is interesting to notice that, in the cases of both plakortin and dihydroplakortin, all the available data seem to indicate that only the rearrangement mechanism operates in the evolution of the oxygen radical to the carbon radical.

This mechanism can allow us to explain the very low *in vitro* antimalarial activity ($IC_{50} > 8$ µg/mL) of some plakortin analogues[20] possessing a *trans*-double bond between the first two carbons of the side chain. Indeed, in spite of their similarity with the plakortin scaffold, when the side chain double bond is directly attached at the dioxane ring, its geometry can almost completely

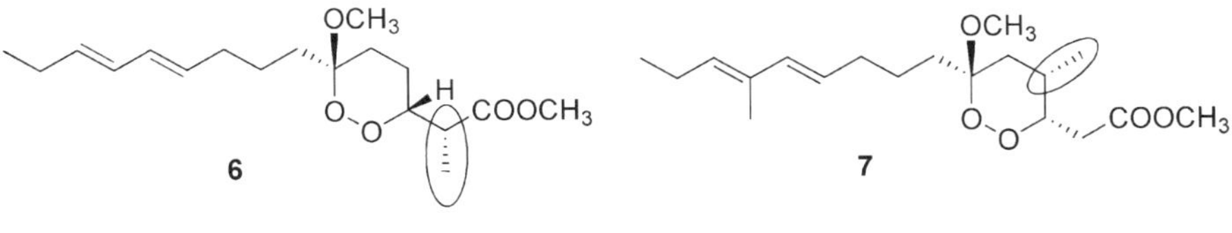

Figure 17.6 The structures of peroxyplakoric ester B3 (left) and manadoperoxide B (right). The change in the position of a methyl group has been highlighted.

prevent the evolution of the oxygen radical to a side chain carbon radical, and, thus, alkylation of the protozoan macromolecules could be not possible.

Another good piece of evidence supporting the proposed mechanism for antimalarial activity of 1,2-dioxanes comes from comparison of *in vitro* antimalarial potencies of peroxyplakoric esters (**6**)[21] and manadoperoxides (**7**) (Figure 17.6).[22] Both these molecules belong to the class of 3-alkoxy-1,2-dioxanes (peroxyketals) derivatives and they differ almost exclusively in the position of a methyl group, which is moved from the short alkyl chain to the dioxane ring. This corresponds to a difference in the last two steps of the hypothesized polyketide biogenesis: for peroxyplakoric derivatives the last two units are, most likely, acetate-propionate, while in the case of manadoperoxides the sequence of the last two incorporated units should be reversed, i.e. propionate-acetate.

Peroxyplakoric derivatives showed a potent *in vitro* activity (IC_{50} = 50 ng/mL against *P. falciparum*) with a good selective toxicity index (about 200). When these compounds were examined *in vivo* against *P. berghei* infection, they showed little antimalarial potency because of the relative instability in mouse serum due to the hydrolysis of the ester function. A better stability in physiological media was reached with the preparation of the monoethyl amide analogue, which showed a good *in vivo* activity.[23]

In spite of the very minor structural change above described, manadoperoxides experienced a dramatic decrease in the antimalarial activity (IC_{50} = 3.70 μM). However, a computational study revealed that this unexpected outcome can be rationalized by analysing the conformational behaviour of the molecule, and, consequently, its ability to interact with heme and produce the toxic C-centred radical.[22] In the reasonably populated conformations of manadoperoxides, the endoperoxide linkage is either not accessible to heme iron or it cannot evolve to produce the C-centred radical. On the contrary, peroxyplakoric derivatives can adopt a conformation which offers no obstacles to heme interaction and evolution to the C-centred radical.

The reaction of peroxyplakoric ester analogues with Fe^{II} species was explored to mimic the interaction with heme,[24] and analysis of the products appeared to suggest that after the homolytic cleavage of the O–O bond, the O-centred radical is almost exclusively located at the oxygen attached to the methoxy-linking carbon. This radical would evolve through a regiospecific carbon–carbon cleavage producing a neutral methyl ester species and a very

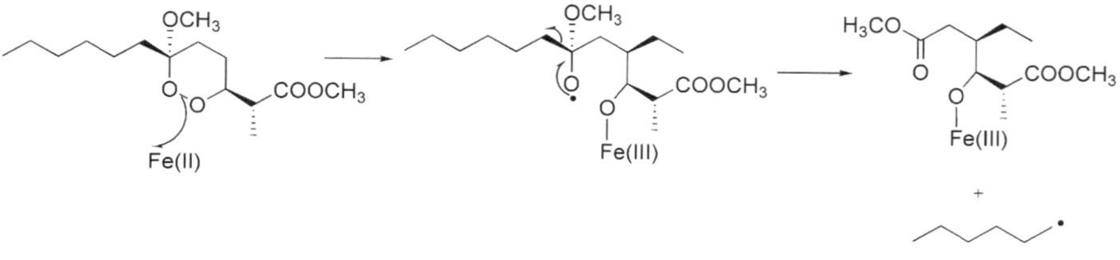

Figure 17.7   The postulated mechanism of action of peroxyplakoric derivatives. Notice that the key step is a cleavage reaction.

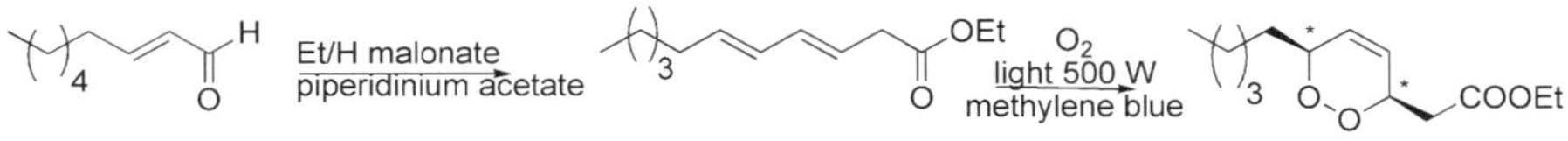

Figure 17.8 Synthetic strategy to the simplified plakortin analogues.

reactive primary carbon radical which should be the toxic species responsible for the death of the parasite (Figure 17.7). This model differs significantly, in the last step, from that proposed for plakortin (Figure 17.5), since in this case the production of the C-radical relies on a cleavage reaction. However, it should be noted that some synthetic analogues of peroxyplakoric derivatives, potentially producing reactive carbon radicals according to the cleavage model, failed to show a significant *in vitro* activity.

Total syntheses have been reported for both peroxyplakoric derivatives[25] and plakortin.[26] In this latter case, the creation of four stereogenic centres required a long and low-yielding strategy, whose application to a large-scale production of the molecule appears absolutely unfeasible. Indeed, low-cost production is a mandatory requirement of an antimalarial drug given the poor market to which these compounds are addressed. Medicine for Malaria Venture (MMV) has published the Target Product Profile (TPP),[27] which is a list of requirements that a new molecule should satisfy in order to be evaluated as clinical candidate. Among these requirements, oral activity and a cost treatment that should be not more than 1$ for simple malaria and 5$ for cerebral malaria are of particular significance. In this context, a synthesis of simplified plakortin analogues based on the sequence of two easy and high-yielding reactions and starting from commercially available materials has been reported very recently.[28] The carbon skeleton was assembled through a modified Knoevenagel condensation, while the singlet oxygen addition to 1,3-butadiene derivatives was exploited for the construction of the endoperoxide ring, the critical step of the synthetic procedure (Figure 17.8). The obtained dioxin derivative showed a good antimalarial activity ($IC_{50} = 1.2$ μM).

However, the chemical synthesis should not be viewed as the only option to access large-scale production of marine antimalarials; as discussed in the preceding paragraph, some of the marine secondary metabolites are actually produced by microbial symbionts and this could pave the way for their biotechnological production. In the case of plakortin, strong evidence has been gathered about the plakortin production by the microbial population hosted by the *Plakortis* sponge.[29] Accurate investigation of the metagenome of the sponge is underway in our lab with the final aim of identifying the genes encoding for plakortin, which could allow the production of these molecules by low-cost bacterial fermentation.

17.4   Isonitriles and Related Compounds

Isonitrile-containing terpenoids are typically marine secondary metabolites and, in particular, they appear to be efficiently produced as feeding deterrent agents by marine sponges of the Axinellidae and Halicondridae families. Interestingly, it has been shown that nudibranchs do sequester isonitriles produced by sponges into their skin and use them to be protected from predation. The parent compound of the isonitrile class is axisonitrile-1 (**8**), isolated about 40 years ago from the marine sponge *Axinella cannabina*, where it co-occurred with the corresponding isothiocyanate and formamide analogues.[30] As for plakortin (see above), at the time of the first isolation the antimalarial potential of these molecules was not disclosed. More recently, the related axisonitrile-3 (**9**),[31] also found in *Axinella cannabina*, has been re-isolated from the sponge *Acanthella klethra* and found to possess a potent antimalarial activity on *P. falciparum* strains with $IC_{50} < 0.1$ µg/mL.[32] Interestingly, the corresponding isothiocyanate derivative was completely inactive, suggesting a crucial role for the isonitrile functional group. On the other hand, the role of the sesquiterpene skeleton of both compounds **8** and **9** was not absolutely clear.

The chemical analysis of *Cymbastela hooperi* (Axinellidae) afforded a series of tetracyclic diterpenes based on amphilectane (and similar) skeletons and bearing isonitrile, isothiocyanate and isocyanate functionalities, which provided some interesting insights on the role played by the terpene scaffold.[33] These molecules displayed a potent (e.g. **10**, $IC_{50} = 4$ ng/mL) and selective *in vitro* antimalarial activity with a marked decrease of activity for isothiocyanate and isocyanate analogues. Moreover, the ten-times lower activity of the neoamphilectane derivative (e.g. **11**) indicated that the carbon skeleton can have a role in the modulation of the antiplasmodial activity of isonitrile derivatives.[34] The most potent isonitrile antimalarials isolated to date are

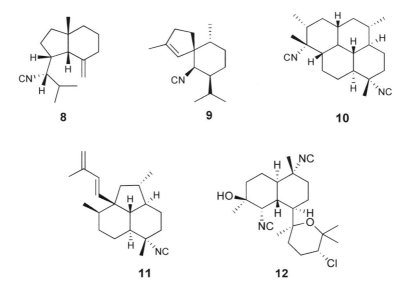

Figure 17.9   Some representatives isonitrile antimalarials.

kalihinols (e.g. **12**, Figure 17.9), isonitrile diterpenes isolated from the Japanese sponge *Acanthella* sp. showing an exceptionally potent activity ($IC_{50} < 1$ ng/mL).[35]

As for the mechanism of action, isonitrile antimalarials have been proposed to interact with free heme through the formation of a coordination complex between the isonitrile group and the iron centre, thus inhibiting the transformation of heme into β-hematin and then hemozoin, a polymer produced by *Plasmodium* in order to neutralize the toxic (detergent-like) free heme produced in the food vacuole. In addition, isonitriles were shown to prevent both the peroxidative and glutathione-mediated destruction of heme under conditions that mimic the environment within the malaria parasite.[36] The pharmacophore must possess an overall lipophilic rigid molecular core carrying an isonitrile group and establishing further hydrophobic interactions above the ring plane.

The ability of these isonitriles to interact with heme iron also gives them anti-photosynthetic activity, which has been very recently investigated by means of an accurate docking analysis,[37] and has been proposed as an important feature for the regulation of the growth of photosynthetic organisms within the sponge tissues.

17.5 Manzamines

Manzamines constitute a unique class of polycyclic (7–8 rings or more) alkaloids characterized by an intricate heterocyclic system attached to a β-carboline moiety, first reported from an Okinawan sponge belonging to the genus *Haliclona*.[38] Since the first report of manzamine A (**13**, Figure 17.10), at least 60 additional manzamine-type alkaloids have been reported from taxonomically unrelated sponges belonging to different genera (e.g. *Xestospongia*, *Ircinia* and *Amphimedon*) and different orders. These findings

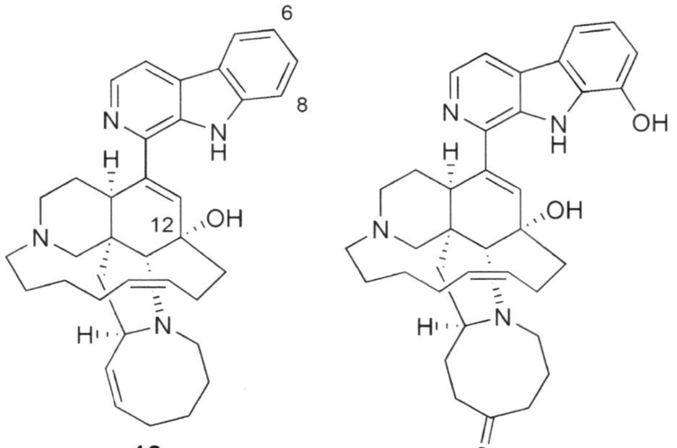

Figure 17.10 The structure of manzamine A (**13**) and manzamine F (**14**).

strengthen the hypothesis that manzamines are not true sponge metabolites but, more likely, they have a symbiotic origin. Accordingly, microbial community analyses for one of the most common manzamine-producing sponges resulted in the identification of *Micronosphora* sp. as the bacterial strain producing manzamines.[39] A common biosynthetic route for cyclostel-lettamines, halicyclamines and manzamines has also been hypothesized, highlighting the existence of a kind of 'nature diversity-oriented syntheses'.[40]

Hamann *et al.* disclosed the antimalarial potential of manzamine A and of its 8-hydroxy derivative, showing that these molecules potently inhibit the growth of *Plasmodium falciparum* both *in vitro* ($IC_{50} \sim 5.0$ ng/mL) and *in vivo*.[41] Oral treatment with manzamine A (2×100 µmol/kg) showed 90% reduction in parasitemia and mice treated with a single dose of manzamine A also showed significant improvements in survival times over mice treated with chloroquine or artemisinin. The mechanism of action of manzamine A is not clear and an extensive structure–activity relationship study, aimed at showing the relative importance of all the single moieties of this complex scaffold, has been carried out. For example, a closely related derivative of manzamine A, manzamine F (**14**), was completely devoid of activity ($IC_{50} > 1000$ ng/mL), thus showing the key role of the eight-membered ring. Similarly, the attachment of the hydroxyl group at position 6 has a deleterious impact on the antimalarial activity, as indicated by the lower potency of 6-hydroxy-manzamine A.[42] The crucial role of a free hydroxy group at position 12 has also been highlighted by a study reporting that acetylation of this group significantly reduced the antimalarial activity.[43]

Unfortunately, a major drawback of manzamine antimalarials is related to the toxicity associated with higher doses. It has been proposed that this toxicity may be due to DNA intercalation operated by the planar β-carboline moiety, which causes arrest of the cell cycle in the S phase. A recent study has been aimed at preparing a series of semi-synthetic derivatives of manzamine A showing modifications in this aromatic moiety, critical for toxicity.[44] It was found that introduction of amide functionalities (with bulky alkyl groups) at positions 6 and 8 of the carboline moiety can completely eliminate the toxicity, with only a marginal reduction of the antimalarial activity. Manzamines have also been reported to be antibacterial and antituberculosis agents and to exhibit activity against AIDS opportunistic pathogens as the protozoan *Toxoplasma gondii*.[45] More recently, manzamine A and derivatives have been shown to possess inhibitory activity against glycogen synthase kinase-3 (GSK-3) with potential application in Alzheimer's disease.[46]

17.6 Lepadins and Salinosporamide

Lepadins are decahydroquinoline derivatives bearing a linear eight-carbon chain obtained from two tunicates of Australian origin, *Clavelina lepadiformis* and *Didemnum* sp.[47,48] Lepadin E (**15**, Figure 17.11) exhibited a significant antimalarial activity ($IC_{50} = 0.4$ µg/mL) while its analogues lacking the 2*E*-

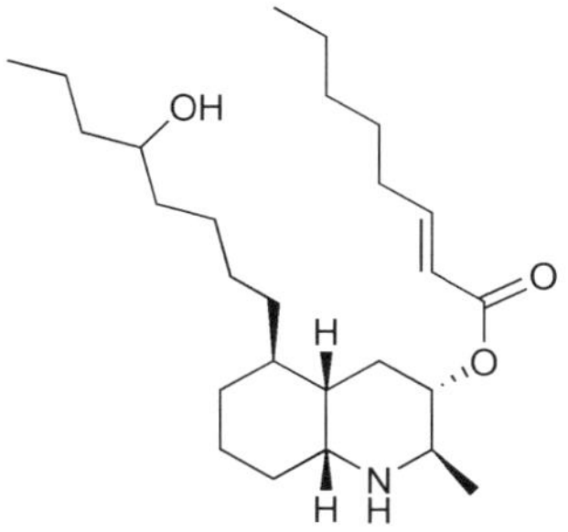

Figure 17.11   The structure of lepadin E (**15**) and salinosporamide (**16**).

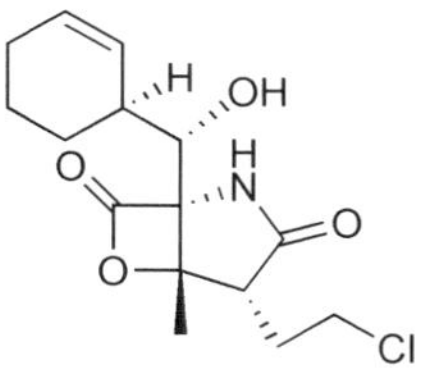

octenoic acid ester functionality were completely inactive. Authors have proposed that this conformationally mobile side chain could serve to stabilize non-bonding interactions with heme, or with any other 'receptor' molecule. No further investigations on the mechanism of action of lepadins have since been reported.

Salinosporamide A (**16**, Figure 17.11), a γ-lactam-β-lactone alkaloid isolated from a marine bacterium of the new genus *Salinispora*,[49] has been found to be parasite proteasome inhibitor and to possess a significant antimalarial activity *in vitro* (IC$_{50}$ = 11.4 nM).[50] This chlorinated molecule dominates a family of natural structural analogues that differ for the substituent on the γ-lactam ring.[51] Salinosporamide was shown to act in the erythrocytic stage and maintained its potent activity in a malaria mouse model with inhibition of the parasite growth in treated mice at very low doses (130 μg/kg). Salinosporamide A (marizomib) is in Phase I clinical trials for cancer treatment and, in an important example of industrial marine microbiology, the supplies of this compound come from a robust saline fermentation process using a *Salinispora tropica* strain.

17.7   Conclusions

In this chapter we have presented four antimalarial leads (and some other related compounds) obtained from marine organisms whose good activities and knowledge of their mechanism of action could foster the development of new classes of antimalarial drugs. The marine antimalarial molecules are a good example of the incredible marine chemodiversity, which hopefully could give a contribution to the humanitarian challenge posed by malaria infections. In many cases, marine antimalarials possess innovative structures and can constitute a valuable contribution to the research in this field.

These lead compounds can be conceived as the result of the drug discovery programs launched by academia and public agencies in recent years with the aim of developing effective alternatives to the old drugs used, whose efficacy is diminishing due to the multi-drug resistance developed by parasites and to severe side effects, sometimes completely preventing their utilization. It should be hoped that in the near future the investigations on these molecules could surpass the initial stage of assessment of *in vitro* (in very few cases *in vivo*) activity and move to the stage of optimization of both activity and pharmacokinetics. A further important issue is given by large-scale production, which is probably the major obstacle to the massive utilization of marine organisms as drug sources. However, the finding that many marine natural products (including some antimalarials) are actually produced by the microbial symbionts promises to open a new era in this field and, together with the recent progress in molecular biology and genomics, predicts exciting developments in the production of marine metabolites.

Acknowledgement

The support from MIUR-PRIN2008 (Leads ad Attività Antimalarica di Origine Naturale: Isolamento, Ottimizzazione e Valutazione Biologica) is acknowledged.

References

1. D. A. Casteel, in *Burger's Medicinal Chemistry and Drug Discovery*ed. M. E. Wolff, John Wiley and Sons, New York, 1997, p 3.
2. J. G. Breman, A. Egan and G. Keusch, *Am. J. Trop. Med. Hyg.*, 2001, **64**, iv.
3. P. Olliaro, J. Cattani and D. Wirth, *J. Am. Med. Ass.*, 1996, **275**, 230.
4. M. J. Gardner, N. Hall, E. Fung, O. White, M. Berriman, R. W. Hyman, J. M. Carlton, A. Pain, K. E. Nelson, S. Bowman, I. T. Paulsen, K. James, J. A. Eisen, K. Rutherford, S. L. Salzberg, A. Craig, S. Kyes, M. S. Chan, V. Nene, S. J. Shallom, B. Suh, J. Peterson, S. Angiuoli, M. Pertea, J. Allen, J. Selengut, D. Haft, M. W. Mather, A. B. Vaidya, D. M. Martin, A. H. Fairlamb, M. J. Fraunholz, D. S. Roos, S. A. Ralph, G. I. McFadden, L. M. Cummings, G. M. Subramanian, C. Mungall, J. C. Venter, D. J. Carucci, S. L. Hoffman, C. Newbold, R. W. Davis, C. M Fraser and B. Barrell, *Nature*, 2002, **419**, 498.
5. W. E. Collins and J. W. Barnwell, *N. Engl. J. Med.*, 2008, **359**, 2599.
6. J. W. Blunt and M. H. G. Munro, eds. *Dictionary of Marine Natural Products*, Chapman & Hall/CRC, Boca Raton, , FL, 2008.
7. U. Hentschel, *ChemBioChem*, 2002, **3**, 1151.
8. C. E. Salomon, N. A. Magarvey and D. H. Sherman, *Nat. Prod. Rep.*, 2004, **21**, 105.
9. A. M. S. Mayer, K. B. Glaser, C. Cuevas, R. S. Jacobs, W. Kem, R. D. Little, J. M. McIntosh, D. J. Newmann, B. C. Potts and D. E. Shuster, *Trends Pharmacol. Sci.*, 2010, **31**, 255.
10. K. R. Watts, K.Tenney, and P. Crews, *Curr. Opin. Biotechnol.*, 2010, **21**, 808.
11. E. Fattorusso and O. Taglialatela-Scafati, *Marine Drugs*, 2009, **7**, 130.

12. E. Fattorusso and O. Taglialatela-Scafati, *Phytochem. Rev.*, 2010, **9**, 515.
13. D. L. Klayman, A. J. Lin, N. Acton, J. P. Scovill, J. M. Hoch, W. K. Milhous, A. D. Theoharides and A. S. Dobek, *J. Nat. Prod.*, 1984, **47**, 715.
14. M. D. Higgs and D. J. Faulkner, *J. Org. Chem.*, 1978, **43**, 3454.
15. F. Cafieri, E. Fattorusso, O. Taglialatela-Scafati and A. Ianaro, *Tetrahedron*, 1999, **55**, 7045.
16. C. Campagnuolo, E. Fattorusso, A. Romano, O. Taglialatela-Scafati, N. Basilico, S. Parapini and D. Taramelli, *Eur. J. Org. Chem.*, 2005, 5077.
17. E. Fattorusso, S. Parapini, C. Campagnuolo, N. Basilico, O. Taglialatela-Scafati and D. Taramelli, *J. Antimicrob. Chemother.*, 2002, **50**, 883.
18. C. Fattorusso, G. Campiani, B. Catalanotti, M. Persico, N. Basilico, S. Parapini, D. Taramelli, C. Campagnuolo, E. Fattorusso, A. Romano and O. Taglialatela-Scafati, *J. Med. Chem.*, 2006, **49**, 7088.
19. O. Taglialatela-Scafati, E. Fattorusso, A. Romano, F. Scala, V. Barone, P. Cimino, E. Stendardo, B. Catalanotti, M. Persico and C. Fattorusso, *Org. Biomol. Chem.*, 2010, **8**, 846.
20. J. Hu, H. Gao, M. Kelly and M. T. Hamann, *Tetrahedron*, 2001, **57**, 9379.
21. M. Kobayashi, K. Kondo and I. Kitagawa, *Chem. Pharm. Bull.*, 1993, **41**, 1324.
22. C. Fattorusso, M. Persico, B. Calcinai, C. Cerrano, S. Parapini, D. Taramelli, E. Novellino, A. Romano, F. Scala, E. Fattorusso and O. Taglialatela-Scafati, *J. Nat. Prod.*, 2010, **73**, 1138.
23. N. Murakami, M. Kawanishi, H. M. Mostaqul, J. Li, S. Itagaki, T. Horii and M. Kobayashi, *Bioorg. Med. Chem. Lett.*, 2003, **13**, 4081.
24. M. Kawanishi, N. Kotoku, S. Itagaki, T. Horii and M. Kobayashi, *Bioorg. Med. Chem.*, 2004, **12**, 5297.
25. N. Murakami, M. Kawanishi, S. Itagaki, T. Horii and M. Kobayashi, *Bioorg. Med. Chem. Lett.*, 2002, **12**, 69.
26. S. Gemma, E. Gabellieri, S. Sanna Coccone, F. Martì, O. Taglialatela-Scafati, E. Novellino, G. Campiani and S. Butini, *J. Org. Chem.*, 2010, **75**, 2333.
27. http://www.MMV.org/IMG/pdf/PRODUCT_PROFILE_with-logo.pdf
28. C. Fattorusso, M. Persico, N. Basilico, D. Taramelli, E. Fattorusso, F. Scala and O. Taglialatela-Scafati, *Bioorg. Med. Chem.*, 2011, **19**, 312.
29. M. Laroche, C. Imperatore, L. Grozdanov, V. Costantino, A. Mangoni, U. Hentschel and E. Fattorusso, *Marine Biol.*, 2006, **151**, 1365.
30. F. Cafieri, E. Fattorusso, S. Magno, C. Santacroce and D. Sica, *Tetrahedron*, 1973, **29**, 4259.
31. B. Di Blasio, E. Fattorusso, S. Magno, L. Mayol, C. Pedone, C. Santacroce and D. Sica, *Tetrahedron*, 1976, **32**, 473.
32. C. K. Angerhofer, J. M. Pezzuto, G. M. Koenig, A. D. Wright and O. Sticher, *J. Nat. Prod.*, 1992, **55**, 1787.
33. G. M. Koenig, A. D. Wright and C. K. Angerhofer, *J. Org. Chem.*, 1996, **61**, 3259.
34. A. D. Wright and N. Lang-Unnasch, *J. Nat. Prod.*, 2009, **72**, 492.

35. H. Miyaoka, M. Shimomura, H. Kimura, Y. Yamada, H. S. Kim and Y. Wataya, *Tetrahedron*, 1998, **54**, 13467.
36. A. D. Wright, H. Wang, M. Gurrath, G. M. Koenig, G. Kocak, G. Neumann, P. Loria, M. Foley and L. Tilley, *J. Med. Chem.*, 2001, **44**, 873.
37. A. D. Wright, A. McCluskey, M. J. Robertson, K. A. MacGregor, C. P. Gordon and J. Guenther, *Org. Biomol. Chem.*, 2011, **9**, 400.
38. R. Sakai, T. Higa, C. W. Jefford and G. Bernardinelli, *J. Am. Chem. Soc.*, 1986, **108**, 6404.
39. M. Yousaf, K. A. El Sayed, K. V. Rao, C. W. Lim, J. Hu, M. Kelly, S. G. Franzblau, F. Zhang, O. Peraud, R. T. Hill and M. T. Hamann, *Tetrahedron*, 2002, **58**, 7397.
40. J. C. Wypych, T. M. Nguyen, P. Nuhant, M. Benechie and C. Marazano, *Angew. Chem., Int. Ed.*, 2008, **47**, 5418.
41. K. K. H. Ang, M. J. Holmes, T. Higa, M. T. Hamann and U. A. K. Kara, *Antimicrob. Agent. Chemother.*, 2000, **44**, 1645.
42. K. V. Rao, M. S. Donia, J. Pen, E. Garcia-Palomero, D. Alonso, A. Martinez, M. Medina, S. G. Franzblau, B. L. Tekwani, S. I. Khan, S. Wayhuono, K. L. Willett and M. T. Hamann, *J. Nat. Prod.*, 2006, **69**, 1034.
43. A. G. Shilabin, N. Kasanah, B. L. Tekwani and M. T. Hamann, *J. Nat. Prod.*, 2008, **71**, 1218.
44. A. E. Wahba, J. Peng, S. Kudrimoti, B. L. Tekwani and M. T. Hamann, *Bioorg. Med. Chem.*, 2009, **17**, 7775.
45. K. V. Rao, B. D. Santarsiero, A. D. Mesecar, R. F. Schinazi, B. L. Tekwani and M. T. Hamann, *J. Nat. Prod.*, 2003, **66**, 823.
46. M. T. Hamann, D. Alonso, E. Martin-Aparicio, A. Fuertes, J. Perez-Puerto, A. Castro, S. Morales, M. L. Navarro, M. del Monte-Millan, M. Medina, H. Pennaka, A. Balaiah, J. Peng, J. Cook, S. Wahyuono and A. Martinez, *J. Nat. Prod.*, 2007, **70**, 1397.
47. B. Steffan, *Tetrahedron*, 1991, **42**, 8729.
48. A. D. Wright, E. Goclik, G. M. Koenig and R. J. Kaminsky, *J. Med. Chem.*, 2002, **45**, 3067.
49. R. H. Feling, G. O. Buchanan, T. J. Mincer, C. A. Kauffman, P. R. Jensen and W. Fenical, *Angew. Chem., Int. Ed.*, 2003, **42**, 355.
50. J. Prudhomme, E. McDaniel, N. Ponts, S. Bertani, W. Fenical, P. Jensen and K. Le Roch, *PLoS One*, 2008, **3**, e2335.
51. T. A. M. Gulder and B. S. Moore, *Angew. Chem., Int. Ed.*, 2010, **49**, 9346.

Future Challenges

Natural products are accepted today as one of the most promising sources for the discovery of future drugs that are urgently needed in many therapeutic areas. Natural products research has experienced an extraordinary evolution in the last decade, with a large and converging cross-disciplinary activity that is supporting innovative ways to approach the discovery of novel scaffolds. Although the process of drug discovery from natural products is still complex and difficult, significant technological progress made in the last years has improved our understanding of the key factors required for success. At present it is widely accepted that natural products are the future of drug discovery – because of their unparalleled structural diversity, their optimized interaction with molecular targets derived from a long evolutionary process, the still unexplored vast microbial diversity, and the existing potential of novel approaches to trigger the expression of silent pathways – and they have been setting the basis for the current research that is nuturing the field of future natural products drug discovery.

Technological advances in the automated purification, de-replication, and structure elucidation of natural products have shortened the timeline of drug discovery similar to that of synthetic molecules. Recent improvements in NMR and mass spectrometry, a reduction in the amounts of materials needed for the detection and identification of new molecules in complex samples, and increasing amounts and better management of analytical data, have removed previous bottlenecks in high throughput screening (HTS) drug discovery programs. Furthermore, this ability to operate at the microscale has also favored the production of natural product libraries that are playing a new role in drug discovery programs.

In addition, bioprospecting in environments that have been only poorly sampled and innovative new methods for culturing microorganisms are likely to significantly increase the number of natural products that can be screened –

RSC Drug Discovery Series No. 25
Drug Discovery from Natural Products
Edited by Olga Genilloud and Francisca Vicente
© The Royal Society of Chemistry 2012
Published by the Royal Society of Chemistry, www.rsc.org

platensimycin and kibdelomycin are good examples. Converging trends from microbial genomics and protein structural biology are continuing to drive new insights into the biosynthetic capacity of bacteria and fungi and the molecular recognition of small molecules with specific macromolecular targets. Genomics is playing a major role in the development of natural products research and next generation sequencers are enabling high coverage of genome sequences. The use of microbial genome mining and the search for natural products in relatively untapped sources will expand our ability to find novel, potent, and selective drug leads. Given that only a tiny fraction of the microbial world has been explored and we are witnessing exponential growth in the microbial genomic database, the outlook for discovering new biologically active natural products remains very promising.

Moreover, success with the heterologous expression of genomic and metagenomic DNA is providing novel insights into alternative ways of accessing uncultured microbes for novel drugs and their potential benefits – and the difficulties derived from these technologies. To date, all reported small molecules from captured biosynthetic pathways have relied on heterologous expression and even today background metabolism, codon usage, or promoter structure, among others, represent major limitations for the development of novel molecules. In this scenario, synthetic biology is bringing completely new avenues for the expression of novel pathways, a rational design of biological systems, and a new potential for combinatorial synthesis and heterologous expression of novel scaffolds.

Once genuine natural product leads are identified, new organic synthetic methodologies, biotransformation, combinatorial biosynthesis, and combinations of these techniques can be applied to modify natural product leads, thereby generating a large number of novel, structurally diverse analogs that can be screened for improved properties or new activities. While most of the attention has now shifted to smaller high-quality libraries, natural product leads are emerging as a legitimate starting template for combinatorial chemistry. Novel biologically active analogs with improved properties or new activities can be discovered by this process. The same combinatorial chemistry that initially caused the decline in natural product screening is now perceived as an essential tool for generating analogs of natural products.

In order for natural product drug discovery to continue to deliver novel leads, new and innovative approaches that have been shown be successful, with very elegant examples reported in literature, must be set up in a high throughput 'industrial format'. Whereas it might be possible to increase the current efficiency in identifying and developing new drugs from natural products by applying these new approaches in a systematic manner to natural product drug discovery, it is still too early to predict whether it will lead to the success observed during the 'golden age' of natural product drug discovery.

On this evidence, efforts to discover (or rediscover) natural product families seem likely to complement and offer new alternatives to the decades-long trend of iterative semisynthetic modification of old scaffolds in the drive for discovering new molecules. At the same time, many of these new drugs will provide valuable clues about currently unexploited targets against which natural product and synthetic chemical libraries should be screened.

Despite the recent vanishing internal resources and the lack of efforts by most of the large pharmaceutical companies, natural product research has maintained a tremendous activity throughout this decade at academic laboratories and small biotechnological companies. This volume has contributed to provide new insights into the substantial impact of natural products on the recent supply of new drugs and the renovated interest and interdisciplinary effort that is being developed across different fields. All these new approaches have shown that they can deliver new drug leads in different therapeutic areas, thereby reinforcing the advantages of screening natural products for drug discovery.

Any new efforts in this area will benefit from an optimal integration of all these disciplines, as well as from the continued implementation of new technologies that will contribute to improving the efficiency of the process. The resources required may appear to be considerable, but the reward for research teams willing to take the risk is worthwhile – the solution to one of the most serious health threats and therapeutic needs that we may be facing in the years to come.

Subject Index

Note: italic page numbers indicate a table, scheme or figure.